T0265386

THE LOGARITHMIC INTEGRAL II

$$\int_{-\infty}^{\infty} \frac{\log M(t)}{1+t^2}\,\mathrm{d}t$$

The logarithmic integral
II

PAUL KOOSIS

McGill University in Montreal

 CAMBRIDGE
UNIVERSITY PRESS

CAMBRIDGE UNIVERSITY PRESS
Cambridge, New York, Melbourne, Madrid, Cape Town, Singapore, São Paulo, Delhi

Cambridge University Press
The Edinburgh Building, Cambridge CB2 8RU, UK

Published in the United States of America by Cambridge University Press, New York

www.cambridge.org
Information on this title: www.cambridge.org/9780521102544

First published 1992
This digitally printed version 2009

A catalogue record for this publication is available from the British Library

Library of Congress Cataloguing in Publication data
(Revised for volume 2)
Koosis, Paul
The logarithmic integral.
(Cambridge studies in advanced mathematics; 21)
Includes bibliographies and indexes.
1. Analytic functions.   2. Harmonic analysis.
3. Integrals, Logarithmic.   I. Title.   II. Series.
QA331.K7393   1988   515.4   85-28018

ISBN 978-0-521-30907-3 hardback
ISBN 978-0-521-10254-4 paperback

Remember

Geneviève Bergeron, age 21,

Hélène Colgan, age 23,

Nathalie Croteau, age 23,

Barbara Daigneault, age 22,

Anne-Marie Edward, age 21,

Maud Haviernick, age 29,

Barbara Maria Klucznik-Widajewicz, age 31,

Maryse Laganière, age 25,

Maryse Leclair, age 23,

Anne-Marie Lemay, age 22,

Sonia Pelletier, age 28,

Michèle Richard, age 21,

Annie St-Arneault, age 23,

Annie Turcotte, age 20;

murdered in the Montreal Ecole Polytechnique on December 6, 1989.

Contents

Foreword to volume II, with an example for the end of volume I

Art is long and life is short. More than four years elapsed between completion of the MS for volume I and its publication; a good deal of that time was taken up with the many tasks, often tedious, called for by the production of any decently printed book on mathematics.

An attempt has been made to speed up the process for volume II. Three quarters of it has been set directly from handwritten MS, with omission of the intermediate preparation of typed copy, so useful for bringing to light mistakes of all kinds. I have tried to detect such deficiencies on the galleys and corrected all the ones I could find there; I hope the result is satisfactory.

Some mistakes did remain in volume I in spite of my efforts to remove them; others crept in during the successive proof revisions. Those that have come to my attention are reported in the *errata* immediately following this foreword.

In volume I the theorem on simultaneous polynomial approximation was incorrectly ascribed to Volberg; it is almost certainly due to T. Kriete, who published it some three years earlier. L. de Branges' name should have been mentioned in connection with the theorem on p. 215, for he gave (with different proof) an essentially equivalent result in 1959. The developments in §§A and C of Chapter VIII have been influenced by earlier work of Akhiezer and Levin. A beautiful paper of theirs made a strong impression on me many years ago. For exact references, see the bibliography at the end of this volume.

I thank Jal Choksi, my friend and colleague, for having frequently helped me to extricate myself from entanglements with the English language while I was writing and revising both volumes.

Suzanne Gervais, maker of animated films, became my friend at a bad time in my life and has constantly encouraged me in my work on this book, from the time I first decided I would write it early in 1983. Although she had visual work enough of her own to think about, she was always willing to examine my drawings of the figures and give me practical advice on how to do them. For that help and for her friendship which I am fortunate to enjoy, I thank her affectionately.

One point raised at the very end of volume I had there to be left unsettled. This concerned the likelihood that Brennan's improvement of Volberg's theorem, presented in article 1 of the addendum, was essentially best possible. An argument to support that claim was made on pp. 578–83; it depended, however, on an example which had been reported, but not described, by Borichev and Volberg. No description was available before Volume I went to press, so the claim about Brennan's improvement could not be fully substantiated.

Now we are able to complete verification of the claim by providing the missing example. Its description is found at the end of a paper by Borichev and Volberg appearing in the very first issue of the new Leningrad periodical *Algebra i analiz*. We continue using the notation of the addendum to volume I.

Two functions have to be constructed. The first, $h(\xi)$, should be decreasing for $0 < \xi < \infty$ and satisfy $\xi h(\xi) \geqslant 1$, together with the relation

$$\int_0^1 \log h(\xi)\, d\xi = \infty.$$

The second, $F(z)$, is to be continuous on the closed unit disk and $\mathscr{C}_\infty$ in its interior, with

$$\left|\frac{\partial F(z)}{\partial \bar{z}}\right| \leqslant \exp(-h(\log(1/|z|))), \qquad |z| < 1,$$

$$|F(e^{i\vartheta})| > 0 \quad \text{a.e.},$$

and

$$\int_{-\pi}^{\pi} \log|F(e^{i\vartheta})|\, d\vartheta = -\infty.$$

The function $F(z)$ we obtain will in fact be analytic in most of the unit disk Δ, ceasing to be so only in the neighborhood of some very small segments on the positive radius, accumulating at 1. The function $h(\log(1/x))$ will be very much larger than $1/\log(1/x)$ for most of the

$x \in (0, 1)$ contiguous to those segments.

Three simple ideas form the basis for the entire construction:

(1) In a domain $\mathscr{E}$ with piecewise analytic boundary having a 90° corner (internal measure) at ζ, say, we have

$$\omega_{\mathscr{E}}(I, z) \leqslant K_z |I|^2, \quad z \in \mathscr{E},$$

for arcs I on $\partial\mathscr{E}$ containing ζ (see volume I, pp 260–1);

(2) The use of a Blaschke product involving factors affected with fractional exponents to 'correct', in an infinitely connected subdomain of Δ, a function analytic there and multiple-valued, but with single valued modulus;

(3) The use of a smoothing operation inside Δ, scaled according to that disk's hyperbolic geometry.

We start by looking at harmonic measure in domains $\mathscr{E} = \Delta \sim [a, 1]$, where $0 < a < 1$. According to (1), if $\eta > 0$ is small (and $< 1 - a$), we have

$$\omega_{\mathscr{E}}(E_{\eta}, 0) \leqslant O(\eta^2)$$

for the sets $E_{\eta} = [1 - \eta, 1] \cup I_{\eta}$, where I_{η} is the arc of length η on the unit circle, centered at 1. This is so because $\partial\mathscr{E}$ has two (internal) square corners at 1 that contribute separately to harmonic measure. (The slit $[a, 1]$ can be opened up by making a conformal mapping of $\mathscr{E}$ given by $z \longrightarrow \sqrt{(a - z)}$; when this is done the two corners at 1 are separated and they remain square.) Suppose that, for some given $a \in (0, 1)$, we fix an $\eta > 0$ small enough to make $\omega_{\mathscr{E}}(E_{\eta}, 0)/\eta$ less than some preassigned amount. Then, if we put $\mathscr{G} = \Delta \sim [a, 1 - \eta]$, we will have, by simple comparison of $\omega_{\mathscr{G}}(I_{\eta}, z)$ and $\omega_{\mathscr{E}}(E_{\eta}, z)$ in $\mathscr{E}$,

$$\omega_{\mathscr{G}}(I_{\eta}, 0)/\eta \leqslant \omega_{\mathscr{E}}(E_{\eta}, 0)/\eta.$$

This relation is taken as the base of an inductive process. Beginning with an $a_1 > 2/(2 + \sqrt{3})$ and < 1 (we shall see presently why the first condition is needed), we take a b_1, $a_1 < b_1 < 1$, so close to 1 as to make

$$\frac{\omega_{\mathscr{G}_1}(I_1, 0)}{|I_1|} < \frac{1}{2}$$

for $\mathscr{G}_1 = \Delta \sim [a_1, b_1]$ and the arc I_1 of length $1 - b_1$ on $\partial\Delta$ centered at 1. One next chooses a_2, $b_1 < a_2 < 1$, in a way to be specified later on (a_2 will in fact be much *closer* to 1 than b_1), and then takes

b_2, $a_2 < b_2 < 1$, near enough to 1 to have

$$\frac{\omega_{\mathscr{G}_2}(I_2, 0)}{|I_2|} < \frac{1}{4}$$

and

$$|I_2| < \tfrac{1}{2}|I_1|$$

for $\mathscr{G}_2 = \Delta \sim [a_2, b_2]$ and the arc I_2 of length $1 - b_2$ on $\partial\Delta$ centered at 1. Continuing this procedure indefinitely, we get a sequence of segments

$$J_n = [a_n, b_n],$$

where $b_n < a_{n+1} < b_{n+1} < 1$, and nested arcs I_n of length $1 - b_n$ on $\partial\Delta$, each centered at 1, with

$$\frac{\omega_{\mathscr{G}_n}(I_n, 0)}{|I_n|} < \frac{1}{2^n}$$

for the corresponding domains $\mathscr{G}_n = \Delta \sim J_n$, and

$$|I_n| < \tfrac{1}{2}|I_{n-1}|.$$

Take now

$$\mathscr{D} = \Delta \sim J_1 \sim J_2 \sim J_3 \sim \cdots;$$

then, since $\mathscr{D}$ is contained in each $\mathscr{G}_n$, the principle of extension of domain tells us that

$$\frac{\omega_{\mathscr{D}}(I_n, 0)}{|I_n|} \leqslant \frac{\omega_{\mathscr{G}_n}(I_n, 0)}{|I_n|} < \frac{1}{2^n}.$$

Our first ingredient in the formation of the desired $F(z)$ is a function $u(z)$ positive and harmonic in $\mathscr{D}$. Let $T_n(\vartheta)$ be periodic of period 2π, with

$$T_n(\vartheta) = \frac{1}{|I_n|}\left(1 - \frac{2|\vartheta|}{|I_n|}\right)^+ \qquad \text{for } -\pi \leqslant \vartheta \leqslant \pi.$$

The graph of $T_n(\vartheta)$ for $|\vartheta| \leqslant \pi$ is an isosceles triangle of height $1/|I_n|$ with its base on the segment $\{|\vartheta| \leqslant |I_n|/2\}$ corresponding to the arc I_n. We have

$$\int_{-\pi}^{\pi} T_n(\vartheta)\,d\vartheta = \frac{1}{2}$$

while

$$\int_{-\pi}^{\pi} T_n(\vartheta)\,d\omega_{\mathscr{D}}(e^{i\vartheta},\,0) \;\leqslant\; \frac{\omega_{\mathscr{D}}(I_n,\,0)}{|I_n|} \;<\; \frac{1}{2^n},$$

so

$$\int_{-\pi}^{\pi} \sum_{n=1}^{\infty} T_n(\vartheta)\,d\omega_{\mathscr{D}}(e^{i\vartheta},\,0) \;<\; \infty.$$

although

$$\int_{-\pi}^{\pi} \sum_{n=1}^{\infty} T_n(\vartheta)\,d\vartheta \;=\; \infty$$

For $z \in \mathscr{D}$, we put

$$u(z) \;=\; \int_{-\pi}^{\pi} \sum_{n=1}^{\infty} T_n(\vartheta)\,d\omega_{\mathscr{D}}(e^{i\vartheta},\,z)\;;$$

the integral on the right is certainly *finite* by the third of the preceding four relations and Harnack's inequality, so $u(z)$ is harmonic in $\mathscr{D}$ and

$$u(z) \;>\; 0$$

there. For $0 < |\vartheta| \leqslant \pi$, $\sum_{n=1}^{\infty} T_n(\vartheta)$ is continuous (and *even* locally Lip 1 !), so at these values of ϑ,

$$u(z) \;\longrightarrow\; \sum_{n=1}^{\infty} T_n(\vartheta) \qquad \text{as } z \longrightarrow e^{i\vartheta}$$

from within $\mathscr{D}$. (It is practically obvious that the corresponding points $e^{i\vartheta}$ are regular for the Dirichlet problem in $\mathscr{D}$ — in fact, *all* points of $\partial\mathscr{D}$ are regular.) Taking $u(e^{i\vartheta})$ *equal* to $\sum_{n=1}^{\infty} T_n(\vartheta)$, we thus get a function $u(z)$ continuous in $\bar{\Delta} \sim \{1\}$, and we have

$$\int_{-\pi}^{\pi} u(e^{i\vartheta})\,d\vartheta \;=\; \infty.$$

The function $u(z)$ has, *locally*, a harmonic conjugate $\tilde{u}(z)$ in $\mathscr{D}$. The latter, of course, need not be *single-valued* in the infinitely connected domain $\mathscr{D}$; we nevertheless put

$$f(z) \;=\; e^{-(u(z)+i\tilde{u}(z))}$$

for $z \in \mathscr{D}$, obtaining a function *analytic and multiple-valued* in $\mathscr{D}$ whose modulus, $e^{-u(z)}$, is *single-valued* there. If $e^{i\vartheta} \neq 1$, *any given branch* of $\tilde{u}(z)$ is *continuous up to* $e^{i\vartheta}$, because $u(e^{it}) = \sum_{n=1}^{\infty} T_n(t)$ is Lip 1 for t near ϑ.

(To verify this, it suffices to look at $u(z)$ and $\tilde{u}(z)$ in the intersection of $\mathscr{D}$ with a small disk about $e^{i\vartheta}$ avoiding the J_n; if there is still any doubt, map that intersection conformally onto Δ.) It therefore makes sense to talk about the *multiple-valued*, but locally continuous boundary value $f(e^{i\vartheta})$ when $e^{i\vartheta} \neq 1$; the modulus $|f(e^{i\vartheta})|$ is again single-valued, being equal to $\exp(-u(e^{i\vartheta}))$. By the previous relation, we have

$$\int_{-\pi}^{\pi} \log|f(e^{i\vartheta})| \, d\vartheta \;=\; -\infty.$$

It is now necessary to cure the multiple-valuedness of $f(z)$; that is where the second of our ideas comes in. In constructing the $J_n = [a_n, b_n]$ and the arcs I_n, there is nothing to prevent our choosing the a_n so as to have

$$\sum_{n=1}^{\infty} (1 - a_n) \;<\; \infty;$$

we henceforth assume that this has been done. (A much *faster* convergence of a_n to 1 will indeed be required later on.) Our condition on the a_n guarantees that the sum

$$\sum_{n=1}^{\infty} \mu_n \log \left| \frac{z - a_n}{1 - a_n z} \right|$$

converges uniformly in the interior of $\Delta \sim \bigcup_n \{a_n\} \supseteq \mathscr{D}$ whenever the coefficients μ_n are bounded. If $0 \leqslant \mu_n \leqslant 1$, that sum is then equal to a function $v(z)$, *harmonic and* $\leqslant 0$ in $\mathscr{D}$. For the latter, there is a multiple-valued harmonic conjugate $\tilde{v}(z)$ defined in $\mathscr{D}$, and we have finally a function

$$b(z) \;=\; e^{v(z) + i\tilde{v}(z)} \;=\; \prod_{n=1}^{\infty} \left(\frac{a_n - z}{1 - a_n z} \right)^{\mu_n},$$

analytic but multiple-valued in $\mathscr{D}$. The *modulus* $|b(z)| = e^{v(z)}$ is single-valued in $\mathscr{D}$.

The points a_n accumulate only at 1, so any branch of $b(z)$ is continuous up to points $e^{i\vartheta} \neq 1$ of the unit circle. For such points, $|b(e^{i\vartheta})| = 1$, and of course $|b(z)| \leqslant 1$ in $\mathscr{D}$, since $v(z) \leqslant 0$ there.

By proper adjustment of the exponents μ_n we can make the *product* $b(z)f(z)$ *single-valued* in $\mathscr{D}$, and hence analytic there in the ordinary sense. Consider what happens when z describes a simple closed path in the counterclockwise sense about just one of the slits $J_n = [a_n, b_n]$. Each given branch of the harmonic conjugate $\tilde{u}(z)$ will then increase by a certain (real) amount λ_n, independent of the branch. At the same time, every branch of $\tilde{v}(z) = \arg b(z)$ will increase by $2\pi\mu_n$. We take μ_n between 0 and 1 so as to make

$$2\pi\mu_n - \lambda_n$$

an integral multiple of 2π; this is clearly possible, and, once it is done, every branch of $\arg(b(z)f(z)) = \tilde{v}(z) - \tilde{u}(z)$ *increases by that amount* when z goes around a path of the kind just mentioned. Then the *product* $b(z)f(z)$ just comes back to its original value! Choosing in this way a value of μ_n, $0 \leqslant \mu_n \leqslant 1$, for *every* n, we ensure that $b(z)f(z)$ is single-valued in $\mathscr{D}$. Note that we have

$$|b(z)f(z)| \leqslant e^{-u(z)} \leqslant 1, \qquad z \in \mathscr{D},$$

and, since $|b(e^{i\vartheta})| = 1$ for $e^{i\vartheta} \neq 1$,

$$|b(e^{i\vartheta})f(e^{i\vartheta})| = |f(e^{i\vartheta})| > 0, \qquad e^{i\vartheta} \neq 1.$$

Because the product $b(z)f(z)$ is analytic in $\mathscr{D}$, we have there

$$\frac{\partial}{\partial\bar{z}}\, b(z)f(z) = 0;$$

the expression on the left may therefore be looked on as a *distribution* in Δ, supported on the slits J_n of $\Delta \sim \mathscr{D}$. In order to obtain a $\mathscr{C}_\infty$ function defined in Δ, we *smooth out* $b(z)f(z)$; that is our *third idea*. The smoothing is scaled according to the *square* of the gauge for the hyperbolic metric in Δ, i.e., like $1/(1-|z|)^2$.

Taking a $\mathscr{C}_\infty$ function $\psi(\rho) \geqslant 0$ supported on the interval $[1/4, 1/2]$ of the real axis, with

$$\int_0^{1/2} \psi(\rho)\rho\,\mathrm{d}\rho = \frac{1}{2\pi},$$

we put, for $z \in \Delta$,

$$G(z) = \iint_\Delta \psi\left(\frac{|z-\zeta|}{(1-|z|)^2}\right) \frac{b(\zeta)f(\zeta)}{(1-|z|)^4}\,\mathrm{d}\xi\,\mathrm{d}\eta$$

(writing, as usual, $\zeta = \xi + i\eta$).

The first thing to observe here is that the expression on the right *makes sense*. Although $b(\zeta)f(\zeta)$ is defined merely in $\mathscr{D}$, the slits J_n making up $\Delta \sim \mathscr{D}$ are of *planar Lebesgue measure zero*, so we *only need* the values of the product in $\mathscr{D}$ in order to do the integral. The *second* observation is that $G(z)$ is $\mathscr{C}_\infty$ in $\mathscr{D}$. As a function of ζ, $\psi(|z-\zeta|/(1-|z|)^2)$ vanishes outside the disk $|\zeta-z| \leqslant \frac{1}{2}(1-|z|)^2$ which, however, lies well within Δ

for $z \in \Delta$, since then $|z| + \frac{1}{2}(1 - |z|)^2 < 1$. We may therefore differentiate under the integral sign with respect to z or $\bar{z}$ as often as we wish, $\psi(\rho)$ being $\mathscr{C}_\infty$ (its identical vanishing for ρ near 0 helps here), and $|b(\zeta)f(\zeta)|$ being < 1 in $\mathscr{D}$. In this way we verify that $G(z)$ is $\mathscr{C}_\infty$ in Δ, and get (practically 'by inspection') the crude estimate

$$\left| \frac{\partial G(z)}{\partial \bar{z}} \right| \leqslant \frac{\text{const.}}{(1 - |z|)^2}, \qquad |z| < 1.$$

As for $G(z)$, just *an average* of the function $b(\zeta)f(\zeta)$, we have

$$|G(z)| < 1, \qquad |z| < 1.$$

The *third* thing to observe is that $G(z)$ is actually *analytic* in a fairly large subset of Δ. Because $\psi(\rho)$ vanishes for $\rho \geqslant 1/2$, the integration in the above formula for $G(z)$ is really over the disk

$$\bar{\Delta}_z = \{\zeta : |\zeta - z| \leqslant \tfrac{1}{2}(1 - |z|)^2\}$$

which, as we have just seen, *lies in* Δ when $|z| < 1$. Suppose that $\bar{\Delta}_z$ touches *none* of the slits J_n. Then $\bar{\Delta}_z \subseteq \mathscr{D}$ where $b(\zeta)f(\zeta)$ is *analytic* and, writing $\zeta = z + re^{i\vartheta}$, we have

$$G(z) = \int_0^{(1-|z|)^2/2} \int_{-\pi}^\pi b(z + re^{i\vartheta})\, f(z + re^{i\vartheta})\, \psi(r/(1 - |z|)^2) \frac{r\,d\vartheta\,dr}{(1 - |z|)^4}.$$

Using Cauchy's theorem to perform the first integration with respect to ϑ and then making the change of variable $r/(1 - |z|)^2 = \rho$, we obtain the value $2\pi b(z)f(z) \int_0^{1/2} \psi(\rho)\rho\,d\rho = b(z)f(z)$, i.e.,

$$G(z) = b(z)f(z) \qquad \text{if} \quad \bar{\Delta}_z \subseteq \mathscr{D}.$$

When $\bar{\Delta}_z \subseteq \mathscr{D}$, the disks $\bar{\Delta}_{z'}$ also lie in $\mathscr{D}$ for the z' belonging to some *neighborhood* of z; we thus have $G(z') = b(z')f(z')$ in that neighborhood, and $G(z')$ (like $b(z')f(z')$) is then *analytic at* z. For the z in Δ such that $\bar{\Delta}_z \subseteq \mathscr{D}$, we therefore have

$$\frac{\partial G(z)}{\partial \bar{z}} = 0$$

although, for the *remaining* z in the unit disk, only the above estimate on $\partial G(z)/\partial \bar{z}$ is available. It is necessary to examine the set of those remaining z.

They are precisely the ones for which $\bar{\Delta}_z$ intersects with some J_n. We proceed to describe the set

$$B_n = \{z \in \Delta : \bar{\Delta}_z \cap J_n \neq \varnothing\}.$$

Write for the moment $J_n = [a, b]$, dropping the subscripts on a_n and b_n. If $\bar{\Delta}_z$ is to intersect with $[a, b]$, we must have $|z| > 2 - \sqrt{3}$. Indeed, a, as one of the a_n, is $\geqslant a_1$ which we initially took $> 2/(2 + \sqrt{3})$, while $\bar{\Delta}_z$ lies in the disk $\{|\zeta| \leqslant |z| + \frac{1}{2}(1 - |z|)^2\}$ whose radius increases with $|z|$. For $|z| = 2 - \sqrt{3}$, that radius works out to $2/(2 + \sqrt{3})$, so if $|z| \leqslant 2 - \sqrt{3}$, $[a, b]$ would lie *outside* the disk containing $\bar{\Delta}_z$; $|z|$ is thus $> 2 - \sqrt{3}$ for $z \in B_n$. Now when $2 - \sqrt{3} < |z| < 1$, $|z| - \frac{1}{2}(1 - |z|)^2 > 0$, so $\bar{\Delta}_z$ is in fact contained in the *ring*

$$|z| - \tfrac{1}{2}(1 - |z|)^2 \;\leqslant\; |\zeta| \;\leqslant\; |z| + \tfrac{1}{2}(1 - |z|)^2$$

(*that's* why a_1 was chosen $> 2/(2 + \sqrt{3})$!). Therefore, if $\bar{\Delta}_z$ intersects with $[a, b]$, we must have

$$|z| - \tfrac{1}{2}(1 - |z|)^2 \;\leqslant\; b,$$
$$|z| + \tfrac{1}{2}(1 - |z|)^2 \;\geqslant\; a.$$

Both left sides are increasing functions of $|z|$ (for $z \in \Delta$), so these relations are equivalent to

$$a' \;\leqslant\; |z| \;\leqslant\; b',$$

where

$$a' + \tfrac{1}{2}(1 - a')^2 \;=\; a,$$
$$b' - \tfrac{1}{2}(1 - b')^2 \;=\; b.$$

In $(0, 1)$ these equations have the solutions

$$a' \;=\; \sqrt{(2a - 1)},$$
$$b' \;=\; 2 - \sqrt{(3 - 2b)};$$

for the first we need $a > 1/2$ but have in fact $a > 2/(2 + \sqrt{3})$. Using differentiation, one readily verifies that $a' < a$ and $b < b' < 1$.

We see that B_n (the set of $z \in \Delta$ for which $\bar{\Delta}_z$ intersects with $[a, b]$) is an oval-shaped region including $[a, b]$ and contained in the ring $a' \leqslant |z| \leqslant b'$; its boundary crosses the x-axis at the points a' and b'. When a is close to 1, B_n is quite thin in the vertical direction because, if $\bar{\Delta}_z$ touches the x-axis at all, we must have $|\Im z| \leqslant \frac{1}{2}(1 - |z|)^2$.

One can specify the a_n and b_n so as to ensure *disjointness* of the oval regions B_n. The preceding description shows that this will be the case if the *rings* $a'_n \leqslant |z| \leqslant b'_n$ are disjoint, where (restoring the subscript n)

$$a'_n \;=\; \sqrt{(2a_n - 1)},$$
$$b'_n \;=\; 2 - \sqrt{(3 - 2b_n)};$$

i.e., if $b'_n < a'_{n+1}$ for $n = 1, 2, 3, \ldots$. It is easy to arrange this in making the successive choices of the a_n and b_n; all we need is to have

$$a_{n+1} = a'_{n+1} + \tfrac{1}{2}(1 - a'_{n+1})^2 > b'_n + \tfrac{1}{2}(1 - b'_n)^2.$$

Here it is certainly true that $b_n < b'_n < 1$ when $0 < b_n < 1$; then, however, the extreme right-hand member of the relation is still < 1, and numbers $a_{n+1} < 1$ satisfying it *are available*. There is obviously no obstacle to our making the a_n increase as rapidly as we like towards 1; we can, in particular, have

$$\sum_{n=1}^{\infty} (1 - a_n) < \infty.$$

We henceforth assume that the last precaution has been heeded in the selection of the a_n. The B_n will then lie in their respective *disjoint* rings $a'_n \leqslant |z| \leqslant b'_n$ besides being all included in the cusp-shaped region $|\Im z| \leqslant \tfrac{1}{2}(1 - |z|)^2$ and, of course, in the right half plane. According to what we have already seen, $G(z)$ is equal to the analytic function $b(z)f(z)$ for $z \in \Delta$ *outside* all of the B_n, so then $\partial G(z)/\partial \bar{z} = 0$. *Within* any of the B_n, we have only the estimate $|\partial G(z)/\partial \bar{z}| \leqslant \text{const.}/(1 - |z|)^2$.

Because of the configuration of the B_n, $G(z)$ is continuous up to the points of $\partial \Delta \sim \{1\}$. Indeed, when $z \in \Delta$ tends to $e^{i\vartheta} \neq 1$, it must eventually *leave* the region $\{\Re z > 0, |\Im z| \leqslant \tfrac{1}{2}(1 - |z|)^2\}$ in which all the B_n lie, and then $G(z)$ becomes equal to $b(z)f(z)$ which has the continuous limit $b(e^{i\vartheta})f(e^{i\vartheta})$ away from 1 on the unit circumference.

If $z \in \Delta$ tends to 1 from *outside* any sector with vertex at 1 of the form $|\arg(1 - z)| \leqslant \alpha$, $0 < \alpha < \pi/2$, we have

$$G(z) \longrightarrow 0.$$

To see this, we argue that such z must leave the region $|\Im z| \leqslant \tfrac{1}{2}(1 - |z|)^2$, making $G(z) = b(z)f(z)$. Then, however,

$$\log|b(z)f(z)| \leqslant -u(z) = -\int_{-\pi}^{\pi} \sum_{n=1}^{\infty} T_n(\vartheta) \, d\omega_{\mathscr{D}}(e^{i\vartheta}, z),$$

and it suffices to show that the expression on the right tends to $-\infty$ whenever $z \longrightarrow 1$ from outside any of the sectors just mentioned. This is so due to the fact that $\sum_{n=1}^{\infty} T_n(\vartheta) \longrightarrow \infty$ for $\vartheta \longrightarrow 0$, as may be verified by taking the region

$$\mathscr{E} = \Delta \sim [1/2, 1] \subseteq \mathscr{D}$$

and comparing harmonic measure for $\mathscr{D}$ with that for $\mathscr{E}$. By the principle of extension of domain, $d\omega_{\mathscr{E}}(e^{i\vartheta}, z) \leqslant d\omega_{\mathscr{D}}(e^{i\vartheta}, z)$ for $z \in \mathscr{E}$, so we need only check that

$$\int_{-\pi}^{\pi} \sum_{n=1}^{\infty} T_n(\vartheta)\, d\omega_{\mathscr{E}}(e^{i\vartheta}, z) \longrightarrow \infty$$

as $z \longrightarrow 1$ from outside any of the sectors in question. That, however, should be *clear*. Let the reader imagine that $\mathscr{E}$ has been mapped conformally onto the upper half plane so as to take the vertices of its two corners at 1 to -2 and 2, say, and then think about how the *ordinary* Poisson integral corresponding to the last expression must behave as one moves towards -2 or 2 from the upper half plane.

We put finally

$$F(z) = c \exp\left(-K\frac{1+z}{1-z}\right) G(z)$$

for $z \in \Delta$, with c a *small* constant > 0 and K a *large* one. The exponential serves two purposes. It is, in the first place, < 1 in modulus in Δ and continuous up to $\partial\Delta \sim \{1\}$ where it has boundary values of modulus 1. When $z \longrightarrow 1$ from *within* any sector $|\arg(1-z)| \leqslant \alpha$, $0 < \alpha < \pi/2$, the exponential tends to *zero*, making $F(z) \longrightarrow 0$, since $|G(z)| < 1$ in Δ. This, however, is also true when $z \longrightarrow 1$ from *outside* such a sector because then $G(z) \longrightarrow 0$ as we have just seen. Thus,

$$F(z) \longrightarrow 0 \quad \text{for } z \longrightarrow 1, \ z \in \Delta.$$

We have already remarked that $G(z)$ is continuous up to $\partial\Delta \sim \{1\}$, where it coincides with $b(z)f(z)$, so we have

$$F(z) \longrightarrow c\, e^{-Ki\cot(\vartheta/2)} b(e^{i\vartheta}) f(e^{i\vartheta})$$

when $z \in \Delta$ tends to $e^{i\vartheta} \neq 1$. Denoting the boundary value on the right by $F(e^{i\vartheta})$, we have $|F(e^{i\vartheta})| = c|f(e^{i\vartheta})| = c\exp(-u(e^{i\vartheta}))$, and this tends to *zero* as $\vartheta \longrightarrow 0$ since $u(e^{i\vartheta}) = \sum_{n=1}^{\infty} T_n(\vartheta)$ then tends to ∞. *The function $F(z)$ thus extends continuously up to the unit circumference* thanks to the factor $\exp(-K(1+z)/(1-z))$. We have $|F(e^{i\vartheta})| = c|f(e^{i\vartheta})| > 0$ for $e^{i\vartheta} \neq 1$, and

$$\int_{-\pi}^{\pi} \log|F(e^{i\vartheta})|\, d\vartheta = 2\pi \log c + \int_{-\pi}^{\pi} \log|f(e^{i\vartheta})|\, d\vartheta = -\infty.$$

Since $G(z)$ is $\mathscr{C}_{\infty}$ inside Δ, so is $F(z)$. The second service rendered by the factor $\exp(-K(1+z)/(1-z))$ is to make $\partial F(z)/\partial\bar{z}$ small near $\partial\Delta$.

Outside the B_n, $F(z)$ (like $G(z)$) is *analytic, so* $\partial F(z)/\partial \bar{z} = 0$. *Within any* of the B_n, we use the formula

$$\frac{\partial F(z)}{\partial \bar{z}} = c \exp\left(-K\frac{1+z}{1-z}\right)\frac{\partial G(z)}{\partial \bar{z}},$$

which holds because the exponential is analytic in Δ. The B_n all lie in the right half plane, and in them,

$$|\Im z| \leqslant \tfrac{1}{2}(1-|z|)^2 < \tfrac{1}{2}(1-|z|),$$

whence

$$\Re\frac{1+z}{1-z} \geqslant \frac{\text{const.}}{1-|z|}.$$

This makes

$$\left|\frac{\partial F(z)}{\partial \bar{z}}\right| \leqslant c \exp\left(-K\frac{\text{const.}}{1-|z|}\right)\left|\frac{\partial G(z)}{\partial \bar{z}}\right|$$

for z belonging to any of the B_n. As we have seen, the last factor on the right is $\leqslant \text{const.}/(1-|z|)^2$ which, for $|z| < 1$ near 1, is *greatly outweighed* by the exponential. Bearing in mind that $\log(1/|z|) \sim 1-|z|$ for $|z| \longrightarrow 1$, we see that the constants c and K can be adjusted so as to have

$$\left|\frac{\partial F(z)}{\partial \bar{z}}\right| \leqslant \exp\left(-\frac{1}{\log(1/|z|)}\right),$$

within the B_n at least. But then this holds *outside* them as well (in Δ, including in the neighborhood of 0), because $\partial F(z)/\partial \bar{z} = 0$ there.

$F(z)$ has now been shown to enjoy all the properties enumerated at the beginning of this exposition except the one involving the function $h(\xi)$, not yet constructed. That construction comes almost as an afterthought. Since the sets B_n lie inside the disjoint rings $a'_n \leqslant |z| \leqslant b'_n$, we start by putting $h(\log(1/|z|)) = 1/\log(1/|z|)$ on each of the latter; in view of the preceding relation, this *already implies* that

$$\left|\frac{\partial F(z)}{\partial \bar{z}}\right| \leqslant \exp(-h(\log(1/|z|)))$$

throughout Δ, no matter *how* $h(\log(1/|z|))$ is defined for the remaining $z \in \Delta$, because the left side is *zero* outside the B_n. To complete the definition of $h(\xi)$ for $0 < \xi < \infty$, we continue to use $h(\log(1/|z|)) = 1/\log(1/|z|)$ on the range $0 < |z| \leqslant a'_1$ and then take $h(\log(1/|z|))$ to be *linear in*

$|z|$ on each of the *complementary* rings

$$b'_n \leqslant |z| \leqslant a'_{n+1}, \qquad n = 1, 2, 3, \ldots .$$

The function $h(\xi)$ we obtain in this fashion is certainly *decreasing* (in ξ); $h(\log(1/|z|))$ is also > 1 for $|z| \geqslant b'_1$, because $b'_1 > a_1 > 2/(2+\sqrt{3}) > 1/e$. $h(\log(1/|z|))$ is moreover $\geqslant 1/\log(1/|z|)$ on the complementary rings, for $1/\log(1/x)$ is a *convex* function of x for $1/e^2 < x < 1$, and $b'_1 > 1/e^2$. In terms of the variable $\xi = \log(1/|z|)$ we therefore have

$$\xi h(\xi) \geqslant 1, \qquad 0 < \xi < \infty.$$

The trick in arranging to have

$$\int_0^1 \log h(\xi)\,d\xi = \infty$$

is to use *linearity* of $h(\log(1/x))$ in x on each interval $b'_n \leqslant x \leqslant a'_{n+1}$ to get lower bounds on the integrals

$$\int_{\log(1/a'_{n+1})}^{\log(1/b'_n)} \log h(\xi)\,d\xi.$$

We have indeed $h(\xi) > 1$ for $\xi \leqslant \log(1/b'_n) \leqslant \log(1/b'_1)$ and $h(\log(1/a'_{n+1})) = 1/\log(1/a'_{n+1})$, so the linearity just mentioned makes $h(\xi) \geqslant 1/2\log(1/a'_{n+1})$ for $(b'_n + a'_{n+1})/2 \leqslant e^{-\xi} \leqslant a'_{n+1}$, i.e., for $\log(1/a'_{n+1}) \leqslant \xi \leqslant \log(2/(b'_n + a'_{n+1}))$. The preceding integral is therefore

$$\geqslant \log\left(\frac{2a'_{n+1}}{a'_{n+1} + b'_n}\right) \cdot \log^+\left(\frac{1}{2\log(1/a'_{n+1})}\right),$$

since $\log h(\xi)$ is > 0 on the whole range of integration. For any *given* value of b'_n, $0 < b'_n < 1$, *the last expression tends to* ∞ *as* $a'_{n+1} \longrightarrow 1$! We can therefore make it $\geqslant 1$ by taking $a'_{n+1} > b'_n$ *close enough to* 1, and that can in turn be achieved by choosing $a_{n+1} = a'_{n+1} + \frac{1}{2}(1 - a'_{n+1})^2$ sufficiently near 1. We therefore *select the successive a_n in accordance with this requirement* in carrying out the inductive procedure followed at the beginning of our construction. That will certainly guarantee that $b'_n < a'_{n+1}$ (which we needed), and may obviously be done so as to have $\sum_{n=1}^\infty (1 - a_n) < \infty$ (by making the a_n tend more rapidly towards 1 we can only improve matters).

Once the a_n have been specified in this way, we will have

$$\int_{\log(1/a'_{n+1})}^{\log(1/b'_n)} \log h(\xi)\,d\xi \geqslant 1$$

for each n, and therefore

$$\int_0^1 \log h(\xi) \, d\xi \;=\; \infty.$$

Our construction of the functions $F(z)$ and $h(\xi)$ with the desired properties is thus complete, and the gap in the second half of article 2 in the addendum to volume I filled in. This means, in particular, that in the hypothesis of Brennan's result (top of p. 574, volume I), *the condition that $M(v)/v^{1/2}$ be increasing cannot be replaced by the weaker one that $M(v)/v^{1/2} \geqslant 2$.*

January 26, 1990
Outremont, Québec.

Errata for volume I

Location	Correction
page 66	At end of the theorem's statement, words in roman should be in italic, and words in italic in roman.
pages 85, 87	In running title, delete bar under second M_n but keep it under first one.
page 102	In heading to §E, delete bar under M_n in first and third $\mathscr{C}_R(\{M_n\})$ but keep it in second one.
page 126, line 8	In statement of theorem, change *determinant* to *determinate*.
page 135, line 11	In displayed formula, change w* to w*.
page 136, line 4 from bottom	In displayed formula $|P(x_0)|^2 v(\{x_0\})$ should stand on the right.
page 177, line 11 from bottom	The sentence beginning 'Since, as we already' should start on a new line, separated by a horizontal space from the preceding one
page 190	In last displayed formula, change x'' to x^n
page 212 and following even numbered pages up to page 232 inclusive	Add to running title: *Comparison of $\mathscr{C}_W(0)$ to $\mathscr{C}_W(0+)$*
page 230	In the last two displayed formulas replace $(1 - \alpha^2)$ throughout by $|1 - \alpha^2|$.
page 241, line 3	Change b_b^2 in denominator of right-hand expression to b_n^2.
page 270, line 10	Change $F(z)$ to $F(Z)$.
page 287	In figure 69, B_1 and B_2 should designate the lower and upper sides of $\mathscr{D}_0$, not $\mathscr{D}$.

page 379, line 8 from bottom	Change comma after 'theorem' to a full stop, and capitalize 'if'.
page 394, line 3	Change y_1 to y_l.
page 466, last line	Delete full stop.
page 563, line 9	Change 'potential' to 'potentials'.
page 574, line 9 from bottom	Delete full stop after 'following'.
page 604	In running title, '*volume*' should not be capitalized.
page 605	In titles of §§C.1 and C.4 change 'Chapter 8' to 'Chapter VIII'.

IX

Jensen's formula again

The derivations of the two main results in this chapter – Pólya's gap theorem and a *lower* bound for the completeness radius of a set of imaginary exponentials – are both based on the same simple idea: application of Jensen's formula with a circle of varying radius and *moving* centre. I learned about this device from a letter that J.-P. Kahane sent me in 1958 or 1959, where it was used to prove the first of the results just mentioned. Let us begin our discussion with an exposition of that proof.

A.   Pólya's gap theorem
Consider a Taylor series expansion

$$f(w) = \sum_0^\infty a_n w^n$$

with radius of convergence equal to 1. The function $f(w)$ must have at least one singularity on the circle $|w| = 1$. It was observed by Hadamard that *if many of the coefficients a_n are zero*, i.e., if, as we say, *the Taylor series has many gaps, $f(w)$ must have lots of singularities on the series' circle of convergence*. In a certain sense, *the more gaps the power series has, the more numerous must be the singularities associated thereto on its circle of convergence*.

This phenomenon was studied by Hadamard and by Fabry; the best result was given by Pólya. In order to formulate it, Pólya invented the maximum density bearing his name which has already appeared in Chapter VI.

In this §, it will be convenient to denote by $\mathbb{N}$ *the set of integers ≥ 0* (and *not just the ones ≥ 1 as is usually done, and as we will do in §B!). If $\Sigma \subseteq \mathbb{N}$, we denote by $n_\Sigma(t)$ the number of elements of Σ in $[0, t]$, $t \geq 0$*. The *Pólya*

1

maximum density of Σ, studied in §E.3 of Chapter VI, is the quantity

$$D_\Sigma^* = \lim_{\lambda \to 1-} \left(\limsup_{r \to \infty} \frac{n_\Sigma(r) - n_\Sigma(\lambda r)}{(1 - \lambda)r} \right).$$

We have shown in the article referred to that the outer limit really does exist for any Σ, and that D_Σ^* is the *minimum of the densities of the measurable sequences containing* Σ. In this §, we use a property of D_Σ^* furnished by the following

Lemma. *Given* $\varepsilon > 0$, *we have, for* $\rho \geqslant \varepsilon r$,

$$\frac{n_\Sigma(r + \rho) - n_\Sigma(r)}{\rho} \leqslant D_\Sigma^* + \varepsilon$$

when r is large enough (depending on ε).

Proof. According to the above formula, if N is large enough and

$$\lambda = (1 + \varepsilon)^{-1/N},$$

we will have

$$\frac{n_\Sigma(r) - n_\Sigma(\lambda r)}{(1 - \lambda)r} < D_\Sigma^* + \frac{\varepsilon}{2}$$

for $r \geqslant R$, say. *Fix* such an N.

When $r \geqslant R$, we certainly have

$$\frac{n_\Sigma(\lambda^{-k-1}r) - n_\Sigma(\lambda^{-k}r)}{(\lambda^{-k-1} - \lambda^{-k})r} < D_\Sigma^* + \frac{\varepsilon}{2}$$

for $k = 0, 1, 2, \ldots$, so

$$\frac{n_\Sigma(\lambda^{-k}r) - n_\Sigma(r)}{(\lambda^{-k} - 1)r} < D_\Sigma^* + \frac{\varepsilon}{2}$$

for $k = 1, 2, 3, \ldots$. Let

$$\rho \geqslant \varepsilon r = (\lambda^{-N} - 1)r.$$

Then, if k is the least integer such that $(\lambda^{-k} - 1)r \geqslant \rho$, we have $k \geqslant N$, so, $n_\Sigma(t)$ being increasing,

$$\frac{n_\Sigma(r + \rho) - n_\Sigma(r)}{\rho} \leqslant \frac{n_\Sigma(\lambda^{-k}r) - n_\Sigma(r)}{(\lambda^{-k} - 1)r} \cdot \frac{(\lambda^{-k} - 1)r}{\rho}$$

$$< \left(D_\Sigma^* + \frac{\varepsilon}{2} \right) \frac{\lambda^{-k} - 1}{\lambda^{-k+1} - 1}$$

$$\leqslant \ \left(D_\Sigma^* + \frac{\varepsilon}{2}\right)\frac{\lambda^{-N} - 1}{\lambda^{-N+1} - 1} \ = \ \frac{\varepsilon}{(1 + \varepsilon)^{(N-1)/N} - 1}\left(D_\Sigma^* + \frac{\varepsilon}{2}\right)$$

when $r \geqslant R$. If N is chosen large enough to begin with, the last number is $\leqslant D_\Sigma^* + \varepsilon$. This does it.

Theorem (Pólya). *Let the power series*

$$f(w) \ = \ \sum_{n \in \Sigma} a_n w^n$$

have radius of convergence 1. *Then, on every arc of* $\{|w| = 1\}$ *with length* $> 2\pi D_\Sigma^*$, $f(w)$ *has at least one singularity.*

Proof (Kahane). Assume that $f(w)$ can be continued analytically through an arc on the unit circle of length $> 2\pi D$, which we may wlog take to be symmetric about -1. We then have to prove that $D \leqslant D_\Sigma^*$. We may of course take $D > 0$. There is also no loss of generality in assuming $D < 1$, for here the power series' circle of convergence, which does include at least one singularity of $f(w)$, has length 2π.

Pick any $\delta > 0$. In the formula

$$a_n \ = \ \frac{1}{2\pi i}\int_{|w| = e^{-\delta}} f(w)w^{-n-1}\,\mathrm{d}w$$

(we are, of course, taking a_n as *zero* for $n \notin \Sigma$, $n \geqslant 0$) one may, thanks to the analyticity of $f(w)$, *deform* the path of integration $\{|w| = e^{-\delta}\}$ to the contour Γ_δ shown here:

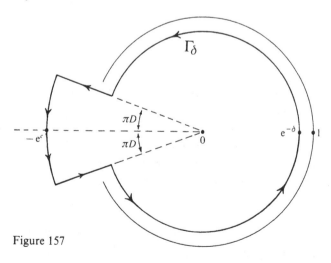

Figure 157

The quantity $c > 0$ is fixed once D is given, and independent of δ.

In the integral around Γ_δ, make the change of variable $w = e^{-s}$, where $s = \sigma + i\tau$ with τ ranging from $-\pi$ to π. Our expression then goes over into

$$\frac{1}{2\pi i} \int_{\gamma_\delta} f(e^{-s}) e^{ns} \, ds \quad = \quad a_n$$

with this path γ_δ:

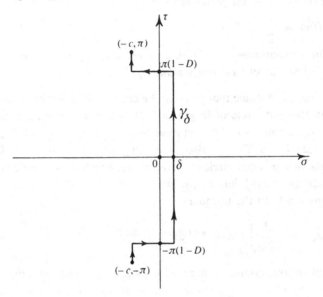

Figure 158

Write

$$F(z) \quad = \quad \frac{1}{2\pi i} \int_{\gamma_\delta} f(e^{-s}) e^{zs} \, ds$$

so that $F(n) = a_n$ for $n \in \mathbb{N}$ (and is hence *zero* for $n \in \mathbb{N} \sim \Sigma$); $F(z)$ is of course entire and of exponential type. We break up the integral along γ_δ into three pieces, I, II and III, coming from the *front vertical, horizontal* and *rear vertical* parts of γ_δ respectively.

On the *front vertical* part of γ_δ, $|f(e^{-s})| \leqslant M_\delta$ and $|e^{sz}| \leqslant e^{\delta x + \pi(1-D)|y|}$ (writing as usual $z = x + iy$); hence

$$|\mathrm{I}| \leqslant M_\delta e^{\delta x + \pi(1-D)|y|}$$

On the *horizontal* parts of γ_δ, $|f(e^{-s})| \leqslant C$ (a number independent of δ, by the way), and $|e^{sz}| \leqslant e^{\delta x + \pi(1-D)|y|}$ for $x > 0$, whence

$$|\mathrm{II}| \leqslant C e^{\delta x + \pi(1-D)|y|}, \quad x > 0.$$

Finally, on the *rear vertical* parts of γ_δ, $|f(e^{-s})| \leqslant C$ and $|e^{sz}| \leqslant e^{-cx+\pi|y|}$ for $x > 0$, making

$$|\mathrm{III}| \leqslant Ce^{-cx+\pi|y|}, \quad x > 0.$$

Adding these three estimates, we get

$$|F(z)| \leqslant (M_\delta + C)e^{\delta x + \pi(1-D)|y|} + Ce^{-cx+\pi|y|}$$

for $x > 0$. Since $c > 0$, the *second* term on the right will be $\leqslant$ the *first* in the sector

$$S = \left\{ z: \ |\Im z| \leqslant \frac{c}{\pi D}\Re z \right\}$$

with opening *independent of δ*. We thus have

$$|F(z)| \leqslant K_\delta e^{\delta x + \pi(1-D)|y|} \quad \text{for } z \in S,$$

K_δ being a constant depending on δ. The idea here is that *the availability, for $f(w)$, of an analytic continuation through the arc* $\{e^{i\vartheta}: |\vartheta - \pi| \leqslant \pi D\}$ *has made it possible for us to diminish the term* $\pi|y|$, *which would normally occur in the exponent on the right, to* $\pi(1-D)|y|$, *thanks to the term* $-cx$ figuring in the previous expression.

Because $\sum_{n\in\Sigma} a_n w^n$ has *radius of convergence* 1, there is a subsequence Σ' of Σ with

$$\frac{\log|a_n|}{n} \longrightarrow 0 \quad \text{for } n \longrightarrow \infty \text{ in } \Sigma'.$$

Let 2α be the opening (*independent of δ*) of our sector S, i.e.,

$$\alpha = \arctan\frac{c}{\pi D}$$

With $n \in \Sigma'$, write Jensen's formula for $F(z)$ and the circle of radius $n\sin\alpha$ about n (this is Kahane's idea). That is just

$$\int_0^{n\sin\alpha} \frac{N(\rho,n)}{\rho}\,d\rho = \frac{1}{2\pi}\int_{-\pi}^{\pi} \log|F(n+n\sin\alpha\,e^{i\vartheta})|\,d\vartheta - \log|a_n|,$$

where $N(\rho,n)$ denotes the number of zeros of $F(z)$ in the disk $\{|z-n| \leqslant \rho\}$.

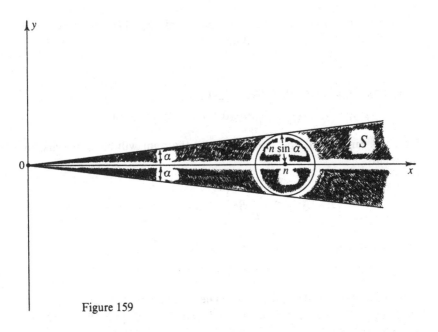

Figure 159

By the above estimate on $|F(z)|$ for $z \in S$, the *right side* of the relation just written is

$$\leqslant \ \log K_\delta \ + \ \frac{1}{2\pi} \int_{-\pi}^{\pi} (\delta n + \delta n \sin \alpha \cos \vartheta + n \sin \alpha \cdot \pi (1 - D) |\sin \vartheta|) \, d\vartheta$$

$$- \ \log |a_n|$$

$$= \ \log K_\delta + \delta n + 2(1 - D) n \sin \alpha - \log |a_n|.$$

The *left side* we estimate *from below*, using the lemma. Since $F(m) = a_m = 0$ for $m \in \mathbb{N} \sim \Sigma$, we have, for $0 < \rho \leqslant n$,

$$N(\rho, n) \ \geqslant \ \text{number of integers in } [n - \rho, \ n + \rho] \ -$$
$$\text{number of elements of } \Sigma \text{ in } [n - \rho, \ n + \rho]$$
$$\geqslant \ 2\rho \ - \ (n_\Sigma(n + \rho) - n_\Sigma(n - \rho)) \ - \ 2.$$

Fix any ε, $0 < \varepsilon < \sin \alpha$. According to the lemma, for n sufficiently large,

$$n_\Sigma(n + \rho) - n_\Sigma(n - \rho) \ \leqslant \ 2\rho(D_\Sigma^* + \varepsilon)$$

when $\varepsilon(1 - \sin \alpha) n \ \leqslant \ 2\rho \ \leqslant \ 2n \sin \alpha$, so, for such ρ (and large n),

$$N(\rho, n) \ \geqslant \ 2(1 - D_\Sigma^* - \varepsilon)\rho \ - \ 2.$$

Hence, since $n_\Sigma(t)$ increases,

$$\int_0^{n\sin\alpha} \frac{N(\rho, n)}{\rho}\, d\rho \;\geqslant\; 2(1 - D_\Sigma^* - \varepsilon)\left(\sin\alpha - \frac{\varepsilon(1 - \sin\alpha)}{2}\right)n$$

$$- \; 2\log\frac{2\sin\alpha}{\varepsilon(1 - \sin\alpha)}.$$

Use this inequality together with the preceding estimate for the right side of the above Jensen formula. After dividing by $2n\sin\alpha$, one finds that

$$(1 - D_\Sigma^* - \varepsilon)\left(1 - \frac{\varepsilon(1 - \sin\alpha)}{2\sin\alpha}\right) \;\leqslant\; \frac{\delta}{2\sin\alpha} + 1 - D$$

$$- \; \frac{\log|a_n|}{2n\sin\alpha} + O\!\left(\frac{1}{n}\right)$$

for large n, whence, making $n \longrightarrow \infty$ in Σ',

$$(1 - D_\Sigma^* - \varepsilon)\left(1 - \frac{\varepsilon(1 - \sin\alpha)}{2\sin\alpha}\right) \;\leqslant\; 1 - D + \frac{\delta}{2\sin\alpha},$$

on account of the behaviour of $\log|a_n|$ for $n \in \Sigma'$.

The quantity ε, $0 < \varepsilon < \sin\alpha$, is arbitrary, and so is $\delta > 0$ with, as we have remarked, the opening 2α of S independent of δ. We thence deduce from the previous relation that $1 - D_\Sigma^* \leqslant 1 - D$, i.e., that

$$D \;\leqslant\; D_\Sigma^*.$$

This, however, is what we had to prove. We are done.

Remark. We see from the proof that it is really the presence in the Taylor series of *many gaps 'near' those* $n \in \Sigma$ *for which* $|a_n|$ *is 'big'* (the $n \in \Sigma'$) that gives rise to *large numbers of singularites* on the circle of convergence. The reader is invited to formulate a precise statement of this observation, obtaining a theorem in which the behaviour of the a_n and that of Σ both figure.

Pólya's gap theorem has.various generalizations to Dirichlet series. For these, the reader should first look in the last chapter of Boas' book, after which the one by Levinson may be consulted. The most useful work on this subject is, however, the somewhat older one of V. Bernstein. Two of Mandelbrojt's books – the one published in 1952 and an earlier Rice Institute pamphlet on Dirichlet series – also contain interesting material, as does J.-P. Kahane's thesis, beginning with part II. There is, in addition, a recent monograph by Leontiev.

B.   Scholium. A converse to Pólya's gap theorem

The quantity D_Σ^* figuring in the result of the preceding § is a kind of *upper density* for sequences Σ of positive integers. Before continuing with the main material of this chapter, it is natural to ask whether D_Σ^* is *the right kind of density measure to use* for a sequence Σ when investigating the distribution of the singularities associated with

$$\sum_{n \in \Sigma} a_n w^n$$

on that series' circle of convergence. Maybe there is always a singularity on each arc of that circle having opening greater than $2\pi d_\Sigma$, with d_Σ a quantity $\leqslant D_\Sigma^*$ associated to Σ which is really $< D_\Sigma^*$ for some sequences Σ. It turns out that *this is not the case*; D_Σ^* is *always* the critical parameter associated with the sequence Σ insofar as distribution of singularities on the circle of convergence is concerned.

This fact, which shows Pólya's gap theorem to be *definitive*, is not well known in spite of its clear scientific importance. It is the content of the following

Converse to Polya's gap theorem *Given any sequence Σ of positive integers with Pólya maximum density $D_\Sigma^* > 0$, there is, for any δ, $0 < \delta < D_\Sigma^*$, a Taylor series*

$$\sum_{n \in \Sigma} a_n w^n$$

with radius of convergence 1, equal, for $|w| < 1$, to a function which can be continued analytically through the arc

$$\{e^{i\vartheta}: \ |\vartheta| < \pi(D_\Sigma^* - \delta)\}.$$

The present § is devoted to the establishment of this result in its full generality.

1.   Special case. Σ measurable and of density $D > 0$.

If $\lim_{t \to \infty} n_\Sigma(t)/t$ exists and equals a number $D > 0$ ($n_\Sigma(t)$ denoting the number of elements of Σ in $[0, \ t]$), the converse* to Pólya's theorem is easy – I think it is due to Pólya himself. The contour integration technique used to study this case goes back to Lindelöf; it was extensive-

---

* in a strengthened version, with analytic continuation through the arc
 $|\vartheta| \ < \ \pi D_\Sigma^* \ = \ \pi D$

ly used by V. Bernstein in his work on Dirichlet series, and later on by L. Schwartz in his thesis on sums of exponentials.

Restricting our attention to sequences Σ of strictly positive integers clearly involves no loss in generality; we do so throughout the present § because that makes certain formulas somewhat simpler. Denote by ℕ *the set of integers* > 0 (N.B. *this is different from the notation of §A, where* ℕ *also included* 0), and by Λ the sequence of *positive integers complementary to* Σ, i.e.,

$$\Lambda = \mathbb{N} \sim \Sigma.$$

For $t \geqslant 0$, we simply write $n(t)$ for the *number of elements of* Λ (N.B.!) in $[0, t]$. Put*

$$C(z) = \prod_{n \in \Lambda} \left(1 - \frac{z^2}{n^2} \right);$$

in the present situation

$$\frac{n(t)}{t} \longrightarrow 1 - D \quad \text{for} \quad t \longrightarrow \infty$$

and on account of this, $C(z)$ turns out to be an entire function of exponential type with quite regular behaviour.

Problem 29

(a) By writing $|\log C(z)|$ as a Stieltjes integral and integrating by parts, show that

$$\frac{\log|C(iy)|}{|y|} \longrightarrow \pi(1 - D)$$

for $y \longrightarrow \pm \infty$

(b) Show that for $x > 0$,

$$\log|C(x)| = 2 \int_0^1 \left(\frac{n(x\tau)}{\tau} - \tau n\left(\frac{x}{\tau}\right) \right) \frac{d\tau}{1 - \tau^2}.$$

(Hint: First write the left side as a Stieltjes integral, then integrate by parts. Make appropriate changes of variable in the resulting expression.)

(c) Hence show that for $x > 0$,

$$\log|C(x)| \leqslant 2n(x)\log\frac{1}{\gamma} + 2\int_0^\gamma \left(\frac{n(x\tau)}{\tau} - \tau n\left(\frac{x}{\tau}\right) \right) \frac{d\tau}{1 - \tau^2},$$

with γ any number between 0 and 1.

---

* When $D = 1$, the complementary sequence Λ has density zero and may even be empty. In the last circumstance we take $C(z) \equiv 1$; the function $f(w)$ figuring in the construction given below then reduces simply to $w/\pi(1 + w)$.

(d) By making an appropriate choice of the number γ in (c), show that $\log|C(x)| \leqslant \varepsilon x$ for large enough x, $\varepsilon > 0$ being arbitrary.

(e) Use an appropriate Phragmén–Lindelöf argument to deduce from (a) and (d) that

$$\limsup_{r \to \infty} \frac{\log|C(re^{i\vartheta})|}{r} \leqslant \pi(1-D)|\sin\vartheta|.$$

(f) Show that in fact

$$\frac{\log|C(n)|}{n} \to 0 \quad \text{for} \quad n \to \infty \text{ in } \Sigma,$$

and that we have *equality* in the result of (e).
(Hint: Form the function

$$K(z) = \prod_{n\in\Sigma}\left(1 - \frac{z^2}{n^2}\right);$$

then, as in (e),

$$\limsup_{r\to\infty}\frac{\log|K(re^{i\vartheta})|}{r} \leqslant \pi D|\sin\vartheta|.$$

Show that the same result holds if $K(re^{i\vartheta})$ is replaced by $K'(re^{i\vartheta})$. Observe that

$$\pi z K(z)C(z) = \sin \pi z.$$

Look at the derivative of the left-hand side at points $n\in\Sigma$.)

We are going to use the function $C(z)$ to construct a power series

$$\sum_{n\in\Sigma} a_n w^n$$

having *radius of convergence* 1, and representing a function which can be *analytically continued into the whole sector* $|\arg w| < \pi D$.
Start by putting

$$f(w) = \frac{1}{2\pi i}\int_{\frac{1}{2}-i\infty}^{\frac{1}{2}+i\infty} \frac{C(\zeta)}{\sin\pi\zeta} w^\zeta \, d\zeta$$

for $|\arg w| < \pi D$. Given any $\varepsilon > 0$, we see, by part (e) of the above problem, that

$$\left|\frac{C(\frac{1}{2}+i\eta)}{\sin\pi(\frac{1}{2}+i\eta)}\right| \leqslant \frac{\text{const.}}{\cosh\pi\eta} e^{(\pi(1-D)+\varepsilon)|\eta|}$$

for real η, where the constant on the right depends on ε.   At the same time,

$$|w^{\frac{1}{2}+i\eta}| \;=\; |w|^{1/2} e^{-\eta \arg w},$$

so the above integral *converges absolutely and uniformly* for w ranging over any bounded part of the sector

$$|\arg w| \;\leqslant\; \pi D - 2\varepsilon.$$

The function $f(w)$ is hence analytic in the interior of that sector, and thus finally for

$$|\arg w| \;<\; \pi D,$$

since $\varepsilon > 0$ was arbitrary.

We proceed now to obtain a series expansion *in powers of w* for $f(w)$, valid for w of small modulus with $|\arg w| < \pi D$. For this purpose the *method of residues* is used. Taking a large *integer R*, let us consider the integral

$$\frac{1}{2\pi i}\int_{\Gamma_R} \frac{C(\zeta)}{\sin \pi \zeta}\, w^{\zeta}\, d\zeta$$

around the following contour Γ_R:

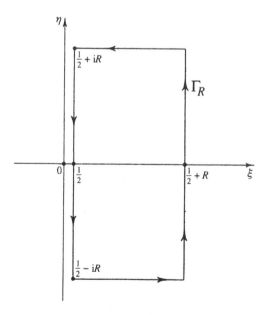

Figure 160

On the *top horizontal* side of Γ_R,

$$\left|\frac{C(\zeta)}{\sin \pi\zeta}\right| \leqslant \text{ const.} \frac{e^{(\pi(1-D)+\varepsilon)R}}{e^{\pi R}} = \text{ const. } e^{(\varepsilon - \pi D)R}$$

by part (e) of our problem. Here, $\varepsilon > 0$ is arbitrary, and the constant depends on it. The same estimate holds on the *lower horizontal* side of Γ_R. Also, if $|w| < 1$ and

$$|\arg w| \quad < \quad \pi D - 2\varepsilon,$$

we have

$$|w^\zeta| \leqslant e^{(\pi D - 2\varepsilon)R}$$

for ζ on the *horizontal sides of* Γ_R; in this circumstance the contribution of the horizontal sides to the contour integral is thus

$$\leqslant \text{ const. } Re^{-\varepsilon R}$$

in absolute value, and *that tends to zero as* $R \longrightarrow \infty$.

Along the *right vertical* side of Γ_R, by part (e) of the problem,

$$\left|\frac{C(\zeta)}{\sin \pi\zeta}\right| \leqslant \text{ const.} \frac{e^{\pi(1-D)|\eta|+\varepsilon R}}{\cosh \pi\eta}$$

with $\varepsilon > 0$ arbitrary as before (we write as usual $\zeta = \xi + i\eta$). For $|\arg w| < \pi D - 2\varepsilon$ and ζ on that side,

$$|w^\zeta| \leqslant |w|^R e^{(\pi D - 2\varepsilon)|\eta|},$$

so, if also $|w| < e^{-2\varepsilon}$, the contribution of the *right vertical* side of Γ_R to the contour integral is in absolute value

$$\leqslant \text{ const. } e^{-\varepsilon R} \int_{-\infty}^{\infty} e^{-2\varepsilon|\eta|} \, d\eta,$$

and *this tends to zero as* $R \longrightarrow \infty$.

Putting together the two results just found, we see that

$$\frac{1}{2\pi i} \int_{\Gamma_R} \frac{C(\zeta)}{\sin \pi\zeta} w^\zeta d\zeta \quad \longrightarrow \quad -\frac{1}{2\pi i} \int_{\frac{1}{2}-i\infty}^{\frac{1}{2}+i\infty} \frac{C(\zeta)}{\sin \pi\zeta} w^\zeta d\zeta = -f(w)$$

as $R \longrightarrow \infty$ for $|w| < 1$ and $|\arg w| < \pi D$, since $\varepsilon > 0$ is arbitrary.

By the residue theorem we have, however (taking $|\arg w| \leqslant \pi$, say),

$$\frac{1}{2\pi i} \int_{\Gamma_R} \frac{C(\zeta)}{\sin \pi\zeta} w^\zeta d\zeta = \frac{1}{\pi} \sum_{n=1}^{R} (-1)^n C(n) w^n$$

Here, $C(n) = 0$ for $n \in \Lambda = \mathbb{N} \sim \Sigma$, and, *by part* (f) *of the above problem,* the power series

$$\frac{1}{\pi} \sum_{n \in \Sigma} (-1)^n C(n) w^n$$

has *radius of convergence* 1. We thus see that

$$\lim_{R \to \infty} \frac{1}{2\pi i} \int_{\Gamma_R} \frac{C(\zeta)}{\sin \pi \zeta} w^{\zeta} \, d\zeta$$

equals the sum of that power series – call it $g(w)$ – for $|w| < 1$ and $|\arg w| < \pi D$. In that region, $g(w)$ must then coincide with $-f(w)$ by the calculation of the preceding limit just made. For $|w| < 1$ and $|\arg w| < \pi D$, we therefore have

$$f(w) = -\frac{1}{\pi} \sum_{n \in \Sigma} (-1)^n C(n) w^n$$

This relation furnishes an *analytic continuation* of $-g(w)$, analytic in $|w| < 1$, *to the whole sector* $|\arg w| < \pi D$ where, as we have seen, $f(w)$ is analytic. The power series on the right has radius of convergence 1.

For our measurable sequence Σ, D_{Σ}^* and D coincide. Hence *Pólya's gap theorem cannot be improved* in the case of such Σ.

2.   General case; Σ not measurable. Beginning of Fuchs' construction

As stated at the beginning of this §, the converse to Pólya's gap theorem holds for *any* sequence Σ of positive integers, and for non-measurable Σ, the critical size for singularity-free arcs on the circle of convergence is $2\pi D_{\Sigma}^*$ radians, where D_{Σ}^* is the *maximum density* of Σ. This remarkable extension of the preceding article's result is not generally known. Malliavin makes passing mention of it in his 1957 *Illinois Journal* paper (one exceedingly difficult to read, by the way), but it really goes back to a publication of W. Fuchs in the 1954 *Proceedings* of the Edinburgh mathematical society, being entirely dependent on the beautiful construction given there. Fuchs, however, does not mention this (almost immediate) application of his construction in that paper.

The treatment for the general case involves a contour integral like the one used in the preceding article. Now, however, we cannot make do with just an entire function of exponential type like $C(z)$, but need another more complicated one besides. The latter, analytic and of exponential type in the *right half plane* (but *not* entire), is obtained by means of Fuchs' construction.

We start with a *non-measurable* sequence Σ of *strictly positive* integers (this last being no real restriction), and assume, throughout the remaining articles of the present §, that

$$D_\Sigma^* = \lim_{\lambda \to 1-} \left(\limsup_{r \to \infty} \frac{n_\Sigma(r) - n_\Sigma(\lambda r)}{(1 - \lambda)r} \right)$$

is > 0. By the second theorem of §E.3, Chapter III, we know that Σ is included in a *measurable* sequence Σ^* of positive numbers with *density* D_Σ^*. *In the present case, we may take* $\Sigma^* \subseteq \mathbb{N}$. Indeed, since $\Sigma \subseteq \mathbb{N}$, D_Σ^* is certainly $\leqslant 1$. If $D_\Sigma^* = 1$, we can just put $\Sigma^* = \mathbb{N}$. A glance at the construction used in proving the theorem referred to shows that *the choice of new elements to be adjoined to Σ so as to make up Σ^* is fairly arbitrary, and that when $D_\Sigma^* < 1$ we may always take them to be distinct positive integers*. Here, this will yield a sequence $\Sigma^* \subseteq \mathbb{N}$ when $D_\Sigma^* < 1$.

Having obtained $\Sigma^* \subseteq \mathbb{N}$, we take the *complement*

$$\Lambda_1 = \mathbb{N} \sim \Sigma^*;$$

since Σ^* is *measurable*, so is Λ_1, and Λ_1 has *density* $1 - D_\Sigma^*$. The complement of Σ in $\mathbb{N}$ consists of Λ_1 together with another sequence

$$\Lambda_0 = \Sigma^* \sim \Sigma$$

*distinct** from Λ_1; *most of the work in the rest of this § will be with Λ_0*. For $t \geqslant 0$ we *denote by $n(t)$ the number of elements of Λ_0 in $[0, t]$*. (We write $n(t)$ instead of $n_{\Lambda_0}(t)$ in order to simplify the notation.) If $n_{\Sigma^*}(t)$ denotes the number of points of Σ^* in $[0, t]$, we have

$$n_{\Sigma^*}(t) = n(t) + n_\Sigma(t),$$

so, since

$$\frac{n_{\Sigma^*}(t)}{t} \longrightarrow D_\Sigma^* \quad \text{for} \quad t \longrightarrow \infty,$$

the relation

$$\lim_{\lambda \to 1-} \left(\liminf_{r \to \infty} \frac{n(r) - n(\lambda r)}{(1 - \lambda)r} \right) = 0$$

---

* The sequence Λ_0 is certainly non-void and indeed infinite since Σ is *non-measurable*, as we are *assuming* throughout this and the next 6 articles. But Λ_1, of density $1 - D_\Sigma^*$, may even be *empty* when $D_\Sigma^* = 1$ (if we then take $\Sigma^* = \mathbb{N}$).

must hold, in view of the above formula for D_Σ^*. We may say that the sequence $\Lambda_0 \subseteq \mathbb{N}$ has *minimum density zero*.

Lemma. *Given $\varepsilon > 0$, there is an increasing sequence of numbers X_j tending to ∞ and an $\alpha > 0$, both depending on ε, such that*

$$n(x) - n(X_j) \leqslant \frac{\varepsilon}{2}(x - X_j) \quad \text{for} \quad X_j \leqslant x \leqslant (1 + \alpha)X_j.$$

Proof. For a certain fixed $c > 0$ we have, with $\lambda = 1/(1 + c)$,

$$\liminf_{r \to \infty} \frac{n(r) - n(\lambda r)}{(1 - \lambda)r} \;<\; \frac{\varepsilon}{4}$$

by the above boxed relation. There are hence *arbitrarily large* numbers R such that

$$\frac{n((1 + c)R) - n(R)}{cR} \;<\; \frac{\varepsilon}{4}.$$

Take such a number R. It is claimed that if the integer M is *large* enough (*independently* of R) and we put

$$1 + \alpha \;=\; (1 + c)^{1/M},$$

there exists an X,

$$R \;\leqslant\; X \;\leqslant\; (1 + \alpha)^{M-1}R,$$

such that

$$n(x) - n(X) \;\leqslant\; \frac{\varepsilon}{2}(x - X) \quad \text{for } X \leqslant x \leqslant (1 + \alpha)X.$$

This assertion, once verified, *will establish the lemma*, for we can then take a sequence of numbers R tending to ∞ and choose a corresponding sequence $\{X_j\}$ of numbers X.

Suppose, for some large integer M and for α related to it by the above formula, that *there is no such X*. There must then be a number x_1, $R \leqslant x_1 \leqslant (1 + \alpha)R$, with

$$n(x_1) - n(R) \;>\; \frac{\varepsilon}{2}(x_1 - R).$$

This certainly makes $n(x_1) \geqslant n(R) + 1$ since $n(t)$ increases by 1 at each of its jumps. By the same token, there is an x_2, $x_1 \leqslant x_2 \leqslant (1 + \alpha)x_1$, with

$$n(x_2) - n(x_1) > \frac{\varepsilon}{2}(x_2 - x_1)$$

(so in particular $n(x_2) \geqslant n(x_1) + 1$). The process continues, yielding x_3, $x_2 \leqslant x_3 \leqslant (1 + \alpha)x_2$, x_4, and so forth, with

$$n(x_{k+1}) - n(x_k) > \frac{\varepsilon}{2}(x_{k+1} - x_k),$$

as long as the number x_k already obtained is $\leqslant (1 + \alpha)^{M-1}R$. Since $n(x_{k+1}) - n(x_k) \geqslant 1$, x_k *cannot remain* $\leqslant (1 + \alpha)^{M-1}R$ *indefinitely* (we must eventually have $n(x_k) > n((1 + \alpha)^{M-1}R))$). Let x_l be the *last* x_k which is $\leqslant (1 + \alpha)^{M-1}R$; then we can still get an x_{l+1} between $(1 + \alpha)^{M-1}R$ and $(1 + \alpha)^M R$, such that

$$n(x_{l+1}) - n(x_l) > \frac{\varepsilon}{2}(x_{l+1} - x_l).$$

Adding to this the corresponding inequalities already obtained, we get

$$n(x_{l+1}) - n(R) > \frac{\varepsilon}{2}(x_{l+1} - R).$$

Since $(1 + \alpha)^M = 1 + c$, $x_{l+1} \leqslant (1 + c)R$, so

$$n((1 + c)R) \geqslant n(x_{l+1}).$$

And

$$x_{l+1} - R \geqslant (1 + \alpha)^{M-1}R - R = \frac{(1 + c)^{(M-1)/M} - 1}{c} \cdot cR.$$

The relation just found therefore implies that

$$\frac{n((1 + c)R) - n(R)}{cR} > \frac{(1 + c)^{(M-1)/M} - 1}{c} \cdot \frac{\varepsilon}{2}.$$

However, if M is *large enough* (depending *only* on c and not on R!), we have

$$\frac{(1 + c)^{(M-1)/M} - 1}{c} > \frac{1}{2}.$$

This would make the *left-hand side* of the previous relation $> \varepsilon/4$, *in contradiction with our choice of the number R*. For such large M, then, a number X with the properties specified above *must exist*. This establishes our claim, and proves the lemma.

Lemma. *Given* $\varepsilon > 0$, *let* $\alpha > 0$ *and* $X = X_j$ *be as in the statement of the previous lemma. There is then a* β, $\alpha/3 \leqslant \beta \leqslant \alpha$, *such that*

$$n((1 + \beta)X) - n(x) \leqslant 2\varepsilon((1 + \beta)X - x) \quad for \quad X \leqslant x \leqslant (1 + \beta)X.$$

Proof. By the argument used to prove the lemma about Bernstein intervals near the beginning of §B.2, Chapter VIII.

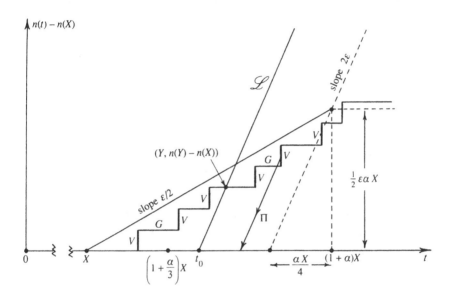

Figure 161

Denote by G the *graph of* $n(t) - n(X)$ vs. t for $X \leqslant t \leqslant (1 + \alpha)X$, and by V the union of the *vertical portions* of G (corresponding to the *jumps* of $n(t)$). Let Π be the operation of *downward projection, along a line of slope* 2ε, *onto the t-axis*. Then, since $2\varepsilon > \varepsilon/2$,

$$\Pi(V) \subseteq [X, (1 + \alpha)X],$$

and we see from the figure that

$$|\Pi(V)| \leqslant \frac{1}{2\varepsilon} \cdot \frac{\varepsilon}{2} \alpha X = \frac{\alpha X}{4}.$$

Therefore $\Pi(V)$ *cannot cover the segment* $[(1 + \alpha/3)X, (1 + 3\alpha/4)X]$ *of length* $(5/12)\alpha X$, *so there is a* t_0 *in that segment not belonging to* $\Pi(V)$.

The line $\mathscr{L}$ of slope 2ε through $(t_0, 0)$ must, from the figure, cut G, say at a point $(Y, n(Y) - n(X))$, with $(1 + \alpha/3)X \leqslant Y \leqslant (1 + \alpha)X$. Since $\mathscr{L}$, passing through that point, does not touch any part of V (t_0

being $\notin \Pi(V)$), we have

$$n(Y) - n(t) \leqslant 2\varepsilon(Y - t) \quad \text{for} \quad X \leqslant t \leqslant Y.$$

Calling $Y/X = 1 + \beta$, we have the lemma.

Let us combine the two results just proved. We see that, given $\varepsilon > 0$, there are *two sequences* $\{X_j\}$ and $\{Y_j\}$ tending to ∞ and an $\alpha > 0$ depending on ε, such that

$$1 + \frac{\alpha}{3} \leqslant \frac{Y_j}{X_j} \leqslant 1 + \alpha$$

and that the *simultaneous relations*

$$\begin{cases} n(x) - n(X_j) \leqslant 2\varepsilon(x - X_j), \\ n(Y_j) - n(x) \leqslant 2\varepsilon(Y_j - x) \end{cases}$$

hold on each of the intervals $[X_j, Y_j]$. (Of course, the first of these relations can be replaced by an even better one!).

We henceforth assume that $\varepsilon \leqslant 1/6$. *That* being granted, we can, at the cost of ending with slightly worse inequalities, *modify* the above constructions *so as to make the* X_j *and* Y_j *half-odd integers*. To see this, we again use an idea from §B.2 of Chapter VIII.

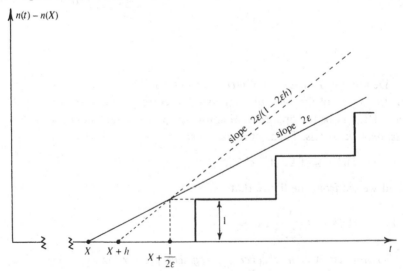

Figure 162

Choosing one of the intervals $[X_j, Y_j]$ described above, we drop the index j, writing simply X for X_j and Y for Y_j. *The function* $n(t)$ *increases by* 1 *at each of its jumps.* Therefore, since

$$n(t) - n(X) \leqslant 2\varepsilon(t - X)$$

for $X \leqslant t \leqslant Y$, we must have

$$n(t) = n(X) \quad \text{for } X \leqslant t < X + \frac{1}{2\varepsilon}.$$

If $0 \leqslant h < 1/2\varepsilon$, the line of slope $2\varepsilon/(1 - 2\varepsilon h)$ through $(X + h, 0)$ must then *lie entirely above* the graph of $n(t) - n(X)$ vs. t for $X + h \leqslant t \leqslant Y$, as the above figure shows. Choosing h, $0 \leqslant h < 1$, *so as to make*

$$X' = X + h$$

a half-odd integer, we thus have

$$n(t) - n(X') \leqslant 3\varepsilon(t - X') \quad \text{for } X' \leqslant t \leqslant Y$$

because for such h,

$$\frac{2\varepsilon}{1 - 2\varepsilon h} \leqslant 3\varepsilon,$$

ε being $< 1/6$.

The same kind of reasoning shows that if we take a *half-odd integer* Y' with $Y - 1 < Y' \leqslant Y$, we will still have

$$n(Y') - n(t) \leqslant 3\varepsilon(Y' - t) \quad \text{for} \quad X' \leqslant t \leqslant Y'.$$

Since

$$1 + \frac{\alpha}{3} \leqslant \frac{Y}{X} \leqslant 1 + \alpha,$$

we have

$$\frac{Y'}{X'} \leqslant 1 + \alpha$$

and also

$$\frac{Y'}{X'} \geqslant \frac{Y - 1}{X + 1} \geqslant 1 + \frac{\alpha}{4}$$

as long as X is large.

From now on, we work with the intervals $[X', Y']$, *and write* $X = X_j$ *instead of* X' *and* $Y = Y_j$ *instead of* Y'. Also, since ε, $0 < \varepsilon < 1/6$, is *arbitrary, we may just as well write* ε *instead of* 3ε. By the above considerations we have then proved the following

Theorem. *Given ε, $0 < ε < 1/2$, there are sequences $\{X_j\}$ and $\{Y_j\}$ of half-odd integers tending to ∞ and an $α > 0$ such that*

$$\left(1 + \frac{α}{4}\right)X_j \;\leqslant\; Y_j \;\leqslant\; (1 + α)X_j$$

and that

$$\left.\begin{array}{l} n(t) - n(X_j) \;\leqslant\; ε(t - X_j) \\[4pt] n(Y_j) - n(t) \;\leqslant\; ε(Y_j - t) \end{array}\right\} \; \textit{for } X_j \leqslant t \leqslant Y_j.$$

When X_j is *large*, $Y_j - X_j \geqslant (α/4)X_j$ is *also* large, so the segment $[X_j, Y_j]$ *contains lots of integers.* Recalling the meaning of $n(t)$, we see by the theorem that if the intervals $[X_j, Y_j]$ are constructed for a *small* value of ε, *most of the integers in them will not belong to our sequence* $Λ_0$.

The purpose of Fuchs' construction is to obtain a function $Φ(z)$, analytic in $\Re z > 0$ and of *small exponential type there*, such that, for large $n \in Λ_0$, $|Φ(n)|^{1/n}$ is *at most* $e^{-δ}$ times the limsup of $|Φ(m)|^{1/m}$ *for m tending to ∞ in* Σ, δ being some constant > 0. The function $Φ(z)$ is constructed so as to *vanish at the points of* $Λ_0$ *belonging to a sparse sequence of the intervals* $[X_j, Y_j]$, and so as to make

$$|Φ(x)| \;\leqslant\; \text{const.}e^{(k - δ)x}$$

for $x > 0$ *outside* of those intervals, while

$$|Φ(m)| \;\geqslant\; \text{const.}e^{km}$$

for most of the integers m *inside* them *that don't belong to* $Λ_0$. As we have just observed, there will be plenty of the latter.

3.   Bringing in the gamma function

For obtaining the function $Φ(z)$ mentioned at the end of the preceding article, procedures yielding entire functions will not work.* Fuchs' idea is to construct $Φ(z)$ by using products of the form

$$\prod \left(\frac{1 - z/n}{1 + z/n}\right)e^{2z/n}$$

taken over *certain* sets of positive integers n; these are analytic in the *right* half plane, but have poles in the left half plane. The exponential factors ensure convergence.

----
* at least, so it seems

The prototype of such a product is

$$\prod_{n=1}^{\infty}\left(\frac{1-z/n}{1+z/n}\right)e^{2z/n};$$

this can be expressed in terms of the gamma function.

$\Gamma(z)$ is the reciprocal of an entire function of order 1 defined by means of a certain infinite product designed to make

$$\Gamma(x+1) = \int_0^{\infty} t^x e^{-t}\,dt$$

for real $x > -1$. Starting from this formula, successive integrations by parts yield

$$\Gamma(x+1) = \frac{1}{x+1}\int_0^{\infty} t^{x+1}e^{-t}\,dt = \frac{1}{(x+1)(x+2)}\int_0^{\infty} t^{x+2}e^{-t}\,dt$$

$$= \cdots = \frac{\displaystyle\int_0^{\infty} t^{x+m}e^{-t}\,dt}{(1+x)(2+x)\cdots(m+x)}.$$

The last expression can be rewritten

$$\frac{\exp\left(-\displaystyle\sum_{k=1}^{m}(x/k)\right)}{\displaystyle\prod_{k=1}^{m}\left(1+\frac{x}{k}\right)e^{-x/k}} \cdot \frac{\displaystyle\int_0^{\infty} t^{x+m}e^{-t}\,dt}{m!}$$

One has, of course, $m! = \int_0^{\infty} t^m e^{-t}\,dt$, and, for real x tending to ∞, Stirling's formula,

$$\int_0^{\infty} t^x e^{-t}\,dt \sim \sqrt{(2\pi x)}\cdot\left(\frac{x}{e}\right)^x$$

is valid. (The latter may be proved by applying Laplace's method to the integral on the left.) Using these relations to simplify the expression just written, we find that

$$\Gamma(x+1) = \lim_{m\to\infty} \frac{m^x\exp\left(-\displaystyle\sum_{k=1}^{m}(x/k)\right)}{\displaystyle\prod_{k=1}^{m}\left(1+\frac{x}{k}\right)e^{-x/k}}$$

We have

$$\exp\left(x\sum_{k=1}^{m}(1/k)\right)m^{-x} = \exp\left\{x\left(\frac{1}{m} + \sum_{k=1}^{m-1}\frac{1}{k} - \log m\right)\right\}.$$

By drawing a picture, one sees that as $m \to \infty$,

$$\sum_{k=1}^{m-1} \frac{1}{k} \quad - \quad \log m$$

increases steadily to a certain finite limit C (called Euler's constant). Therefore, by the preceding formula,

$$\Gamma(x+1) \;=\; 1 \bigg/ \exp(Cx) \prod_{m=1}^{\infty}\left(1+\frac{x}{k}\right)e^{-x/k}.$$

For general complex z, one just *defines*

$$\Gamma(z+1) \;=\; 1 \bigg/ \exp(Cz) \prod_{n=1}^{\infty}\left(1+\frac{z}{n}\right)e^{-z/n}.$$

By a slight adaptation of the work in Chapter III, §§A, B one easily shows using this formula that

$$|1/\Gamma(z+1)| \;\leqslant\; K_\varepsilon \exp(|z|^{1+\varepsilon})$$

for each $\varepsilon > 0$. ($1/\Gamma(z+1)$ is NOT, by the way, *of exponential type*, on account of *Lindelöf's theorem* if for no other reason!)
Since

$$\frac{\sin \pi z}{\pi z} \;=\; \prod_{n=1}^{\infty}\left(1-\frac{z^2}{n^2}\right),$$

we have

$$\prod_{n=1}^{\infty}\left(\frac{1-z/n}{1+z/n}\right)e^{2z/n} \;=\; e^{2Cz}\,\frac{\sin \pi z}{\pi z}\,(\Gamma(z+1))^2.$$

Use of this relation together with Stirling's formula for complex z enables us to get a good grip on the behaviour of the right-hand product.

Problem 30
To extend Stirling's formula to complex values of z in the right half plane. Write

$$g(z) \;=\; \sqrt{(2\pi z)}\cdot\left(\frac{z}{e}\right)^z \bigg/ \Gamma(z+1).$$

(a) Show that $g(z)$ is of order 1 – i.e., that

$$|g(z)| \;\leqslant\; M_\varepsilon \exp(|z|^{1+\varepsilon})$$

for each $\varepsilon > 0$ – in any open sector of the form $|\arg z| \leqslant \pi - \delta, \ \delta > 0$, and is continuous up to the boundary of such a sector.

(b) Show that for real y,

$$|g(iy)| = \sqrt{2} \cdot e^{-(\pi|y|/2)} \sqrt{(\sinh \pi |y|)}.$$

(c) Hence show that $g(z)$ is bounded for $\Re z \geqslant 0$. (Hint: Use Stirling's formula to estimate $g(x)$ for $x > 0$. Then use Phragmén–Lindelöf in the *first* and *fourth* quadrants.)

(d) Hence show that $g(z) \longrightarrow 1$ uniformly for z tending to ∞ in any sector $|\arg z| \leqslant \pi/2 - \delta, \ \delta > 0$. (Hint: $g(x) \longrightarrow 1$ for $x \longrightarrow \infty$ by Stirling's formula. In view of (c), a theorem of Lindelöf may be applied.)

(e) Show that $g(re^{\pm(2\pi i/3)}) \longrightarrow 1$ as $r \longrightarrow \infty$. (Hint: First show that $\Gamma(1 + z)\Gamma(1 - z) = \sin \pi z/\pi z$. Use this in conjunction with the result from (d), noting that $e^{\pm(2\pi i/3)} = -e^{\mp(\pi i/3)}$.)

(f) Hence show that $g(z) \longrightarrow 1$ uniformly for z tending to ∞ in any sector of the form

$$|\arg z| \leqslant \tfrac{2}{3}\pi - \delta, \ \delta > 0.$$

(Hint: Use Phragmén–Lindelöf and the theorem of Lindelöf referred to in the hint to part (d) again.)

From part (f) of this problem we have in particular

$$\Gamma(z + 1) \ \sim \ \sqrt{(2\pi z)} \cdot \left(\frac{z}{e}\right)^{z}$$

for $|\arg z| \leqslant \pi/2$ and $|z|$ large. This means that

$$z^{-2z}e^{2(1-C)z} \prod_{n=1}^{\infty} \left(\frac{1 - z/n}{1 + z/n}\right) e^{2z/n} \ \sim \ 2\sin \pi z$$

for $\Re z \geqslant 0$ when $|z|$ is large. The expression on the left is thus certainly of exponential type π in the right half plane.

Fuchs takes* the intervals $[X_j, Y_j]$ constructed in the previous article,

---

* His construction is, of course, needed by us only for the case of *non-measurable* Σ, when the sequence Λ_0 is certainly *available*, and indeed *infinite*.

corresponding to a *small* value of $\varepsilon > 0$. He then fixes a *large* integer L with, however, $L < 1/\varepsilon$, and forms the function

$$F(z/L) = \left(\frac{z}{L}\right)^{-2z/L} e^{2(1-C)z/L} \prod_{n=1}^{\infty} \left(\frac{1-z/nL}{1+z/nL}\right) e^{2z/nL}.$$

According to the above boxed formula, this has very regular behaviour in the right half plane, and is of exponential type π/L there.

Fuchs' idea is now to *modify* the product on the right side of this last relation by *throwing away* the factors

$$\left(\frac{1-z/nL}{1+z/nL}\right) e^{2z/nL}$$

corresponding to the n for which nL belongs to *certain* of the intervals $[X_j, Y_j]$. *Those* factors are *replaced* by *others* of the form

$$\left(\frac{1-z/\lambda}{1+z/\lambda}\right) e^{2z/\lambda}$$

corresponding to the $\lambda \in \Lambda_0$ belonging to the *same* intervals $[X_j, Y_j]$. This alteration of $F(z/L)$ produces a new function, vanishing at the points of Λ_0 lying in certain of the $[X_j, Y_j]$. We have to see how much the behaviour of the latter differs from that of the former.

4.   Formation of the group products $R_j(z)$

We want, then, to *remove* from the product

$$\prod_{n=1}^{\infty} \left(\frac{1-z/nL}{1+z/nL}\right) e^{2z/nL}$$

the group of factors

$$\prod_{nL\in[X_j,Y_j]} \left(\frac{1-z/nL}{1+z/nL}\right) e^{2z/nL}$$

and to *insert*

$$\prod_{\lambda\in\Lambda_0\cap[X_j,Y_j]} \left(\frac{1-z/\lambda}{1+z/\lambda}\right) e^{2z/\lambda}$$

in their place, doing this for infinitely many of the intervals $[X_j, Y_j]$ constructed in article 2, corresponding to some fixed small $\varepsilon > 0$. This amounts to multiplying our original product by expressions of the form

$$\prod_{nL\in[X_j,Y_j]} \left(\frac{nL+z}{nL-z}\right) e^{-2z/nL} \prod_{\lambda\in\Lambda_0\cap[X_j,Y_j]} \left(\frac{\lambda-z}{\lambda+z}\right) e^{2z/\lambda}.$$

As we said at the end of the preceding article, Fuchs takes the integer $L < 1/\varepsilon$. Therefore, since

$$n(Y_j) - n(X_j) \leqslant \varepsilon(Y_j - X_j)$$

by the theorem of article 2, there are *fewer* $\lambda \in \Lambda_0$ than integral multiples of L in $[X_j, Y_j]$, and the exponential factors in the above expression do not multiply out to 1. Their presence would cause difficulties later on, and we would like to get rid of them.

For this reason, Fuchs brings in a small multiple a_j of X_j, chosen so as to make

$$\sum_{nL \in [X_j, Y_j]} \frac{1}{nL} - \sum_{\lambda \in \Lambda_0 \cap [X_j, Y_j]} \frac{1}{\lambda} = \frac{q_j}{a_j}$$

with a *positive integer* q_j, and then *adjoins* to the previous expression an *additional dummy factor* of the form

$$\left(\frac{a_j - z}{a_j + z} \right)^{q_j} e^{2q_j z / a_j}$$

Once this is done, the exponential factors $e^{2q_j z / a_j}$, $e^{-2z/nL}$ and $e^{2z/\lambda}$ figuring in the resulting product cancel each other out.

The details in this step involve some easy estimates. *In this and the succeeding articles, when concentrating on any particular interval* $[X_j, Y_j]$, *we will simplify the notation by dropping the subscript j, writing just X for* X_j, *Y for* Y_j, *a for* a_j, *and so forth.*
By the theorem of article 2, $n(t) - n(X) \leqslant \varepsilon(t - X)$ for $X \leqslant t \leqslant Y$, and X and Y are *half-odd*, while the $\lambda \in \Lambda_0$ are *integers*. Hence,

$$\sum_{\lambda \in \Lambda_0 \cap [X,Y]} \frac{1}{\lambda} = \int_X^Y \frac{dn(t)}{t} = \frac{n(Y) - n(X)}{Y} + \int_X^Y \frac{n(t) - n(X)}{t^2} dt$$

$$\leqslant \frac{\varepsilon(Y - X)}{Y} + \varepsilon \int_X^Y \frac{t - X}{t^2} dt = \varepsilon \log\left(\frac{Y}{X} \right).$$

In the theorem just mentioned, $Y - X \geqslant (\alpha/4)X$ with a constant $\alpha > 0$ depending on ε. Therefore, if X is a *very large* X_j (which we *always assume henceforth*), there *will be* numbers $nL \in [X, Y]$, $n \in \mathbb{N}$, and then, as a simple picture shows,

$$\sum_{nL \in [X,Y]} \frac{1}{nL} \geqslant \frac{1}{L} \log \frac{Y}{X + L} \geqslant \frac{1}{L} \log \frac{Y}{X} - \frac{1}{X}.$$

Thus, when X is large,

$$\sum_{nL \in [X,Y]} \frac{1}{nL} - \sum_{\lambda \in \Lambda_0 \cap [X,Y]} \frac{1}{\lambda} \geqslant \left(\frac{1}{L} - \varepsilon\right) \log \frac{Y}{X} - \frac{1}{X}.$$

Here, $1/L > \varepsilon$ and $Y \geqslant (1 + \alpha/4)X$, so the right side is bounded below by a constant > 0 depending on L, ε and α for X large enough (again depending on those parameters).

Take now a small parameter $\eta > 0$ which is to remain *fixed* throughout all the following constructions – later on we will see how η is to be chosen. *Here, we observe that if X is large, there is a number a between $\eta X/2$ and ηX whose product with the left side of the preceding inequality is an integer.* Picking such an a, we call the corresponding integer q, and we have

$$\frac{q}{a} = \sum_{nL \in [X,Y]} \frac{1}{nL} - \sum_{\lambda \in \Lambda_0 \cap [X,Y]} \frac{1}{\lambda}.$$

Because $\eta X/2 \leqslant a \leqslant \eta X$, the inequality

$$\sum_{X \leqslant nL \leqslant Y} \frac{1}{nL} \leqslant \frac{1}{X}\left(\frac{Y-X}{L} + 1\right)$$

gives us the useful upper estimate

$$q \leqslant \frac{\eta}{L}(Y - X) + \eta.$$

Definition. We write

$$R(z) = \left(\frac{a-z}{a+z}\right)^q \prod_{nL \in [X,Y]} \left(\frac{nL+z}{nL-z}\right) \prod_{\lambda \in \Lambda_0 \cap [X,Y]} \left(\frac{\lambda-z}{\lambda+z}\right).$$

When using the subscript j with X, Y, a and q, we also write $R_j(z)$ instead of $R(z)$.

Let us establish some simple properties of the group product $R(z)$. In the first place, *we can put* the aforementioned *exponential factors* (needed for convergence) *back* into $R(z)$ *if we want to*:

$$R(z) = \left(\frac{1-z/a}{1+z/a}\right)^q e^{2qz/a} \prod_{nL\in[X,Y]}\left(\frac{1+z/nL}{1-z/nL}\right)e^{-2z/nL}$$

$$\times \prod_{\lambda\in\Lambda_0\cap[X,Y]}\left(\frac{1-z/\lambda}{1+z/\lambda}\right)e^{2z/\lambda}.$$

This is so by the above boxed relation involving q/a.

In the second place, we have the

Lemma. *If $\eta > 0$ is chosen small enough (depending only on L and ε) and X is large,*

$$|R(x)| \leqslant 1 \quad \text{for } 0 \leqslant x < a.$$

Proof.

$$\log|R(x)| = q\log\left|\frac{1-x/a}{1+x/a}\right| + \sum_{nL\in[X,Y]}\log\left|\frac{1+x/nL}{1-x/nL}\right|$$

$$+ \sum_{\lambda\in\Lambda_0\cap[X,Y]}\log\left|\frac{1-x/\lambda}{1+x/\lambda}\right|.$$

Since $a \leqslant \eta X < X$ we can, for $0 \leqslant x < a$, expand the logarithms in powers of x. Collecting terms, we find, thanks to the above boxed formula for q/a, that the *coefficient* of x *vanishes*, and we get

$$\log|R(x)| = \sum_{\substack{N=3\\N\text{ odd}}}^{\infty}\left\{\sum_{nL\in[X,Y]}\frac{1}{(nL)^N} - \sum_{\lambda\in\Lambda_0\cap[X,Y]}\frac{1}{\lambda^N} - \frac{q}{a^N}\right\}\cdot\frac{2x^N}{N}.$$

By a previous inequality for the *right side* of the boxed formula for q/a,

$$\frac{q}{a} \geqslant \left(\frac{1}{L}-\varepsilon\right)\log\frac{Y}{X} - \frac{1}{X}.$$

Hence, since $a \leqslant \eta X$,

$$\frac{q}{a^N} \geqslant \frac{\left(\frac{1}{L}-\varepsilon\right)X^{-(N-1)}}{\eta^{N-1}}\log\frac{Y}{X} - \frac{1}{\eta^{N-1}X^N}$$

for $N > 1$.

At the same time, we already have

$$\sum_{nL \in [X,Y]} \frac{1}{(nL)^N} \;\leqslant\; \frac{1}{X^N}\left(\frac{Y-X}{L}+1\right).$$

Thus, since $(1 + \alpha/4)X \leqslant Y \leqslant (1 + \alpha)X$,

$$\sum_{nL \in [X,Y]} \frac{1}{(nL)^N} \;\leqslant\; \left(\frac{\alpha}{L}+\frac{1}{X}\right)\cdot\frac{1}{X^{N-1}},$$

while, by the preceding calculation,

$$\frac{q}{a^N} \;\geqslant\; \left[\left(\frac{1}{L}-\varepsilon\right)\log\left(1+\frac{\alpha}{4}\right) - \frac{1}{X}\right]\cdot\frac{1}{(\eta X)^{N-1}}.$$

Since $1/L - \varepsilon > 0$, *for sufficiently small values of* η, *the second of these quantities exceeds the first for every* $N \geqslant 2$, as long as X is large enough. The required smallness of η is determined here by α, L and ε, and therefore really by the *last two* of these quantities, for α itself depends on ε.

Under the circumstances just described, *all the coefficients in the above power series expansion of* $\log|R(x)|$ *will be negative.* This makes $\log|R(x)| \leqslant 0$ for $0 \leqslant x < a$,

Q.E.D.

Another result goes in the opposite direction.

Lemma. *If* $\eta > 0$ *is taken* small enough (*depending* only on L and ε) and X is large,

$$|R(X/2)| \;>\; 1.$$

Proof.

$$\log|R(X/2)| \;=\; q\log\left|\frac{\frac{1}{2}X-a}{\frac{1}{2}X+a}\right| + \sum_{nL \in [X,Y]}\log\left|\frac{nL+\frac{1}{2}X}{nL-\frac{1}{2}X}\right|$$

$$+ \sum_{\lambda \in \Lambda_0 \cap [X,Y]}\log\left|\frac{\lambda-\frac{1}{2}X}{\lambda+\frac{1}{2}X}\right|.$$

Since $a \leqslant \eta X$, the first term on the right is

$$\geqslant \; \eta\left(\frac{Y-X}{L}+1\right)\log\frac{1-2\eta}{1+2\eta}$$

by the previous boxed estimate on q.

Recall that in Fuchs' construction, L is *an integer,* Λ_0 consists of integers, and (by the theorem of article 2) X and Y are *half-odd integers.* The sum of the second and third terms on the right in the previous relation can therefore be written as

$$\int_X^Y \log\left|\frac{t+\frac{1}{2}X}{t-\frac{1}{2}X}\right| d([t/L]-n(t))$$

$$= \left\{[Y/L]-[X/L]-(n(Y)-n(X))\right\}\log\left(\frac{2Y+X}{2Y-X}\right)$$

$$+ \int_X^Y \frac{4X}{4t^2-X^2}([t/L]-[X/L]-(n(t)-n(X)))dt$$

(we are using the symbol $[p]$ to denote the *biggest integer* $\leqslant p$). We have

$$[t/L]-[X/L] \geqslant [(t-X)/L].$$

Also, $n(t)-n(X) \leqslant \varepsilon(t-X)$ for $X \leqslant t \leqslant Y$, with the left side integer-valued, so *in fact*

$$n(t)-n(X) \leqslant [\varepsilon(t-X)], \quad X \leqslant t \leqslant X.$$

Here $1/L > \varepsilon$, so surely $[(t-X)/L] \geqslant [\varepsilon(t-X)]$, and the right-hand integral in the last formula is positive. Since $(1+\alpha/4)X \leqslant Y \leqslant (1+\alpha)X$, we thus have

$$\int_X^Y \log\left|\frac{t+\frac{1}{2}X}{t-\frac{1}{2}X}\right| d([t/L]-n(t))$$

$$\geqslant \left(\left(\frac{1}{L}-\varepsilon\right)(Y-X)-2\right)\log\frac{3+2\alpha}{1+2\alpha}.$$

Combining this estimate with the one previously obtained, we see that

$$\log|R(X/2)| \geqslant (Y-X)\left\{\left(\frac{1}{L}-\varepsilon\right)\log\frac{3+2\alpha}{1+2\alpha} - \frac{\eta}{L}\log\frac{1+2\eta}{1-2\eta}\right\}$$

$$- 2\log\frac{3+2\alpha}{1+2\alpha} - \eta\log\frac{1+2\eta}{1-2\eta}.$$

Because $Y-X \geqslant (\alpha/4)X$ and $1/L-\varepsilon > 0$, the right side will be *positive for all large X provided that* $\eta > 0$ *is sufficiently small* (depending on L, ε and α, hence on L and ε). This does it.

5.   Behaviour of $(1/x)\log|(x-\lambda)/(x+\lambda)|$.

We are going to have to study $(1/x)\log|R_j(x)|$ for the products $R_j(z)$ constructed in the preceding article. For this purpose, frequent use will be made of the

Lemma. If $\lambda > 0$,

$$\frac{\partial}{\partial x}\left(\frac{1}{x}\log\left|\frac{x-\lambda}{x+\lambda}\right|\right) \quad \text{is} \; < 0 \; \text{for} \; 0 < x < \lambda \; \text{and} \; > 0 \; \text{for} \; x > \lambda.$$

Also,

$$\frac{\partial^2}{\partial\lambda\partial x}\left(\frac{1}{x}\log\left|\frac{x-\lambda}{x+\lambda}\right|\right) > 0 \quad \text{for} \; x > 0 \; \text{different from} \; \lambda.$$

Proof.

$$\frac{\partial}{\partial x}\left(\frac{1}{x}\log\left|\frac{x-\lambda}{x+\lambda}\right|\right) = \frac{1}{x^2}\log\left|\frac{x+\lambda}{x-\lambda}\right| + \frac{2\lambda}{x(x^2-\lambda^2)}.$$

The right side is > 0 for $x > \lambda$. When $0 < x < \lambda$ we rewrite the right side as

$$\frac{1}{x^2}\left\{\log\left(\frac{1+\xi}{1-\xi}\right) - \frac{2\xi}{1-\xi^2}\right\}$$

with $\xi = x/\lambda$, and then expand the quantity in curly brackets in powers of ξ. This yields

$$\frac{2}{x^2}\left(\xi + \frac{\xi^3}{3} + \frac{\xi^5}{5} + \cdots - \xi - \xi^3 - \xi^5 - \cdots\right),$$

which is < 0 since $0 < \xi < 1$.

Finally,

$$\frac{\partial^2}{\partial\lambda\partial x}\left(\frac{1}{x}\log\left|\frac{x-\lambda}{x+\lambda}\right|\right) = \frac{4x}{(\lambda^2-x^2)^2} > 0$$

for $x > 0$, $x \neq \lambda$.

We are done.

Corollary. If $0 < \lambda < \lambda'$,

$$\frac{1}{x}\log\left|\frac{x-\lambda}{x+\lambda}\right| - \frac{1}{x}\log\left|\frac{x-\lambda'}{x+\lambda'}\right|$$

is a decreasing *function of* x *for* $0 < x < \lambda$ *and for* $x > \lambda'$.

Proof. By the second derivative inequality from the lemma.

In like manner, we have the

Corollary. If $0 < x < x'$,

$$\frac{1}{x'}\log\left|\frac{x'-\lambda}{x'+\lambda}\right| - \frac{1}{x}\log\left|\frac{x-\lambda}{x+\lambda}\right|$$

is an increasing *function of* λ *for* $\lambda > x'$ *and for* $0 < \lambda < x$.

6.   **Behaviour of $(1/x)\log|R_j(x)|$ outside the interval $[X_j, Y_j]$**

Turning now to the group products $R_j(z)$ constructed in article 4, we have the

Lemma. *If the parameter $\eta > 0$ is taken sufficiently small (depending only on L and ε), $(1/x)\log|R_j(x)|$ is decreasing for $x \geqslant Y_j$ provided that X_j is large enough.*

Proof. Dropping the subscript j, we have, for $x \geqslant Y$,

$$\frac{1}{x}\log|R(x)| = \frac{q}{x}\log\left(\frac{x-a}{x+a}\right) + \sum_{\lambda \in \Lambda_0 \cap [X,Y]}\frac{1}{x}\log\left(\frac{x-\lambda}{x+\lambda}\right)$$

$$- \sum_{nL \in [X,Y]}\frac{1}{x}\log\left(\frac{x-nL}{x+nL}\right).$$

We are going to make essential use of the property

$$n(Y) - n(t) \leqslant \varepsilon(Y-t), \quad X \leqslant t \leqslant Y$$

(see theorem of article 2). Since $1/L > \varepsilon$, we have a picture like the following:

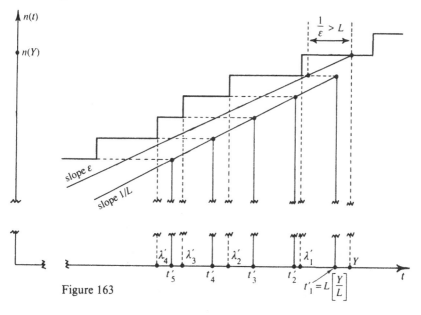

Figure 163

Number the members of Λ_0 in $[X, Y]$ *downwards*, calling the *largest* of those λ'_1, the *next largest* λ'_2, and so forth. We also denote by $t'_1 = L[Y/L]$ the *largest integral multiple of* L in $[X, Y]$, by t'_2 the *next largest one*, and so on. Since L and the members of Λ_0 are *integers* while Y is *half-odd* (theorem, article 2), we in fact have $\lambda_1' < Y$ and $t_1' < Y$.

By the above property of $n(t)$,

$$\lambda_1' \leqslant Y - 1/\varepsilon,$$
$$\lambda_2' \leqslant Y - 2/\varepsilon,$$

etc. Since, however, $L < 1/\varepsilon$, it is also true that

$$t_1' > Y - 1/\varepsilon,$$
$$t_2' > Y - 2/\varepsilon,$$

and so on.

We can *pair off* each of the λ_k', $X \leqslant \lambda_k' \leqslant Y$, with t_k', and *still have some* t_k'*s left over after all the* λ_k' *are taken care of* in this way. Indeed, there are *at most* $n(Y) - n(X) \leqslant \varepsilon(Y - X)$ of the $\lambda \in \Lambda_0$ in $[X, Y]$, X being half-odd – in fact there are at most $[\varepsilon(Y - X)]$ of them, since $n(Y) - n(X)$ is integer-valued. At the same time, there are at least $[(Y - X)/L]$ integral multiples of L in that interval. Since $1/L > \varepsilon$, there are *more* of the latter than the former, and, after pairing off each λ_k' with t_k', there will *still be* at least

$$[(Y - X)/L] - [\varepsilon(Y - X)] \;\;\geqslant\;\; \left[\left(\frac{1}{L} - \varepsilon\right)(Y - X)\right]$$

integral multiples of L *left over* in $[X, Y]$.

Now from article 4,

$$q \leqslant \frac{\eta}{L}(Y - X) + \eta.$$

Since $Y - X \geqslant (\alpha/4)X$, we see that if $\eta/L < 1/L - \varepsilon$, $[(1/L - \varepsilon)(Y - X)]$ is *larger than* q, provided only that X is big enough. Under these circumstances, *there will be more than enough of the points* t_k' *left over to pair off with the q-fold point a, after each* λ_k' *has been paired with the* t_k' corresponding to it.

Write $n(Y) - n(X) = N$. Then, after the pairings just described, the above formula for $(1/x)\log|R(x)|$ can be rewritten thus:

$$\frac{1}{x}\log|R(x)| \;\;=\;\; \sum_{k=1}^{N}\left(\frac{1}{x}\log\left(\frac{x - \lambda_k'}{x + \lambda_k'}\right) - \frac{1}{x}\log\left(\frac{x - t_k'}{x + t_k'}\right)\right)$$

$$+ \sum_{k=N+1}^{N+q}\left(\frac{1}{x}\log\left(\frac{x - a}{x + a}\right) - \frac{1}{x}\log\left(\frac{x - t_k'}{x + t_k'}\right)\right) \;\;+\;\; \sum_{k>N+q}\frac{1}{x}\log\left(\frac{x + t_k'}{x - t_k'}\right).$$

As we have seen, for $1 \leqslant k \leqslant N$,

$$\lambda'_k \;\leqslant\; Y - \frac{k}{\varepsilon} \;<\; t'_k.$$

Therefore, according to the *first corollary* from the preceding article, each of the terms in the *first right-hand sum* is a *decreasing function of x for* $x \geqslant Y$. The same holds good for the terms of the *second right-hand sum*, because $a < X < t'_k$. The terms of the *third* sum are decreasing for $x \geqslant Y$ by the lemma of article 5, since $t'_k < Y$.

All in all, then, $(1/x)\log|R(x)|$ decreases for $x \geqslant Y$. That is what we had to prove.

Lemma. *Under the hypothesis of the preceding lemma,* $(1/x)\log|R_j(x)|$ *is increasing for* $a_j < x \leqslant X_j$.

Proof. We drop the subscript j and argue as in the last proof.

Here, we number the elements of Λ_0 and the whole multiples of L belonging to $[X, Y]$ in *increasing* order, calling the *smallest* of the former λ_1, the *next smallest* λ_2, and so on, and similarly denoting by t_1 the *smallest* integral multiple of L in $[X, Y]$, by t_2 the *next smallest* one, etc. In the present situation, the property

$$n(t) - n(X) \;\leqslant\; \varepsilon(t - X), \quad X \leqslant t \leqslant Y,$$

ensured by the theorem of article 2, is the relevant one for us. By its help we see, remembering that $1/L > \varepsilon$, that

$$t_k \;<\; X + \frac{k}{\varepsilon} \;\leqslant\; \lambda_k.$$

(The reader may wish to make a diagram like the one accompanying the proof of the preceding lemma.)

There are of course *fewer* λ_ks than t_ks, just as in the proof referred to. Denoting by N the number of the *former*, we can write

$$\frac{1}{x}\log|R(x)| \;=\; \frac{q}{x}\log\frac{x-a}{x+a} + \sum_{k=1}^{N}\left(\frac{1}{x}\log\!\left(\frac{\lambda_k - x}{\lambda_k + x}\right) - \frac{1}{x}\log\!\left(\frac{t_k - x}{t_k + x}\right)\right)$$

$$+ \sum_{k>N}\frac{1}{x}\log\!\left(\frac{t_k + x}{t_k - x}\right)$$

for $a < x \leqslant X$. Using the lemma from the preceding article and its first corollary, we readily see that the right-hand expression is an increasing function of x for $a < x \leqslant X$. Done.

Combining the above two lemmas with those from article 4 we now get the

Theorem. *If $\eta > 0$ is taken small enough (depending only on the values of L and ε), and if X_j is large, the maximum value of $(1/x)\log|R_j(x)|$ for $x \geqslant 0$ outside of (X_j, Y_j) is attained for $x = X_j$ or $x = Y_j$, and it is positive.*

Proof. Under the hypothesis, we have, by the results in article 4,

$$\frac{1}{x}\log|R(x)| \;\leqslant\; 0, \quad 0 \leqslant x < a,$$

whereas

$$\frac{2}{X}\log|R(X/2)| \;>\; 0$$

(as usual, we suppress the subscript j). The maximum in question is hence > 0, and it is attained for $a < x \leqslant X$ or for $x \geqslant Y$. Now use the preceding two lemmas. The theorem follows.

7.   Behaviour of $(1/x)\log|R_j(x)|$ inside $[X_j, Y_j]$

The essential step in Fuchs' construction consists in showing that $(1/x)\log|R_j(x)|$ *really gets larger inside* $[X_j, Y_j]$ than *outside* that interval.

Lemma. *If $\kappa > 0$ is sufficiently small (depending only on L and ε) and ξ, $(1 + (\kappa/5))X_j \leqslant \xi \leqslant (1 + (4\kappa/5))X_j$, is an integer not in Λ_0, we have*

$$\frac{1}{\xi}\log|R_j(\xi)| \;\geqslant\; \frac{1}{X_j}\log|R_j(X_j)| + \frac{1}{2}\left(\frac{1}{L} - \varepsilon\right)\sigma\log\frac{1}{\sigma},$$

where

$$1 + \sigma \;=\; \frac{\xi}{X_j},$$

provided that X_j is large enough (depending on κ, L and ε).

Proof. As usual, we suppress the subscript j. For $X < \xi < Y$, we have

$$\frac{1}{\xi}\log|R(\xi)| - \frac{1}{X}\log|R(X)| \;=\; \frac{q}{\xi}\log\left(\frac{\xi - a}{\xi + a}\right) - \frac{q}{X}\log\left(\frac{X - a}{X + a}\right)$$

$$+ \sum_{\lambda\in\Lambda_0\cap[X,Y]}\left(\frac{1}{\xi}\log\left|\frac{\xi - \lambda}{\xi + \lambda}\right| - \frac{1}{X}\log\left|\frac{X - \lambda}{X + \lambda}\right|\right)$$

$$+ \sum_{nL\in[X,Y]}\left(\frac{1}{\xi}\log\left|\frac{\xi + nL}{\xi - nL}\right| - \frac{1}{X}\log\left|\frac{X + nL}{X - nL}\right|\right).$$

Let us start by looking at the *third* term on the right – the summation over λ.

Taking a $\kappa > 0$ which is fairly small in relation to α (the number > 0 depending on ε such that $(1 + (\alpha/4))X \leqslant Y \leqslant (1 + \alpha)X$), we break up the sum in question as

$$\sum_{\substack{X \leqslant \lambda < (1+\kappa)X \\ \lambda \in \Lambda_0}} + \sum_{\substack{(1+\kappa)X \leqslant \lambda \leqslant Y \\ \lambda \in \Lambda_0}},$$

and look initially at the *second* of these terms, which, in fact, gives the main contribution.

Number the elements of Λ_0 belonging to $[X, Y]$ in *increasing* order, calling them $\lambda_1 (> X)$, $\lambda_2 > \lambda_1$, $\lambda_3 > \lambda_2$ and so on. As in the proof of the second lemma from the preceding article, we have

$$\lambda_k \geqslant X + \frac{k}{\varepsilon}.$$

It is, in the first place, possible that some of the λ_k with $1 \leqslant k < \varepsilon\kappa X + 1$ are $\geqslant (1 + \kappa)X$. Denoting by S the set of indices k for which this occurs (if there are any), we can write

$$\sum_{\substack{(1+\kappa)X \leqslant \lambda \leqslant Y \\ \lambda \in \Lambda_0}} = \sum_{\lambda_k, k \in S} + \sum_{\lambda_k, k \geqslant \varepsilon\kappa X + 1}.$$

The first sum on the right is

$$\sum_{k \in S} \left(\frac{1}{\xi} \log \left| \frac{\xi - \lambda_k}{\xi + \lambda_k} \right| - \frac{1}{X} \log \left| \frac{X - \lambda_k}{X + \lambda_k} \right| \right).$$

For $X \leqslant \xi < (1 + \kappa)X$, each of the differences in this expression is *negative* by the lemma of article 5, and each is made *smaller* (*more* negative) when the corresponding λ_k is *moved downwards* to $(1 + \kappa)X$, by the *second* corollary to that lemma. Since it involves at most $\varepsilon\kappa X + 1$ of the λ_k, the sum just written is therefore

$$\geqslant (\varepsilon\kappa X + 1) \left\{ \frac{1}{\xi} \log \left(\frac{(1+\kappa)X - \xi}{(1+\kappa)X + \xi} \right) - \frac{1}{X} \log \left(\frac{(1+\kappa)X - X}{(1+\kappa)X + X} \right) \right\}$$

Writing $\xi = (1 + \sigma)X$, this last works out to

$$\frac{\varepsilon\kappa + (1/X)}{1 + \sigma} \left\{ \log \left(\frac{\kappa - \sigma}{\kappa} \right) + \log \left(\frac{2 + \kappa}{2 + \kappa + \sigma} \right) - \sigma \log \left(\frac{\kappa}{2 + \kappa} \right) \right\},$$

and this is $\geqslant \varepsilon O(\kappa)$ for $\kappa/5 \leqslant \sigma \leqslant 4\kappa/5$ and $X \geqslant 1/\varepsilon\kappa$ when κ is small. The $O(\kappa)$ factor involved here depends *only* on κ and *not* on any of the *other* parameters.

We now turn to the *second* right-hand sum in the above decomposition, namely

$$\sum_{k \geqslant \varepsilon \kappa X + 1} \left(\frac{1}{\xi} \log\left(\frac{\lambda_k - \xi}{\lambda_k + \xi}\right) - \frac{1}{X} \log\left(\frac{\lambda_k - X}{\lambda_k + X}\right) \right).$$

Since $\lambda_k \geqslant X + k/\varepsilon$, the summand involving λ_k is

$$\geqslant \varepsilon \int_{X+(k-1)/\varepsilon}^{X+k/\varepsilon} \left(\frac{1}{\xi} \log\left(\frac{t - \xi}{t + \xi}\right) - \frac{1}{X} \log\left(\frac{t - X}{t + X}\right) \right) dt,$$

because it is $\geqslant$ *the integrand here for each $t \in [X + (k-1)/\varepsilon,\ X + k/\varepsilon]$* according to the *second* corollary of article 5. In our present *sum*, the index k ranges from the smallest integer $\geqslant \varepsilon \kappa X + 1$ up to $n(Y) - n(X)$ (which we assume is not less than the former quantity; otherwise the sum is just *zero*). That sum is thus

$$\geqslant \varepsilon \int_{X+\kappa X}^{X+(n(Y)-n(X))/\varepsilon} \left\{ \frac{1}{\xi} \log\left(\frac{t - \xi}{t + \xi}\right) - \frac{1}{X} \log\left(\frac{t - X}{t + X}\right) \right\} dt$$

where, by the lemma of article 5, *the integrand is negative* (for $X \leqslant \xi < X + \kappa X$). We have $n(Y) - n(X) \leqslant \varepsilon(Y - X)$. The last expression is therefore

$$\geqslant \varepsilon \int_{(1+\kappa)X}^{Y} \left\{ \frac{1}{\xi} \log\left(\frac{t - \xi}{t + \xi}\right) - \frac{1}{X} \log\left(\frac{t - X}{t + X}\right) \right\} dt.$$

This is worked out by partial integration. It is convenient to write $Y = (1 + \beta)X$ (so that $\alpha/4 \leqslant \beta \leqslant \alpha$) and make the substitution $t = X + \tau X$. For $x = (1 + s)X$ with $0 \leqslant s < \kappa$, we find that

$$\frac{1}{x} \int_{(1+\kappa)X}^{Y} \log\left(\frac{t - x}{t + x}\right) dt = \frac{1}{1+s} \int_{\kappa}^{\beta} \log\left(\frac{\tau - s}{2 + \tau + s}\right) d\tau$$

$$= \frac{1}{1+s} \left\{ (\beta - s)\log(\beta - s) - (\beta + s + 2)\log(\beta + s + 2) \right.$$

$$\left. - (\kappa - s)\log(\kappa - s) + (\kappa + s + 2)\log(\kappa + s + 2) \right\}.$$

Putting first $s = \sigma$, then $s = 0$ and subtracting, and afterwards multiplying the result by ε, we get for the previous integral the value

$$\varepsilon \left\{ \frac{\beta - \sigma}{1 + \sigma} \log(\beta - \sigma) - \beta \log \beta - \frac{\beta + \sigma + 2}{1 + \sigma} \log(\beta + \sigma + 2) + \right.$$

$$+ (\beta + 2)\log(\beta + 2) + \frac{\kappa + \sigma + 2}{1 + \sigma}\log(\kappa + \sigma + 2)$$

$$- (\kappa + 2)\log(\kappa + 2) - \frac{\kappa - \sigma}{1 + \sigma}\log(\kappa - \sigma) + \kappa \log \kappa \Big\}.$$

Here, $\alpha/4 \leqslant \beta \leqslant \alpha$, so if $\kappa/5 \leqslant \sigma \leqslant 4\kappa/5$ with κ quite small in relation to α, the *first two terms* in curly brackets are readily seen to amount to $O(\sigma)$, taken together. The same is true for the *third* and *fourth* terms. The *fifth* and *sixth* terms again come to $O(\sigma)$. (The first two quantities $O(\sigma)$ obtained in this manner depend on α, which depends on ε.) There remain the *last two* terms. Those work out to $\sigma \log \sigma + O(\sigma)$. The whole expression thus equals $\varepsilon(\sigma \log \sigma + O(\sigma))$; *this*, then, is the value of the above integral, which, in turn, is a *lower bound* for

$$\sum_{k \geqslant \varepsilon\kappa X + 1}\left(\frac{1}{\xi}\log\left(\frac{\lambda_k - \xi}{\lambda_k + \xi}\right) - \frac{1}{X}\log\left(\frac{\lambda_k - X}{\lambda_k + X}\right)\right).$$

That sum is thus

$$\geqslant -\varepsilon\left(\sigma \log \frac{1}{\sigma} + O(\sigma)\right).$$

We combine this estimate with the one for the sum in which k ranges over S, obtained previously. That yields, for $X \geqslant 1/\varepsilon\kappa$,

$$\sum_{\substack{(1 + \kappa)X \leqslant \lambda \leqslant Y \\ \lambda \in \Lambda_0}}\left(\frac{1}{\xi}\log\left(\frac{\lambda - \xi}{\lambda + \xi}\right) - \frac{1}{X}\log\left(\frac{\lambda - X}{\lambda + X}\right)\right)$$

$$\geqslant -\varepsilon\sigma \log \frac{1}{\sigma} - \varepsilon O(\sigma) - \varepsilon O(\kappa) = -\varepsilon\left(\sigma \log \frac{1}{\sigma} + O(\sigma)\right),$$

where the $O(\sigma)$ term depends on α (and thus on ε).

We lay the sum

$$\sum_{\substack{X \leqslant \lambda < (1 + \kappa)X \\ \lambda \in \Lambda_0}}$$

aside for the moment, and proceed to the examination of

$$\sum_{X \leqslant nL \leqslant Y}\left(\frac{1}{\xi}\log\left|\frac{\xi + nL}{\xi - nL}\right| - \frac{1}{X}\log\left|\frac{X + nL}{X - nL}\right|\right),$$

which we break up as

$$\sum_{X \leqslant nL \leqslant (1 + \kappa)X} + \sum_{(1 + \kappa)X < nL \leqslant Y}.$$

The *second* of these sums can be handled just as the one over λ_k with $k \geqslant \varepsilon \kappa X + 1$ was treated above. Since the numbers nL are *already equally spaced*, that procedure furnishes an *approximate equality* in the present situation, namely*

$$\sum_{(1+\kappa)X < nL \leqslant Y} \left(\frac{1}{\xi} \log\left(\frac{nL+\xi}{nL-\xi}\right) - \frac{1}{X} \log\left(\frac{nL+X}{nL-X}\right) \right)$$

$$= \frac{1}{L}\left(\sigma \log \frac{1}{\sigma} + O(\sigma) \right) + O\left(\frac{1}{X} \right).$$

For the *first* sum,

$$\sum_{X \leqslant nL \leqslant (1+\kappa)X} \left(\frac{1}{\xi} \log\left|\frac{\xi+nL}{\xi-nL}\right| - \frac{1}{X} \log\left|\frac{X+nL}{X-nL}\right| \right),$$

we proceed to find a *lower bound, assuming that ξ is an integer between* $(1+(\kappa/5))X$ and $(1+(4\kappa/5))X$. (Up to now, we have *not used* the fact that $\xi \in \mathbb{N}$, which is part of the hypothesis.) In case ξ is *divisible by the integer* L, the expression in question is *infinite*, for X is half-odd. We therefore need do the computation only when $\xi \in \mathbb{N}$ is *not divisible by L*. In that case, we rewrite the sum as

$$\int_X^{(1+\kappa)X} \left(\frac{1}{\xi} \log\left|\frac{\xi+t}{\xi-t}\right| - \frac{1}{X} \log\left|\frac{X+t}{X-t}\right| \right) \mathrm{d}[t/L],$$

and begin by working out

$$\int_X^{(1+\kappa)X} \log\left|\frac{1}{\xi-t}\right| \mathrm{d}[t/L].$$

(As usual, $[t/L]$ denotes the largest integer $\leqslant t/L$.)

Since $\xi \in \mathbb{N}$ is *not divisible by L*, the last integral is equal to

$$\left(\int_X^{\xi-\gamma} + \int_{\xi+\gamma}^{(1+\kappa)X} \right) \log\left|\frac{1}{t-\xi}\right| \mathrm{d}[t/L]$$

for all sufficiently small $\gamma > 0$. Integrate by parts, using $[t/L] - [\xi/L]$ as

---

* Thanks to article 5's second corollary, the sum in question can differ from a corresponding integral by at most *twice* the value of $(1/\xi)\log((nL+\xi)/(nL-\xi)) - (1/X)\log((nL+X)/(nL-X))$ for $n = [(1+\kappa)X/L]$. This is $O(1/X)$ for large X when $\kappa/5 \leqslant \sigma \leqslant 4\kappa/5$.

a primitive for $d[t/L]$. After making $\gamma \to 0$, we obtain the value

$$\left([\xi/L] - [X/L]\right)\log\frac{1}{\xi - X} + \left([(1 + \kappa)X/L] - [\xi/L]\right)\log\frac{1}{(1 + \kappa)X - \xi}$$

$$+ \int_X^{(1 + \kappa)X}\frac{[t/L] - [\xi/L]}{t - \xi}\,dt.$$

The *integral* in this expression is *positive*, and

$$\log\frac{1}{\xi - X} = \log\frac{1}{X} + \log\frac{1}{\sigma} = \log\frac{1}{X} + \log\frac{1}{\kappa} + O(1),$$

$$\log\frac{1}{(1 + \kappa)X - \xi} = \log\frac{1}{X} + \log\frac{1}{\kappa - \sigma} = \log\frac{1}{X} + \log\frac{1}{\kappa} + O(1),$$

because $\kappa/5 \leqslant \sigma \leqslant 4\kappa/5$. Therefore,

$$\int_X^{(1 + \kappa)X}\log\left|\frac{1}{\xi - t}\right|d[t/L]$$

$$\geqslant \left([(1 + \kappa)X/L] - [X/L]\right)\left(\log\frac{1}{X} + \log\frac{1}{\kappa} + O(1)\right).$$

On the other hand, since $X < \xi < (1 + \kappa)X$, we clearly have

$$\int_X^{(1 + \kappa)X}\log(t + \xi)\,d[t/L]$$

$$= \left([(1 + \kappa)X/L] - [X/L]\right)(\log X + O(1)).$$

This, together with the calculation just made, gives

$$\frac{1}{\xi}\int_X^{(1 + \kappa)X}\log\left|\frac{t + \xi}{t - \xi}\right|d[t/L]$$

$$\geqslant \frac{[(1 + \kappa)X/L] - [X/L]}{(1 + \sigma)X}\left(\log\frac{1}{\kappa} + O(1)\right).$$

Turning to the similar integral involving X in place of ξ, we first get

$$\int_X^{(1 + \kappa)X}\log\left(\frac{1}{t - X}\right)d[t/L] = \left([(1 + \kappa)X/L] - [X/L]\right)\log\frac{1}{\kappa X}$$

$$+ \int_X^{(1 + \kappa)X}\frac{[t/L] - [X/L]}{t - X}\,dt \quad \leqslant$$

$$\leqslant \left([(1+\kappa)X/L] - [X/L] \right)\left(\log\frac{1}{X} + \log\frac{1}{\kappa} \right) + \frac{\kappa X}{L} + O(\log X)$$

after taking account of the fact that $[t/L] - [X/L] \leqslant (t-X)/L + 1$ actually vanishes for $0 \leqslant t - X < 1/2$, X being half-odd and L an integer. Thence, it is readily seen that

$$\frac{1}{X}\int_X^{(1+\kappa)X} \log\left(\frac{t+X}{t-X}\right) d[t/L]$$

$$\leqslant \frac{[(1+\kappa)X/L] - [X/L]}{X}\left(\log\frac{1}{\kappa} + O(1) \right) + \frac{\kappa}{L} + O\left(\frac{\log X}{X}\right).$$

Combining this with the estimate already obtained for the similar integral involving ξ, we find that

$$\sum_{X \leqslant nL \leqslant (1+\kappa)X} \left(\frac{1}{\xi}\log\left|\frac{\xi+nL}{\xi-nL}\right| - \frac{1}{X}\log\left|\frac{X+nL}{X-nL}\right| \right)$$

$$\geqslant \frac{[(1+\kappa)X/L] - [X/L]}{X}\left\{ \left(\frac{1}{1+\sigma} - 1 \right)\log\frac{1}{\kappa} + O(1) \right\}$$

$$- \frac{\kappa}{L} - O\left(\frac{\log X}{X}\right)$$

for integers $\xi = (1+\sigma)X$ between $(1+(\kappa/5))X$ and $(1+(4\kappa/5))X$. When $X/\log X$ is *large in relation to* L/κ, the right side of this inequality reduces to

$$- \frac{\kappa\sigma}{L}\log\frac{1}{\kappa} - O\left(\frac{\kappa}{L}\right) = \frac{O(\sigma)}{L}$$

for small κ. Referring to the estimate for

$$\sum_{(1+\kappa)X < nL \leqslant Y}$$

given above, we thus obtain

$$\boxed{\sum_{X \leqslant nL \leqslant Y} \left(\frac{1}{\xi}\log\left|\frac{\xi+nL}{\xi-nL}\right| - \frac{1}{X}\log\left|\frac{X+nL}{X-nL}\right| \right) \geqslant \frac{1}{L}\sigma\log\frac{1}{\sigma} - \frac{O(\sigma)}{L},}$$

valid for *small* values of $\kappa > 0$ with *integers* $\xi = (1+\sigma)X$ such that $\kappa/5 \leqslant \sigma \leqslant 4\kappa/5$, and for *large* values of X (depending on L and κ).

Let us return to the sum

$$\sum_{\substack{X \leq \lambda < (1+\kappa)X \\ \lambda \in \Lambda_0}} \left(\frac{1}{\xi} \log \left| \frac{\xi - \lambda}{\xi + \lambda} \right| - \frac{1}{X} \log \left| \frac{X - \lambda}{X + \lambda} \right| \right)$$

which was set aside earlier; *here we assume that* ξ *is an integer not belonging to* Λ_0, with $(1 + (\kappa/5))X \leq \xi \leq (1 + (4\kappa/5))X$. The calculation is like the one just made for the similar sum over nL with $X \leq nL \leq (1 + \kappa)X$. In order to keep our notation simple, we assume that $(1 + \kappa)X \notin \Lambda_0$ – actually, whether this is true or not makes no real difference.

Consider the integral

$$\int_X^{(1 + \kappa)X} \log|t - \xi|\, dn(t).$$

Following the procedure used above for the similar integral involving $d[t/L]$ instead of $dn(t)$, this works out to

$$(n((1 + \kappa)X) - n(X))(\log X + \log \kappa + O(1))$$

$$- \int_X^{(1 + \kappa)X} \frac{n(t) - n(\xi)}{t - \xi}\, dt.$$

We have $\Lambda_0 \subseteq \mathbb{N}$, so, if $\xi \in \mathbb{N}$ does *not* belong to Λ_0, $|n(t) - n(\xi)| \leq |t - \xi|$. The last expression therefore *differs* from

$$(n((1 + \kappa)X) - n(X))(\log X + \log \kappa + O(1))$$

by at most κX in absolute value.

From here on, the work goes just as that done above for the integrals involving $d[t/L]$. One finds without trouble that when κ is small,

$$\sum_{\substack{X \leq \lambda < (1+\kappa)X \\ \lambda \in \Lambda_0}} \left(\frac{1}{\xi} \log \left| \frac{\xi - \lambda}{\xi + \lambda} \right| - \frac{1}{X} \log \left| \frac{X - \lambda}{X + \lambda} \right| \right) = O(\sigma)$$

for *integers* $\xi = (1 + \sigma)X \notin \Lambda_0$, $\kappa/5 \leq \sigma \leq 4\kappa/5$, and *sufficiently large* X (depending on κ). (This calculation *really gives* just $O(\sigma)$ and *not* $\varepsilon O(\sigma)$, because the relation $n(t) - n(\xi) \leq |t - \xi|$ was all we had available for our choice of ξ.)

The result just obtained is now combined with the earlier one for the sum

$$\sum_{\substack{(1 + \kappa)X \leq \lambda \leq Y \\ \lambda \in \Lambda_0}}.$$

We get

$$
\sum_{\substack{X \leqslant \lambda \leqslant Y \\ \lambda \in \Lambda_0}} \left(\frac{1}{\xi} \log \left| \frac{\xi - \lambda}{\xi + \lambda} \right| - \frac{1}{X} \log \left| \frac{X - \lambda}{X - \lambda} \right| \right) \;\geqslant\; -\varepsilon\sigma \log \frac{1}{\sigma} \;-\; O(\sigma),
$$

valid for ξ of the kind described above, provided that X is *large enough* (depending on κ).

We still have to look at the difference

$$
\frac{q}{\xi} \log \left(\frac{\xi - a}{\xi + a} \right) \;-\; \frac{q}{X} \log \left(\frac{X - a}{X + a} \right);
$$

it, however, is clearly $\geqslant 0$ by the lemma of article 5, *since* $a < X < \xi$!*

The relation last noted and two boxed estimates above are finally plugged into the formula from the very beginning of this proof. That gives, *for $\kappa > 0$ small in relation to α,*

$$
\frac{1}{\xi} \log |R(\xi)| \;-\; \frac{1}{X} \log |R(X)| \;\geqslant\; \left(\frac{1}{L} - \varepsilon \right) \sigma \log \frac{1}{\sigma} \;-\; O(\sigma),
$$

provided that the *integer* $\xi = (1 + \sigma)X$ with $\kappa/5 \leqslant \sigma \leqslant 4\kappa/5$ is $\notin \Lambda_0$ and that X is *sufficiently large* (depending on κ, L and ε). *Formally*, the $O(\sigma)$ term on the right depends on L, ε and α; *in fact*, however, since ε is *small* and L *large*, it is *essentially* dependent *on α alone* (that is, on ε), as one sees by looking again at how the individual $O(\sigma)$ terms arise in the above computations.

We are using $L < 1/\varepsilon$ in the present construction. Therefore, if $\kappa > \sigma > 0$ is *small enough* (depending on L and ε), the term $(1/L - \varepsilon)\sigma \log 1/\sigma$ in the right-hand side of the preceding inequality *will greatly outweigh* the $O(\sigma)$ term appearing there. Then, for *sufficiently large* X (depending on κ, L and ε) we will have

$$
\frac{1}{\xi} \log |R(\xi)| \;-\; \frac{1}{X} \log |R(X)| \;\geqslant\; \frac{1}{2} \left(\frac{1}{L} - \varepsilon \right) \sigma \log \frac{1}{\sigma}
$$

when the integer $\xi = (1 + \sigma)X$ with $\kappa/5 \leqslant \sigma \leqslant 4\kappa/5$ lies *outside* Λ_0. The lemma is proved.

---

* Using the relation $a \leqslant \eta X$ and the boxed estimate on q from article 4, the difference in question is readily worked out to be $\leqslant \text{const.} \, \sigma\eta^2((\alpha/L) + (1/X))$.

Remark. Since $n(t) - n(X_j) \leqslant \varepsilon(t - X_j)$ for $X_j \leqslant t \leqslant Y_j$, there are *at most* $\frac{4}{5}\kappa\varepsilon X_j$ members of Λ_0 in the interval $[(1 + (\kappa/5))X_j, \ (1 + (4\kappa/5))X_j]$ which, however, contains *at least* $[(3\kappa/5)X_j]$ integers. In our construction, $\varepsilon > 0$ is *small*. Hence, given $\kappa > 0$ there are *at least* $(1 - \frac{4}{3}\varepsilon)\cdot(3\kappa/5)X_j - 1$ *integers* ξ *in* $[(1 + (\kappa/5))X_j, \ (1 + (4\kappa/5))X_j]$ *to which the lemma applies* when X_j is large.

It is important that the ratio $(1/x)\log|R_j(x)|$ has behaviour similar to that described by the preceding lemma when x is near the *right* endpoint Y_j of $[X_j, Y_j]$. Arguing very much as in the long proof just given, but using the inequality

$$n(Y_j) - n(t) \ \leqslant \ \varepsilon(Y_j - t), \quad X_j \leqslant t \leqslant Y_j,$$

instead of

$$n(t) - n(X_j) \ \leqslant \ \varepsilon(t - X_j), \quad X_j \leqslant t \leqslant Y_j,$$

one establishes the

Lemma. *If* $\kappa > 0$ *is* sufficiently small (*depending on* L *and* ε) *and* ξ, $(1 - (4\kappa/5))Y_j \leqslant \xi \leqslant (1 - (\kappa/5))Y_j$, *is an integer not in* Λ_0, *we have*

$$\frac{1}{\xi}\log|R_j(\xi)| \ \geqslant \ \frac{1}{Y_j}\log|R_j(Y_j)| \ + \ \frac{1}{2}\left(\frac{1}{L} - \varepsilon\right)\sigma\log\frac{1}{\sigma}$$

with $1 - \sigma = \xi/Y_j$, *provided that* X_j *is large enough* (*depending on* κ, L *and* ε).

It is recommended that the reader *think through* how the steps in the proof of the previous result can be adapted to that of the present one, without actually writing out the details.

8. Formation of Fuchs' function Φ(z). Discussion

$\Phi(z)$, which actually involves the parameters ε, L and η, is constructed as follows. One starts by taking a *small* $\varepsilon > 0$ and then getting a sequence of intervals $[X_j, Y_j]$ with $X_j \xrightarrow[j]{} \infty$, corresponding to ε in the manner described by the theorem at the end of article 2. One then picks a *large integer* $L < 1/\varepsilon$ and finally, choosing a small value > 0 for η, takes for each j a number a_j,

$$\tfrac{1}{2}\eta X_j \ \leqslant \ a_j \ \leqslant \ \eta X_j,$$

and an integer q_j, according to the procedure of article 4. The parameter

$\eta > 0$ is chosen *small enough* (depending on L and ε) for the results of articles 4 and 6 to apply.

We next take an *exceedingly sparse* sequence of the intervals $[X_j, Y_j]$ – by this we mean that the ratios X_{j+1}/X_j are to *increase very rapidly*, in a way to be determined presently. *The rest of the construction uses only the $[X_j, Y_j]$ from this sparse sequence.* In terms of *these*, we write

$$\Omega \;=\; (0, \infty) \;\sim\; \bigcup_j [X_j, Y_j],$$

and finally put

$$\Phi(z) \;=\; \left(\frac{z}{L}\right)^{-2z/L} e^{(2-2C)z/L} \prod_{\substack{nL\in\Omega \\ n\in\mathbb{N}}} \left(\frac{1 - z/nL}{1 + z/nL}\right) e^{2z/nL}$$

$$\times \prod_j \left\{ \left(\frac{1 - z/a_j}{1 + z/a_j}\right)^{q_j} e^{2q_j z/a_j} \cdot \prod_{\lambda\in\Lambda_0\cap[X_j, Y_j]} \left(\frac{1 - z/\lambda}{1 + z/\lambda}\right) e^{2z/\lambda} \right\}.$$

Here, C is Euler's constant (see article 3).

The function $\Phi(z)$ is analytic in $\Re z > 0$ and vanishes at the positive integral multiples of L *outside* our sparse sequence of intervals $[X_j, Y_j]$, as well as at *the points of* Λ_0 lying *within the latter*. It also has a q_j-fold zero at each a_j. Using the group products defined in article 4 and studied in the previous two articles, we can write

$$\Phi(z) \;=\; F(z/L) \prod_j R_j(z),$$

where

$$F(z) \;=\; z^{-2z} e^{(2-2C)z} \prod_{n=1}^{\infty} \left(\frac{1 - z/n}{1 + z/n}\right) e^{2z/n}$$

$$= \; z^{-2z} e^{2z} \frac{\sin \pi z}{\pi z} (\Gamma(z+1))^2,$$

a function already looked at in article 3. (In this last formula for $\Phi(z)$, it is of course the product of the $R_j(z)$ *corresponding to our sparse sequence of intervals* $[X_j, Y_j]$ that is understood.)

Lemma. *If* $Y_j < \frac{1}{2}\eta X_{j+1}$ *we have*

$$|\Phi(z)| \;\leqslant\; \text{const.} \exp\!\left(\frac{\pi}{L}|\Im z| + A\Re z\right)$$

for $\Re z \geqslant 0$, A *being a constant depending on* L, ε *and* η.

Proof. We have

$$\Phi(z) = \left(\frac{z}{L}\right)^{-2z/L} e^{(2-2C)z/L} \prod_{k=1}^{\infty} \left(\frac{1-z/\mu_k}{1+z/\mu_k}\right) e^{2z/\mu_k},$$

where $\{\mu_k\}$ is a certain increasing sequence consisting of the numbers nL in Ω, $n \in \mathbb{N}$, the points of Λ_0 in our intervals $[X_j, Y_j]$, and the q_j-fold repeated points a_j.

If z, $\Re z > 0$, is given, we have, for $\mu_k \geqslant 2|z|$,

$$\left(\frac{1-z/\mu_k}{1+z/\mu_k}\right) e^{2z/\mu_k} = \exp\left(-\frac{2z^3}{3\mu_k^3} - \frac{2z^5}{5\mu_k^5} - \cdots\right)$$

$$= \exp\left(-\frac{2z^3}{3\mu_k^3}\left(1 + o\left(\frac{z}{\mu_k}\right)\right)\right).$$

Clearly,

$$\sum_{\mu_k \geqslant 2|z|} \frac{1}{\mu_k^3} = O\left(\frac{1}{|z|^2}\right).$$

Hence

$$\left|\prod_{\mu_k \geqslant 2|z|} \left(\frac{1-z/\mu_k}{1+z/\mu_k}\right) e^{2z/\mu_k}\right| \leqslant e^{O(|z|)}.$$

For $0 < \mu_k < 2|z|$, since $x = \Re z > 0$,

$$\left|\left(\frac{1-z/\mu_k}{1+z/\mu_k}\right) e^{2z/\mu_k}\right| \leqslant e^{2x/\mu_k} \ (!),$$

whence

$$\left|\prod_{\mu_k < 2|z|} \left(\frac{1-z/\mu_k}{1+z/\mu_k}\right) e^{2z/\mu_k}\right| \leqslant \exp\left(2x \sum_{\mu_k < 2|z|} \frac{1}{\mu_k}\right).$$

Suppose now that $Y_j \leqslant 2|z| < a_{j+1}$ for some j – note that our condition $Y_j < \frac{1}{2}\eta X_{j+1}$ does make $Y_j < a_{j+1}$ for each j. Then, since the a_l and q_l were chosen so as to make

$$\frac{q_l}{a_l} + \sum_{\lambda \in \Lambda_0 \cap [X_l, Y_l]} \frac{1}{\lambda} = \sum_{nL \in [X_l, Y_l]} \frac{1}{nL}$$

for each l (see article 4), we have

$$\sum_{\mu_k < 2|z|} \frac{1}{\mu_k} = \sum_{\substack{nL < 2|z| \\ n \in \mathbb{N}}} \frac{1}{nL} = \frac{1}{L}\left(\log\left(\frac{2|z|}{L}\right) + O(1)\right)$$

$$= \frac{1}{L}\log|z| - \frac{O(\log L)}{L}.$$

This formula *remains true* when $a_{j+1} \leqslant 2|z| < Y_{j+1}$. For then $\sum_{\mu_k < 2|z|}(1/\mu_k)$ can differ from $\sum_{nL < 2|z|, n \in \mathbb{N}}(1/nL)$ by *at most*

$$\frac{q_{j+1}}{a_{j+1}} + \sum_{\substack{X_{j+1} \leqslant \lambda \leqslant Y_{j+1} \\ \lambda \in \Lambda_0}} \frac{1}{\lambda} + \sum_{\substack{X_{j+1} \leqslant nL \leqslant Y_{j+1} \\ n \in \mathbb{N}}} \frac{1}{nL},$$

and by the above relation this equals

$$2\sum_{X_{j+1} \leqslant nL \leqslant Y_{j+1}} \frac{1}{nL} \quad \leqslant \quad \frac{2}{X_{j+1}} + \frac{2}{L}\log\frac{Y_{j+1}}{X_{j+1}} \quad \leqslant \quad \frac{2}{X_{j+1}} + \frac{2\log(1+\alpha)}{L},$$

since

$$\left(1 + \frac{\alpha}{4}\right)X_{j+1} \quad \leqslant \quad Y_{j+1} \quad \leqslant \quad (1+\alpha)X_{j+1}.$$

There is of course no loss of generality in assuming all the X_l to be $> L$, so, this being granted, we have

$$\sum_{\mu_k < 2|z|} \frac{1}{\mu_k} \quad = \quad \frac{1}{L}\log|z| + \frac{O(\log L)}{L}$$

in the present case also; the formula is thus true generally.

Using the relation just found with the preceding estimate and then combining the result with the one obtained previously we get, for $\Re z > 0$,

$$\left| \prod_{k=1}^{\infty} \left(\frac{1 - z/\mu_k}{1 + z/\mu_k}\right) e^{2z/\mu_k} \right| \quad \leqslant \quad \exp\left(\frac{2x}{L}\log|z| + O(|z|)\right).$$

Hence, in the right half plane,

$$|\Phi(z)| \quad \leqslant \quad \left|\frac{z}{L}\right|^{-2x/L} e^{(2y/L)\arg z} e^{(2-2C)x/L} e^{(2x/L)\log|z| + O(|z|)} \quad = \quad e^{O(|z|)},$$

the exponential in $(2x/L)\log|z|$ being cancelled by $|z|^{-2x/L}$. The function $\Phi(z)$ is thus of *exponential type* in the half plane $\{\Re z > 0\}$.

Once this is known, we can use Φ's obvious continuity up to the imaginary axis and apply the *second* Phragmén–Lindelöf theorem from §C of Chapter III. For $z = iy$ *pure imaginary*, the product over the μ_k in the above formula for $\Phi(z)$ *has modulus* 1, and we see that

$$|\Phi(iy)| \quad = \quad e^{(2y/L)\arg(iy)} \quad = \quad e^{\pi|y|/L}$$

On the other hand, $|\Phi(x)| \leqslant \text{const.} \, e^{Ax}$ with a certain constant A, by what has just been shown. The function

$$e^{\pi i z/L} e^{-Az} \Phi(z)$$

is thus *bounded* on the *sides* of the *first quadrant*, and *hence within it*, by Phragmén–Lindelöf. Similarly,

$$e^{-\pi i z/L}e^{-Az}\,\Phi(z)$$

is *bounded* in the *fourth quadrant*. Thus,

$$|\Phi(z)| \;\leqslant\; \text{const. } e^{(\pi|y|/L)+Ax} \quad \text{for } x \geqslant 0,$$

Q.E.D.

Remark. Paying a little more attention to the computation at the beginning of the proof just given, one sees that the constant A can be taken to be *small if L is large*. Our function $\Phi(z)$ is thus of *small exponential type in* $\Re z \geqslant 0$. This fact will not be used in our application.

We now return to our *original non-measurable sequence* $\Sigma \subseteq \mathbb{N}$ with Polya maximum density $D_\Sigma^* > 0$. At the beginning of article 2, the complement $\mathbb{N} \sim \Sigma$ was broken up into two disjoint sequences: Λ_0, infinite and of *minimum density zero*, which has figured in the constructions of articles 2–7, and a *measurable sequence* Λ_1 of *density* $1 - D_\Sigma^*$. *The main purpose of all the above work has been to arrive at the function* $\Phi(z)$, *having properties described by the preceding lemma and by the*

Theorem. *If* $\varepsilon > 0$ *is small enough and the integer L,* $0 < L < 1/\varepsilon$ *is large, if, moreover,*

$$\frac{1}{L} + 2\varepsilon \;<\; D_\Sigma^*$$

and the sequence of intervals $[X_j, Y_j]$ *used in the construction of* $\Phi(z)$ *is sparse enough, we have*

$$\limsup_{\substack{m\to\infty\\ m\in\Lambda_0}} \frac{\log \Phi(m)}{m} \;\leqslant\; \limsup_{\substack{n\to\infty\\ n\in\Sigma}} \frac{\log |\Phi(n)|}{n} \;-\; \delta(L,\varepsilon),$$

where $\delta(L,\varepsilon)$ *is a quantity* > 0 *depending on L and ε. The quantity*

$$\limsup_{\substack{n\to\infty\\ n\in\Sigma}} \frac{\log |\Phi(n)|}{n}$$

is finite *and* > 0.

Proof. Using the group products $R_j(z)$ constructed in article 4, we have

$$\Phi(z) \;=\; F(z/L)\prod_j R_j(z)$$

with the function F studied in article 3. For each fixed j, the modulus of

$$R_j(z) \;=\; \left(\frac{a_j - z}{a_j + z}\right)^{q_j} \prod_{nL\in[X_j,Y_j]}\left(\frac{nL+z}{nL-z}\right) \prod_{\lambda\in\Lambda_0\cap[X_j,Y_j]}\left(\frac{\lambda - z}{\lambda + z}\right)$$

tends obviously to 1 when $z\to\infty$. $R_j(z)$ also tends to 1 when $z\to 0$, and in a manner dependent only on the ratio $|z|/X_j^{2/3}$, while otherwise independent of j. To see this, recall that $a_j \geqslant \frac{1}{2}\eta X_j$ so that, for $|z| \leqslant \frac{1}{4}\eta X_j$, say, we can expand $\log R_j(z)$ in powers of z, as in the proof of the *first* lemma in article 4. As we saw there, the first degree term in z is absent from this expansion, and we can readily deduce from the latter that

$$|R_j(z)-1| \;\leqslant\; \text{const.}\,|z|^3\left\{\frac{q_j}{a_j^3} \;+\; \sum_{\lambda\in\Lambda_0\cap[X_j,Y_j]}\frac{1}{\lambda^3} \;+\; \sum_{nL\in[X_j,Y_j]}\frac{1}{(nL)^3}\right\}$$

for $|z| \leqslant \frac{1}{4}\eta X_j$. The sum in curly brackets is clearly $\leqslant \text{const.}/X_j^2$ so we have

$$|R_j(z)-1| \;\leqslant\; \text{const.}\frac{|z|^3}{X_j^2}, \qquad |z| \leqslant \tfrac{1}{4}\eta X_j,$$

verifying our claim.

Thanks to this behaviour of the $R_j(z)$, we can select a sequence of the numbers X_j *increasing sufficiently rapidly* so that, for $x > 0$, the product $\prod_j|R_j(x)|$ *will be sensibly equal to* 1 *unless* x *is much nearer to one of the intervals* $[X_j, Y_j]$ – *to* $[X_l, Y_l]$ *say* – *than to any of the others. In the latter situation,* $\prod_{j\neq l}|R_j(x)|$ *will be practically equal to* 1 *and the whole product* $\prod_j|R_j(x)|$ *essentially equal to* $|R_l(x)|$.

By the asymptotic behaviour of $F(z)$ obtained in article 3 and the formula at the beginning of this proof,

$$\Phi(x) \;\sim\; \left(2\sin\frac{\pi x}{L}\right)\prod_j R_j(x)$$

for large $x > 0$. For $m \in \mathbb{N}$ *not* divisible by L, the asymptotic behaviour of $\log|\Phi(m)|$ is thus governed by that of $\prod_j|R_j(m)|$. And, as we have just seen, the latter is practically 1 unless m is *much closer* to some $[X_l, Y_l]$ than to *any of the other* $[X_j, Y_j]$, in which case the product is nearly equal to $|R_l(m)|$.

Suppose, first of all, that $m \in \Lambda_0$. If also $m \in [X_l, Y_l]$, then $R_l(m) = 0$

by our definition of the $R_j(z)$. If, on the other hand, $m \notin [X_l, Y_l]$ we have, by the theorem at the end of article 6,

$$\frac{\log|R_l(m)|}{m} \leqslant \max\left(\frac{\log|R_l(X_l)|}{X_l}, \frac{\log|R_l(Y_l)|}{Y_l}\right),$$

a strictly positive quantity. Thus, if $m \in \Lambda_0$ is *near* the interval $[X_l, Y_l]$, we have

$$\frac{\log|\Phi(m)|}{m} \leqslant \frac{\text{const.}}{m} + \max\left(\frac{\log|R_l(X_l)|}{X_l}, \frac{\log|R_l(Y_l)|}{Y_l}\right).$$

Fix now a number $\kappa > 0$ small enough to ensure the conclusions of the two lemmas in article 7. By the remark following the *first* of those lemmas, the interval $[(1 + \kappa/5)X_l, (1 + 4\kappa/5)X_l]$ contains *at least*

$$\left(1 - \frac{4}{3}\varepsilon\right)\cdot\frac{3\kappa}{5}X_l - 1$$

integers not belonging to Λ_0, when X_l is *large*. Since Λ_1 is *measurable and of density* $1 - D_\Sigma^*$, *at most*

$$\left(1 - D_\Sigma^* + \frac{2\varepsilon}{3}\right)\cdot\frac{3\kappa}{5}X_l$$

of the integers just mentioned can belong to Λ_1, *when X_l is large.* And at most

$$\frac{1}{L}\cdot\frac{3\kappa}{5}X_l + 1$$

of them can be divisible by L. We are, however, assuming that $D_\Sigma^* > 1/L + 2\varepsilon$. Hence

$$1 - D_\Sigma^* + \frac{2\varepsilon}{3} + \frac{1}{L} < 1 - \frac{4}{3}\varepsilon,$$

so, if X_l is large, *there are at least*

$$\left\{\left(1 - \frac{4}{3}\varepsilon\right) - \left(1 - D_\Sigma^* + \frac{2\varepsilon}{3} + \frac{1}{L}\right)\right\}\frac{3\kappa}{5}X_l - 2$$

$$= \left(D_\Sigma^* - \frac{1}{L} - 2\varepsilon\right)\cdot\frac{3\kappa}{5}X_l - 2$$

integers ξ in the above interval not divisible by L, and belonging neither to Λ_0 *nor to* Λ_1. *Such ξ are thus in* Σ. For them, by the first lemma of article

7, we have

$$\frac{\log|R_l(\xi)|}{\xi} \geqslant \frac{\log|R_l(X_l)|}{X_l} + \frac{1}{2}\left(\frac{1}{L} - \varepsilon\right)\sigma\log\frac{1}{\sigma}$$

with $1 + \sigma = \xi/X_l$, if X_l is large enough. Here, $\kappa/5 \leqslant \sigma \leqslant 4\kappa/5$, so (wlog $\kappa < 1/e$!)

$$\frac{1}{2}\left(\frac{1}{L} - \varepsilon\right)\sigma\log\frac{1}{\sigma} \geqslant \frac{1}{10}\left(\frac{1}{L} - \varepsilon\right)\kappa\log\frac{5}{\kappa}.$$

The *choice* of our small *fixed* number $\kappa > 0$ depended on L and ε (refer to the first lemma in article 7). The *right side* of the last relation is therefore a certain *strictly positive quantity* $\delta(L, \varepsilon)$ dependent on L and ε. The integers $\xi \in \Sigma$ now under consideration are not divisible by L, so

$$\left|\sin\frac{\pi\xi}{L}\right| \geqslant \sin\frac{\pi}{L},$$

and it thence follows from the above inequality that

$$\frac{\log|\Phi(\xi)|}{\xi} \geqslant \frac{\text{const.}}{\xi} + \frac{\log|R_l(X_l)|}{X_l} + \delta(L, \varepsilon)$$

for them when X_l is large.

The ξ satisfying this relation are in Σ and also in the interval

$$\left[\left(1 + \frac{\kappa}{5}\right)X_l, \left(1 + \frac{4\kappa}{5}\right)X_l\right].$$

An argument just like the one used to get them, but based on the *second* lemma of article 7 instead of the *first*, will similarly give us *other* $\xi \in \Sigma$, this time in

$$\left[\left(1 - \frac{4\kappa}{5}\right)Y_l, \left(1 - \frac{\kappa}{5}\right)Y_l\right],$$

such that

$$\frac{\log|\Phi(\xi)|}{\xi} \geqslant \frac{\text{const.}}{\xi} + \frac{\log|R_l(Y_l)|}{Y_l} + \delta(L, \varepsilon),$$

provided that X_l is sufficiently large.

From this and the preceding inequality we see in the first place that

$$\limsup_{\substack{\xi \to \infty \\ \xi \in \Sigma}} \frac{\log|\Phi(\xi)|}{\xi}$$

is certainly > 0 by the theorem of article 6 – it is, on the other hand, finite by the preceding lemma. The *second statement* of our theorem is thus verified.

For the *first statement*, we confront the two inequalities just obtained with the previous estimate on $(\log|\Phi(m)|)/m$ for $m \in \Lambda_0$ close to $[X_l, Y_l]$. In view of the behaviour of the product $\prod_j |R_j(m)|$ described earlier, we see in that way that

$$\limsup_{\substack{\zeta \to \infty \\ \zeta \in \Sigma}} \frac{\log|\Phi(\zeta)|}{\zeta} \geqslant \limsup_{\substack{m \to \infty \\ m \in \Lambda_0}} \frac{\log|\Phi(m)|}{m} + \delta(L, \varepsilon).$$

The theorem is now completely proved. We are done.

Discussion. Let us look back and try to grasp the idea behind this and the preceding 6 articles, taken as a whole. On the sequence Λ_0, $|\Phi(m)|$ is smaller by a factor of roughly $e^{-\delta m}$ than on a certain sequence Σ in the complement $\mathbb{N} \sim \Lambda_0$. It seems at first glance as though we had succeeded in 'controlling' the magnitude of $\Phi(m)$ on Λ_0 by causing it to have zeros at the points of the latter contained in an *extremely sparse* sequence of intervals $[X_j, Y_j]$, that is, by using only an *insignificantly small part* of Λ_0. This is hard to believe. What is going on?

The truth is that we are *not so much controlling* $\Phi(m)$ *on* Λ_0 *as making it large at the points of* $\mathbb{N} \sim \Lambda_0$ *in the intervals* $[X_j, Y_j]$; $|\Phi(m)|$ is of about *the same* order of magnitude *outside* those intervals *whether* $m \in \Lambda_0$ or not, as long as L does not divide m. $|\Phi(m)|$ is made large *inside* the $[X_j, Y_j]$ by what amounts to the *removal of some of the zeros that* $F(z/L)$ *has in them*. The latter function vanishes at the points nL, $n \in \mathbb{N}$, and behaves like $2\sin((\pi/L)z)$ on the real axis; $\Phi(z)$ is obtained from it by essentially *replacing* its zeros in each $[X_j, Y_j]$, which are about $(1/L)(Y_j - X_j)$ in number, by the elements of Λ_0 therein, of which there are at most $\varepsilon(Y_j - X_j)$. Since $1/L > \varepsilon$, we are in effect just *throwing away* some of the zeros that $F(z/L)$ has in each interval $[X_j, Y_j]$ in order to arrive at $\Phi(z)$, and the result of this is to make $|\Phi(m)|$ *considerably larger than* $|F(m/L)|$ at the integers $m \notin \Lambda_0$ therein. *Outside* the $[X_j, Y_j]$ (where the modification has taken place), this effect is *less pronounced*. Its evaluation in the two cases (*m inside* one of the intervals or *outside all of them*) depends ultimately on the *behaviour of factorials* – that is the real origin (somewhat disguised by the use of integrals) of the (crucial) terms in $\sigma \log 1/\sigma$ appearing in the lemmas of article 7.

It is the simple monotoneity properties of $(1/x)\log|(x - \lambda)/(x + \lambda)|$ given in article 5 that make the computations work out the way they do; those properties form the basis for Fuchs' construction with the factors

$$\left(\frac{1-z/\lambda}{1+z/\lambda}\right)e^{2z/\lambda}$$

and the resulting appearance of the gamma function. The use of such factors leads of course to *functions analytic in the right half plane rather than to entire functions.* Analogous constructions with entire functions of exponential type would involve the somewhat more complicated monotoneity properties of $(1/x)\log|1 - x/\lambda|$ or of $(1/x)\log|1 - x^2/\lambda^2|$; for such work one should consult Rubel's 1955 paper and especially the one of Malliavin and Rubel published in 1961.

Malliavin's very difficult 1957 paper is *also* based on use of the factors

$$\left(\frac{1-z/\lambda}{1+z/\lambda}\right)e^{2z/\lambda}$$

(he works mainly with the logarithms of their absolute values), and is thus in part a generalization of Fuchs' work. Keeping this in mind should help anyone who wishes to understand Malliavin's article.

9.   **Converse of Pólya's gap theorem in general case**

Based on Fuchs' construction, we can now establish the

Theorem. *Let* $\Sigma \subseteq \mathbb{N}$ *have Pólya maximum density* $D_\Sigma^* > 0$. *Given any* $D < D_\Sigma^*$, *there is an analytic function*

$$f(w) = \sum_{n\in\Sigma} a_n w^n$$

whose expansion in powers of w has radius of convergence 1, *and which can be analytically continued across the arc*

$$\{e^{i\vartheta}: -\pi D < \vartheta < \pi D\}$$

of the unit circle.

Proof. The method is from the end of Malliavin's 1957 paper. We start by picking a small $\varepsilon > 0$ and a large integer $L < 1/\varepsilon$ with

$$\frac{1}{L} + 2\varepsilon < D_\Sigma^*.$$

Our result has already been established for *measurable* sequences Σ in article 1, so here we may as well assume Σ to be *non-measurable*. Then, as described in article 2, the complement $\mathbb{N} \sim \Sigma$ can be split up into two *disjoint* sequences: a *measurable* one Λ_1 of *density* $1 - D_\Sigma^*$ and *another*, Λ_0, of *minimum density zero*. Using Λ_0 which, in the present circumstances, is really infinite, we form the Fuchs function described in the preceding article, corresponding to the parameters L and ε. From Λ_1 we construct the entire function of exponential type

$$C(z) = \prod_{\lambda \in \Lambda_1} \left(1 - \frac{z^2}{\lambda^2}\right)$$

already considered in article 1.*

According to the theorem of the preceding article, the quantity

$$\gamma = \limsup_{\substack{n \to \infty \\ n \in \Sigma}} \frac{\log|\Phi(n)|}{n}$$

is *finite*. To get our function $f(w)$, we first look at

$$g(w) = \frac{1}{2\pi i} \int_{\frac{1}{2} - i\infty}^{\frac{1}{2} + i\infty} \frac{e^{-\gamma\zeta}\Phi(\zeta)C(\zeta)}{\sin \pi\zeta} w^\zeta \, d\zeta$$

for $|\arg w| < \pi(D_\Sigma^* - 1/L - 2\varepsilon)$; it is claimed that the integral converges *absolutely and uniformly* for w in that sector, making $g(w)$ *analytic* there.

To check this, observe that

$$|C(\zeta)| \leqslant \text{const.} \, e^{\pi((1 - D_\Sigma^*)|\eta| + \varepsilon|\zeta|)}$$

by problem 29 (article 1) – as usual, we are writing $\zeta = \xi + i\eta$. Again by the lemma of article 8,

$$|\Phi(\tfrac{1}{2} + i\eta)| \leqslant \text{const.} \, e^{\pi|\eta|/L}$$

for real η. Hence, for $\zeta = \frac{1}{2} + i\eta$, $\eta \in \mathbb{R}$,

$$\left|\frac{w^\zeta e^{-\gamma\zeta}\Phi(\zeta)C(\zeta)}{\sin \pi\zeta}\right| \leqslant \text{const.} \exp\left\{\left(|\arg w| + \frac{\pi}{L} + \pi\varepsilon - \pi D_\Sigma^*\right)|\eta|\right\},$$

and the asserted convergence is manifest.

We now proceed as in article 1, approximating the above integral by others taken around rectangles. For $\Re\zeta \geqslant 0$, by the lemma of the preceding article,

$$|\Phi(\zeta)| \leqslant \text{const.} \exp\left(\frac{\pi}{L}|\eta| + A\xi\right),$$

---

* Should Λ_1 be empty, $C(z)$ is taken equal to 1.

A being a certain constant. From this and our estimate on $|C(\zeta)|$, we thus have

$$\left|\frac{w^\zeta e^{-\gamma \zeta}\Phi(\zeta)C(\zeta)}{\sin \pi\zeta}\right| \;\leqslant\; \text{const.}\exp\Big\{\xi\log|w| \;+\; |\eta||\arg w| \;-\; \gamma\xi \;+\frac{\pi}{L}|\eta|$$

$$+\; A\xi \;+\; \pi(1-D^*_\Sigma)|\eta| \;+\; \pi\varepsilon|\zeta| \;-\; \pi|\eta|\Big\}$$

$$\leqslant\; \text{const.}\, e^{(A-\gamma+\pi\varepsilon+\log|w|)\xi}\, e^{(|\arg w|+\pi/L+\pi\varepsilon-\pi D^*_\Sigma)|\eta|}$$

for $\Re\zeta \geqslant 0$, *as long as the distance between ζ and the integers (zeros of* $\sin \pi\zeta$ *) stays bounded away from 0.* With the help of this relation we now easily see as in article 1 that if R is a *large integer* and Γ_R the contour

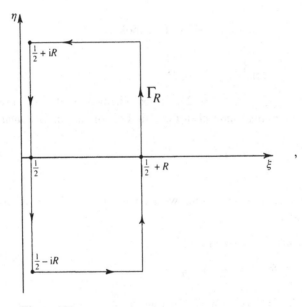

Figure 164

the contributions from the *right-hand, top,* and *bottom* parts of Γ_R to the value of

$$\frac{1}{2\pi i}\int_{\Gamma_R} \frac{e^{-\gamma\zeta}\Phi(\zeta)C(\zeta)}{\sin \pi\zeta}\, w^\zeta\, d\zeta$$

will be *very small* when

$$|w| \;<\; e^{-(A-\gamma+2\pi\varepsilon)}$$

and

$$|\arg w| < \pi\left(D_\Sigma^* - \frac{1}{L} - 2\varepsilon\right).$$

This means that for *such w,*

$$\frac{1}{2\pi i}\int_{\Gamma_R}\frac{e^{-\gamma\zeta}\Phi(\zeta)C(\zeta)}{\sin\pi\zeta}w^\zeta\,d\zeta \longrightarrow -\frac{1}{2\pi i}\int_{\frac{1}{2}-i\infty}^{\frac{1}{2}+i\infty}\frac{e^{-\gamma\zeta}\Phi(\zeta)C(\zeta)}{\sin\pi\zeta}w^\zeta\,d\zeta$$

as the integer *R* tends to infinity.

As in article 1, by the residue theorem

$$\frac{1}{2\pi i}\int_{\Gamma_R}\frac{e^{-\gamma\zeta}\Phi(\zeta)C(\zeta)}{\sin\pi\zeta}w^\zeta\,d\zeta = \frac{1}{\pi}\sum_{n=1}^{R}(-1)^n e^{-\gamma n}\Phi(n)C(n)w^n$$

for integers *R*. Here,

$$|\Phi(n)C(n)| \leqslant e^{O(n)}$$

for large *n*, so the power series

$$\frac{1}{\pi}\sum_{1}^{\infty}(-1)^n e^{-\gamma n}\Phi(n)C(n)w^n$$

certainly has a *positive radius of convergence*. For $|w| > 0$ sufficiently small and

$$|\arg w| < \pi\left(D_\Sigma^* - \frac{1}{L} - 2\varepsilon\right),$$

its *sum* must then be equal to

$$-\frac{1}{2\pi i}\int_{\frac{1}{2}-i\infty}^{\frac{1}{2}+i\infty}\frac{e^{-\gamma\zeta}\Phi(\zeta)C(\zeta)}{\sin\pi\zeta}w^\zeta\,d\zeta = -g(w)$$

by the preceding two relations.

Let us look more carefully at the power series just written. The function

$$C(z) = \prod_{\lambda\in\Lambda_1}\left(1 - \frac{z^2}{\lambda^2}\right)$$

vanishes at the points of Λ_1. Therefore, since $\mathbb{N} \sim \Lambda_1 = \Sigma \cup \Lambda_0$ our series can be written as

$$\frac{1}{\pi}\left(\sum_{n\in\Sigma} + \sum_{n\in\Lambda_0}\right)(-1)^n e^{-\gamma n}\Phi(n)C(n)w^n.$$

By our choice of γ,

$$\limsup_{\substack{n \to \infty \\ n \in \Sigma}} |e^{-\gamma n} \Phi(n)|^{1/n} \;=\; 1,$$

and $|C(n)|^{1/n} \longrightarrow 1$ as $n \longrightarrow \infty$ with $n \notin \Lambda_1$, according to problem 29, part (f) (article 1). Hence

$$\frac{1}{\pi} \sum_{n \in \Sigma} (-1)^n e^{-\gamma n} \Phi(n) C(n) w^n,$$

the *first* of the two power series into which our original one was split, *has radius of convergence* 1 *and is equal, in* $\{|w| < 1\}$, *to a certain function* $f(w)$, *analytic there*.

It is at this point that we apply the main part of the theorem from the preceding article. According to that theorem, if the sequence of intervals $[X_j, Y_j]$ used in constructing the Fuchs function $\Phi(z)$ is sparse enough,

$$\limsup_{\substack{n \to \infty \\ n \in \Lambda_0}} \frac{\log |\Phi(n)|}{n} \;\leqslant\; \limsup_{\substack{m \to \infty \\ m \in \Sigma}} \frac{\log |\Phi(m)|}{m} - \delta \;=\; \gamma - \delta$$

with a certain constant $\delta > 0$ depending on L and ε. In view of our previous relation involving $C(n)$ we thus have

$$\limsup_{\substack{n \to \infty \\ n \in \Lambda_0}} |e^{-\gamma n} \Phi(n) C(n)|^{1/n} \;\leqslant\; e^{-\delta},$$

and the radius of convergence of

$$\frac{1}{\pi} \sum_{n \in \Lambda_0} (-1)^n e^{-\gamma n} \Phi(n) C(n) w^n$$

(the *second* of the series into which our original one was broken) is $\geqslant e^{\delta} > 1$. There is thus a function $h(w)$, *analytic for* $|w| < e^{\delta}$ and *equal* there to the *sum* of this *second series*.

For $|\arg w| < \pi(D_\Sigma^* - 1/L - 2\varepsilon)$ and $|w| > 0$ small enough,

$$f(w) + h(w) \;=\; \frac{1}{\pi} \sum_1^{\infty} (-1)^n e^{-\gamma n} \Phi(n) C(n) w^n$$

is, as we have just seen, equal to $-g(w)$, a function *analytic in the whole sector*

$$|\arg w| < \pi \left(D_\Sigma^* - \frac{1}{L} - 2\varepsilon \right).$$

The formula

$$f(w) \;=\; -g(w) \;-\; h(w)$$

thus furnishes *an analytic continuation* of $f(w)$ from the unit disk into the *intersection* of our sector with the disk $\{|w| < e^{\delta}\}$:

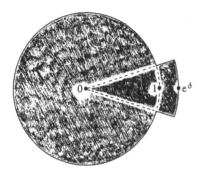

Figure 165

The function given, for $|w| < 1$, by the power series

$$\frac{1}{\pi} \sum_{n \in \Sigma} (-1)^n e^{-\gamma n} \Phi(n) C(n) w^n,$$

having convergence radius 1 can, in other words, be *continued analytically across the arc*

$$\left\{ e^{i\vartheta} \colon \; |\vartheta| < \pi \left(D_{\Sigma}^{*} - \frac{1}{L} - 2\varepsilon \right) \right\}$$

of the unit circle. Here, $\varepsilon > 0$ can be as small as we like and L is any large integer $< 1/\varepsilon$. Hence $D = D_{\Sigma}^{*} - 1/L - 2\varepsilon$ can be made as close as we like to D_{Σ}^{*}.

Our theorem is proved.

C. A Jensen formula involving confocal ellipses instead of circles

Suppose that we are only interested in the *real zeros* of a function $f(z)$ analytic in some disk $\{|z| < R\}$ with $f(0) \neq 0$. If we denote by $n(r)$ the number of zeros of f on the segment $[-r, r]$, Jensen's formula implies that

$$\int_{0}^{r} \frac{n(\rho)}{\rho} \, d\rho \;\leqslant\; \frac{1}{2\pi} \int_{0}^{2\pi} \log|f(re^{i\vartheta})| \, d\vartheta \;-\; \log|f(0)|$$

for $0 < r < R$. This relation can be used to *estimate* $n(r)$ for certain values of *r*, and that application has been frequently made in the present book: in Chapter III, for instance, and in §A of this one. Such use of it does, however, involve a drawback – it furnishes a kind of average of $n(\rho)$ for $\rho \leqslant r$ rather than $n(r)$ itself. In order to alleviate this shortcoming, we proceed to derive a similar formula by working with confocal ellipses instead of concentric circles.

The standard Joukowski mapping

$$w \longrightarrow z = \frac{1}{2}\left(w + \frac{1}{w}\right)$$

takes $\{|w| > 1\}$ conformally onto the *complement* (in $\mathbb{C}$) of the real segment $[-1, 1]$, and each of the circles $|w| = R > 1$ onto an *ellipse*

$$\frac{4x^2}{(R + R^{-1})^2} + \frac{4y^2}{(R - R^{-1})^2} = 1$$

with foci at 1 and -1:

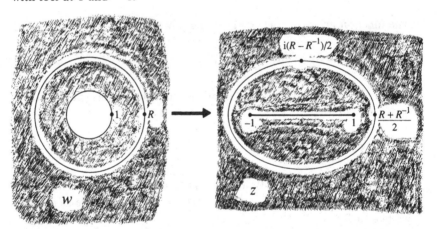

Figure 166

For such an ellipse we have the parametric representation

$$z = \frac{1}{2}\left(Re^{i\vartheta} + \frac{e^{-i\vartheta}}{R}\right),$$

and as *R* increases, the ellipse gets bigger.

Theorem. *Let* $f(z)$ *be analytic inside and on the ellipse*

$$z = \frac{1}{2}\left(Re^{i\vartheta} + \frac{e^{-i\vartheta}}{R}\right),$$

$R > 1$, *and, for* $1 < r \leqslant R$, *denote by* $N(r)$ *the* number of zeros *of* f (*counting multiplicities*) inside or on the ellipse

$$z = \frac{1}{2}\left(re^{i\vartheta} + \frac{e^{-i\vartheta}}{r}\right).$$

Then

$$\int_1^R N(r)\frac{dr}{r} = \frac{1}{2\pi}\int_0^{2\pi} \log\left|f\left(\frac{1}{2}\left(Re^{i\vartheta} + \frac{e^{-i\vartheta}}{R}\right)\right)\right|d\vartheta - \frac{1}{\pi}\int_{-1}^1 \frac{\log|f(x)|}{\sqrt{(1-x^2)}}dx.$$

Proof. Like that of a theorem of Littlewood given at the end of Chapter III in Titchmarsh's *Theory of Functions*. (Our result can in fact be derived from that theorem.)

Suppose that $1 \leqslant r_0 < r_1 \leqslant R$, and that $f(z)$ *has no zeros* inside or on the boundary of the ring-shaped open region bounded by the two ellipses

$$z = \frac{1}{2}\left(r_0 e^{i\vartheta} + \frac{e^{-i\vartheta}}{r_0}\right), \qquad z = \frac{1}{2}\left(r_1 e^{i\vartheta} + \frac{e^{-i\vartheta}}{r_1}\right).$$

Then $\log f(z)$ can be defined so as to be *analytic and single-valued* in the *simply connected* domain obtained by removing the segment.

$$\left(-\tfrac{1}{2}(r_1 + r_1^{-1}), \quad -\tfrac{1}{2}(r_0 + r_0^{-1})\right)$$

from that region:

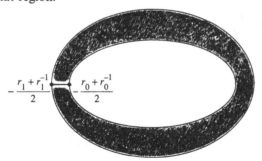

Figure 167

Along the *upper and lower sides* of the removed segment, arg f will generally have *two different* determinations. Indeed, by the principle of argument,

$$\arg f(x + i0) - \arg f(x - i0) = 2\pi N(r_0)$$

for

$$-\tfrac{1}{2}(r_1 + r_1^{-1}) \leqslant x \leqslant -\tfrac{1}{2}(r_0 + r_0^{-1}).$$

Let us parametrize the *boundary* of our simply connected domain by putting $z = \frac{1}{2}(w + 1/w)$ and then having w go around the path γ shown here:

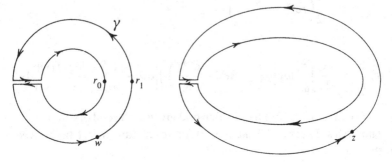

Figure 168

Then, using appropriate determinations of $\log f(\frac{1}{2}(u + 1/u))$ along the upper and lower horizontal stretches of γ, we have by Cauchy's theorem

$$\int_\gamma \log f\left(\frac{1}{2}\left(w + \frac{1}{w}\right)\right)\frac{dw}{w} = 0.$$

Taking the imaginary part of this relation, we get

$$\int_{-r_1}^{-r_0}\left\{\arg f\left(\frac{u + u^{-1}}{2} + i0\right) - \arg f\left(\frac{u + u^{-1}}{2} - i0\right)\right\}\frac{du}{u}$$

$$- \int_{-\pi}^{\pi} \log\left|f\left(\frac{1}{2}\left(r_0 e^{i\vartheta} + \frac{e^{-i\vartheta}}{r_0}\right)\right)\right| d\vartheta$$

$$+ \int_{-\pi}^{\pi} \log\left|f\left(\frac{1}{2}\left(r_1 e^{i\vartheta} + \frac{e^{-i\vartheta}}{r_1}\right)\right)\right| d\vartheta = 0,$$

i.e.,

$$2\pi N(r_0)\log\frac{r_1}{r_0} = \int_{-\pi}^{\pi} \log\left|f\left(\frac{1}{2}\left(r_1 e^{i\vartheta} + \frac{e^{-i\vartheta}}{r_1}\right)\right)\right| d\vartheta$$

$$- \int_{-\pi}^{\pi} \log\left|f\left(\frac{1}{2}\left(r_0 e^{i\vartheta} + \frac{e^{-i\vartheta}}{r_0}\right)\right)\right| d\vartheta,$$

in view of the previous formula.

The integral $\int_{-\pi}^{\pi} \log|f(\frac{1}{2}(re^{i\vartheta} + e^{-i\vartheta}/r))| d\vartheta$ is, however, a *continuous function* of r for $1 \leqslant r \leqslant R$, even at values of r for which the ellipse $z = \frac{1}{2}(re^{i\vartheta} + e^{-i\vartheta}/r)$ has zeros of f lying on it – this is immediate by an argument like the one used in Chapter I. The result just found therefore remains valid when f has zeros on one or both of the ellipses

$$z = \frac{1}{2}\left(r_0 e^{i\vartheta} + \frac{e^{-i\vartheta}}{r_0}\right), \qquad z = \frac{1}{2}\left(r_1 e^{i\vartheta} + \frac{e^{-i\vartheta}}{r_1}\right),$$

as long as it has *none in the region between them.* Hence, if all the zeros of $f(z)$ inside or on the ellipse $z = \frac{1}{2}(Re^{i\vartheta} + e^{-i\vartheta}/R)$ occur on the ellipses (or segment!)

$$z = \frac{1}{2}\left(\rho_k e^{i\vartheta} + \frac{e^{-i\vartheta}}{\rho_k}\right)$$

with

$$1 \leqslant \rho_1 < \rho_2 < \cdots < \rho_m \leqslant R,$$

we have

$$N(\rho_{k-1})\log\frac{\rho_k}{\rho_{k-1}} = \frac{1}{2\pi}\int_0^{2\pi}\log\left|f\left(\frac{1}{2}\left(\rho_k e^{i\vartheta} + \frac{e^{-i\vartheta}}{\rho_k}\right)\right)\right|d\vartheta$$

$$- \frac{1}{2\pi}\int_0^{2\pi}\log\left|f\left(\frac{1}{2}\left(\rho_{k-1}e^{i\vartheta} + \frac{e^{-i\vartheta}}{\rho_{k-1}}\right)\right)\right|d\vartheta$$

for $k = 2, 3, \ldots, m$. If $\rho_1 > 1$, the same relation holds with $\rho_0 = 1$ and ρ_1, and, if $\rho_m < R$, with ρ_m and $\rho_{m+1} = R$. Since $N(r) = N(\rho_{k-1})$ for $\rho_{k-1} \leqslant r < \rho_k$, addition of all these formulas yields

$$\int_1^R N(r)\frac{dr}{r} = \frac{1}{2\pi}\int_0^{2\pi}\log\left|f\left(\frac{1}{2}\left(Re^{i\vartheta} + \frac{e^{-i\vartheta}}{R}\right)\right)\right|d\vartheta$$

$$- \frac{1}{2\pi}\int_0^{2\pi}\log|f(\cos\vartheta)|d\vartheta.$$

Putting $\cos\vartheta = x$ in the second integral on the right now gives us the theorem. Done.

Our desired amelioration of the information obtainable from Jensen's formula is provided by the following

Corollary. *Let a and $\gamma > 0$, and suppose that $F(z)$ is analytic inside and on the ellipse*

$$z = \frac{a}{2}\left(e^{\gamma}e^{i\vartheta} + \frac{e^{-i\vartheta}}{e^{\gamma}}\right) = a\cosh(\gamma + i\vartheta),$$

$0 \leqslant \vartheta \leqslant 2\pi$. *If $0 \leqslant \eta < \gamma$ and $F(z)$ has at least N zeros (counting* *multiplicities) on the segment* $[-a\cosh\eta,\ a\cosh\eta]$ *of the real axis, then*

$$(\gamma - \eta)N \;\leqslant\; \frac{1}{2\pi}\int_0^{2\pi}\log|F(a\cosh(\gamma + i\vartheta))|\,\mathrm{d}\vartheta \;-\; \frac{1}{\pi}\int_{-a}^{a}\frac{\log|F(t)|}{\sqrt{(a^2 - t^2)}}\,\mathrm{d}t.$$

Proof. Apply the theorem to $f(z) = F(az)$ with $R = \mathrm{e}^{\gamma}$.

D.   A condition for completeness of a collection of imaginary exponentials on a finite interval

Suppose we are given a one-way or two-way strictly increasing sequence of *real* numbers λ_n, with $\lambda_n > 0$ for $n > 0$, $\lambda_n \longrightarrow \infty$ as $n \longrightarrow \infty$, and, if there are λ_n with negative indices, $\lambda_n \leqslant 0$ for $n \leqslant 0$ and $\lambda_n \longrightarrow -\infty$ as $n \longrightarrow -\infty$. Considering the corresponding set of *imaginary exponentials* $\mathrm{e}^{i\lambda_n t}$, we take any number $L > 0$ and ask whether the finite linear combinations of these exponentials are *uniformly dense* in $\mathscr{C}(-L, L)$. If they *are*, the exponentials $\mathrm{e}^{i\lambda_n t}$ are said to be *complete* on $[-L,\ L]$; otherwise, they are *incomplete* on that interval.

If the $\mathrm{e}^{i\lambda_n t}$ are complete on $[-L,\ L]$, they are obviously complete on $[-L',\ L']$ for any L' with $0 < L' < L$. There is thus a certain number A associated with those exponentials, $0 \leqslant A \leqslant \infty$, such that the former are *complete* on $[-L,\ L]$ if $0 < L < A$ and *incomplete* on $[-L,\ L]$ if $L > A$. We are, of course, not limited here to consideration of intervals centred at the origin; when $0 < A < \infty$ it is immediate (by translation!) that the $\mathrm{e}^{i\lambda_n t}$ will in fact be complete on any real interval of length $< 2A$ and incomplete on any one of length $> 2A$. In the extreme case where $A = 0$, the given exponentials are incomplete on any real interval of length > 0, and, when $A = \infty$, they are complete on all finite intervals. Simple examples show that both of these extreme cases are possible.

Regarding completeness of the exponentials on intervals of length *exactly equal* to $2A$, nothing can be said a priori. There are examples in which the $\mathrm{e}^{i\lambda_n t}$ are *complete* on $[-A, A]$ and others where they are *incomplete* thereon. Without going into the matter at all, it seems clear that the outcome in this borderline situation must depend in very delicate and subtle fashion on the sequence of frequencies λ_n. We will not consider that particular question in this book; various fragmentary results concerning it may be found in Levinson's monograph and in Redheffer's expository article.

What interests us is the more basic problem of finding out how the number A – L. Schwartz called it the *completeness radius* associated with the λ_n – actually depends on those frequencies. We would like, if possible, to get a *formula* relating A to the distribution of the λ_n.

This important question was investigated by Paley and Wiener, Levinson, L. Schwartz and others. A complete solution was obtained around 1960 by Beurling and Malliavin, whose work involved two main steps:

(i) The determination of a certain *lower bound* for A,

(ii) Proof that the lower bound found in (i) is also an *upper bound* for A.

The first of these can be presented quite simply using the formula from the preceding §; that is what we will do presently. The second step is much more difficult; its completion required a deep existence theorem established expressly for that purpose by Beurling and Malliavin. That part of the solution will be given in Chapter X, with proof of the existence theorem itself deferred until Chapter XI.

The first step amounts to a *proof of completeness* of the $e^{i\lambda_n t}$ on intervals $[-L, L]$ with L small enough (depending on the λ_n). The idea for this goes back to Szasz and to Paley and Wiener.

Reasoning by contradiction, we take an $L > 0$ and assume *incompleteness* of the $e^{i\lambda_n t}$ on $[-L, L]$. *Duality* (Hahn–Banach theorem) then gives us a *non-zero complex measure* μ on $[-L, L]$ with

$$\int_{-L}^{L} e^{i\lambda_n t} d\mu(t) = 0$$

for each λ_n, i.e., $\hat{\mu}(\lambda_n) = 0$ for the Fourier–Stieltjes transform

$$\hat{\mu}(z) = \int_{-L}^{L} e^{izt} d\mu(t).$$

The function $\hat{\mu}(z)$ is entire, of exponential type $\leq L$, and bounded on the real axis. Using a familiar result from Chapter III, §G.2, together with the one from the preceding §, one now shows that for *small enough $L > 0$*, the *zeros λ_n* of $\hat{\mu}(z)$ *cannot* (in some suitable sense) *be too dense without forcing* $\hat{\mu}(z) \equiv 0$, contrary to our choice of μ.

The details of this argument are given in the following article. Before proceeding to it, we should observe how duality can be used to demonstrate one very important fact: *the completeness radius A associated with a sequence $\{\lambda_n\}$ is not really specific to the topology of uniform convergence and the spaces $\mathscr{C}(-L, L)$. If, in place of $\mathscr{C}(-L, L)$, we take any of the spaces $L_p(-L, L)$, $1 \leq p < \infty$, the value of A corresponding to a given sequence of frequencies λ_n turns out to be the same.*

Suppose indeed that $0 < L < A$. The space $\mathscr{C}(-L, L)$ is contained in each of the $L_p(-L, L)$ and *dense* in the latter. Therefore, since uniform convergence on $[-L, L]$ implies L_p convergence thereon, the linear combinations of the $e^{i\lambda_n t}$, being uniformly dense in $\mathscr{C}(-L, L)$, will be $\| \ \|_p$ dense in $L_p(-L, L)$.

Let, on the other hand, $L > A$. *Then the finite linear combinations of the $e^{i\lambda_n t}$ are not $\| \ \|_p$ dense in any of the spaces $L_p(-L, L)$.* To see this, we can take an L', $A < L' < L$, and apply duality as above to get a non-zero complex measure μ on $[-L', L']$ *(sic!)* with $\hat{\mu}(\lambda_n) = 0$ for each n. If $h > 0$ is sufficiently small, the function

$$\varphi(t) = \frac{1}{2h} \int_{t-h}^{t+h} d\mu(\tau)$$

is supported on $[-L, L]$; φ is clearly *bounded*, hence in each of the *duals* $L_q(-L, L)$, $1 < q \leqslant \infty$, to our L_p spaces. We have

$$\hat{\varphi}(z) = \frac{\sin hz}{hz} \hat{\mu}(z),$$

so in particular $\hat{\varphi}(z) \not\equiv 0$, hence $\varphi(t)$ *cannot vanish a.e. on* $[-L, L]$. By the same token, however,

$$\int_{-L}^{L} e^{i\lambda_n t} \varphi(t) \, dt = \hat{\varphi}(\lambda_n) = 0$$

for each n, so finite linear combinations of the $e^{i\lambda_n t}$ cannot be dense in $L_p(-L, L)$.

A considerable refinement of the preceding observation is due to L. Schwartz – Levinson also certainly knew of it:

Theorem. *If, for $L > 0$, the finite linear combinations of the $e^{i\lambda_n t}$ are not uniformly dense in $\mathscr{C}(-L, L)$, removal of any one of the exponentials $e^{i\lambda_n t}$ leaves us with a collection whose finite linear combinations are not dense in $L_1(-L, L)$ (hence not dense in any of the $L_p(-L, L)$, $1 \leqslant p < \infty$).*

Taking the exponentials e^{int}, $n \in \mathbb{Z}$, on $[-\pi, \pi]$, we see that this result is sharp. Finite linear combinations of the former *are* dense in each of the $L_p(-\pi, \pi)$, $1 \leqslant p < \infty$, but can *uniformly* approximate *only* those $f \in \mathscr{C}(-\pi, \pi)$ for which $f(-\pi) = f(\pi)$. In order to have (uniform) completeness on $[-\pi, \pi]$ we must use *one additional* imaginary exponential $e^{i\lambda z}$, $\lambda \notin \mathbb{Z}$ (*any* such one will do!).

Problem 31

Prove the above theorem. (Hint: Assuming a non-zero measure μ on

$[-L, L]$ with $\hat{\mu}(\lambda_n) = 0$, take, for instance, λ_1, and look at the function φ supported on $[-L, L]$ given by

$$\varphi(t) = e^{-i\lambda_1 t} \int_{-L}^{t} e^{i\lambda_1 \tau} d\mu(\tau)$$

for $-L \leqslant t \leqslant L$. Compute $\hat{\varphi}(z)$.)

1. Application of the formula from §C

Let us, without further, ado, proceed to this chapter's basic result about completeness.

Theorem. *Given the sequence of distinct frequencies $\lambda_n > 0$, suppose that for some $D > 0$ there are disjoint half-open intervals $(a_k, b_k]$ in $(0, \infty)$ such that, for each k,*

$$\frac{\text{number of } \lambda_n \text{ in } (a_k, b_k]}{b_k - a_k} \;\geqslant\; D,$$

and that

$$\sum_k \left(\frac{b_k - a_k}{a_k} \right)^2 = \infty.$$

Then, if $0 < L < \pi D$, the exponentials $e^{i\lambda_n t}$ are complete on $[-L, L]$.

Remark. The *second* condition on the intervals $(a_k, b_k]$ has already figured in Beurling's gap theorem (§A.2, Chapter VII).

Proof of Theorem. Assume that the $e^{i\lambda_n t}$ are *not* complete on $[-L, L]$, where $0 < L < \pi D$. Then, as in the discussion immediately preceding the present article, there is a *non-zero complex measure μ* on $[-L, L]$ with $\hat{\mu}(\lambda_n) = 0$. The function $\hat{\mu}(z)$ is entire, of exponential type $\leqslant L$, and bounded on the real axis, indeed, wlog,

(*) $|\hat{\mu}(z)| \leqslant e^{L|\Im z|}$,

as one sees by direct inspection of the Fourier–Stieltjes integral used to define $\hat{\mu}(z)$. Our aim is to show that

$$\int_{-\infty}^{\infty} \frac{\log^{-} |\hat{\mu}(x)|}{1 + x^2} dx = \infty,$$

which, by §G.2 of Chapter III, implies that $\hat{\mu}(z) \equiv 0$, contrary to μ's being non-zero.

If $\hat{\mu}(z)$ is not to vanish identically, we must have $b_k \xrightarrow[k]{} \infty$, for each interval $(a_k, b_k]$ contains at least *one* zero λ_n of $\hat{\mu}$. We may therefore re-enumerate the $(a_k, b_k]$ so as to ensure that

$$0 < a_1 < b_1 \leqslant a_2 < b_2 \leqslant a_3 < \cdot \cdot \cdot \text{(with } a_k \xrightarrow[k]{} \infty); \text{ this we}$$

henceforth suppose done.

Following the idea mentioned at the beginning of this chapter, we proceed to apply the Jensen formula from the preceding § to certain ellipses whose centres have been *moved* from the origin to the midpoints of the $(a_k, b_k]$. Let us fix our attention on *any one* of the latter which, for the moment, we designate as $(c - R, c + R]$. We take a fixed small number $\gamma > 0$ (whose value will be assigned presently) and, with

$$\frac{R}{\cosh \gamma} < r < R,$$

apply the corollary at the end of §C to $\hat{\mu}(c + z)$ (*sic!*) in the ellipse

$$z = r \cosh (\gamma + i\vartheta), \quad 0 \leqslant \vartheta \leqslant 2\pi;$$

we are, in other words, looking at $\hat{\mu}(z)$ in an ellipse whose major axis is $[c - r \cosh \gamma, \ c + r \cosh \gamma]$:

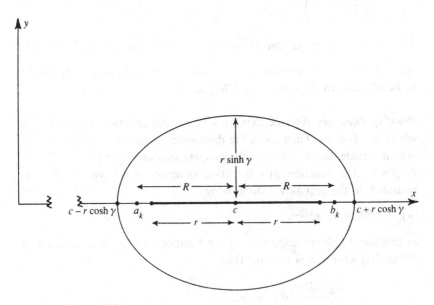

Figure 169

If η, $0 < \eta < \gamma$, is a number such that $r \cosh \eta \geqslant R$ and N denotes the number of λ_n in $(c - R, \ c + R]$, we find that

$$N(\gamma - \eta) \quad \leqslant \quad \frac{1}{2\pi} \int_0^{2\pi} \log |\hat{\mu}(c + r \cosh(\gamma + i\vartheta))| \, d\vartheta$$

$$- \frac{1}{\pi} \int_{-r}^r \frac{\log |\hat{\mu}(c + t)|}{\sqrt{(r^2 - t^2)}} \, dt.$$

By (∗), $\log |\hat{\mu}(c + r \cosh(\gamma + i\vartheta))| \leqslant Lr \sinh \gamma |\sin \vartheta|$, and this, substituted in the first integral on the right, yields

$$\frac{1}{\pi} \int_{-r}^r \frac{\log |\hat{\mu}(c + t)|}{\sqrt{(r^2 - t^2)}} \, dt \quad \leqslant \quad \frac{2L}{\pi} r \sinh \gamma \; - \; N(\gamma - \eta).$$

However, the number N of λ_n in $(a_k, b_k]$ is by hypothesis $\geqslant 2RD \geqslant 2Dr$. Hence

$$(†) \qquad \frac{1}{\pi} \int_{-r}^r \frac{\log |\hat{\mu}(c + t)|}{\sqrt{(r^2 - t^2)}} \, dt \quad \leqslant \quad \frac{2}{\pi} \Big(L \sinh \gamma \; - \; \pi D(\gamma - \eta) \Big) r$$

for $\dfrac{R}{\cosh \gamma} < \dfrac{R}{\cosh \eta} \leqslant r \leqslant R.$

Now we are assuming that $\pi D > L$. We can therefore *fix $\gamma > 0$ small enough (independently of k !)* so that

$$\pi D \frac{\gamma}{\sinh \gamma} \quad > \quad L$$

and *then fix $\eta > 0$ much smaller than the small number γ so as to still have*

$$\pi D \frac{\gamma - \eta}{\sinh \gamma} \quad > \quad L.$$

With such values of γ and η *the right side of (†) is negative.*

Multiply both sides of (†) by $r \, dr$ (the purpose of the factor r being to make our computation a little easier) and then integrate r from $R/\cosh \eta$ to R. After changing the order of integration on the left, we get

$$\int_{-R}^R \varphi_R(t) \log |\hat{\mu}(c + t)| \, dt \quad \leqslant \quad \frac{2}{3\pi} \Big(L \sinh \gamma \; - \; \pi D(\gamma - \eta) \Big) \frac{\cosh^3 \eta - 1}{\cosh^3 \eta} R^3,$$

where

$$\pi \varphi_R(t) \quad = \quad \int_{\max(|t|, R/\cosh \eta)}^R \frac{r \, dr}{\sqrt{(r^2 - t^2)}}$$

$$= \quad \begin{cases} \sqrt{(R^2 - t^2)} - \sqrt{(R^2/\cosh^2 \eta - t^2)} & \text{for } |t| < R/\cosh \eta, \\ \sqrt{(R^2 - t^2)} & \text{for } R/\cosh \eta \leqslant |t| \leqslant R. \end{cases}$$

The function $\pi\varphi_R(t)$ assumes its maximum on $[-R,\ R]$ for $t = R/\cosh\eta$, where it equals $R\tanh\eta$. Hence, since $\log|\hat{\mu}(x)| \leqslant 0$ by (∗),

$$R\tanh\eta \int_{-R}^{R} \log|\hat{\mu}(c+t)|\,dt \;\leqslant\; \frac{2}{3}\left(L\sinh\gamma \;-\; \pi D(\gamma-\eta)\right)\frac{\cosh^3\eta - 1}{\cosh^3\eta}\,R^3,$$

i.e.

$$(\overset{*}{*})\qquad \int_{a_k}^{b_k} \log|\hat{\mu}(x)|\,dx \;\leqslant\; -\frac{\pi D(\gamma-\eta) - L\sinh\gamma}{2}\cdot\frac{\tanh(\eta/2)}{\cosh^2\eta}(b_k - a_k)^2,$$

a relation holding for all the intervals $(a_k,\ b_k]$, with our fixed γ and η for which

$$\pi D(\gamma-\eta) \;-\; L\sinh\gamma \;>\; 0.$$

As we saw at the beginning of this discussion, $a_k \xrightarrow[k]{} \infty$. If $\hat{\mu}(z) \not\equiv 0$, we must also have

$$\frac{b_k}{a_k} \longrightarrow 1 \quad\text{as } k \longrightarrow \infty.$$

Indeed, such a function $\hat{\mu}(z)$ satisfies the hypothesis of *Levinson's theorem* (Chapter III, §H.3), after being multiplied by a suitable exponential $e^{i\alpha z}$ (see the observation at the beginning of §H.2, Chapter III). According to that result, if we denote the number of zeros of $\hat{\mu}(z)$ with modulus $\leqslant r$ *in the right half plane* by $n_+(r)$, we have

$$\frac{n_+(r)}{r} \longrightarrow \frac{L'}{\pi} \quad\text{for } r \longrightarrow \infty,$$

where (here) $0 \leqslant L' \leqslant L$. Therefore, if $\varepsilon > 0$, we will certainly have

$$n_+(a_k) \;\geqslant\; \left(\frac{L'}{\pi} - \varepsilon^2\right)a_k$$

and

$$n_+(b_k) \;\leqslant\; \left(\frac{L'}{\pi} + \varepsilon^2\right)b_k$$

for all sufficiently large k, since a_k and b_k tend to ∞ with k. If now $a_k < (1-\varepsilon)b_k$ for some large enough k, the previous inequalities yield

$$n_+(b_k) - n_+(a_k) \;\leqslant\; \frac{L'}{\pi}(b_k - a_k) + \varepsilon^2(a_k + b_k) \;\leqslant\; \left(\frac{L}{\pi} + 2\varepsilon\right)(b_k - a_k).$$

In particular, the interval $(a_k,\ b_k]$ can contain at most

$$\left(\frac{L}{\pi} + 2\varepsilon\right)(b_k - a_k)$$

of the points λ_n, since $\hat{\mu}(\lambda_n) = 0$. By hypothesis, however, that interval contains at least $D(b_k - a_k)$ of those points. Hence, since $L < \pi D$, we have a contradiction if $\varepsilon > 0$ is small enough.

Once we know that $a_k \longrightarrow \infty$ and $b_k/a_k \longrightarrow 1$ for $k \longrightarrow \infty$, we can be sure that the quantity

$$M = \sup_k \frac{b_k^2 + 1}{a_k^2}$$

is finite. Then, since $\log|\hat{\mu}(x)| \leqslant 0$, $\binom{*}{*}$ yields

$$\sum_{k=1}^{\infty} \int_{a_k}^{b_k} \frac{\log|\hat{\mu}(x)|}{1 + x^2} \, dx \;\leqslant\; -\frac{\pi D(\gamma - \eta) - L \sinh\gamma}{2M} \cdot \frac{\tanh(\eta/2)}{\cosh^2\eta} \sum_{k=1}^{\infty} \left(\frac{b_k - a_k}{a_k}\right)^2,$$

and the right side equals $-\infty$ *by hypothesis.* That is,

$$\int_{-\infty}^{\infty} \frac{\log^-|\hat{\mu}(x)|}{1 + x^2} \, dx = \infty,$$

the relation sought. The proof is complete.

Remark 1. Beurling and Malliavin originally proved this result using the ordinary Jensen formula (for circles). For such a proof, a covering lemma for intervals on $\mathbb{R}$ is required.

Remark 2. In case the λ_n are *not* distinct but $\hat{\mu}(z)$, having the other properties assumed in the proof has, at each of the former, a zero of order equal to that point's number of occurences in the sequence, we still conclude that $\hat{\mu}(z) \equiv 0$ by reasoning the same as above.

Remark 3. If we assume the apparently stronger condition

$$\sum_{k=1}^{\infty} \left(\frac{b_k - a_k}{b_k}\right)^2 = \infty$$

on the intervals $(a_k, b_k]$, the appeal to Levinson's theorem in the above argument can be avoided. In that way one arrives at what looks like a weaker criterion for completeness.

That criterion is in fact *not* weaker, for the condition just written is *implied* by the divergence of $\sum_k ((b_k - a_k)/a_k)^2$. Convergence of *either* series is actually equivalent to that of the other; see the top of p. 81.

Be that as it may, we *are* in possession of Levinson's theorem. It was therefore just as well to use it.

2.   **Beurling and Malliavin's effective density $\tilde{D}_\Lambda$**

A certain notion of *density* for positive real sequences, different from the one used in the first two §§ of the present chapter, is suggested by the result proved in the preceding article.

Starting with a sequence Λ of numbers $\lambda_n > 0$ tending to ∞, we denote by $n_\Lambda(t)$ the number* of λ_n in $[0, t]$ when $t \geq 0$ (as in §A), and take $n_\Lambda(t)$ as *zero* for $t < 0$. Fixing a $D > 0$, we then consider the set $\mathcal{O}_D$ of $t > 0$ such that

$$\frac{n_\Lambda(\tau) - n_\Lambda(t)}{\tau - t} > D$$

for at least one $\tau > t$. Since $n_\Lambda(t) = n_\Lambda(t+)$, $\mathcal{O}_D$ is *open*, and hence the union of a sequence of disjoint open intervals $(a_k, b_k) \subseteq (0, \infty)$, perhaps only finite in number. It is convenient in the present article to have the *index k start* from the value *zero*.

The (a_k, b_k) are yielded by a geometric construction reminiscent of that of the Bernstein intervals made at the beginning (first stage) of §B.2, Chapter VIII, but different from the latter. Imagine *light shining downwards and from the right, in a direction of slope D, onto the graph* of $n_\Lambda(t)$ vs. t for $t \geq 0$. The intervals (a_k, b_k) will then lie *under those portions of that graph which are left in shadow*:

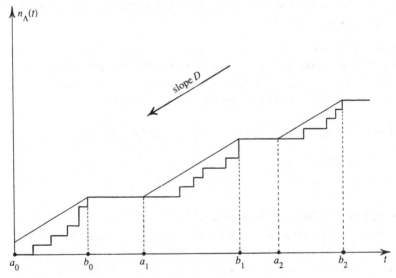

Figure 170

* We allow *repetitions* in the sequence Λ; $n_\Lambda(t)$ thus counts the points in Λ *with their appropriate multiplicities.*

It is clear that *each of the points* λ_n (where $n_\Lambda(t)$ has a jump) *must lie in one of the half-open intervals* $(a_k, \ b_k]$. Therefore since $\lambda_n \xrightarrow[n]{} \infty$ we may enumerate those intervals so as to have

$$0 \leqslant a_0 < b_0 \leqslant a_1 < b_1 \leqslant a_2 < b_2 \leqslant \cdots,$$

with $a_k \xrightarrow[k]{} \infty$ (see beginning of proof of the theorem in the preceding article). This having been done, we see from the diagram that when $b_k < \infty$,

$$\frac{n_\Lambda(b_k) - n_\Lambda(a_k)}{b_k - a_k} \ = \ D,$$

except perhaps for $k = 0$, where the left side may be $> D$.

Let us, for the work of this article, agree to *call the set* $\mathscr{O}_D$ *substantial* if either $b_0 = \infty$ or

$$\sum_{k \geqslant 1} \left(\frac{b_k - a_k}{a_k} \right)^2 \ = \ \infty.$$

Lemma. *If, for some* $D > 0$, $\mathscr{O}_D$ *is substantial, then* $\mathscr{O}_{D'}$ *is substantial for* $0 < D' < D$.

Proof. It is clear from our definition that $\mathscr{O}_D \subseteq \mathscr{O}_{D'}$ for $0 < D' < D$. If, then, we write $\mathscr{O}_D$ and $\mathscr{O}_{D'}$ as disjoint unions

$$\mathscr{O}_D \ = \ \bigcup_{k \geqslant 0} (a_k, \ b_k),$$

$$\mathscr{O}_{D'} \ = \ \bigcup_{l \geqslant 0} (a'_l, \ b'_l)$$

of intervals enumerated in the way just described, *each of the* $(a_k, \ b_k)$ *is contained in some* $(a'_l, \ b'_l)$, and in particular $(a_0, \ b_0) \subseteq (a'_0, \ b'_0)$. Hence, if $b_0 = \infty$, surely $b'_0 = \infty$, and $\mathscr{O}_{D'}$ is substantial.

When $b_0 < \infty$ and $\mathscr{O}_D$ is substantial, we have

(∗) $\qquad \sum_{k \geqslant 1} \left(\frac{b_k - a_k}{a_k} \right)^2 \ = \ \infty.$

If there are only *finitely many* $(a_k, \ b_k)$, b_k must be infinite for the *last one* of those by (∗), so, if that one is contained in $(a'_l, \ b'_l)$, say, surely $b'_l = \infty$, making $\mathscr{O}_{D'}$ substantial. Otherwise, (∗) consists of *infinitely many terms*, each of which is *finite*. Then, in case $b'_0 = \infty$, we are done. When $b'_0 < \infty$, however, there must be intervals $(a'_l, \ b'_l)$ with $l \geqslant 1$ since $\lambda_n \xrightarrow[n]{} \infty$ and each λ_n is contained in some interval of $\mathscr{O}_{D'}$. Taking, then, any $l \geqslant 1$ and denoting by

$$N_l, \ N_l + 1, \ . \ . \ . \ , \ M_l$$

the indices k – there *are* some – for which

$$(a_k, \ b_k) \subseteq (a'_l, \ b'_l),$$

we get

$$b'_l - a'_l \ \geqslant \ \sum_{k=N_l}^{M_l} (b_k - a_k),$$

so that

$$(b'_l - a'_l)^2 \ \geqslant \ \sum_{k=N_l}^{M_l} (b_k - a_k)^2$$

(the possibility that $M_l = \infty$ is not excluded here). Because $0 < a'_l \leqslant a_k$ for $k \geqslant N_l$, we see that

$$\left(\frac{b'_l - a'_l}{a'_l} \right)^2 \ \geqslant \ \sum_{k=N_l}^{M_l} \left(\frac{b_k - a_k}{a_k} \right)^2 .$$

Adding both sides for the values of $l \geqslant 1$, we obtain *on the right* a sum which differs from the one in (∗) *by at most a finite number of terms* (those, if any, for which $1 \leqslant k < N_1$). The former sum must thus *diverge*, making

$$\sum_{l \geqslant 1} \left(\frac{b'_l - a'_l}{a'_l} \right)^2 \ = \ \infty,$$

and $\mathcal{O}_{D'}$ is substantial.
  We are done.

Definition. If a sequence Λ of (perhaps repeated) strictly positive numbers has no finite limit point, its *effective density* $\tilde{D}_\Lambda$ is the *supremum* of the $D > 0$ for which the sets $\mathcal{O}_D$ corresponding to Λ in the way described above *are substantial*. If *none* of the $\mathcal{O}_D$ with $D > 0$ are substantial, we put $\tilde{D}_\Lambda = 0$. Finally, if Λ *has* a finite limit point, we put $\tilde{D}_\Lambda = \infty$.

  The density $\tilde{D}_\Lambda$ was brought into the investigation of completeness for sets of exponentials $e^{i\lambda_n t}$ by Beurling and Malliavin; its rôle there turns out to be analogous to the one played by the Pólya maximum density D^*_Λ in studying singularities of Taylor series on their circles of convergence. We will see at the end of this article that $\tilde{D}_\Lambda$ is *a kind of upper density*, being the *infimum of the* (ordinary) *densities of those measurable sequences containing Λ that enjoy a certain definite property*, to be described presently.

It is convenient to extend our definition of $\tilde{D}_\Lambda$ to *arbitrary real sequences* Λ.

Definition. If the real sequence Λ includes 0 infinitely often, $\tilde{D}_\Lambda = \infty$. Otherwise, $\tilde{D}_\Lambda$ is the *greater* of $\tilde{D}_{\Lambda+}$ and $\tilde{D}_{\Lambda-}$ for the positive sequences

$$\Lambda_+ = \Lambda \cap (0, \infty),$$
$$\Lambda_- = (-\Lambda) \cap (0, \infty) \quad (sic!).$$

$(-\Lambda$ denotes the sequence $\{-\lambda_n\}$ when $\Lambda = \{\lambda_n\}$.)

The result from the preceding article can then be reformulated as follows:

Theorem. *Let Λ be a sequence of distinct real numbers λ_n with $\tilde{D}_\Lambda > 0$. Then, if $0 < L < \pi D_\Lambda$, the exponentials $e^{i\lambda_n t}$ are complete on $[-L, L]$.*

Proof. Consists mainly of reductions to the result referred to.

Suppose in the first place that the λ_n have a finite limit point, making $\tilde{D}_\Lambda = \infty$. We see then as at the beginning of the proof of the theorem from the preceding article that the $e^{i\lambda_n t}$ are complete on *any* finite interval. Having disposed of this trivial case, we look at $\Lambda_+ = \Lambda \cap (0, \infty)$ and $\Lambda_- = (-\Lambda) \cap (0, \infty)$. Assume, wlog, that $\tilde{D}_\Lambda = \tilde{D}_{\Lambda+}$; in that case we re-enumerate Λ so as to make Λ_+ consist of the λ_n with $n \geq 1$, and then claim that the $e^{i\lambda_n t}$ with $n \geq 1$ are already complete on $[-L, L]$ for

$$0 < L < \pi\tilde{D}_{\Lambda+} = \pi\tilde{D}_\Lambda.$$

Fix a number D with

$$\frac{L}{\pi} < D < \tilde{D}_{\Lambda+} = \tilde{D}_\Lambda,$$

and form the open set

$$\mathcal{O}_D = \bigcup_{k \geq 0} (a_k, b_k)$$

in the manner described above. By definition of $\tilde{D}_{\Lambda+}$ there must be a D',

$$D < D' < \tilde{D}_{\Lambda+}$$

such that the set $\mathcal{O}_{D'}$ corresponding to it is substantial; $\mathcal{O}_D$ is *therefore substantial by the above lemma.* We thus either have $b_k = \infty$ for some k, or else there are infinitely many finite intervals (a_k, b_k) with

$$\sum_{k \geq 1} \left(\frac{b_k - a_k}{a_k} \right)^2 = \infty.$$

Let us consider the first possibility. If, say, $b_{k_0} = \infty$, there must be arbitrarily large $\tau_j > a_{k_0}$ for which

$$\frac{n_\Lambda(\tau_j) - n_\Lambda(a_{k_0}+)}{\tau_j - a_{k_0}} \geqslant D.$$

There is, indeed, a $\tau_1 > a_{k_0}$ such that

$$\frac{n_\Lambda(\tau_1) - n_\Lambda(a_{k_0}+)}{\tau_1 - a_{k_0}} \geqslant D$$

(see the above diagram, and keep in mind that $n_\Lambda(t) = n_\Lambda(t+)$). Then, however, $\tau_1 \in (a_{k_0}, \infty) \subseteq \mathcal{O}_D$, so there is a $\tau_2 > \tau_1$ with

$$\frac{n_\Lambda(\tau_2) - n_\Lambda(\tau_1)}{\tau_2 - \tau_1} > D,$$

and similarly a $\tau_3 > \tau_2$ with

$$\frac{n_\Lambda(\tau_3) - n_\Lambda(\tau_2)}{\tau_3 - \tau_2} > D,$$

and so forth. Since $n_\Lambda(t)$ *increases by at least unity at each of its discontinuities* λ_n, we must have $n_\Lambda(\tau_j) \underset{j}{\longrightarrow} \infty$. But then $\tau_j \underset{j}{\longrightarrow} \infty$ since we are in the case where $\lambda_n \underset{n}{\longrightarrow} \infty$.

Putting together the inequalities from the chain just obtained, we see that

$$\frac{\text{number of } \lambda_n \text{ in } (a_{k_0}, \tau_j]}{\tau_j - a_{k_0}} \geqslant D$$

with $\tau_j \underset{j}{\longrightarrow} \infty$. If now the $e^{i\lambda_n t}$ were *incomplete* on $[-L, L]$ for $0 < L < \pi D$, we would as in the previous article get a non-zero complex measure μ on $[-L, L]$ with $\hat{\mu}(\lambda_n) = 0$ for $n \geqslant 1$, and the zeros of $\hat{\mu}(z)$ in the right half plane would have density $\leqslant L/\pi < D$ by Levinson's theorem (Chapter III, §H.3). This, however, is incompatible with the previous relation. The $e^{i\lambda_n t}$ with $n \geqslant 1$ must hence be complete on $[-L, L]$ in the event that one of the b_k is infinite.

There remains the case where $\mathcal{O}_D$ consists of infinitely many finite intervals (a_k, b_k) with

$$\sum_{k \geqslant 1} \left(\frac{b_k - a_k}{a_k} \right)^2 = \infty.$$

Here, however,

$$\frac{\text{number of } \lambda_n \text{ in } (a_k, b_k]}{b_k - a_k} = D$$

for $k \geqslant 1$, and the completeness of the $e^{i\lambda_n t}$ on $[-L, L]$ for $0 < L < \pi D$ is an immediate consequence of the preceding article's result.

We are done.

Corollary. *The completeness radius associated with* Λ *is* $\geqslant \pi \tilde{D}_\Lambda$.

Remark. Work in the next chapter will show that in the corollary we actually have *equality*. That is the real reason for $\tilde{D}_\Lambda$'s having been defined as it was. This extension, due also to Beurling and Malliavin, lies much deeper than the results of the present §.

We proceed to look at how $\tilde{D}_\Lambda$ can be regarded as an upper density. The following lemma and corollary will be used in Chapter X.

Lemma. *If* $\tilde{D}_\Lambda < \infty$ *for a sequence of (perhaps repeated) strictly positive numbers* Λ *there is, corresponding to any* $D > \tilde{D}_\Lambda$, *a sequence* $\Sigma \supseteq \Lambda$ *of strictly positive numbers for which*

$$\int_0^\infty \frac{|n_\Sigma(t) - Dt|}{1 + t^2} \, dt \;\; < \;\; \infty.$$

Remark. Such a sequence Σ does not differ by much from the straight *arithmetic progression* $1/D$, $2/D$, $3/D$, ... About this more later on.

Proof of lemma. Is based on a very simple geometric construction.

Starting with a fixed $D > \tilde{D}_\Lambda$, we form the set

$$\mathcal{O}_D \;=\; \bigcup_{k \geq 0} (a_k, \, b_k)$$

corresponding to Λ in the manner described above, with the intervals (a_k, b_k) enumerated from left to right. By choice of D, $\mathcal{O}_D$ *cannot be substantial*, hence $b_0 < \infty$ and

$$\sum_{k \geq 1} \left(\frac{b_k - a_k}{a_k} \right)^2 \;<\; \infty.$$

Let us first find a continuous increasing function $\mu(t)$ such that

(†) $\qquad \int_0^\infty \frac{|n_\Lambda(t) + \mu(t) - Dt|}{1 + t^2} \, dt \;\; < \;\; \infty.$

This is not difficult: take $\mu(t)$ to be the piecewise linear continuous function having *slope* D on *each interval* of $[0, \infty)$ *complementary* to $\mathcal{O}_D$ and *slope zero* on each of the *components* (a_k, b_k) of $\mathcal{O}_D$, with $\mu(0) = 0$:

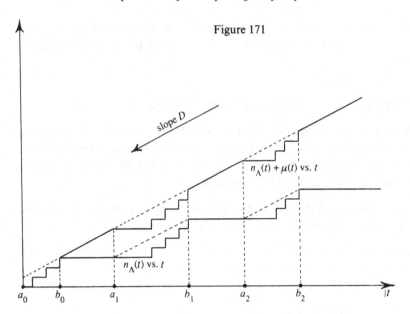

Figure 171

Outside of $\mathscr{O}_D$ on the positive real axis, $n_\Lambda(t) + \mu(t)$ is simply *equal* to $n_\Lambda(b_0) - D(b_0 - a_0) + Dt$, so

$$\int_{[0,\infty) \sim \mathscr{O}_D} \frac{|n_\Lambda(t) + \mu(t) - Dt|}{1 + t^2}\, dt \quad < \quad \infty.$$

Writing $n_\Lambda(b_0) - D(b_0 - a_0) = c$, we see from the figure that *on each of the intervals* (a_k, b_k),

$$Da_k + c = n_\Lambda(a_k) + \mu(a_k) \leqslant n_\Lambda(t) + \mu(t) \leqslant n_\Lambda(b_k) + \mu(b_k) = Db_k + c,$$

so

$$|n_\Lambda(t) + \mu(t) - Dt| \quad \leqslant \quad D(b_k - a_k) + c$$

for $a_k < t < b_k$, and

$$\int_{a_k}^{b_k} \frac{|n_\Lambda(t) + \mu(t) - Dt|}{1 + t^2}\, dt \quad \leqslant \quad D \frac{(b_k - a_k)^2}{a_k^2 + 1} + c \int_{a_k}^{b_k} \frac{dt}{t^2 + 1}.$$

The convergence of

$$\int_{\mathscr{O}_D} \frac{|n_\Lambda(t) + \mu(t) - Dt|}{1 + t^2}\, dt$$

thus follows from that of the sum $\sum_{k \geqslant 1} (b_k - a_k)^2 / a_k^2$, and, referring to the previous relation, we get (†).

In virtue of (†) we have also

$$\int_0^\infty \frac{|n_\Lambda(t) + [\mu(t)] - Dt|}{1 + t^2}\, dt \quad < \quad \infty,$$

where, as usual, $[\mu(t)]$ denotes the greatest integer $\leqslant \mu(t)$. The increasing function $n_\Lambda(t) + [\mu(t)]$ takes, however, only *integral values*, and it vanishes at 0. It is therefore equal to $n_\Sigma(t)$ for some strictly positive sequence Σ. Σ clearly consists of *the points where $n_\Lambda(t)$ jumps together with those where* $[\mu(t)]$ *jumps*, so $\Sigma \supseteq \Lambda$. And

$$\int_0^\infty \frac{|n_\Sigma(t) - Dt|}{1 + t^2}\, dt \;\; < \;\; \infty,$$

as required.

Remark. For the sequence Σ actually furnished by the construction we have

$$n_\Sigma(t) \;\leqslant\; Dt + c.$$

Corollary. *If Λ is a real sequence for which $\tilde{D}_\Lambda < \infty$ and $D > \tilde{D}_\Lambda$, there is a real sequence Σ including Λ such that*

$$\int_{-\infty}^\infty \frac{|n_\Sigma(t) - Dt|}{1 + t^2}\, dt \;\; < \;\; \infty.$$

▶ N.B. Here, $n_\Sigma(t)$ has its usual meaning for $t \geqslant 0$, *but denotes the negative of the number of members of Σ in $[t, 0)$ when $t < 0$* (convention of Chapter III, §H.2).

Proof of corollary. Write $\Lambda_+ = \Lambda \cap (0, \infty)$ and $\Lambda_- = (-\Lambda) \cap (0, \infty)$. Given $D >$ both $\tilde{D}_{\Lambda_+}$ and $\tilde{D}_{\Lambda_-}$ we apply the lemma to Λ_+ and Λ_- separately, and then put the two results together to get Σ, adjoining thereto, if needed, the point 0^* so as to ensure that $\Lambda \subseteq \Sigma$.

The preceding lemma has a *converse* whose proof requires somewhat more work.

Lemma. *If $A \geqslant 0$ and Σ is a strictly positive sequence such that*

$$\int_0^\infty \frac{|n_\Sigma(t) - At|}{1 + t^2}\, dt \;\; < \;\; \infty,$$

we have $\tilde{D}_\Sigma \leqslant A$.

Proof. Let us take any $D > A$ and form a set

$$\mathcal{O}_D \;=\; \bigcup_{k \geqslant 0} (a_k,\, b_k)$$

corresponding to the sequence Σ, following the procedure used up to now with positive sequences Λ. According to the definition of $\tilde{D}_\Sigma$, it is enough

---

* with appropriate multiplicity

to show that $\mathcal{O}_D$ is *not substantial*. In the following discussion, we assume that $A > 0$. When $A = 0$, the treatment is similar (and easier).

Order the interval components (a_k, b_k) of $\mathcal{O}_D$ in the now familiar fashion:

$$0 \leqslant a_0 < b_0 \leqslant a_1 < b_1 \leqslant a_2 < \cdots .$$

It is claimed first of all that $b_0 < \infty$. Suppose indeed that $b_0 = \infty$. Then, as in the proof of the previous lemma, we obtain a sequence $\tau_j \xrightarrow[j]{} \infty$ such that

$$\frac{n_\Sigma(\tau_j) - n_\Sigma(a_0)}{\tau_j - a_0} \geqslant D,$$

i.e.,

$$n_\Sigma(\tau_j) \geqslant D(\tau_j - a_0),$$

since of course $n_\Sigma(a_0) = 0$ (see figure near the beginning of this article). This means that $n_\Sigma(t) \geqslant D(\tau_j - a_0)$ for $t \geqslant \tau_j$, $n_\Sigma(t)$ being increasing. Therefore, if τ_j is large enough to make

$$\frac{D}{A}\tau_j - \frac{D}{A}a_0 > \tau_j$$

(we *are taking* $D > A$!), we have

$$n_\Sigma(t) - At \geqslant D\tau_j - At - Da_0 > 0$$

for $\tau_j \leqslant t < (D/A)\tau_j - (D/A)a_0$:

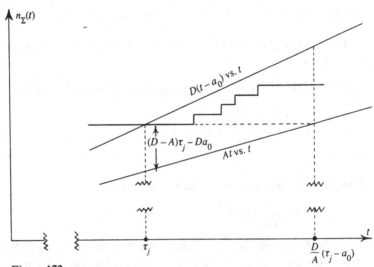

Figure 172

Looking at a τ_j larger than $(2D/(D-A))a_0$, we see from the figure that

$$\int_{\tau_j}^{(D/A)\tau_j-(D/A)a_0} \frac{n_\Sigma(t) - At}{1+t^2}\,dt$$

$$\geqslant \frac{\frac{1}{2}\left\{(D-A)\tau_j - Da_0\right\}\left\{\frac{D}{A}\tau_j - \frac{D}{A}a_0 - \tau_j\right\}}{1 + \left(\frac{D}{A}\tau_j - \frac{D}{A}a_0\right)^2} \geqslant \frac{(D-A)^2}{8A}\frac{\tau_j^2}{1 + \frac{D^2}{A^2}\tau_j^2},$$

and *this is* $\geqslant A(D-A)^2/16D^2$ (say) for large enough τ_j. Since the τ_j tend to ∞, selection of a suitable subsequence of them shows that

$$\int_0^\infty \frac{|n_\Sigma(t) - At|}{1+t^2}\,dt = \infty,$$

a contradiction.

Having thus proved that $b_0 < \infty$, we are assured of the existence of intervals (a_k, b_k) with $k \geqslant 1$, and need to show that

$$\sum_{k\geqslant 1}\left(\frac{b_k - a_k}{a_k}\right)^2 < \infty.$$

Considering any one of the intervals (a_k, b_k) in question,* we denote by $\mathscr{L}_k$ the straight line of slope D through $(a_k, n_\Sigma(a_k))$ and $(b_k, n_\Sigma(b_k))$, and look at the *abscissa* c_k of the *point where* $\mathscr{L}_k$ *and the line of slope* A *through the origin intersect:*

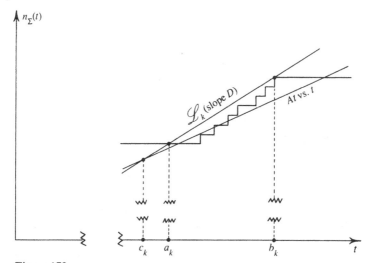

Figure 173

* $b_k < \infty$ by the argument just made for b_0

If c_k *lies to the right of the midpoint,* $(a_k + b_k)/2$, of (a_k, b_k), we say that the index $k \geqslant 1$ *belongs to the set* R. Otherwise, when $c_k < (a_k + b_k)/2$ (as in the last picture), we say that $k \geqslant 1$ *belongs to the set* S.

Let us first show that

$$\sum_{k \in R} \left(\frac{b_k - a_k}{a_k} \right)^2 < \infty.$$

When $k \in R$, the situation is as follows:

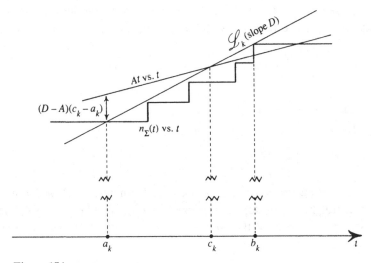

Figure 174

It may, of course, happen that $c_k > b_k$. In order to allow for that possibility, we work with

$$c'_k = \min(c_k, b_k).$$

The preceding figure shows that here

$$\int_{a_k}^{c'_k} \frac{At - n_\Sigma(t)}{1 + t^2} \, dt \geqslant \frac{\frac{1}{2}(D - A)(c_k - a_k)(c'_k - a_k)}{1 + (c'_k)^2} \geqslant \frac{D - A}{8} \cdot \frac{(b_k - a_k)^2}{1 + b_k^2},$$

since $c'_k - a_k \geqslant \frac{1}{2}(b_k - a_k)$ for $k \in R$. However, $(a_k, c'_k) \subseteq (a_k, b_k)$ with the latter intervals *disjoint*. On adding the previous inequalities for $k \in R$ it thus follows by the hypothesis that

$$\sum_{k \in R} \frac{(b_k - a_k)^2}{1 + b_k^2} < \infty,$$

whence

$$\sum_{k \in R} \left(\frac{b_k - a_k}{b_k} \right)^2 < \infty,$$

since we are dealing with numbers $b_k \geqslant a_1 > 0$. The last relation certainly implies that there cannot be infinitely many $k \in R$ with

$$\frac{b_k}{a_k} > 4, \quad \text{say,}$$

so we must also have

$$\sum_{k \in R} \left(\frac{b_k - a_k}{a_k} \right)^2 < \infty.$$

We now show that

$$\sum_{k \in S} \left(\frac{b_k - a_k}{a_k} \right)^2 < \infty;$$

this involves a covering argument. Given $k \in S$, we put

$$b'_k = \min\left(b_k + \frac{D - A}{A}(b_k - c_k), \ b_k + \frac{D - A}{A}(b_k - a_k) \right),$$

so that

$$\frac{D - A}{2A}(b_k - a_k) < b'_k - b_k \leqslant \frac{D - A}{A}(b_k - a_k),$$

and observe that $n_\Sigma(t) > At$ for $b_k \leqslant t < b'_k$:

Figure 175

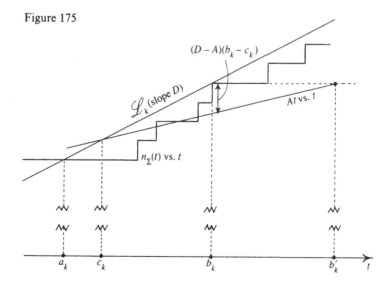

We see that in the present case

$$\int_{b_k}^{b'_k} \frac{n_\Sigma(t) - At}{1+t^2}\, dt \;\geqslant\; \frac{\frac{1}{2}(D-A)(b_k - c_k)(b'_k - b_k)}{1+(b'_k)^2}$$

$$\geqslant\; \frac{(D-A)^2}{8A} \cdot \frac{(b_k - a_k)^2}{1 + \dfrac{D^2}{A^2} b_k^2}.$$

What prevents us now from reasoning as we did when examining the sum $\sum_{k\in R}(b_k - a_k)^2/b_k^2$ is that the intervals $(b_k,\, b'_k)$, $k \in S$, *may overlap*, although of course the $(a_k,\, b_k)$ do not.

To deal with this complication, we fix for the moment *any finite subset* S' of S, and set out to obtain a bound *independent of S'* on the sum

$$\sum_{k\in S'} \frac{(b_k - a_k)^2}{1 + (Db_k/A)^2}.$$

For this purpose, we *select certain of the intervals* $(a_k,\, b_k)$, $k \in S'$, in the following manner.

First of all, we take the *leftmost of the* $(a_k,\, b_k)$, $k \in S'$, and *denote it by* $(\alpha_1,\, \beta_1)$. If $(\alpha_1,\, \beta_1)$ is $(a_{k_1},\, b_{k_1})$, say, *we denote* b'_{k_1} *by* β'_1; thus,

$$\frac{D-A}{2A}(\beta_1 - \alpha_1) \;<\; \beta'_1 - \beta_1 \;\leqslant\; \frac{D-A}{A}(\beta_1 - \alpha_1).$$

Having picked $(\alpha_1,\, \beta_1)$, we *skip over* any of the *remaining* $(a_k,\, b_k)$, $k \in S'$, which happen to be *entirely contained* in $(\alpha_1,\, \beta'_1)$ (sic!), and then, if there are *still any* $(a_k,\, b_k)$ left over for $k \in S'$, choose $(\alpha_2,\, \beta_2)$ as the *leftmost of those*. If $(\alpha_2,\, \beta_2) = (a_{k_2},\, b_{k_2})$, say, we write β'_2 for b'_{k_2}, which makes

$$\frac{D-A}{2A}(\beta_2 - \alpha_2) \;<\; \beta'_2 - \beta_2 \;\leqslant\; \frac{D-A}{A}(\beta_2 - \alpha_2).$$

It is important that $(\beta_1,\, \beta'_1)$ *cannot overlap with* $(\beta_2,\, \beta'_2)$, *even though* $(\alpha_1,\, \beta'_1)$ *may well overlap with* $(\alpha_2,\, \beta_2)$:

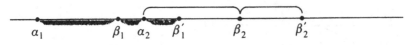

Figure 176

Otherwise, $(\alpha_2,\, \beta_2)$ would certainly be *included* in $(\alpha_1,\, \beta'_1)$, contrary to the way it was chosen. It is *also true* that *any of the intervals* $(a_k,\, b_k)$, $k \in S'$, *skipped over* in going from $(\alpha_1,\, \beta_1)$ to $(\alpha_2,\, \beta_2)$ *must lie in* $(\beta_1,\, \beta'_1)$. The former are indeed included in (α_1, β_1'), but certainly have no intersection with (α_1, β_1), which, as a *particular* (a_k, b_k) is *disjoint from all the others*.

If, after choosing (α_2, β_2), there are still some (a_k, b_k) left over with $k \in S'$, we go on in the same fashion, *first skipping over any that may be entirely included in* (α_2, β_2') – by the argument just made, those must in fact lie in (β_2, β_2') – and *then* taking (α_3, β_3) as the *leftmost of the remaining* (a_k, b_k), $k \in S'$, if such there be. Defining β_3' as we did β_1' and β_2' above, we see that the three intervals

$$(\beta_1, \beta_1'), \qquad (\beta_2, \beta_2'), \qquad (\beta_3, \beta_3')$$

must be *disjoint*.

This process can be continued as long as there are any (a_k, b_k) left with $k \in S'$. After a finite number of steps we finish, ending with certain intervals

$$(\alpha_1, \beta_1), \qquad (\alpha_2 \ \beta_2), \ \ldots \ , (\alpha_p, \beta_p),$$

selected from among the (a_k, b_k) with $k \in S'$, and having the following two properties:

(i) the intervals (β_l, β_l') are *disjoint* for $1 \leqslant l \leqslant p$;

(ii) *each* of the *remaining* (a_k, b_k), $k \in S'$, is entirely contained in one of the (β_l, β_l'), $1 \leqslant l \leqslant p$.

For each l, $1 \leqslant l \leqslant p$, the inequality proved above can be rewritten

$$\int_{\beta_l}^{\beta_l'} \frac{n_\Sigma(t) - At}{1 + t^2}\, dt \ \geqslant \ \frac{(D - A)^2}{8A} \cdot \frac{(\beta_l - \alpha_l)^2}{1 + (D\beta_l/A)^2}.$$

Denoting by S_l the set of $k \in S'$ for which $(a_k, b_k) \subseteq (\beta_l, \beta_l')$, we have

$$\sum_{k \in S_l} (b_k - a_k) \ \leqslant \ \beta_l' - \beta_l \ \leqslant \ \frac{D - A}{A}(\beta_l - \alpha_l),$$

the (a_k, b_k) being disjoint, and b_k is of course $> \beta_l$ for $k \in S_l$. Hence,

$$\sum_{k \in S_l} \frac{(b_k - a_k)^2}{1 + (Db_k/A)^2} \ \leqslant \ \left(\frac{D - A}{A}\right)^2 \cdot \frac{(\beta_l - \alpha_l)^2}{1 + (D\beta_l/A)^2}$$

which is

$$\leqslant \ \frac{8}{A}\int_{\beta_l}^{\beta_l'} \frac{n_\Sigma(t) - At}{1 + t^2}\, dt$$

by the previous relation, and finally

$$\frac{(\beta_l - \alpha_l)^2}{1 + (D\beta_l/A)^2} + \sum_{k \in S_l} \frac{(b_k - a_k)^2}{1 + (Db_k/A)^2} \ \leqslant \ \left(\frac{8A}{(D - A)^2} + \frac{8}{A}\right)\int_{\beta_l}^{\beta_l'} \frac{n_\Sigma(t) - At}{1 + t^2}\, dt.$$

By property (i) we find, summing over l, that

$$\sum_{l=1}^{p} \frac{(\beta_l - \alpha_l)^2}{1 + (D\beta_l/A)^2} + \sum_{l=1}^{p} \sum_{k \in S_l} \frac{(b_k - a_k)^2}{1 + (Db_k/A)^2}$$

$$\leq \left(\frac{8A}{(D-A)^2} + \frac{8}{A} \right) \int_0^{\infty} \frac{|n_\Sigma(t) - At|}{1 + t^2} \, dt$$

According to property (ii), the sum on the left is just

$$\sum_{k \in S'} \frac{(b_k - a_k)^2}{1 + (Db_k/A)^2};$$

that quantity is therefore *bounded by the right hand member, obviously independent* of S', of the relation just written.

Since S' was *any* finite subset of S, we thus have

$$\sum_{k \in S} \frac{(b_k - a_k)^2}{1 + (Db_k/A)^2} \leq \left(\frac{8A}{(D-A)^2} + \frac{8}{A} \right) \int_0^{\infty} \frac{|n_\Sigma(t) - At|}{1 + t^2} \, dt,$$

and from this point we may argue just as during the consideration of $\sum_{k \in R} (b_k - a_k)^2 / a_k^2$ to show that

$$\sum_{k \in S} \left(\frac{b_k - a_k}{a_k} \right)^2 < \infty.$$

Knowing that the corresponding sum over R is finite, we conclude that

$$\sum_{k \geq 1} \left(\frac{b_k - a_k}{a_k} \right)^2 < \infty.$$

That, however, was what we needed to establish in order to finish showing the *non-substantiability* of $\mathcal{O}_D$, from which it follows that $\tilde{D}_\Sigma \leq D$. Thus $\tilde{D}_\Sigma \leq A$, since $D > A$ was arbitrary.   Q.E.D.

Putting together this and the preceding lemma, we immediately obtain the

Theorem. *Let Λ be a strictly positive sequence.* Then $\tilde{D}_\Lambda$ is the infimum of the positive numbers A such that there exist positive sequences $\Sigma \supseteq \Lambda$ with*

$$\int_0^{\infty} \frac{|n_\Sigma(t) - At|}{1 + t^2} \, dt < \infty.$$

---

* perhaps with repetitions.

Corollary. *If* Λ *is a real sequence,* $\tilde{D}_\Lambda$ *is the* infimum *of the positive numbers* A *for which there exist real sequences* $\Sigma \supseteq \Lambda$ *with*

$$\int_{-\infty}^{\infty} \frac{|n_\Sigma(t) - At|}{1 + t^2} \, dt \quad < \quad \infty.$$

This corollary follows from the theorem in the same way that the one to the *first* of the preceding two lemmas does from that lemma.

A positive sequence Σ such that

$$\int_0^{\infty} \frac{|n_\Sigma(t) - At|}{1 + t^2} \, dt \quad < \quad \infty$$

is *measurable* according to the definition in §E.3, Chapter VI.

Lemma. *If the relation just written holds for the positive sequence* Σ *and some* $A \geqslant 0$, *we have*

$$\frac{n_\Sigma(t)}{t} \longrightarrow A \quad for \ t \longrightarrow \infty.$$

Proof. If, for some $\eta > 0$, we have $n_\Sigma(t_0) > (A + \eta)t_0$ with $t_0 > 1$, we see from the following diagram that

$$\int_{t_0}^{(1 + \eta/A)t_0} \frac{n_\Sigma(t) - At}{1 + t^2} \, dt \quad \geqslant \quad \frac{\dfrac{1}{2}\eta t_0 \cdot \dfrac{\eta}{A} t_0}{1 + \left(1 + \dfrac{\eta}{A}\right)^2 t_0^2} \quad \geqslant \quad \frac{\eta^2}{4A\left(1 + \dfrac{\eta}{A}\right)^2} :$$

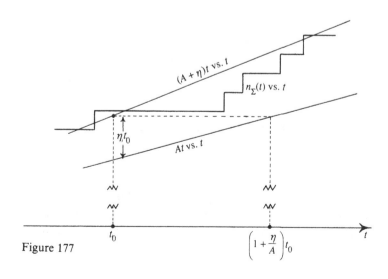

Figure 177

If, then, this happens for *arbitrarily large* t_0, we can get a sequence $\{t_k\}$,

$$t_{k+1} > \left(1 + \frac{\eta}{A}\right)t_k,$$

such that

$$\int_{t_k}^{(1+\eta/A)t_k} \frac{n_\Sigma(t) - At}{1 + t^2}\, dt \;\geq\; \frac{\eta^2 A}{4(A + \eta)^2}$$

for each k, making

$$\int_0^\infty \frac{|n_\Sigma(t) - At|}{1 + t^2}\, dt \;=\; \infty,$$

contrary to hypothesis. Therefore $n_\Sigma(t)/t$ must be $\leq A + \eta$ for all sufficiently large t.

By working with an integral over $((1 - \eta/A)t_0,\ t_0)$, one shows in like manner that $n_\Sigma(t_0)$ cannot be $< (A - \eta)t_0$ for arbitrarily large values of t_0.
The lemma is proved.

This result shows that the (ordinary) *density* of any positive sequence Σ for which

$$\int_0^\infty \frac{|n_\Sigma(t) - At|}{1 + t^2}\, dt \;<\; \infty$$

is *defined* (in the sense of §E.3, Chapter VI) *and equal to A*. The previous theorem thus furnishes our *characterization of $\tilde{D}_\Lambda$ as an upper density*:

For a positive sequence Λ, $\tilde{D}_\Lambda$ is the infimum of the densities D_Σ of the positive measurable sequences $\Sigma \supseteq \Lambda$ such that

$$\int_0^\infty \frac{|n_\Sigma(t) - D_\Sigma t|}{1 + t^2}\, dt \;<\; \infty.$$

Referring to the definition of *Pólya's maximum density* D_Λ^* given in §E.3 of Chapter VI, we also see that

$$D_\Lambda^* \;\leq\; \tilde{D}_\Lambda$$

for positive sequences Λ. Simple examples show that $\tilde{D}_\Lambda$ can be *really bigger* than D_Λ^*; it is recommended, for instance, that the reader construct a *measurable* sequence Λ of *positive integers* for which the *ordinary density* D_Λ is *zero*, while $\tilde{D}_\Lambda = 1$.

E.   Extension of the results in §D to the zero distribution of entire functions $f(x)$ of exponential type, with $\int_{-\infty}^{\infty}(\log^+|f(x)|/(1+x^2))\,dx$ convergent

In the proof of the theorem from §D.1 we may, thanks to the third Phragmén–Lindelöf theorem of §C, Chapter III, replace the Fourier transform $\hat{\mu}(z)$ by any entire function $f(z)$ of exponential type $\leqslant L$, *bounded on the real axis*, and vanishing* at the points $\lambda_n > 0$. This yields a result about the real zeros of such functions which is best formulated in terms of the effective density $\tilde{D}_\Lambda$ introduced in §D.2:

>   If $f(z)$, a non-zero entire function of exponential type $\leqslant L$, bounded on the real axis, vanishes at the points of the real sequence Λ, then $\tilde{D}_\Lambda \leqslant L/\pi$.

In cases where $\Lambda \subseteq \mathbb{R}$ consists of *all* the zeros of $f(z)$ (each counted according to its multiplicity) and where the two quantities

$$\limsup_{y \to \infty} \frac{\log|f(iy)|}{y}, \qquad \limsup_{y \to -\infty} \frac{\log|f(iy)|}{|y|}$$

are both *equal* to L, we in fact have $\tilde{D}_{\Lambda_+} = \tilde{D}_{\Lambda_-} = L/\pi$ for the two 'halves' $\Lambda_+ = \Lambda \cap (0, \infty)$, $\Lambda_- = (-\Lambda) \cap (0, \infty)$ of Λ, which is a considerable amelioration of Levinson's theorem. This is so because under the stated conditions Λ_+ and Λ_- will both be *measurable* and of (ordinary) density L/π by the Levinson theorem from §H.2, Chapter III. We know on the other hand that $\tilde{D}_{\Lambda_+}$ and $\tilde{D}_{\Lambda_-}$ must be $\geqslant$ the respective ordinary densities of Λ_+ and Λ_-, according to the work at the end of §D.2.

The requirement that $f(x)$ be *bounded* on the real axis can be relaxed. From the *Beurling–Malliavin multiplier theorem*, to be proved in Chapter XI, it readily follows that the boundedness can be replaced by the milder condition that

$$\int_{-\infty}^{\infty} \frac{\log^+|f(x)|}{1+x^2}\,dx < \infty,$$

and the above result, together with its consequences, *will still hold*. The multiplier theorem is, however, a deep result about *existence*, so an argument which depends on it for merely refining the simple estimation procedure of §D.1 hardly seems satisfactory. That is why Beurling and Malliavin gave a *direct* proof of the more general result in their 1967 *Acta*

---

* With the appropriate multiplicity at any *repeated* value λ_n; see *Remark 2* at the end of §D.1

paper. Their work is presented in article 2 below. In it, an estimate for harmonic measure going back to Ahlfors and Carleman is used in somewhat unusual fashion. That estimate is derived in article 1.

1.   **Introduction to extremal length and to its use in estimating harmonic measure**

The notion of extremal length, due to Beurling, is a natural development of a more special idea already appearing in his thesis, and is closely related to material in Grötzsch's work. Ahlfors also is closely associated with the early study of it.

The use of extremal length (or rather of its *reciprocal*) is very helpful and convenient in the investigation of various problems involving analytic functions, due partly to the strong appeal such use makes to our geometric intuition. This technique, based on a simple and beautiful idea, has been valuable in the study of quasiconformal mappings and even of problems in $\mathbb{R}^n$, as well as in ordinary function theory. It is really a pity that familiarity with extremal length is not more widespread among analysts, and that most textbooks on analytic functions do not discuss it. The most accessible introduction is in W. Fuchs' little book; material is also contained in the one by Ahlfors on conformal invariants. Hersch's *Commentarii Helvetici* paper from the 1950s has a longer (and somewhat pedantic) development, as does the book of Ohtsuka. This last is *not* the place for beginning one's study of the subject.*

Extremal length is defined for a *given family G of curves in a domain* $\mathscr{D}$. One usually requires the curves belonging to G to be at least locally rectifiable. Once G and $\mathscr{D}$ are prescribed, we look at certain *positive Borel functions* ('*weights*') $p(z)$ *defined on* $\mathscr{D}$. We say that one of these is *admissible for G* if

$$\int_{\gamma} p(z)|dz| \geqslant 1$$

for every curve γ *in the family G.* The *reciprocal extremal length* $\Lambda(\mathscr{D}, G)$ (often called the *modulus*) associated with G and $\mathscr{D}$ is simply the *infimum of*

$$\iint_{\mathscr{D}} (p(z))^2 \,dx\,dy$$

for all the p admissible for G.

---

* The beautiful outline in Beurling's *Collected Works* (vol. I, pp. 361–85) has now appeared (*Collected Works of Arne Beurling*; 2 vols, edited by L. Carleson, P. Malliavin, J. Neuberger and J. Wermer; Birkhäuser, Boston, 1989).

The idea here is natural and straightforward. We think of $p(z)$ as some kind of varying *gauge* or *conversion factor* which must be used *to the first power* to get (infinitesimal) *lengths* and *to the second power* to get *areas*. Saying that $\int_\gamma p(z)|dz| \geqslant 1$ for all the curves γ of the family G means that *we require our p to make each of those have* (gauged) *length* $\geqslant 1$. We then look to see *how small* the (gauged) *areas* $\iint_\mathscr{D}(p(z))^2\,dx\,dy$ can come out *using the different conversion factors p fulfilling that requirement. The infimum of those gauged areas is our quantity* $\Lambda(\mathscr{D}, G)$.

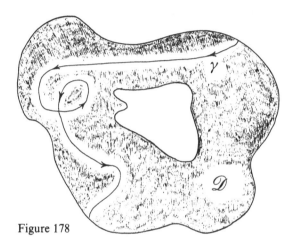

Figure 178

WARNING Most authors work with the *actual extremal length* $\lambda(\mathscr{D}, G)$ equal to $1/\Lambda(\mathscr{D}, G)$, although at least *one* uses $\lambda(\mathscr{D}, G)$ to denote *our* $\Lambda(\mathscr{D}, G)$ and calls *it* extremal length. Some write $\Lambda(\mathscr{D}, G)$ where we have $1/\Lambda(\mathscr{D}, G)$. *Care must therefore be taken* when consulting the formulas in other publications *not to confound what we call* $\Lambda(\mathscr{D}, G)$ *with its reciprocal.*

Here are some practically obvious properties of reciprocal extremal length:

1. *If* $G' \subseteq G$, $\Lambda(\mathscr{D}, G') \leqslant \Lambda(\mathscr{D}, G)$.

Indeed, there are certainly *at least as many* weights $p \geqslant 0$ admissible for G' as there are for G.

2. $\Lambda(\mathscr{D}, G)$ *is a conformal invariant; in other words, if φ is a conformal mapping of $\mathscr{D}$ onto $\tilde{\mathscr{D}}$, say, and $\tilde{G}$ consists of the images under φ of the*

curves belonging to G, we have

$$\Lambda(\tilde{\mathcal{D}},\tilde{G}) = \Lambda(\mathcal{D},G).$$

To verify this, observe that if $\gamma \in G$ and $\varphi(\gamma) = \tilde{\gamma}$, then

$$\int_\gamma p(z)|dz| = \int_{\tilde{\gamma}} \tilde{p}(\zeta)|d\zeta|$$

where, for $\zeta = \varphi(z) \in \tilde{\mathcal{D}}$, $\tilde{p}(\zeta) = p(z)/|\varphi'(z)|$. Thus, to each $p \geqslant 0$ defined on $\mathcal{D}$ and admissible for G corresponds a $\tilde{p} \geqslant 0$ defined on $\tilde{\mathcal{D}}$, admissible for $\tilde{G}$; we obviously can get all weights on $\tilde{\mathcal{D}}$ admissible for $\tilde{G}$ in this fashion. Since φ is conformal, we also have (with $\zeta = \xi + i\eta$):

$$\iint_{\tilde{\mathcal{D}}} |\tilde{p}(\zeta)|^2\,d\xi\,d\eta = \iint_{\mathcal{D}} \left|\frac{p(z)}{\varphi'(z)}\right|^2 |\varphi'(z)|^2\,dx\,dy = \iint_{\mathcal{D}} (p(z))^2\,dx\,dy$$

for each such p.

It is *property 2* that makes extremal length so useful in the study of analytic functions. For those applications, we can go a long way using just the two properties and the result of the following simple

Calculation. To find $\Lambda(\mathcal{D}, G)$ when $\mathcal{D}$ is a *rectangle* of *height* h and *length* l, and G consists of the *curves* in $\mathcal{D}$ *joining the two vertical sides* of $\mathcal{D}$.

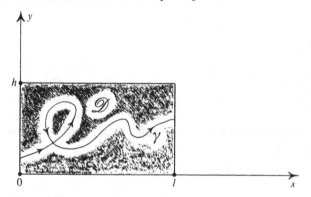

Figure 179

Choosing coordinates in the manner shown, we look at any function $p(z) \geqslant 0$ defined on $\mathcal{D}$ for which $\int_\gamma p(z)|dz| \geqslant 1$ when γ is *any* curve like the one in the diagram. That relation must in particular hold when γ is a *line parallel to* the x-axis, so we must have

$$\int_0^l p(x + iy)\,dx \geqslant 1 \quad \text{for } 0 < y < h.$$

From this, by Schwarz' inequality,

$$\int_0^l (p(x+iy))^2 \, dx \cdot l \;\; \geqslant \;\; \left(\int_0^l p(x+iy) \, dx \right)^2 \;\; \geqslant \;\; 1,$$

so

$$\int_0^h \int_0^l (p(x+iy))^2 \, dx \, dy \;\; \geqslant \;\; \frac{h}{l},$$

and $\Lambda(\mathscr{D}, G) \geqslant h/l$.

However, the function $p(z) \equiv 1/l$ *is* admissible for G, because if γ is *any* curve like the one shown,

$$\int_\gamma \frac{1}{l} |dz| \;\; = \;\; \frac{\text{length } \gamma}{l} \;\; \geqslant \;\; \frac{l}{l} \;\; = \;\; 1 \; (!)$$

This p gives us exactly *the value h/l for $\iint_{\mathscr{D}} p^2 \, dx \, dy$.* Therefore

$$\Lambda(\mathscr{D}, G) \;\; = \;\; \frac{h}{l}$$

for this particular situation.

It is important to note that the computation just made *goes through in the same way, and yields the same result, when we except a finite number of values of y from the requirement* (on $p(z)$) *that*

$$\int_0^l p(x+iy) \, dx \;\; \geqslant \;\; 1$$

for $0 < y < h$. This means that we obtain the same value, h/l, for $\Lambda(\mathscr{D}, G)$ when $\mathscr{D}$ is a rectangle of height h and length l with a finite number of horizontal slits in it, and G consists of the curves in $\mathscr{D}$ joining $\mathscr{D}$'s vertical sides (and avoiding those horizontal slits):

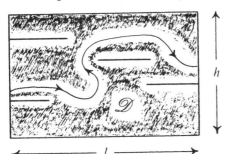

Figure 180

We can now show how extremal length can be used to *express* the *harmonic measure of a single arc* on the boundary of a *simply connected domain* $\mathscr{D}$. Given such a domain $\mathscr{D}$ with a Jordan curve boundary $\partial\mathscr{D}$ and an arc σ on $\partial\mathscr{D}$, we take *any fixed* $z_0 \in \mathscr{D}$ and consider the *family G of curves in $\mathscr{D}$ which start out from σ, loop around z_0, and then* (eventually) *go back to σ*:

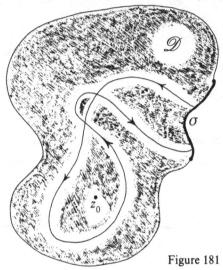

Figure 181

(The curves belonging to G are not *required* to *intersect themselves*, although they are *allowed* to do so.) There is a *precise relation* between $\Lambda(\mathscr{D}, G)$ and $\omega_{\mathscr{D}}(\sigma, z_0)$, the harmonic measure of σ in $\mathscr{D}$, as seen from z_0.

This is due to the *conformal invariance* enjoyed by both $\Lambda(\mathscr{D}, G)$ and $\omega_{\mathscr{D}}(\sigma, z_0)$. Let us first *map $\mathscr{D}$ conformally* onto the *unit disk* Δ in such a way that z_0 *goes to* 0 and σ to an arc $\tilde{\sigma}$ of the unit circle *with midpoint at* -1:

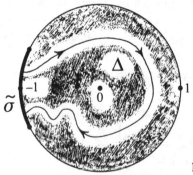

Figure 182

Then

$$\omega_{\mathscr{D}}(\sigma, z_0) \ = \ \omega_{\Delta}(\tilde{\sigma}, 0) \ = \ |\tilde{\sigma}|/2\pi,$$

and, if $\tilde{G}$ denotes the *family of curves* in Δ which *leave $\tilde{\sigma}$, loop around* 0, and *then come back to* $\tilde{\sigma}$ (see the figure),

$$\Lambda(\mathcal{D}, G) = \Lambda(\Delta, \tilde{G})$$

according to property 2. From this, we already see that $\Lambda(\mathcal{D}, G)$ is a *function of* $|\tilde{\sigma}| = 2\pi\omega_{\mathcal{D}}(\sigma, z_0)$, because the *whole configuration* used to *define* $\Lambda(\Delta, \tilde{G})$ is *completely determined* by the *size* of the arc $\tilde{\sigma}$. Calling that function ψ, we have

$$\Lambda(\mathcal{D}, G) = \psi(2\pi\omega_{\mathcal{D}}(\sigma, z_0)),$$

the relation referred to above.

If the boxed formula is to be of any use, we need some information about ψ. With that in mind, we look first at the way $\Lambda(\Delta, \tilde{G})$, equal to $\Lambda(\mathcal{D}, G)$, is obtained. The *reflection, $\tilde{\gamma}^*$, of any curve $\tilde{\gamma} \in \tilde{G}$ in the real axis also belongs to $\tilde{G}$, because the arc $\tilde{\sigma}$ is symmetric with respect to the real axis,* due to our having *chosen* it that way:

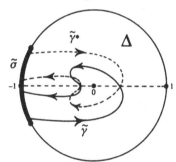

Figure 183

Therefore, if $p \geqslant 0$ (defined on Δ) is *admissible for* $\tilde{G}$, we *not only have* $\int_{\tilde{\gamma}} p(z) |dz| \geqslant 1$ for any $\tilde{\gamma} \in \tilde{G}$, but *also*

$$\int_{\tilde{\gamma}^*} p(z) |dz| \geqslant 1$$

for such curves $\tilde{\gamma}$, i.e.,

$$\int_{\tilde{\gamma}} p(\bar{z}) |dz| \geqslant 1.$$

This means that $p(\bar{z})$ is *also admissible* for $\tilde{G}$, from which it follows that

$$\tfrac{1}{2}(p(z) + p(\bar{z}))$$

is admissible for $\tilde{G}$, whenever p is. However,

$$\iint_{\Delta} \left(\frac{p(z) + p(\bar{z})}{2} \right)^2 dx\,dy \;\leqslant\; \frac{1}{2} \iint_{\Delta} (p(z))^2\,dx\,dy \;+\; \frac{1}{2} \iint_{\Delta} (p(\bar{z}))^2\,dx\,dy$$

$$= \;\iint_{\Delta} (p(z))^2\,dx\,dy,$$

so it follows that *for the computation of* $\Lambda(\Delta, \tilde{G})$, *we need only look at the functions p admissible for $\tilde{G}$ such that*

$$p(z) \;=\; p(\bar{z}).$$

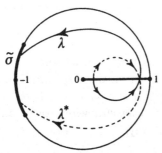

Figure 184

For *such* admissible weights p,

$$\int_{\lambda} p(z)|dz| \;\geqslant\; \frac{1}{2}$$

when λ is *any curve, going from the segment* $[0, 1)$ *to the arc $\tilde{\sigma}$ and lying in* Δ. Indeed, *given such a curve λ, it and its reflection λ^* in the real axis together make up a curve $\tilde{\gamma}$ belonging to* $\tilde{G}$ (when λ^* is traversed in the reverse direction), so that

$$\int_{\lambda} p(z)|dz| \;+\; \int_{\lambda^*} p(z)|dz| \;\geqslant\; 1,$$

i.e.,

$$\int_{\lambda} (p(z) + p(\bar{z}))|dz| \;\geqslant\; 1,$$

which implies the relation in question. *The weight $2p(z)$ is thus admissible for the family H consisting of the curves λ just described.* It is, moreover, clear that *all the $q \geqslant 0$ admissible for H with $q(z) = q(\bar{z})$ are of the form $2p$ where $p(z) = p(z)$ is admissible for* $\tilde{G}$.

To *obtain* $\Lambda(\Delta, H)$ we may, however, *limit our attention to the q admissible for H with $q(z) = q(\bar{z})$*. This is seen by arguing as we did for

$\Lambda(\Delta, \tilde{G})$. In view of the preceding observation, we therefore have

$$\Lambda(\Delta, H) = 4\Lambda(\Delta, \tilde{G}).$$

If we now *restrict* the family H *so as to only have in it curves λ lying entirely in the slit disk* $\Omega = \Delta \sim [0, 1)$ *except for their endpoints,* $\Lambda(\Delta, H)$, according to property 1, *will not be augmented. It will not be diminished either,* for such restriction of H does not give us any *new* admissible functions q.* *We may therefore take H to consist only of curves λ of the kind just mentioned* without affecting the last relation. Once this is done, $\Lambda(\Delta, H)$ becomes identical with $\Lambda(\Omega, H)$,[†] so we have finally

$$\Lambda(\Omega, H) = 4\Lambda(\Delta, \tilde{G}),$$

and if ψ is the function introduced above,

$$\psi(|\tilde{\sigma}|) = \frac{1}{4}\Lambda(\Omega, H).$$

Another conformal mapping will enable us to identify $\Lambda(\Omega, H)$ with the reciprocal extremal length already worked out in the above special calculation. *Let it be granted* for the moment that $\Omega = \Delta \sim [0, 1)$ can be *mapped conformally* onto a certain *rectangle* in such a way as to make $\tilde{\sigma}$ *go onto one side of that rectangle,* while the *slit* $[0, 1]$ *goes onto the opposite side.* We shall see presently *why* there always *is* such a mapping.

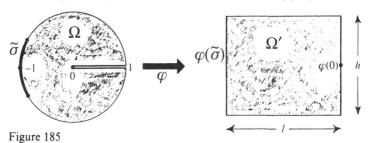

Figure 185

The rectangle may evidently be oriented so that the images of $\tilde{\sigma}$ and the slit are *vertical*; the *mapping* – call it φ – will then take the curves of our (restricted) family H to the *ones in the rectangle joining its two vertical sides.* Denoting the latter family of curves by H' and the rectangle itself by Ω', we see by property 2 that

$$\Lambda(\Omega, H) = \Lambda(\Omega', H').$$

---

* That's because any curve λ in Δ running from $[0, 1)$ to $\tilde{\sigma}$ has on it *an arc lying in the smaller domain Ω and joining $[0, 1)$ to $\tilde{\sigma}$.*

[†] The slit $[0, 1]$ has *zero area!*

If Ω' has height h and length l, we know by the special computation* that $\Lambda(\Omega', H') = h/l$. Hence

$$\Lambda(\Omega, H) = h/l.$$

As we have already seen, however,

$$\Lambda(\Omega, H) = 4\psi(|\tilde\sigma|),$$

ψ being the function presently under investigation. We have therefore arrived at two conclusions:

(i) that the *side ratio h/l* of the *rectangle Ω'* (assuming always that a mapping φ exists!) is *determined by* $|\tilde\sigma|$ even though, for the *same* arc $\tilde\sigma$, *different* mappings φ of the kind described onto *different* rectangles Ω' may be (and in fact *are*) possible;

(ii) that the *function value $\psi(|\tilde\sigma|)$* can be *evaluated*, once a mapping φ is available, by the formula

$$\psi(|\tilde\sigma|) = h/4l$$

With a little more work we can show that a mapping φ *really does exist* and, at the same time, obtain a simpler description of the function ψ. The idea here is to get at φ by *going backwards*.

We start by taking an arbitrary $h > 0$ and *mapping* the *rectangle* $\{z: 0 < \Re z < 1, \ 0 < \Im z < h/2\}$ *conformally* onto the *quarter circle* $\{z: |z| < 1, \ \Re z > 0, \ \Im z > 0\}$ in such a way as to take 1 to 1, $hi/2$ to i, and 0 to 0:

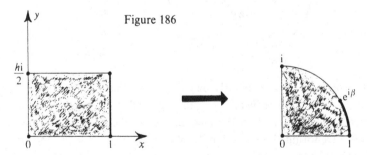

Figure 186

Under this mapping, the *upper right-hand corner* of the rectangle goes to a certain point $e^{i\beta}$, $0 < \beta < \pi/2$, where β evidently depends on h. Successive Schwarz reflections in the x and y axes will now yield a conformal mapping of the *enlarged* rectangle $\{z: -1 < \Re z < 1, \ -h/2 < \Im z < h/2\}$

---

* The curves in H' join the left vertical side of Ω' to its right one *with endpoints omitted*. That does not affect the computation; see the observation following it.

onto the unit disk Δ, which takes 0 to 0, the *right vertical side of the new rectangle* to the *arc* $(\overgroup{e^{-i\beta},\ e^{i\beta}})$, and its *left vertical side* to the *opposite arc* $(\overgroup{-e^{i\beta},\ -e^{-i\beta}})$:

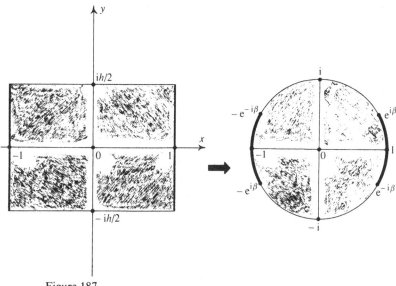

Figure 187

From this we see that $2\beta/\pi$ is equal to the *harmonic measure of the rectangle's two vertical sides* relative to that rectangle, as seen from 0. It is, however, obvious by the principle of extension of domain that *this harmonic measure increases when h does*, in fact, *grows steadily from 0 towards 1 as h increases from 0 to ∞*:

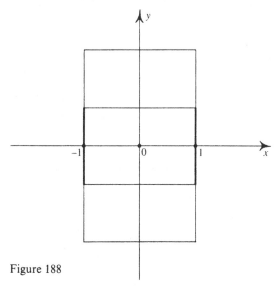

Figure 188

Given α, $0<\alpha<2\pi$, we may therefore *adjust h so as to make $\beta\;=\;\alpha/4$*, and there will be *only one value* of h for which this happens.

Taking that value of h (which depends on α), we denote by F_α the *inverse* of the *last* of the above conformal mappings (the one from the enlarged rectangle to Δ). If α is the *length* of our arc $\tilde{\sigma}$ on the boundary of the slit disk Ω, the transformation

$$z\;\longrightarrow\;F_\alpha(\sqrt{(-z)})$$

maps Ω conformally onto the *rectangle*

$$\left\{z:\;0<\Re z<1,\;-\frac{h}{2}\;<\;\Im z\;<\;\frac{h}{2}\right\}$$

and sends $\tilde{\sigma}$ to the right vertical side of that *rectangle*, taking, at the same time, the slit $[0,1]$ to its *left vertical side*. Indeed, the simple conformal mapping $z\;\longrightarrow\;\sqrt{(-z)}$ does take Ω onto the *right half* of Δ, and sends the slit $[0,1]$ to the *vertical diameter* of that semi-circle, while the arc $\tilde{\sigma}$ of length α (having midpoint at -1) goes to the *arc* $(e^{-i\alpha/4},\,e^{i\alpha/4})$:

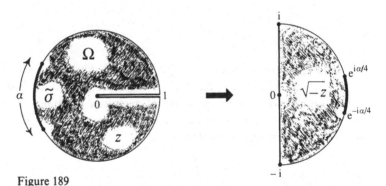

Figure 189

And, since $\beta\;=\;\alpha/4$ for our value of h, F_α does, according to one of the preceding diagrams, take the *right half* of Δ onto the above mentioned *rectangle*, making the *diameter* and *arc* go onto that rectangle's *vertical sides*. We *have*, in other words, *obtained* a mapping φ of the required sort:

$$\varphi(z)\;=\;F_\alpha(\sqrt{(-z)}).$$

The rectangle onto which this φ takes Ω has *height h* and *length unity*. Therefore, with h corresponding to $\alpha\;=\;|\tilde{\sigma}|$ in the manner described,

$$\psi(\alpha)\;=\;h/4.$$

We see by the preceding discussion that $\psi(\alpha)$ is *strictly increasing* (from 0 to ∞) when α increases from 0 to 2π; this means, of course, that the *inverse function* ψ^{-1} *exists.* In fact, if β is related to h through the conformal mapping of a *rectangle onto a quarter disk* used above, we simply have

$$\psi^{-1}(h/4) = 4\beta:$$

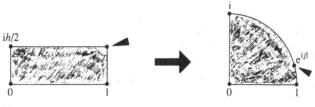

Figure 190

Let us now return to our original boxed relation between $\Lambda(\mathcal{D}, G)$ and $\omega_{\mathcal{D}}(\sigma, z_0)$. From it and the above reasoning we obtain without further ado the following

Theorem. *Given a simply connected domain $\mathcal{D}$ with Jordan curve boundary $\partial\mathcal{D}$, we have, for any arc σ on $\partial\mathcal{D}$ and any $z_0 \in \mathcal{D}$,*

$$\omega_{\mathcal{D}}(\sigma, z_0) = \frac{1}{2\pi}\psi^{-1}(\Lambda(\mathcal{D}, G)),$$

where G is the family of curves in $\mathcal{D}$ beginning and ending on σ and looping around z_0. Here, ψ^{-1} is the strictly increasing function just described.

Remark. The relation between $\omega_{\mathcal{D}}(\sigma, z_0)$ and $\Lambda(\mathcal{D}, G)$ is thus *one–one*; either of these quantities determines the other.

As we have seen, the description of ψ^{-1} is based on a certain conformal mapping of a *rectangle onto a disk*. Since elliptic functions are needed for the precise expression of such mappings, those must be required for the *explicit formula* for ψ^{-1}, which in fact involves *elliptic modular functions*. Fortunately, the *exact value* of $\psi^{-1}(\Lambda(\mathcal{D}, G))$ is *hardly ever needed* in applications, and an *approximation, asymptotically correct for small values of* $\Lambda(\mathcal{D}, G)$, *suffices.* That can be obtained in completely elementary fashion.

Lemma. *For small values of $\alpha > 0$,*

$$\psi(\alpha) = \frac{\pi}{2\log(1/Q\alpha^2)},$$

with a quantity Q tending to a limit $\neq 0$ as $\alpha \to 0$

Proof. The mapping φ used above to express ψ is obtained by putting together a chain of simpler conformal mappings. The whole construction can be most easily presented using the following diagram. In it, $C.\alpha$ *denotes a quantity asymptotic to some non-zero constant multiple of* α *for* $\alpha \longrightarrow 0$. *That constant need not be the same in the different steps.*

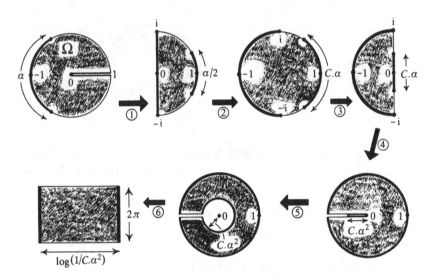

Figure 191

$$\therefore \; \psi(\alpha) \;=\; \frac{1}{4}\times\frac{2\pi}{\log(1/C.\alpha^2)} \;=\; \frac{\pi}{2\log(1/C.\alpha^2)}, \qquad \text{Q.E.D.}$$

Theorem. *In the preceding theorem,* $\omega_{\mathscr{D}}(\sigma,\,z_0)$ *lies between two constant multiplies of*

$$e^{-\pi/4\Lambda(\mathscr{D},G)}$$

(*with* absolute *constants*).

Proof. For *small* values of $\alpha = 2\pi\omega_{\mathscr{D}}(\sigma,\,z_0)$, the statement follows immediately from the lemma in view of the relation $\Lambda(\mathscr{D},G) = \psi(\alpha)$.

Because ψ is *strictly* increasing, when *either* of the quantities $\omega_{\mathscr{D}}(\sigma,\,z_0)$, $\Lambda(\mathscr{D},G)$, is *not* small, the *other* is *not small either*. Hence, since $0 \leqslant \omega_{\mathscr{D}}(\sigma,\,z_0) \leqslant 1$, the statement holds generally.

Remark. Usually what is used is the inequality

$$\omega_{\mathscr{D}}(\sigma,\,z_0) \;\leqslant\; Ce^{-\pi/4\Lambda(\mathscr{D},G)} \;;$$

for most applications the precise value of the numerical constant C does not matter.

Problem 32

Show that one may replace $\leqslant$ in the above boxed formula by $=$ and C by $(8/\pi) + o(1)$ for values of $\Lambda(\mathcal{D}, G)$ tending to zero (which covers just about all the situations where the formula is ever used).

(Hint: In the proof of the lemma, all the mappings of the chain shown are elementary except ⑤. Approximate the latter by a Joukowski transformation which takes the *small inner slit* onto a *circle* about 0 and the *outer circle* onto a curve that is *nearly* a circle about 0 when α is small.)

According to the preceding boxed inequality, *any upper bound for* $\Lambda(\mathcal{D}, G)$ *yields one for* $\omega_{\mathcal{D}}(\sigma, z_0)$. An upper bound for $\Lambda(\mathcal{D}, G)$ is obtained, however, *as soon as we are able to specify a weight* $p(z) \geqslant 0$ *on* $\mathcal{D}$ *admissible for the family* G. To this feature is due the great practical value of the inequality.

Suppose we have a *simply connected unbounded domain* $\mathcal{D}$, with reasonably nice boundary. We *fix* some $z_0 \in \mathcal{D}$ and take any $R > \text{dist}(z_0, \partial\mathcal{D})$. The *intersection* of $\mathcal{D}$ with the disk $|z - z_0| < R$ will then have a *connected component containing* z_0, which we denote by $\mathcal{O}_R$:

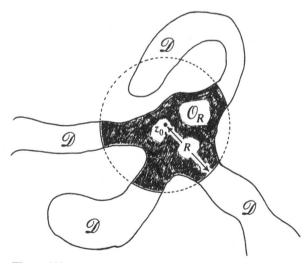

Figure 192

The boundary of $\mathcal{O}_R$ consists of part of $\partial\mathcal{D}$ and a certain number of *arcs* on the circle $|z - z_0| = R$. Some of those separate $\mathcal{O}_R$ from *unbounded* components of $\mathscr{D} \sim \bar{\mathscr{O}}_R$ (which must be present, $\mathscr{D}$ being *assumed* unbounded); we call the former *distinguished arcs.* Let us denote by $\mathscr{D}_R$ the set of points in $\mathscr{D}$ which can be joined to z_0 by paths *lying entirely in $\mathscr{D}$ and not crossing any distinguished arc.* If $\partial\mathscr{D}$ has *sufficient regularity,* which we are assuming, $\mathscr{D}_R$ will be a *bounded domain.* It may, however, contain points z with $|z - z_0| > R$, and hence *include $\mathscr{O}_R$ properly:*

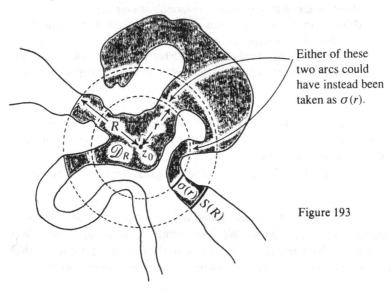

Either of these
two arcs could
have instead been
taken as $\sigma(r)$.

Figure 193

We *now single out at pleasure one* of the *distinguished arcs* on $|z - z_0| = R$ and call it $S(R)$. $S(R)$ is part of $\partial\mathscr{D}_R$. *We are interested in estimating the harmonic measure*

$$\omega_{\mathscr{D}_R}(S(R),\ z_0)$$

from above.

For each r, $\mathrm{dist}(z_0, \partial\mathscr{D}) < r < R$, the circle of radius r about z_0 intersects $\mathscr{D}_R$ (*sic!*) in a number of *open arcs. One or more* of these must *separate z_0 from $S(R)$* in $\mathscr{D}_R$; in other words, any *path in $\mathscr{D}_R$* from z_0 to $S(R)$ *must pass through* it (or them). *We choose such an arc and call it $\sigma(r)$* (see the preceding figure). When several choices are possible, this may be done in *fairly arbitrary fashion; we do require,* however, that the selection be done in such a way as to make the *union of the $\sigma(r)$ at least a Borel set in $\mathbb{C}$.* As long as $\partial\mathscr{D}$ is *decent* (which we are assuming), this is certainly possible.

Put $\vartheta(r) = |\sigma(r)|/r$ for $\mathrm{dist}(z_0, \partial\mathscr{D}) < r < R$; $\vartheta(r)$ is simply the *angle subtended by the arc $\sigma(r)$ at z_0. For $0 < r < \mathrm{dist}(z_0, \partial\mathscr{D})$ we take* $\vartheta(r) = \infty$ (*sic!*). In our present set-up, we then have the

Theorem. (due essentially to Ahlfors and Carleman)

$$\omega_{\mathscr{D}_R}(S(R),\, z_0) \;\leqslant\; C \exp\!\left(-\pi \int_0^R \frac{dr}{r\,\vartheta(r)}\right),$$

C being an absolute constant.

Proof. We use the preceding theorem, specifying a suitable weight p on $\mathscr{D}_R$ admissible for the family G of curves in $\mathscr{D}_R$ that loop around z_0 and have both ends on $S(R)$. In this we are guided by the special computation for a rectangle made earlier.

Denote by S the *union* of the $\sigma(r)$, $\operatorname{dist}(z_0, \partial\mathscr{D}) < r < R$. Then $S \subseteq \mathscr{D}_R$; note that S need not be connected! We are *assuming* that it is a Borel set. For $z \in S$, we put

$$p(z) \;=\; \frac{k}{|z-z_0|\,\vartheta(|z-z_0|)}$$

with a constant k to be presently determined. When $z \in \mathscr{D}_R$ lies *outside the set* S, we put $p(z) = 0$; this is consistent with our having taken $\vartheta(r) = \infty$ for $0 < r < \operatorname{dist}(z_0, \partial\mathscr{D})$. Our weight p will be admissible for G provided that

$$\int_\gamma p(z)|dz| \;\geqslant\; 1$$

for each curve γ in $\mathscr{D}_R$ starting from $S(R)$, looping around z_0, and then returning to $S(R)$.

In terms of the polar coordinates

$$re^{i\varphi} \;=\; z - z_0$$

with origin at z_0, we have

$$|dz| \;=\; \sqrt{((dr)^2 + r^2(d\varphi)^2)} \;\geqslant\; |dr|$$

(with possible equality) along the curves γ; we thus require that

$$\int_\gamma p(z)|dr| \;\geqslant\; 1$$

for $\gamma \in G$. Any of these γ must, however, *pass at least twice through each of the arcs* $\sigma(r)$, $\operatorname{dist}(z_0, \partial\mathscr{D}) < r < R$; *going and coming back:*

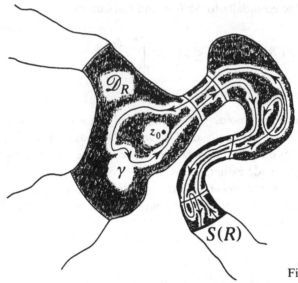

Figure 194

Our condition on $p(z) \; = \; k/r\,\vartheta(r)$ will therefore be *met* if k is adjusted so as to make

$$2 \int_{\text{dist}(z_0,\partial\mathscr{D})}^{R} \frac{k\,dr}{r\,\vartheta(r)} \;=\; 1,$$

i.e., if

$$k \int_0^R \frac{dr}{r\,\vartheta(r)} \;=\; \frac{1}{2}.$$

Choosing the value of k which satisfies this last relation, we then have

$$\iint_{\mathscr{D}_R} (p(z))^2 \, dx\,dy \;=\; \iint_S \frac{k^2}{r^2(\vartheta(r))^2}\, r\,d\varphi\,dr$$

$$=\; \int_{\text{dist}(z_0,\partial\mathscr{D})}^{R} \int_{\sigma(r)} \frac{k^2}{r^2(\vartheta(r))^2}\, r\,d\varphi\,dr$$

$$=\; \int_{\text{dist}(z_0,\partial\mathscr{D})}^{R} \frac{k^2}{r\,\vartheta(r)}\,dr \;=\; k \int_0^R \frac{k}{r\,\vartheta(r)}\,dr \;=\; \frac{k}{2}.$$

Here, the *very first* of the above integrals is by definition $\geqslant \Lambda(\mathscr{D}, G)$, so

$$\Lambda(\mathscr{D}, G) \;\leqslant\; \frac{k}{2} \;=\; 1 \Big/ 4 \int_0^R \frac{dr}{r\,\vartheta(r)}.$$

Our result now follows by the previous theorem.

Remark 1. There is a version of this result for *horizontal curvilinear strips* in which *vertical crosscuts* of those strips play the rôle of the arcs $\sigma(r)$. That version, obtainable formally from ours by a change of variable, is best derived *ab initio* by again reasoning as above. From it and the principle of extension of domain one immediately gets the harmonic measure estimate used in the scholium at the end of §D.5, Chapter VII. The reader should go through the verification of this because that estimate has many practical applications.

Remark 2. If there are several candidates for $\sigma(r)$ as in the preceding two diagrams whenever $r < R$ belongs to a *set of positive measure*, an *improvement* of the estimate provided by the theorem is available. This is pointed out in an *Arkiv* article by K. Haliste.

Let the candidates in question be denoted by $\sigma_l(r)$ with $1 \leqslant l \leqslant n(r)$, and write $r\vartheta_l(r)$ for the length of each $\sigma_l(r)$, taking $\vartheta_1(r) = \infty$ when $r < \text{dist}(z_0, \partial\mathscr{D})$. *Then the integral* $\int_0^R (1/r\,\vartheta(r))\,dr$ *figuring in the theorem's statement can be replaced by*

$$\int_0^R \sum_{l=1}^{n(r)} (1/r\,\vartheta_l(r))\,dr.$$

The proof of this better result is just like that of the theorem. One takes for S the union of *all* the $\sigma_l(r)$ for $\text{dist}(z_0, \partial\mathscr{D}) < r < R$ and then works with a weight $p(z)$, equal to *zero* outside S, and to

$$\frac{k}{|z - z_0|\,\vartheta_l(|z - z_0|)}$$

for $z \in S$ if $\sigma_l(|z - z_0|)$ is the arc on which it lies.

Remark 3. If $\Sigma(R)$ denotes the *union of the arcs* on $|z - z_0| = R$ bounding $\mathscr{O}_R$ (the component of $\mathscr{D} \cap \{|z - z_0| < R\}$ containing z_0), it is possible to get an estimate for

$$\omega_{\mathscr{O}_R}(\Sigma(R),\ z_0)$$

similar to the one for $\omega_{\mathscr{D}_R}(S(R), z_0)$ furnished by the last theorem. One form of that estimate usually goes under the name of *Tsuji's inequality*, although a better version of it can already be found in Beurling's thesis. This matter will be taken up in §F.3 below.

A celebrated application of the preceding result is in the proof, also due to Ahlfors and Carleman, of the *Denjoy conjecture*, which should be part of every analyst's general background.

The conjecture deals with the number of limiting values that an entire function of finite order may tend to when its argument moves out to ∞ along various continuous paths. We say that an entire function $f(z)$ is of *order p* if, for *each* $\varepsilon > 0$, there is a constant K_ε such that

$$|f(z)| \leqslant K_\varepsilon e^{|z|^{p+\varepsilon}}.$$

(The entire functions of exponential type considered so often in this book are thus *of order* 1.) A *finite* number a is called an *asymptotic value* for $f(z)$ if there is a *curve* γ going out to ∞ with

$$f(\zeta) \longrightarrow a \quad \text{as} \quad \zeta \longrightarrow \infty \quad \text{along } \gamma.$$

For example, 0 is an asymptotic value for the function e^z.

An entire function may have *more than one* asymptotic value. Let us for instance take *any integer* $p > 1$. Then the functions $(\sin z^p)/z^p$ and

$$f(z) = \int_0^z \frac{\sin \zeta^p}{\zeta^p}\,d\zeta$$

are both entire and of order p. When z goes out to ∞ along any one of the $2p$ rays

$$\arg z = \frac{\pi}{p}k, \quad k = 1, 2, 3, \ldots, 2p,$$

$f(z)$ tends to the finite value

$$e^{\pi i k/p} \int_0^\infty \frac{\sin t^p}{t^p}\,dt.$$

Since the integral itself is $\neq 0$,* these values are all *different*. The function $f(z)$ thus has $2p$ asymptotic values.

Denjoy made the conjecture that the example just given represents an *extreme case*, and that indeed an *entire function of order p (integral or not) cannot have more than $[2p]$ asymptotic values. To prove this, Ahlfors and Carleman argued as follows.*

Let us assume that $f(z)$ has n different asymptotic values, where $n > 1$. The case where $n = 1$ needs also to be considered when $f(z)$ has order $< \frac{1}{2}$; its treatment is like that of the one for $n > 1$, and somewhat easier

---

* It is equal to $\cos(\pi/2p)\Gamma(1/p)/(p-1)$, as is readily seen on putting $t^p = x$, integrating by parts and then taking the integral of $z^{1/p-1}e^{iz}$ around a contour consisting of the positive real and imaginary axes. Here, $\Gamma(1/p)$ is clearly > 0 — look at the integral representation for $\Gamma(x+1)$ in §B.3!

than the latter. Taking, then, $n > 1$, we have certain curves

$$\gamma_1, \ \gamma_2, \ \cdots, \ \gamma_n$$

going out to ∞ with $f(\zeta)$ tending to some limit as $\zeta \longrightarrow \infty$ along any one of them, *these limits being all different*. We may obviously take each of the γ_k to be *polygonal*, and *without self-intersections* (just cut off any closed loops that γ_k may have:

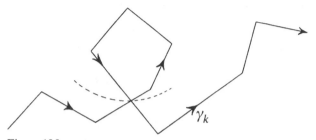

Figure 195

Since $f(\zeta)$ tends to different limits as $\zeta \longrightarrow \infty$ along the different curves γ_k, *two* of those *cannot intersect* at points *arbitrarily far out from* 0. There is thus no loss of generality in taking the γ_k disjoint, and in assuming that *the origin does not lie on any one or them*. The γ_k then bound n separate *channels*, or *tracts*, starting from a common central neighbourhood of 0 and going out to ∞:

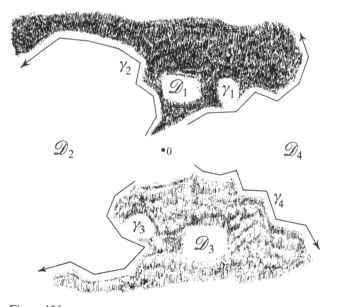

Figure 196

We can index the γ_k in such fashion that for $k = 1, 2, \ldots, n-1$, γ_k and γ_{k+1} together bound one of these tracts, denoted by $\mathcal{D}_k$, and that γ_n and γ_1 bound one, called $\mathcal{D}_n$. The preceding figure shows how things could look when $n = 4$; in it, the tracts $\mathcal{D}_1$ and $\mathcal{D}_3$ are shaded.

The function $f(z)$ cannot be bounded in any of the tracts $\mathcal{D}_k$. Suppose, for instance, that $f(z)$ is bounded in $\mathcal{D}_1$. Closing up the 'base' of $\mathcal{D}_1$ in any convenient fashion then gives us a simply connected region, part of whose boundary consists of the curves γ_1 and γ_2, both going out to ∞:

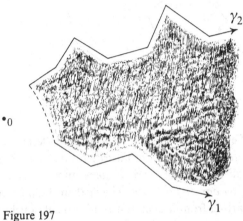

Figure 197

$f(z)$, *bounded* in that region and continuous up to γ_1 and γ_2, then tends to *two limits*, say a_1 and a_2, according as $z \longrightarrow \infty$ along γ_1 or along γ_2. In this circumstance, a well known theorem of Lindelöf implies that $a_1 = a_2$. Since, however, a_1 and a_2 are *two different asymptotic values* of f, we have a contradiction.

Problem 33

Prove Lindelöf's theorem. (Hint: By means of a conformal mapping, one may convert the region in question to the upper half plane and the function f to a new one, $F(z)$, *analytic* and *bounded* for $\Im z > 0$, *continuous* up to $\mathbb{R}$, and having the property that $F(x) \longrightarrow a_2$ for $x \longrightarrow -\infty$ while $F(x) \longrightarrow a_1$ for $x \longrightarrow \infty$. Apply the Poisson representation to $G(z) = (F(z) - a_1)(F(z) - a_2)$ (*sic!*), thus showing that $G(z) \longrightarrow 0$ *uniformly* for $z \longrightarrow \infty$ in $\{\Im z \geq 0\}$.)

Having established that $f(z)$ *cannot be bounded* in any of the $\mathcal{D}_k$, it suffices, in order to prove the Denjoy conjecture, to *assume that $n > [2p]$* (with $f(z)$ of order p) and *deduce* that *then $f(z)$ must be bounded in some $\mathcal{D}_k$*. For this purpose, we take large values of r and look at the intersections

$\Sigma_k(r)$ of each of the tracts $\mathcal{D}_k$ with the circle $|z| = r$. Each $\Sigma_k(r)$ is the union of one or more arcs; we single out one of them, called $\sigma_k(r)$, in such a way as to ensure that *any path from 0 to $\sigma_k(r)$ which touches neither γ_k nor γ_{k+1}* (and hence *stays in $\mathcal{D}_k$ after once entering that channel*) must *necessarily cut every $\sigma_k(r')$ with $r' < r$* (the latter being *defined for all sufficiently large r'*).

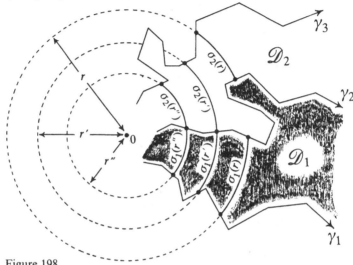

Figure 198

This we do for each k, taking care to select the $\sigma_k(r)$ for different values of r in such fashion as to *make their union a Borel set*, which is clearly *possible* since the γ_k are *polygonal curves*.

Calling $\vartheta_k(r) = |\sigma_k(r)|/r$, it is then evident that

$$\vartheta_1(r) + \vartheta_2(r) + \cdots + \vartheta_n(r) \leqslant 2\pi.$$

The above picture shows that the sum on the left may actually be $< 2\pi$.

Problem 34

(a) Show that if r_0 is *fixed and large enough* and $R > r_0$, we have

$$\sum_{k=1}^{n} \int_{r_0}^{R} \frac{dr}{r\,\vartheta_k(r)} \geqslant \frac{n^2}{2\pi} \log \frac{R}{r_0}.$$

(Hint: $\sum_{k=1}^{n} (1/\sqrt{(\vartheta_k(r))})\cdot\sqrt{(\vartheta_k(r))} = n$ (!).)

(b) Hence show that for some (fixed) k there must be arbitrarily large values of R for which

$$\int_{r_0}^{R} \frac{dr}{r\,\vartheta_k(r)} \geqslant \frac{n}{2\pi} \log \frac{R}{r_0}.$$

(c) Wlog, let the index k in (b) be unity. Take, then, the tract $\mathscr{D}_1$ and attach to it a bounded region containing 0 so as to obtain a *simply connected unbounded domain $\mathscr{D}$*:

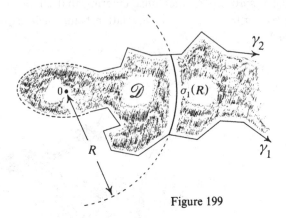

Figure 199

For large $R > 0$, denote by $\mathscr{D}_R$ the *set of points in $\mathscr{D}$ which can be reached by paths in $\mathscr{D}$ starting at 0 and not crossing $\sigma_1(R)$.* Show that for each $z \in \mathscr{D}$ there is a number C_z such that, for large enough R,

$$\omega_{\mathscr{D}_R}(\sigma_1(R),\, z) \;\leqslant\; C_z \exp\!\left(-\pi \int_{r_0}^{R} \frac{dr}{r\,\vartheta_1(r)}\right).$$

(Hint: First do this for $z = 0$. Then use Harnack.)

(d) Assuming that $n \geqslant [2p] + 1$, show that $f(z)$ is *bounded in $\mathscr{D}$*, and thus bounded in $\mathscr{D}_1$, yielding a *contradiction that proves the Denjoy conjecture*. (Hint: $f(z)$ is bounded on $\partial\mathscr{D}$ since the part of that boundary lying outside some large circle consists of points either on γ_1 or on γ_2. Fix any $z \in \mathscr{D}$, take large values of R for which the conclusion of (b) holds (with, as we are assuming, the index $k = 1$), and use the *theorem on harmonic estimation* (Chapter VII, §B.1) to estimate $\log|f(z)|$ in the domains $\mathscr{D}_R$. Note that on $\partial\mathscr{D}_R \cap \mathscr{D} \;=\; \sigma_1(R)$,

$$\log|f(\zeta)| \;\leqslant\; O(1) + R^{p+\varepsilon}$$

with $\varepsilon > 0$ arbitrary. Apply the conclusion of (c).)

2.   **Real zeros of functions $f(z)$ of exponential type with $\int_{-\infty}^{\infty}(\log^+|f(x)|/(1+x^2))\,dx < \infty$**

Now that the Ahlfors–Carleman estimate for harmonic measure is at our disposal, we are ready to carry out the extension of the results from the preceding § described at the beginning of the present one. With that in mind, we turn again to the *proof* of the theorem in §D.1, considering, instead of the Fourier–Stieltjes transform $\hat{\mu}(z)$, an entire function $f(z)$ of exponential type $\leqslant L$ with

$$(*) \qquad \int_{-\infty}^{\infty} \frac{\log^+ |f(x)|}{1+x^2}\,dx \quad < \quad \infty.$$

Taking $f(z)$ to *vanish** at each point of a certain positive sequence $\{\lambda_n\}$, we *assume* as in §D.1 that for some number $D > L/\pi$ there is a sequence of *disjoint half-open intervals* $(a_k, b_k]$, $a_k > 0$, such that

$$\frac{\text{number of } \lambda_n \text{ in } (a_k, b_k]}{b_k - a_k} \quad \geqslant \quad D$$

and

$$\sum_k \left(\frac{b_k - a_k}{a_k} \right)^2 \quad = \quad \infty.$$

Our *object* is to prove that *then*

$$\int_{-\infty}^{\infty} \frac{\log^- |f(x)|}{1+x^2}\,dx \quad = \quad \infty,$$

from which it will follow by §G.2 of Chapter III that $f(z) \equiv 0$.

The argument starts out exactly as in §D.1, and proceeds as it did there until we arrive at the examination of $f(z)$ in the ellipses

$$z \;=\; c + r\cosh(\gamma + i\vartheta)$$

about the midpoint

$$c \;=\; (a_k + b_k)/2$$

of one of the intervals $(a_k, b_k]$, where

$$\frac{b_k - a_k}{2\cosh\gamma} \quad < \quad r \quad < \quad \frac{b_k - a_k}{2}.$$

Here, γ is a *small fixed number* > 0, *the same for all of the intervals* $(a_k, b_k]$. We continue to write

$$R \;=\; (b_k - a_k)/2$$

as we did in §D.1, so that $(a_k, b_k] \;=\; (c - R, c + R]$.

---

* with the appropriate multiplicity at any repeated point of the sequence – in the next displayed formula, such points are counted according to the multiplicity of their repetition.

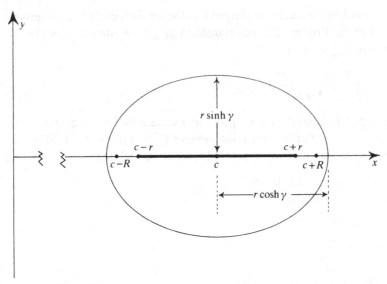

Figure 200

Picking a *fixed $\eta > 0$ much smaller than* γ, and denoting the number of λ_n in $(c - R, \ c + R]$ by N (a quantity $\geqslant 2RD$), we find as before that

$$N(\gamma - \eta) \ \leqslant \ \frac{1}{2\pi}\int_0^{2\pi} \log|f(c + r\cosh(\gamma + i\vartheta))|\, d\vartheta$$

$$- \ \frac{1}{\pi}\int_{-r}^r \frac{\log|f(c + t)|}{\sqrt{(r^2 - t^2)}}\, dt$$

for $r \geqslant R/\cosh\eta$. *Here*, however, the simple inequality

$$\log|f(c + r\cosh(\gamma + i\vartheta))| \ \leqslant \ Lr\sinh\gamma|\sin\vartheta|$$

is *no longer available* for the estimation of the first integral on the right, because *f is no longer assumed to be bounded on the real axis. Instead of boundedness, (∗) is all we have to work with.*

Our adaptation of the earlier reasoning to the present circumstances is nevertheless not altogether thwarted. In the passage from the previous relation to what corresponds to (†) of §D.1, there is a certain amount of leeway. *Provided that the constant $\delta > 0$ is small enough, it is sufficient to have*

$$\text{(∗)} \qquad \log|f(c + r\cosh(\gamma + i\vartheta))| \ \leqslant \ Lr\sinh\gamma|\sin\vartheta| \ + \ \delta R$$

$$\text{for } 0 \leqslant r \leqslant R$$

in place of the stronger inequality written above. Indeed, substitution of this into

$$\frac{1}{2\pi} \int_0^{2\pi} \log |f(c + r\cosh(\gamma + i\vartheta))|\,d\vartheta$$

yields, for $R/\cosh\eta \leqslant r \leqslant R$,

$$\frac{1}{\pi} \int_{-r}^{r} \frac{\log |f(c + t)|}{\sqrt{(r^2 - t^2)}}\,dt \;\leqslant\; \frac{2(L\sinh\gamma - \pi D(\gamma - \eta) + \frac{1}{2}\pi\delta\cosh\eta)}{\pi}r,$$

instead of (†), §D.1, so we certainly have*

$$-\frac{1}{\pi} \int_{-r}^{r} \frac{\log^- |f(c + t)|}{\sqrt{(r^2 - t^2)}}\,dt \;\leqslant\; \frac{2(L\sinh\gamma - \pi D(\gamma - \eta) + \frac{1}{2}\pi\delta\cosh\gamma)}{\pi}r,$$

for the values of r just indicated. (*Remember*: our *convention in this book is that* $\log^- |f| \geqslant 0$!)

Our *assumption* here is that $\pi D > L$, from which it follows that the *coefficient of* r *in the preceding relation is surely* (strictly) *negative* when the constants γ, η and δ (all > 0) are *chosen properly, as we henceforth suppose they are.* This being the case we may now deduce from that relation the inequality

$$(\S) \qquad \int_{a_k}^{b_k} \log^- |f(x)|\,dx \;\geqslant\; \frac{(\pi D(\gamma - \eta) - L\sinh\gamma - \frac{1}{2}\pi\delta\cosh\gamma)\tanh(\eta/2)}{2\cosh^2\eta}$$

$$\times (b_k - a_k)^2$$

by arguing as in §D.1. *This is valid, then, for any of the intervals* $(a_k, b_k]$ for which (*) holds with $c = (a_k + b_k)/2$ and $R = (b_k - a_k)/2$.

▶   Let us denote by S the set of indices k such that

$$\log |f(z)| \;\leqslant\; L|\Im z| + \frac{\delta}{2}(b_k - a_k)$$

on the whole rectangle

$$a_k - (b_k - a_k) \;\leqslant\; \Re z \;\leqslant\; b_k + (b_k - a_k)$$

$$|\Im z| \;\leqslant\; (b_k - a_k)/2.$$

---

* remember, $0 < \eta < \gamma$!

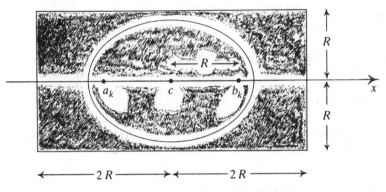

Figure 201

(Do not be misled into looking at the term $\delta(b_k - a_k)/2$ on the right side of the inequality just written as a *majorant* for the other right-hand term in the rectangle. The constant δ is usually *very much smaller* than L.) Since we are working with a *small* constant $\gamma > 0$, there is no loss of generality in *assuming – which we do from now on –* that $\cosh \gamma < 2$ and $\sinh \gamma < 1$. Then the ellipses $z = c + r \cosh(\gamma + i\vartheta)$, $0 < r \leqslant R$, *are all contained in the rectangle* (we are writing $c = (a_k + b_k)/2$ and $R = (b_k - a_k)/2$), so, if $k \in S$, $\binom{*}{*}$ holds on those ellipses, and (§) *is therefore true.* From this we see, reasoning as near the end of §D.1, that

$$\sum_{k \in S} \int_{a_k}^{b_k} \frac{\log^{-} |f(x)|}{1 + x^2} dx$$

$$\geqslant \frac{(\pi D(\gamma - \eta) - L \sinh \gamma - \frac{1}{2} \pi \delta \cosh \gamma) \tanh(\eta/2)}{2M \cosh^2 \eta} \sum_{k \in S} \left(\frac{b_k - a_k}{a_k} \right)^2,$$

where M is a certain finite constant.

The last relation shows that *if*

$$\sum_{k \in S} ((b_k - a_k)/a_k)^2 = \infty,$$

we will have

$$\int_{-\infty}^{\infty} \frac{\log^{-} |f(x)|}{1 + x^2} dx = \infty,$$

whence $f(z) \equiv 0$, the conclusion sought. It was, however, *given* that

$$\sum_{k=1}^{\infty} ((b_k - a_k)/a_k)^2 = \infty.$$

Our extension will thus be *fully established* if it can be proved that

$$\sum_{k\notin S}^{\infty}((b_k-a_k)/a_k)^2\;<\;\infty.$$

We set out to verify this convergence.

Our starting point here is just the simple inequality from §E of Chapter III, which can be applied to any of the functions $f(z-ih)$, $h\in\mathbb{R}$, in either the *upper* or *lower* half plane, f being *entire*. After making a change of variable, this gives

$$(\ddagger)\qquad \log|f(z)|\;\leqslant\;L|\Im z+h|\;+\;\frac{1}{\pi}\int_{-\infty}^{\infty}\frac{|\Im z+h|\log^{+}|f(t-ih)|}{|z-t+ih|^2}\,\mathrm{d}t,$$

because f is of exponential type $\leqslant L$. Picking any k, we write $c=(a_k+b_k)/2$, $R=(b_k-a_k)/2$ as before, and fix our attention on the rectangle

$$|\Re z-c|\;\leqslant\;2R\qquad |\Im z|\;\leqslant\;R,$$

wishing to see whether or not

$$\log|f(z)|\;\leqslant\;L|\Im z|\;+\;\delta R$$

therein. For this purpose we use $(\ddagger)$ twice, making a *hall of mirrors* argument like the one resorted to several times in Chapter VI (cf. §§A.3, B.1 and E.4 there).

Let us look in the *top half*, $\mathcal{O}$, of the rectangle in question:

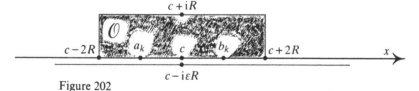

Figure 202

Fix a small $\varepsilon>0$, *the same for all the indices* k – in a moment we will see how small ε must be taken. Putting first $h=\varepsilon R$ in $(\ddagger)$ we get, for $z\in\mathcal{O}$:

$$\log|f(z)|\;\leqslant\;L\Im z+\varepsilon LR+\frac{1}{\pi}\int_{-\infty}^{\infty}\frac{(\Im z+\varepsilon R)\log^{+}|f(s-i\varepsilon R)|}{|z-s+i\varepsilon R|^2}\,\mathrm{d}s.$$

Using then $(\ddagger)$ with $h=0$, we find that

$$\log^{+}|f(s-i\varepsilon R)|\;\leqslant\;\varepsilon LR+\frac{1}{\pi}\int_{-\infty}^{\infty}\frac{\varepsilon R\log^{+}|f(t)|}{(s-t)^2+(\varepsilon R)^2}\,\mathrm{d}t,$$

which, substituted into the previous, yields

$$\log|f(z)| \leqslant L\mathfrak{J}z + 2\varepsilon LR + \frac{1}{\pi}\int_{-\infty}^{\infty}\frac{(\mathfrak{J}z + 2\varepsilon R)\log^+|f(t)|}{|z - t + 2i\varepsilon R|^2}\,dt, \quad z\in\mathcal{O},$$

by Fubini's theorem and the reproducing property of the Poisson kernel. The desired estimate for $\log|f(z)|$ will therefore *hold in $\mathcal{O}$, provided that*

$$2\varepsilon L = \frac{\delta}{4}, \quad\text{say,}$$

and also

(§§) $$\frac{1}{\pi}\int_{-\infty}^{\infty}\frac{(\mathfrak{J}z + 2\varepsilon R)\log^+|f(t)|}{|z - t + 2i\varepsilon R|^2}\,dt < \frac{3\delta}{4}R \quad\text{for } z\in\mathcal{O}.$$

Now, however, the simplest form of Harnack's inequality shows that for $z\in\mathcal{O}$,

$$\frac{1}{\pi}\int_{-\infty}^{\infty}\frac{(\mathfrak{J}z + 2\varepsilon R)\log^+|f(t)|}{|z - t + 2i\varepsilon R|^2}\,dt \leqslant C_\varepsilon\cdot\frac{1}{\pi}\int_{-\infty}^{\infty}\frac{R\log^+|f(t)|}{(c - t)^2 + R^2}\,dt$$

with a constant C_ε depending only on ε – the exact form of the dependence is not important here. (Actually, C_ε acts like $1/\varepsilon$.) Therefore, picking ε equal to $\delta/8L$, we may then determine a small constant $\alpha > 0$ such that (§§) *is ensured as long as*

$$\frac{1}{\pi}\int_{-\infty}^{\infty}\frac{R\log^+|f(t)|}{(c - t)^2 + R^2}\,dt < \alpha R.$$

If this relation holds, we will have

$$\log|f(z)| \leqslant L|\mathfrak{J}z| + \delta R$$

for $|c - \mathfrak{R}z| \leqslant 2R$ and $0 \leqslant \mathfrak{J}z \leqslant R$ and then, by obvious symmetry, for such $\mathfrak{R}z$ and $-R \leqslant \mathfrak{J}z \leqslant 0$.

We now bring the index k back into our notation, writing

$$(a_k + b_k)/2 = c_k \quad\text{(instead of } c)^*$$

and

$$(b_k - a_k)/2 = R_k \quad\text{(instead of } R).$$

What we have just shown is that *k belongs to the set S introduced above, provided that*

---

* Note that wlog, $c_k \xrightarrow[k]{} \infty$; see top of p. 66

$$\frac{1}{\pi} \int_{-\infty}^{\infty} \frac{R_k \log^+ |f(t)|}{(c_k - t)^2 + R_k^2}\, dt \; < \; \alpha R_k.$$

Regarding this condition there is, however, the important

Lemma (Beurling and Malliavin, 1967). *Let* $\alpha > 0$, *and suppose that* $c_k \xrightarrow{\;k\;} \infty$ *with the intervals* $(c_k - R_k,\; c_k + R_k)$ *lying on* $(0, \infty)$ *and disjoint. If f satisfies* (∗) *and S′ is the set of indices k for which* the preceding boxed relation fails to hold,

$$\sum_{k \in S'} \left(\frac{R_k}{c_k} \right)^2 \; < \; \infty.$$

Proof. Is based on the Ahlfors–Carleman inequality for harmonic measure derived in the preceding article.
   The function

$$U(z) \;=\; \frac{1}{\pi} \int_{-\infty}^{\infty} \frac{\Im z \, \log^+ |f(t)|}{|z - t|^2}\, dt,$$

harmonic and *positive* in $\Im z > 0$, is available thanks to (∗). For each fixed real x,

$$\frac{U(x + iy)}{y} \;=\; \frac{1}{\pi} \int_{-\infty}^{\infty} \frac{\log^+ |f(t)|}{(x - t)^2 + y^2}\, dt$$

tends to zero as $y \longrightarrow \infty$ and is a *decreasing* function of y for $y > 0$, indeed, *strictly decreasing* unless $\log^+ |f(t)| = 0$ a.e., i.e., $|f(t)| \leq 1$ on $\mathbb{R}$, in which case the lemma is obviously true anyway. Wlog, then, there is a certain value $Y(x)$ of y (perhaps equal to zero), such that

$$\frac{\alpha y}{2} \;-\; U(x + iy) \quad \begin{cases} = 0 \text{ for } y = Y(x), \\ > 0 \text{ for } y > Y(x). \end{cases}$$

Continuity of $Y(x)$ as a function of x follows easily from that of $U(z)$ for $\Im z > 0$; the set

$$\mathscr{D} \;=\; \{(x, y):\; y > Y(x)\}$$

is therefore *open*, besides being obviously *simply connected*.

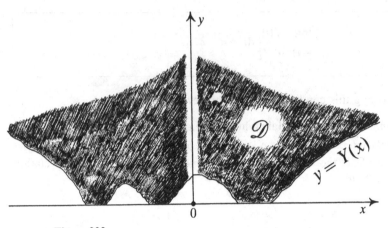

Figure 203

We work with the function

$$V(z) \;=\; \frac{\alpha}{2}\Im z \;-\; U(z)$$

in the domain $\mathscr{D}$; $V(z)$ is *harmonic* and (by construction) *strictly positive* there, and *zero* on $\partial\mathscr{D}$. Wlog, $i \in \mathscr{D}$, so that

$$V(i) \;>\; 0.$$

The idea now is to *assume* that

$$\sum_{k \in S'} (R_k/c_k)^2 \;=\; \infty$$

and then *derive* therefrom the *contradictory* conclusion that $V(i) = 0$. That will prove the lemma.

To say that $k \in S'$ means that

$$U(c_k + iR_k) \;\geqslant\; \alpha R_k,$$

whence, $U(z)$ being $\geqslant 0$ in $\{\Im z > 0\}$,

$$U(x + iR_k) \;\geqslant\; \frac{\alpha}{2} R_k \quad \text{for } c_k - \tfrac{1}{3}R_k \leqslant x \leqslant c_k + \tfrac{1}{3}R_k$$

by Harnack, in other words,

$$Y(x) \;\geqslant\; R_k \quad \text{for } c_k - \tfrac{1}{3}R_k \leqslant x \leqslant c_k + \tfrac{1}{3}R_k \text{ when } k \in S'.$$

This observation we will use presently.

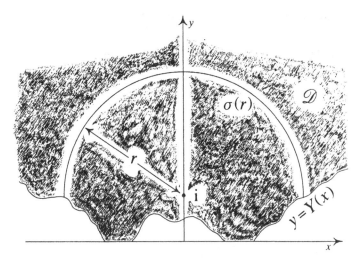

Figure 204

For large r, let $\sigma(r)$ be the circular arc of radius r about i (*sic!*), *lying entirely in $\mathscr{D}$ save for its two endpoints* which are on the curve $y = Y(x)$. Given R, the arc $\sigma(R)$ divides $\mathscr{D}$ into at least two simply connected regions, one of which contains i and is denoted by $\mathscr{D}_R$. We apply the *theorem on harmonic estimation* (Chapter VII, §B.1) to $V(z)$ in $\mathscr{D}_R$. Since

$$V(z) \leqslant \frac{\alpha}{2}\Im z \leqslant \frac{\alpha}{2}(R+1)$$

for z on $\sigma(R)$ and $V(z) = 0$ on $\partial\mathscr{D}$, we find that

$$V(\mathrm{i}) \leqslant \frac{\alpha}{2}(R+1)\omega_{\mathscr{D}_R}(\sigma(R),\,\mathrm{i}).$$

To estimate the harmonic measure appearing on the right, we now apply the *last theorem* of the preceding article (Ahlfors–Carleman). *Fixing any large r_0* (more or less at pleasure) and writing $\vartheta(r) = |\sigma(r)|/r$ as usual, we have by that result

$$\omega_{\mathscr{D}_R}(\sigma(R),\,\mathrm{i}) \leqslant C_0 \exp\left(-\pi\int_{r_0}^{R}\frac{dr}{r\,\vartheta(r)}\right)$$

for $R > r_0$; here C_0 is a constant, depending, of course, on the choice of r_0, but independent of R.

In the case where $Y(x)$ is *zero at both ends* of an arc $\sigma(r)$, we have $\vartheta(r) = \pi + 2\arcsin(1/r)$. In the general situation, one may bring in the angles $\varphi(r)$ and $\psi(r)$ shown in the following figure, and in terms of them,

$$\mathcal{G}(r) \;\; = \;\; \pi \;+\; 2\arcsin\frac{1}{r} \;-\; \varphi(r) \;-\; \psi(r):$$

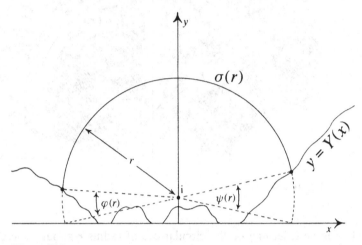

Figure 205

To investigate the integral occurring on the right side of the previous inequality, it is better to first work directly with $\varphi(r)$ and $\psi(r)$, bringing in the quantities c_k, R_k (with $k \in S'$) only towards the end.

We have:

$$\frac{\pi}{\pi + 2\arcsin\dfrac{1}{r} - \varphi(r) - \psi(r)}$$

$$\geqslant \;\; \frac{\pi}{\pi + 2\arcsin\dfrac{1}{r}} \;+\; \frac{\pi}{\left(\pi + 2\arcsin\dfrac{1}{r}\right)^2}\Big(\varphi(r) + \psi(r)\Big)$$

$$= \;\; 1 \;-\; O\!\left(\frac{1}{r}\right) \;+\; \frac{\varphi(r) + \psi(r)}{\pi}.$$

Hence

$$\pi \int_{r_0}^{R} \frac{dr}{r\,\mathcal{G}(r)} \;\; \geqslant \;\; \log\frac{R}{r_0} \;-\; O(1) \;+\; \frac{1}{\pi}\int_{r_0}^{R} \frac{\varphi(r) + \psi(r)}{r}\,dr.$$

In terms of the *polar coordinates about* i

$$re^{i\theta} = z - i,$$

the *integral on the right* has a simple interpretation. It is none other than

$$\frac{1}{\pi} \iint_{\Omega \cap \{r_0 < |z-i| < R\}} \frac{r \, d\theta \, dr}{r^2},$$

where Ω is the *complement*, in

$$\{z: \Im z > 0 \text{ and } |z-i| > r_0\},$$

of the union of the arcs $\sigma(r)$ *for* $r > r_0$. Ω *certainly includes the complement of* $\mathscr{D}$ in the region just mentioned, and may, *indeed*, include the latter *properly*:

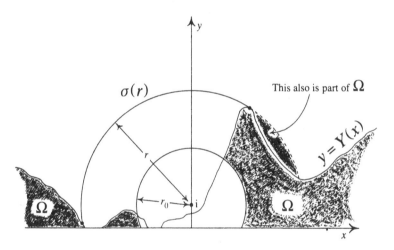

Figure 206

Writing

$$\Omega_R = \Omega \cap \{z: r_0 < |z-i| < R\},$$

we see, going over to rectangular coordinates, that

$$\frac{1}{\pi} \int_{r_0}^{R} \frac{\varphi(r) + \psi(r)}{r} dr = \frac{1}{\pi} \iint_{\Omega_R} \frac{dx \, dy}{x^2 + (y-1)^2},$$

whence finally

$$\pi \int_{r_0}^{R} \frac{dr}{r \vartheta(r)} \geq \log \frac{R}{r_0} - O(1) + \frac{1}{\pi} \iint_{\Omega_R} \frac{dx \, dy}{1 + x^2 + y^2}.$$

Substitution of this into the above estimate for harmonic measure yields

$$\omega_{\mathscr{D}_R}(\sigma(R),\,i) \;\leqslant\; \frac{\text{const.}}{R}\exp\left(-\frac{1}{\pi}\iint_{\Omega_R}\frac{dx\,dy}{1+x^2+y^2}\right),$$

from which

$$V(i) \;\leqslant\; \text{const.}\,\frac{R+1}{R}\exp\left(-\frac{1}{\pi}\iint_{\Omega_R}\frac{dx\,dy}{1+x^2+y^2}\right).$$

The constant here depends on r_0, but is independent of R.

Let us now make $R \longrightarrow \infty$. By the last relation, *divergence* of

$$\iint_{\Omega}\frac{dx\,dy}{1+x^2+y^2}$$

will then force $V(i) = 0$. As we have seen, however,

$$Y(x) \geqslant R_k \quad \text{for } c_k - \tfrac{1}{3}R_k \leqslant x \leqslant c_k + \tfrac{1}{3}R_k$$

when $k \in S'$, so, when $k \in S'$ is large, the *rectangle* of *height* R_k with *base* on $[c_k - \tfrac{1}{3}R_k,\, c_k + \tfrac{1}{3}R_k]$ *must lie in* Ω *(recall that the intervals* $(c_k - R_k,\, c_k + R_k) \subseteq (0,\infty)$ *are disjoint and that* $c_k \xrightarrow[k]{} \infty$!):

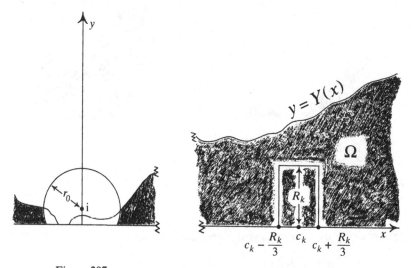

Figure 207

Therefore, fixing a suitable k_0,

$$\iint_{\Omega}\frac{dx\,dy}{1+x^2+y^2} \;\geqslant\; \sum_{\substack{k\in S'\\ k\geqslant k_0}}\frac{R_k^2}{3(1+c_k^2+R_k^2)}.$$

Again, since $(c_k - R_k,\ c_k + R_k) \subseteq (0, \infty)$, $c_k \geqslant R_k$, and, since $c_k \xrightarrow{k} \infty$, $c_k \geqslant$ some number $A > 0$ for $k \geqslant k_0$. From the preceding relation we thus get

$$\iint_\Omega \frac{dx\,dy}{1 + x^2 + y^2} \geqslant \frac{A^2}{6A^2 + 3} \sum_{\substack{k\in S' \\ k \geqslant k_0}} \left(\frac{R_k}{c_k}\right)^2.$$

Divergence of the sum on the right therefore makes the integral on the left infinite, which, as we have just shown, implies that $V(\mathrm{i}) = 0$.

But $V(\mathrm{i}) > 0$. The right-hand sum must therefore *converge*, so that

$$\sum_{k\in S'} (R_k/c_k)^2 < \infty.$$

This is what we had to prove, however. We are done.

Let us return to the discussion preceding the lemma, where we saw that *the index k belongs to the set S provided that*

$$\frac{1}{\pi} \int_{-\infty}^{\infty} \frac{R_k \log^+ |f(t)|}{(t - c_k)^2 + R_k^2} dt < \alpha R_k.$$

Knowing that

$$\sum_{k=1}^{\infty} \left(\frac{b_k - a_k}{a_k}\right)^2 = \infty,$$

we wished to conclude that

$$\sum_{k\in S} \left(\frac{b_k - a_k}{a_k}\right)^2 = \infty,$$

for, as had already been deduced then, this would imply that

$$\int_{-\infty}^{\infty} \frac{\log^- |f(x)|}{1 + x^2} dx = \infty$$

and hence that $f(z) \equiv 0$ (our desired result) by §G.2 of Chapter III.

The lemma shows, however, that the above condition on the numbers $c_k = (a_k + b_k)/2$ and $R_k = (b_k - a_k)/2$ *certainly holds* except for those k belonging to a set S' for which*

$$\sum_{k\in S'} \left(\frac{R_k}{c_k}\right)^2 < \infty.$$

---

* see footnote, p. 116

As in §D.1, we may, wlog, assume that

$$\frac{b_k}{a_k} \xrightarrow[k]{} 1,$$

so, since the $a_k > 0$, the preceding relation becomes

$$\sum_{k \in S'} \left(\frac{b_k - a_k}{a_k} \right)^2 \;<\; \infty.$$

Thus, we surely have

$$\sum_{k \notin S} \left(\frac{b_k - a_k}{a_k} \right)^2 \;<\; \infty,$$

from which it follows that

$$\sum_{k \in S} \left(\frac{b_k - a_k}{a_k} \right)^2 \;=\; \infty$$

in view of the divergence of the corresponding sum over all k.

In this way, we have *proved* our generalization of the result in §D.1:

Theorem. *Let* $f(z)$, *entire and of exponential type* $\leqslant L$, *satisfy*

$$\int_{-\infty}^{\infty} \frac{\log^+ |f(x)|}{1 + x^2}\,dx \;<\; \infty$$

and vanish* at each of the numbers $\lambda_n > 0$. *If, for some* $D > L/\pi$, *there is a sequence of disjoint intervals* $(a_k, b_k]$, $a_k > 0$, *with*

$$\sum_{k} \left(\frac{b_k - a_k}{a_k} \right)^2 \;=\; \infty$$

and

$$\frac{\text{number of } \lambda_n \text{ in } (a_k, b_k]}{b_k - a_k} \;\geqslant\; D$$

for each k, *then* $f(z) \equiv 0$.

Remark. The number L enters into this theorem *solely on account* of (‡). We may therefore replace the condition, figuring in the hypothesis, that f be of exponential type $\leqslant L$ by the simpler requirement that f be of

---

* with appropriate multiplicity at any repeated value λ_n

(some) exponential type, with

$$\limsup_{y \to \infty} \frac{\log |f(iy)|}{y} \qquad \text{and} \qquad \limsup_{y \to -\infty} \frac{\log |f(iy)|}{|y|}$$

both $\leqslant L$ (see Chapter III, §E). Of course, the latter in fact *implies* that $f(z)$ is of exponential type $\leqslant L$ (see the last theorem in §E.2, Chapter VI and especially the discussion in §B.2 there), so that, *logically*, we have gained *nothing*. It is nevertheless often *easier in practice* to estimate the two limsups than to obtain a good upper bound on f's exponential type by direct examination of that function.

Once the result just stated is available, it is useful to bring in the effective density $\tilde{D}_\Lambda$ discussed in §D.2. Arguing as at the very beginning of the present §, we then obtain the following propositions:

Theorem. *Let Λ be a sequence of real numbers on which a non-zero entire function f of exponential type vanishes,* with*

$$\limsup_{y \to \infty} \frac{\log |f(iy)|}{y} \qquad \text{and} \qquad \limsup_{y \to -\infty} \frac{\log |f(iy)|}{|y|}$$

both $\leqslant L$ *and*

$$\int_{-\infty}^{\infty} \frac{\log^+ |f(x)|}{1 + x^2} dx \; < \; \infty.$$

Then $\tilde{D}_\Lambda \leqslant L/\pi$.

Theorem. *Let the real sequence Λ consist of all the zeros (with repetitions according to multiplicities) of the entire function f figuring in the previous result, and suppose that the two lim sups occurring there are each* exactly *equal to L. Then for $\Lambda_+ = \Lambda \cap (0, \infty)$ and $\Lambda_- = (-\Lambda) \cap (0, \infty)$, we have*

$$\tilde{D}_{\Lambda_+} \; = \; \tilde{D}_{\Lambda_-} \; = \; \frac{L}{\pi}.$$

As remarked at the beginning of this §, the second of these two theorems is a considerable improvement of the one of Levinson for functions with real zeros (§H.2, Chapter III). It can be used to *replace Pólya maximum density by effective density* in some of the earlier results in this book, thus

---

* with appropriate multiplicity at any repeated member of Λ

strengthening them. For instance, the *first* theorem from §E.3, Chapter VI, can be extended as follows (keeping the notation of the place where it appears originally):

Theorem. *Given a weight* $W(x) \geqslant 1$ *tending to* ∞ *as* $x \longrightarrow \pm\infty$ *and a number* $A > 0$, *suppose that*

$$\int_{-\infty}^{\infty} \frac{\log W_A(x)}{1 + x^2}\, dx \;<\; \infty$$

and that $W(x_k) < \infty$ *for each of the positive numbers* x_k. *If the* effective density $\tilde{D}$ *of the sequence* $\{x_k\}$ *is* $> A/\pi$, $\mathscr{E}_A$ *is not* $\|\ \|_W$-*dense in* $\mathscr{C}_W(\mathbb{R})$.

To compare this improved result with the original one, it suffices to note that there exist positive sequences *having ordinary density equal to zero* for which the *effective density is infinite*.

Again, we can now see that the entire function $\Phi(z)$ of exponential type A with

$$\int_{-\infty}^{\infty} \frac{\log^+ |\Phi(x)|}{1 + x^2}\, dx \;<\; \infty$$

whose existence was treated during the discussion of de Branges' theorem (near the beginning of §F.3, Chapter VI) in fact satisfies $|\Phi(x_n)| \geqslant W(x_n)$ on a sequence Λ of real x_n for which

$$\tilde{D}_{\Lambda_+} \;=\; \tilde{D}_{\Lambda_-} \;=\; \frac{A}{\pi}.$$

This represents a certain improvement over the asymptotic relation

$$x_n \;\sim\; \frac{\pi}{A} n, \qquad n \longrightarrow \pm\infty,$$

obtained for that sequence by use of Levinson's theorem.

F.   Scholium. Extension of results in §E.1. Pfluger's theorem and Tsuji's inequality

Extremal length is not only useful for finding the harmonic measure of *single arcs* on the boundaries of simply connected domains (as in §E.1); it can also be applied when examining the harmonic measure of *fairly general sets* lying on those boundaries. In the latter situation one obtains *an estimate* instead of the *exact relation* holding for arcs. The main result there is based on work of Pfluger which was published in the same volume of *Commentarii Helvetici* as the one where Hersch's article appeared. Another form of the result, involving, however, a more special notion than extremal length (or rather one *formulated* in more particular fashion), can already be found in Beurling's thesis.

Logarithmic capacity and the logarithmic conductor potential play an important rôle in Pfluger's work; these are explained in the first of the following three articles. In order to apply his result so as to obtain an analogue of the Ahlfors–Carleman estimate given near the end of §E.1, some elementary theorems about univalent functions are needed. Those may be found in many standard texts on complex variable theory.

1   Logarithmic capacity and the conductor potential

If $\mathscr{D}$ is a *bounded* domain having reasonably decent boundary $\partial\mathscr{D}$, we have, for the Green's function $G_{\mathscr{D}}(z, w)$ associated thereto,

$$G_{\mathscr{D}}(z, w) = \log\frac{1}{|z - w|} + \int_{\partial\mathscr{D}} \log|\zeta - w|\, d\omega_{\mathscr{D}}(\zeta, z); \quad z, w \in \mathscr{D}.$$

We saw in §C.1 of Chapter VIII that this formula remains valid for certain *unbounded* domains $\mathscr{D}$ whose boundary *includes* the point at ∞ and is *not too sparse* there. It *cannot*, however, be true for *unbounded* domains with *compact* boundary. Indeed, when $\partial\mathscr{D}$ is compact but $\mathscr{D}$ unbounded, the *right side* of the relation tends to $-\infty$ as $z \longrightarrow \infty$ if $w \in \mathscr{D}$ is fixed and finite. At the same time, $G_{\mathscr{D}}(z, w)$ is supposed *by definition* to stay > 0.

In complex variable theory, there is a standard procedure for adapting the various notions of local behaviour originally defined for the points of $\mathbb{C}$ to the point at ∞ on the Riemann sphere. One first uses a linear fractional transformation to map ∞ to a finite point a, and then says that a function defined near and at ∞ has such and such behaviour there if the one related to it through the transformation behaves thus (in the usual accepted sense) at a. The Green's function for a domain $\mathscr{D}$ on the Riemann sphere having reasonable boundary, with $\infty \in \mathscr{D}$, is defined in accordance with that convention: taking a linear fractional transformation φ which maps $\mathscr{D}$ onto some domain $\mathscr{E} \subseteq \mathbb{C}$, we put

$$G_{\mathscr{D}}(z, w) = G_{\mathscr{E}}(\varphi(z), \varphi(w)), \quad z, w \in \mathscr{D}.$$

In this extension of the definition of $G_{\mathscr{D}}(z, w)$ to domains $\mathscr{D}$ containing ∞, all of the usual general properties of that function holding for domains $\mathscr{D} \subseteq \mathbb{C}$ are preserved. That is, in particular, true of the important *symmetry* relation

$$G_{\mathscr{D}}(z, w) = G_{\mathscr{D}}(w, z),$$

established at the end of §A.2, Chapter VIII. We see also that when $w \in \mathscr{D}$ is *not* equal to ∞, $G_{\mathscr{D}}(z, w)$ is described by just the ordinary specification:

As a function of z, $G_{\mathscr{D}}(z, w)$ is continuous and $\geqslant 0$ on $\bar{\mathscr{D}} \sim \{w\}$, harmonic in $\mathscr{D} \sim \{w\}$ (including at ∞), and zero on $\partial \mathscr{D}$. Near w, it equals $\log(1/|z - w|)$ plus a harmonic function of z.

For $w = \infty$, however, our definition leads to the following description:

$G_{\mathscr{D}}(z, \infty)$ is continuous and $\geqslant 0$ on $\bar{\mathscr{D}} \sim \{\infty\}$, harmonic in $\mathscr{D} \sim \{\infty\}$, and zero on $\partial \mathscr{D}$. Near ∞, it equals $\log|z|$ (*sic!*) plus a function harmonic in z (including at ∞).

Keeping these characterizations in mind, we easily obtain an integral representation for $G_{\mathscr{D}}(z, \infty)$ akin to the one given above for $G_{\mathscr{D}}(z, w)$ and *bounded* domains $\mathscr{D}$. Fix, for the moment, a *finite* $w \in \mathscr{D}$, and consider the function

$$h_w(z) \;=\; \log|z - w| \;+\; G_{\mathscr{D}}(z, w) \;-\; G_{\mathscr{D}}(z, \infty).$$

According to the descriptions just made $h_w(z)$ is *bounded* and *harmonic* in $\mathscr{D}$ (including at $z = w$ and at $z = \infty$), and *continuous* up to $\partial \mathscr{D}$. Therefore, as long as $\partial \mathscr{D}$ is *decent* (which we are *assuming throughout this* §!),

$$h_w(z) \;=\; \int_{\partial \mathscr{D}} h_w(\zeta) \, d\omega_{\mathscr{D}}(\zeta,\, z) \;=\; \int_{\partial \mathscr{D}} \log|\zeta - w| \, d\omega_{\mathscr{D}}(\zeta,\, z).$$

As $z \longrightarrow \infty$, $\log|z - w| - G_{\mathscr{D}}(z, \infty) \;=\; \mathrm{o}(1) + \log|z| - G_{\mathscr{D}}(z, \infty)$ tends to *a certain finite limit*, which we denote by $-\gamma_{\mathscr{D}}$, so then

$$h_w(z) \;\longrightarrow\; G_{\mathscr{D}}(\infty, w) \;-\; \gamma_{\mathscr{D}}.$$

Making $z \longrightarrow \infty$ in the previous relation, we thus obtain

$$-\gamma_{\mathscr{D}} \;+\; G_{\mathscr{D}}(\infty, w) \;=\; \int_{\partial \mathscr{D}} \log|\zeta - w| \, d\omega_{\mathscr{D}}(\zeta,\, \infty)$$

After using the above mentioned symmetry property and then replacing w by z, this becomes

$$G_{\mathscr{D}}(z, \infty) \;=\; \gamma_{\mathscr{D}} \;+\; \int_{\partial \mathscr{D}} \log|z - \zeta| \, d\omega_{\mathscr{D}}(\zeta,\, \infty),$$

the integral representation sought.

From the last formula, we see in particular that

$$\int_{\partial \mathscr{D}} \log \frac{1}{|z - \zeta|} \, d\omega_{\mathscr{D}}(\zeta, \, \infty) \; = \; \gamma_{\mathscr{D}} \quad \text{for } z \in \partial \mathscr{D}.$$

The positive measure $\omega_{\mathscr{D}}(\, , \, \infty)$ of total mass 1 *supported on $\partial \mathscr{D}$ has constant logarithmic potential equal to $\gamma_{\mathscr{D}}$ thereon.* $\gamma_{\mathscr{D}}$ is called the *Robin constant* for $\mathscr{D}$.

If we imagine a very long metallic cylinder perpendicular to the $x - y$ plane, cutting the latter, near its own middle, precisely in $\partial \mathscr{D}$, and having its different pieces joined to each other by thin perfectly conducting wires, an electric charge placed on it will distribute itself so as to make the electrostatic potential constant thereon. Near the $x - y$ plane, that equilibrium distribution will depend mainly on $z = x + iy$ and hardly at all on distance measured perpendicularly to the cylinder's cross section; the same is true of the corresponding electrostatic potential. To within a constant factor, the latter*, near the $x - y$ plane, is practically equal to a *logarithmic potential* in $z = x + iy$ corresponding to a *measure* giving the *amount of electric charge per unit of cylinder generator length*. The *second* of the aboved boxed formulas shows that if the whole cylinder carries *one unit of electric charge per unit of length* (measured along a generator), the electrostatic potential (see footnote) at equilibrium is

$$\int_{\partial \mathscr{D}} \log \frac{1}{|z - \zeta|} \, d\omega_{\mathscr{D}}(\zeta, \, \infty)$$

(near the $x - y$ plane); this is called the *logarithmic conductor potential* (or *equilibrium potential*) for $\partial \mathscr{D}$. The *constant value* that this potential assumes on $\partial \mathscr{D}$ is equal to the Robin constant $\gamma_{\mathscr{D}}$.

In physics, the *capacity* of a conductor is the *quantity of electricity* which must be placed on it *in order to raise its (equilibrium) electrostatic potential to unity*. That potential is there taken as $\log L \, + \, \gamma_{\mathscr{D}}$ instead of $\gamma_{\mathscr{D}}$ when dealing with a long cylinder of length L bearing one unit of electric charge per unit of length (see footnote); in this way one arrives at a value $L/(\log L \, + \, \gamma_{\mathscr{D}})$ for the capacity of the cylinder. Even after division by L

---

* measured from a certain reference value depending on the cylinder's length and net electric charge, but *not* on its cross-sectional form

(in order to obtain a *capacity per unit length*), this quantity shows practically *no dependence* on $\gamma_{\mathscr{D}}$, i.e., *on the cylinder's cross section*, because the length L is assumed to be very large. That is why mathematicians have agreed on a different specification of logarithmic capacity.

Definition. The *logarithmic capacity*, Cap $\partial\mathscr{D}$, of the compact boundary $\partial\mathscr{D}$ is equal to

$$e^{-\gamma_{\mathscr{D}}},$$

where $\gamma_{\mathscr{D}}$ is the Robin constant for $\mathscr{D}$.

The logarithmic conductor potential and measure $\omega_{\mathscr{D}}(\ ,\ \infty)$ corresponding to it are characterized by an important extremal property. From physics, we expect that the *equilibrium charge distribution* $\omega_{\mathscr{D}}(\ ,\ \infty)$ on $\partial\mathscr{D}$ should be the positive measure μ of total mass 1 carried thereon for which the *energy* (cf. Chapter VIII, §B.5)

$$\int_{\partial\mathscr{D}}\int_{\partial\mathscr{D}}\log\frac{1}{|z-\zeta|}\,d\mu(\zeta)\,d\mu(z)$$

is as small as possible. That is true.

Lemma (goes back to Gauss). *If μ is a positive measure on $\partial\mathscr{D}$ with* $\mu(\partial\mathscr{D}) = 1$,

$$\int_{\partial\mathscr{D}}\int_{\partial\mathscr{D}}\log\frac{1}{|z-\zeta|}\,d\mu(\zeta)\,d\mu(z)\ \geqslant\ \gamma_{\mathscr{D}},$$

and equality is realized for $\mu = \omega_{\mathscr{D}}(\ ,\ \infty)$.

Remark. It is not too hard to show that equality holds *only* for $\mu = \omega_{\mathscr{D}}(\ ,\ \infty)$. We will not require that fact.

Proof of lemma. Since

$$\int_{\partial\mathscr{D}}\log\frac{1}{|z-\zeta|}\,d\omega_{\mathscr{D}}(\zeta,\ \infty)\ =\ \gamma_{\mathscr{D}},\qquad z\in\partial\mathscr{D},$$

we have

$$\int_{\partial\mathscr{D}}\int_{\partial\mathscr{D}}\log\frac{1}{|z-\zeta|}\,d\omega_{\mathscr{D}}(\zeta,\infty)\{d\mu(z)-d\omega_{\mathscr{D}}(z,\infty)\}\ =\ 0$$

for any measure μ on $\partial\mathscr{D}$ with $\mu(\partial\mathscr{D}) = \omega_{\mathscr{D}}(\partial\mathscr{D}, \infty) = 1$. Now*

$$\int_{\partial\mathscr{D}}\int_{\partial\mathscr{D}}\log\frac{1}{|z-\zeta|}\,d\mu(\zeta)\,d\mu(z) \quad - \quad \gamma_{\mathscr{D}}$$

$$= \int_{\partial\mathscr{D}}\int_{\partial\mathscr{D}}\log\frac{1}{|z-\zeta|}\big\{d\mu(\zeta)\,d\mu(z) - d\omega_{\mathscr{D}}(\zeta, \infty)\,d\omega_{\mathscr{D}}(z, \infty)\big\}$$

$$= \int_{\partial\mathscr{D}}\int_{\partial\mathscr{D}}\log\frac{1}{|z-\zeta|}\big(d\mu(\zeta) + d\omega_{\mathscr{D}}(\zeta, \infty)\big)\big(d\mu(z) - d\omega_{\mathscr{D}}(z, \infty)\big)$$

$$= \int_{\partial\mathscr{D}}\int_{\partial\mathscr{D}}\log\frac{1}{|z-\zeta|}\big(d\mu(\zeta) - d\omega_{\mathscr{D}}(\zeta, \infty)\big)\big(d\mu(z) - d\omega_{\mathscr{D}}(z, \infty)\big)$$

$$+ 2\int_{\partial\mathscr{D}}\int_{\partial\mathscr{D}}\log\frac{1}{|z-\zeta|}\,d\omega_{\mathscr{D}}(\zeta, \infty)\big\{d\mu(z) - d\omega_{\mathscr{D}}(z, \infty)\big\}.$$

In the last expression, the *second* double integral is *zero*, as we have just seen. However, $\int_{\partial\mathscr{D}}(d\mu(\zeta) - d\omega_{\mathscr{D}}(\zeta, \infty)) = 0$, so the *first* double integral in the last expression is *positive*, according to the *scholium and warning* just past the middle of §B.5, Chapter VIII. (The argument alluded to there works at least for sufficiently *smooth* signed measures of total mass zero supported on compact sets. The positivity thus established can be extended to our present signed measures $\mu - \omega_{\mathscr{D}}(\ ,\ \infty)$, with μ *not* necessarily *smooth*, by an appropriate limiting argument (regularization).) We see in this way that the *first* member in the above chain of equalities is *positive*. That's what we had to prove.

The following exercise gives us an alternative procedure for verifying that

$$\int_{\partial\mathscr{D}}\int_{\partial\mathscr{D}}\log\frac{1}{|z-\zeta|}\big(d\mu(\zeta) - d\omega_{\mathscr{D}}(\zeta, \infty)\big)\big(d\mu(z) - d\omega_{\mathscr{D}}(z, \infty)\big) \quad \geqslant \quad 0$$

when $\mu(\partial\mathscr{D}) = 1$.

Problem 35. (Ahlfors)
(a) Show that for large R,

$$\int\int_{|z|<R}\frac{dx\,dy}{|z-\zeta||z-w|} = 2\pi\log\frac{1}{|\zeta-w|} + 2\pi\log R + C + \delta(\zeta, w, R),$$

where C is a certain numerical constant (its value will not be needed) and $\delta(\zeta, w, R) \longrightarrow 0$ *uniformly* for ζ and w ranging over any compact set in $\mathbb{C}$, as $R \longrightarrow \infty$. (Hint: Wlog, $w - \zeta = a > 0$. Use the polar coordinates $re^{i\vartheta} = z - \zeta$ and, in the double integral thus obtained, *integrate r first*.)

---

* note that $\int_{\partial\mathscr{D}}\int_{\partial\mathscr{D}}\log(1/|z-\zeta|)d\omega_{\mathscr{D}}(\zeta, \infty)d\mu(z) = \gamma_{\mathscr{D}}$, a *finite* quantity!

(b) Hence show that

$$\int_{\partial\mathscr{D}}\int_{\partial\mathscr{D}} \log\frac{1}{|\zeta - w|}(d\mu(\zeta) - d\omega_{\mathscr{D}}(\zeta,\infty))(d\mu(w) - d\omega_{\mathscr{D}}(w,\infty)) \;\geqslant\; 0$$

for any positive measure μ on $\partial\mathscr{D}$ having total mass 1 (the double integral may be infinite). Pay attention to the problems of convergence and of possibly getting $\infty - \infty$. (Note that under our assumptions on $\partial\mathscr{D}$, $\gamma_{\mathscr{D}}$ is certainly *finite*.)

2   A conformal mapping. Pfluger's theorem

Let us fix any finite union E of closed arcs on $\{|z| = 1\}$ and a simple closed curve Γ about 0 lying in the unit disk Δ. The unit circumference and Γ bound a certain ring domain Δ_Γ lying in Δ, and we denote by G *the family of curves λ lying in Δ_Γ and going from the set E to Γ*:

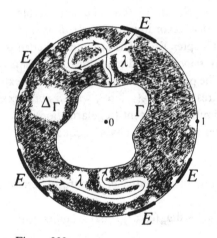

Figure 208

In this article we will be mainly concerned with the *reciprocal extremal length*

$$\Lambda(\Delta_\Gamma, G)$$

(see §E.1 for the definition and elementary properties of reciprocal extremal length).

Pfluger found a *relation* between $\Lambda(\Delta_\Gamma, G)$ and the *logarithmic capacity* of E defined in the preceding article. That relation also involves the curve Γ of course, but in fairly straightforward fashion. To arrive at Pfluger's result, we construct a *conformal mapping* of Δ onto a certain *disk with radial slits*.

Let $\mathscr{D} = (\mathbb{C} \cup \{\infty\}) \sim E$, so that $\partial\mathscr{D} = E$. With the harmonic measure $\omega_{\mathscr{D}}(\ ,z)$ for this domain, we form the *conductor potential* for E,

$$U_E(z) = \int_E \log\frac{1}{|z-\zeta|}\,d\omega_{\mathscr{D}}(\zeta,\ \infty),$$

described in the last article. Our conformal mapping will be constructed from the function

$$V_E(z) = U_E(z) + U_E\!\left(\frac{1}{\bar{z}}\right).$$

We have, first of all,

$$V_E(z) = V_E(1/\bar{z}).$$

Explicitly, since $E \subseteq \{|\zeta| = 1\}$,

$$V_E(z) = \int_E \log\left|\frac{z}{(z-\zeta)(1-\bar{\zeta}z)}\right|\,d\omega_{\mathscr{D}}(\zeta,\ \infty)$$

$$= \log|z| + 2\int_E \log\frac{1}{|z-\zeta|}\,d\omega_{\mathscr{D}}(\zeta,\ \infty),$$

from which we see in particular that

$$V_E(z) \longrightarrow -\infty \quad \text{for } z \longrightarrow 0$$

and hence also, that

$$V_E(z) \longrightarrow -\infty, \quad z \longrightarrow \infty.$$

Since $V_E(z)$ is *harmonic* in $\mathbb{C} \sim E \sim \{0\}$ and $V_E(z) = 2\gamma_{\mathscr{D}}$ for $z \in E$, the previous relations and the maximum principle imply that

$$V_E(z) < 2\gamma_{\mathscr{D}} \quad \text{for } z \notin E.$$

The function $V_E(z)$ has a *harmonic conjugate* $\tilde{V}_E(z)$ in $\{0 < |z| < 1\}$. $\tilde{V}_E(z)$ is of course *multiple valued* there; we proceed to investigate its behaviour. For $0 < r < 1$, we have

$$V_E(re^{i\vartheta}) = \int_E \log\!\left(\frac{r}{1+r^2 - 2r\cos(\vartheta-\tau)}\right)\,d\omega_{\mathscr{D}}(e^{i\tau},\ \infty),$$

from which

$$\frac{\partial V_E(re^{i\vartheta})}{\partial r} = \frac{1}{r}\int_E \frac{1-r^2}{1+r^2 - 2r\cos(\vartheta-\tau)}\,d\omega_{\mathscr{D}}(e^{i\tau},\ \infty),$$

whence, by the Cauchy–Riemann equations,

$$\frac{\partial \tilde{V}_E(re^{i\vartheta})}{\partial \vartheta} \;=\; \int_E \frac{1-r^2}{1+r^2-2r\cos(\vartheta-\tau)}\,\mathrm{d}\omega_{\mathscr{D}}(e^{i\tau},\,\infty).$$

From this we see, *firstly,* that $\tilde{V}_E(re^{i\vartheta})$ is a *strictly increasing* function of ϑ for each fixed r, $\;0<r<1$, and, *secondly,* that $V_E(re^{i\vartheta})$ *increases by*

$$\int_0^{2\pi}\!\!\int_E \frac{1-r^2}{1+r^2-2r\cos(\vartheta-\tau)}\,\mathrm{d}\omega_{\mathscr{D}}(e^{i\tau},\,\infty)\,\mathrm{d}\vartheta$$

$$=\;\; 2\pi\int_E \mathrm{d}\omega_{\mathscr{D}}(e^{i\tau},\,\infty)\;\;=\;\;2\pi$$

when ϑ goes from 0 to 2π. The determinations of $\tilde{V}_E(z)$ are, however, well defined and *single valued* in the simply connected region

$$\{z:\;\;0<\arg z<2\pi,\;0<|z|<1\}.$$

It therefore follows from the calculation just made that

$$\tilde{V}_E(z)\;-\;\arg z$$

is single valued (and harmonic) in $\{0<|z|<1\}$, and hence that

$$f_E(z)\;=\;\exp(V_E(z)+i\tilde{V}_E(z))\;=\;z\exp(V_E(z)+i\tilde{V}_E(z)-\log z)$$

is *single valued and analytic* there. Since $V_E(z)\;=\;\log|z|+O(1)$ near 0, $f_E(z)$ is *bounded* in a punctured neighbourhood of that point and thus analytic at 0 also (with $f_E(0)\;=\;0$). The function $f_E(z)$ is therefore *analytic* in Δ. We are going to show that it maps Δ conformally onto a disk with radial slits.

As we have seen,

$$V_E(e^{i\vartheta})\;=\;2\gamma_{\mathscr{D}}\quad\text{for }e^{i\vartheta}\in E$$

while

$$V_E(e^{i\vartheta})\;<\;2\gamma_{\mathscr{D}}\quad\text{when }e^{i\vartheta}\notin E.$$

The boundary value

$$\tilde{V}_E(e^{i\vartheta})\;=\;\lim_{r\to1-}\tilde{V}_E(re^{i\vartheta})$$

must, on the other hand, be an *increasing* function of ϑ like each of the $\tilde{V}_E(re^{i\vartheta})$, augmenting by 2π when ϑ does. In fact, $\tilde{V}_E(e^{i\vartheta})$ increases *only* on the arcs making up E; on *each* of the *complementary* arcs of $\{|z|=1\}$ it is *constant*. To see this, it suffices to observe that

$$V_E(z) = U_E(z) + U_E(1/\bar{z})$$

is *harmonic*, hence *infinitely differentiable*, along each of those complementary arcs, on which, by *symmetry*,

$$\frac{\partial V_E(z)}{\partial r} = 0.$$

Thence, by the Cauchy–Riemann equations,

$$\frac{\partial \tilde{V}_E(e^{i\vartheta})}{\partial \vartheta} = 0 \quad \text{for } e^{i\vartheta} \notin E.$$

We see that when ϑ increases from 0 to 2π, $f_E(e^{i\vartheta})$ describes a closed path of the following form, going through it *exactly once*:

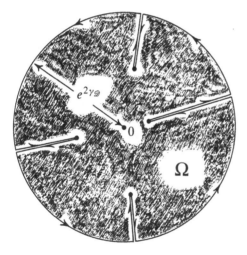

Figure 209

The arcs of E are taken onto those on the circle $\{|w| = e^{2\gamma_{\mathscr{D}}}\}$, and the ones *complementary* to E go onto the *radial slits*. Each of the *latter*, by the way, *really is* traversed *exactly once in each direction* (inward and then outwards) as $e^{i\vartheta}$ runs through the complementary arc corresponding to it. This is due to the *convexity* of $V_E(e^{i\vartheta})$ in ϑ on each of those complementary arcs, which may be verified by noting that

$$V_E(e^{i\vartheta}) = \int_E \log \left(\frac{1}{4\sin^2 \dfrac{(\vartheta - \tau)}{2}} \right) d\omega_{\mathscr{D}}(e^{i\tau}, \infty)$$

for $e^{i\vartheta} \notin E$, whence

$$\frac{\partial^2 V_E(e^{i\vartheta})}{\partial \vartheta^2} = \int_E \frac{1}{2 \sin^2\left(\dfrac{\vartheta - \tau}{2}\right)} d\omega_{\mathscr{D}}(e^{i\tau},\, \infty) > 0$$

for such ϑ.

Let Ω denote the (bounded) region enclosed by the path traced out by $f_E(e^{i\vartheta})$ as ϑ goes from 0 to 2π. Then, if $w \in \Omega$, $\arg(f_E(e^{i\vartheta}) - w)$ *increases by exactly 2π when ϑ does, and $f_E(z)$ must assume the value w exactly once in Δ, by the principle of argument.* We see in the same way that *none* of the values w *outside of $\bar{\Omega}$ are assumed by $f_E(z)$ in Δ; f_E thus maps Δ conformally onto Ω.*

It is good to have a more compact formula for $f_E(z)$ at our disposal. Since

$$V_E(z) = \log|z| + 2\int_E \log\frac{1}{|1 - \bar{\zeta}z|} d\omega_{\mathscr{D}}(\zeta,\, \infty),$$

we may take the harmonic conjugate $\tilde{V}_E(z)$ equal to

$$\arg z + 2\int_E \arg\left(\frac{1}{1 - \bar{\zeta}z}\right) d\omega_{\mathscr{D}}(\zeta,\, \infty)$$

for $0 < |z| < 1$, and hence

$$f_E(z) = z \exp\left\{2\int_E \log\left(\frac{1}{1 - \bar{\zeta}z}\right) d\omega_{\mathscr{D}}(\zeta,\, \infty)\right\}$$

for $z \in \Delta$.

These results are important enough to be summarized in the following

Theorem. *Let E be a finite union of arcs on $\{|z| = 1\}$, and denote by $\omega_{\mathscr{D}}(\;,\, z)$ the harmonic measure for*

$$\mathscr{D} = (\mathbb{C} \cup \{\infty\}) \sim E.$$

The function

$$f_E(z) = z \exp\left\{2\int_E \log\left(\frac{1}{1 - \bar{\zeta}z}\right) d\omega_{\mathscr{D}}(\zeta,\, \infty)\right\}$$

maps the unit disk Δ conformally onto a domain Ω obtained by removing certain radial sits from the disk $\{|w| < e^{2\gamma_{\mathscr{D}}}\}$. Under the mapping, the arcs of E go onto others, lying on the circumference $\{|w| = e^{2\gamma_{\mathscr{D}}}\}$ and precisely covering it, *and the* complementary arcs *of the unit circumference go onto the* radial slits.

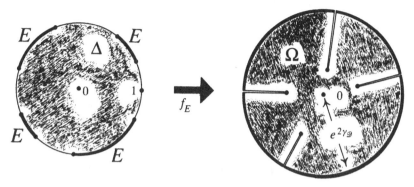

Figure 210

Once the conformal mapping f_E is available, it is easy to obtain estimates for the reciprocal extremal length $\Lambda(\Delta_\Gamma, G)$ specified at the beginning of the present article. To do that, we need two lemmas.

Lemma. $e^{\gamma_{\mathscr{D}}} \geqslant 1$.

Proof. Uses the minimum property of the conductor potential. Denoting the *unit circumference* by K, let us write

$$\mathscr{E} = \{|z| > 1\} \cup \{\infty\},$$

so that $\partial \mathscr{E} = K$. If $E \subseteq K$, any positive measure μ of total mass 1 supported on E is *certainly* supported on K, so, by the extremal property in question (lemma at the end of the preceding article),

$$\gamma_{\mathscr{E}} \leqslant \int_E \int_E \log \frac{1}{|z - \zeta|} \, d\mu(\zeta) d\mu(z)$$

Choosing $\mu = \omega_{\mathscr{D}}(\ , \infty)$, we get

$$\gamma_{\mathscr{E}} \leqslant \gamma_{\mathscr{D}}.$$

Referring to the formula

$$\frac{1}{2\pi} \int_0^{2\pi} \log \frac{1}{|z - e^{i\vartheta}|} \, d\vartheta = \log \frac{1}{|z|}, \qquad |z| \geqslant 1,$$

we see that $\gamma_{\mathscr{E}} = 0$. Therefore $\gamma_{\mathscr{D}} \geqslant 0$, as required.

Lemma. *For* $|z| < 1$,

$$\frac{|z|}{(1+|z|)^2} \;\leqslant\; |f_E(z)| \;\leqslant\; \frac{|z|}{(1-|z|)^2}.$$

Proof. By the boxed formula from the theorem,

$$\log\left|\frac{f_E(z)}{z}\right| \;=\; 2\int_E \log\frac{1}{|1-\bar{\zeta}z|}\,\mathrm{d}\omega_{\mathscr{D}}(\zeta,\,\infty).$$

The integral on the right evidently lies between $2\log(1/(1+|z|))$ and $2\log(1/(1-|z|))$. The lemma follows.

Consider now any simple closed curve Γ about 0 lying in Δ, and write

$$M_\Gamma \;=\; \sup_{z\in\Gamma}\frac{|z|}{(1-|z|)^2},$$

$$m_\Gamma \;=\; \inf_{z\in\Gamma}\frac{|z|}{(1+|z|)^2}.$$

For a finite union E of arcs on the unit circumference, $\mathrm{Cap}\,E \;=\; \mathrm{e}^{-\gamma_{\mathscr{D}}}$ and the quantity $\Lambda(\Delta_\Gamma, G)$ are then related through

Pfluger's Theorem. *If* $|z| \;<\; \frac{1}{2}(3 - \sqrt{5})$ *for* $z\in\Gamma$ *(so as to make* $M_\Gamma \;<\; 1$*), the logarithmic capacity of E satisfies the double inequality*

$$M_\Gamma^{-1/2}\,\mathrm{e}^{-\pi/\Lambda(\Delta_\Gamma,G)} \;\leqslant\; \mathrm{Cap}\,E \;\leqslant\; m_\Gamma^{-1/2}\,\mathrm{e}^{-\pi/\Lambda(\Delta_\Gamma,G)}$$

Proof. The conformal mapping f_E described in the preceding theorem takes the ring domain Δ_Γ bounded by the unit circumference and Γ onto another, $\Omega_{\tilde{\Gamma}}$, bounded by $\partial\Omega$ and the curve $\tilde{\Gamma} \;=\; f_E(\Gamma)$:

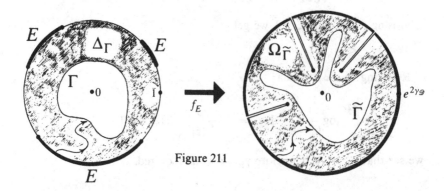

Figure 211

Under this mapping the curves of the family G – those lying in Δ_Γ and joining E to Γ – are taken to the ones lying in $\Omega_{\tilde{\Gamma}}$ which join *the circumference* $\{|w| = e^{2\gamma_{\mathscr{D}}}\}$ to $\tilde{\Gamma}$. Denoting the family of the latter curves by $\tilde{G}$, we have

$$\Lambda(\Delta_\Gamma, G) = \Lambda(\Omega_{\tilde{\Gamma}}, \tilde{G})$$

on account of property 2 of extremal length (§E.1).

By the *second* of the preceding two lemmas, $\tilde{\Gamma}$ lies inside the circle of radius M_Γ about 0, and in the present circumstances, $M_\Gamma < 1 \leqslant e^{2\gamma_{\mathscr{D}}}$ thanks to the *first* of those lemmas. The picture is thus as follows:

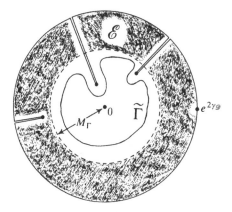

Figure 212

For the moment, let us denote the ring

$$\{M_\Gamma < |w| < e^{2\gamma_{\mathscr{D}}}\}$$

by $\mathscr{E}$, and the *family of curves* in $\mathscr{E}$ joining its *inner* to its *outer* boundary by H. Then, if $p(z) \geqslant 0$ is any weight on $\mathscr{E}$ *admissible for the family* H (in the parlance of §E.1), the weight $p^*(z)$ on $\Omega_{\tilde{\Gamma}}$ equal to $p(z)$ on $\Omega_{\tilde{\Gamma}} \cap \mathscr{E}$ and to *zero* on $\Omega_{\tilde{\Gamma}} \cap \sim \mathscr{E}$ *is certainly admissible* for $\tilde{G}$, so, by definition (§E.1),

$$\Lambda(\Omega_{\tilde{\Gamma}}, \tilde{G}) \leqslant \iint_{\Omega_{\tilde{\Gamma}}} (p^*(z))^2 \, dx \, dy = \iint_{\mathscr{E}} (p(z))^2 \, dx \, dy$$

(the presence of radial slits in Ω makes no difference here). Since the infimum of the right hand integral for the weights p in question is just $\Lambda(\mathscr{E}, H)$, we have $\Lambda(\Omega_{\tilde{\Gamma}}, \tilde{G}) \leqslant \Lambda(\mathscr{E}, H)$; that is, in view of the previous relation,

$$\Lambda(\Delta_\Gamma, G) \leqslant \Lambda(\mathscr{E}, H).$$

The right side of this inequality can easily be calculated explicitly by a procedure much like that of the special computation for a rectangle in §E.1, *reducible*, in fact, to the latter by logarithmic substitution. In that way one finds without difficulty that

$$\Lambda(\mathcal{E}, H) = \frac{2\pi}{\log(e^{2\gamma_\mathcal{D}}/M_\Gamma)}.$$

Plugging this into the preceding relation we get

$$\Lambda(\Delta_\Gamma, G) \leqslant \frac{\pi}{\gamma_\mathcal{D} - (\log M_\Gamma/2)}.$$

The second of the above two lemmas also implies that the circle of radius m_Γ about 0 lies entirely *inside* the curve $\tilde{\Gamma}$. From this we see by an argument like the one just made that

$$\frac{\pi}{\gamma_\mathcal{D} - (\log m_\Gamma/2)} \leqslant \Lambda(\Delta_\Gamma, G)$$

(again the radial slits of Ω cause no trouble).

Combining the last inequalities, we find that

$$\gamma_\mathcal{D} - \tfrac{1}{2}\log M_\Gamma \leqslant \frac{\pi}{\Lambda(\Delta_\Gamma, G)} \leqslant \gamma_\mathcal{D} - \tfrac{1}{2}\log m_\Gamma,$$

or, since $\operatorname{Cap} E = e^{-\gamma_\mathcal{D}}$,

$$M_\Gamma^{-1/2} e^{-\pi/\Lambda(\Delta_\Gamma, G)} \leqslant \operatorname{Cap} E \leqslant m_\Gamma^{-1/2} e^{-\pi/\Lambda(\Delta_\Gamma, G)}. \qquad \text{Q.E.D.}$$

3.   **Application to the estimation of harmonic measure.**
      Tsuji's inequality

The use of Pfluger's theorem in estimating harmonic measure is made possible by the fact that $\operatorname{Cap} E$ is a majorant for $|E|$ when E is a closed subset of the unit circumference. We restrict our attention to *finite unions of closed arcs*; results valid for such sets are general enough for most purposes.*

A sharp (i.e., *best possible*) relation between $\operatorname{Cap} E$ and $|E|$ (for E on the unit circumference) is known; its derivation is set as problem 36, given further on. For us a less precise result, having, however, a more straightforward proof, suffices:

Lemma. *If E is a finite union of closed arcs on the unit circumference,*

$$|E| \leqslant 4\pi \operatorname{Cap} E.$$

---

* See below, at end of the proof of Tsuji's inequality.

Proof. Is based on the extremal property of the conductor potential.

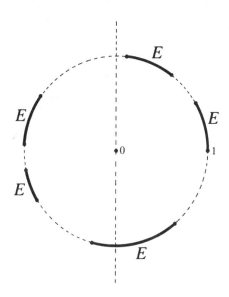

Figure 213

The part of E lying to *one of the two sides* of the imaginary axis has measure $\geqslant |E|/2$; suppose, wlog, that for E_+, the part lying to the *right* of that axis, we have

$$|E_+| \;\geqslant\; \tfrac{1}{2}|E|.$$

Arguing as in the proof of the *first* of the two lemmas in article 2, we see that

$$\operatorname{Cap} E_+ \;\leqslant\; \operatorname{Cap} E.$$

We have, however, $\operatorname{Cap} E_+ = e^{-\gamma}$, where, by the extremal property referred to (lemma, end of article 1), γ is the *minimum value* of the expressions

$$\int_{E_+}\int_{E_+} \log\frac{1}{|z-\zeta|}\,\mathrm{d}\mu(z)\mathrm{d}\mu(\zeta)$$

formed using positive measures μ on E_+ of total mass 1.

Suppose that E_+ consists of the arcs $I_1, I_2, \ldots, I_n$. By means of different rotations about the origin, we may *move* the I_j to *new arcs* I'_j on the *right half of the unit circumference*, arranged in the same order as the I_j but just touching each other, with $|I'_j| = |I_j|$ for $j = 1, 2, \ldots, n$.

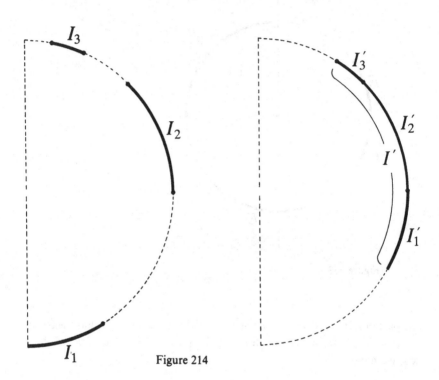

Figure 214

We denote $\bigcup_{j=1}^n I'_j$ by I'; I' is a *single arc* and $|I'| = |E_+|$.

A mapping ψ from E_+ to I' can now be defined in the following way: if $\zeta \in I_j$, we let $\psi(\zeta)$ be the point on I'_j *having the same position, relative to the endpoints of that arc, that ζ has, relative to the endpoints of I_j*. We then have

$$|\psi(z) - \psi(\zeta)| \leqslant |z - \zeta| \quad \text{for } z \text{ and } \zeta \in E_+.$$

Indeed, when z and ζ belong to the *same* arc I_j, there is *equality*, and, if z and ζ belong to *different* arcs, strict inequality, the effect of ψ being to move *each* arc of E_+ *closer to all the others* (since E_+ lies on a *semi-circumference*):

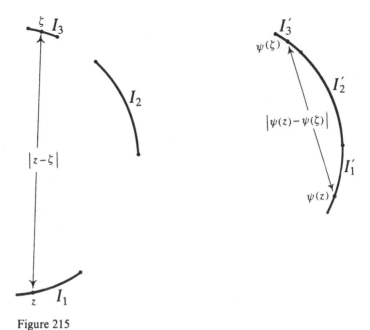

Figure 215

From the last inequality, we get

$$\int_{E_+}\int_{E_+} \log\frac{1}{|z-\zeta|}\,d\mu(\zeta)d\mu(z) \;\leqslant\; \int_{E_+}\int_{E_+} \log\frac{1}{|\psi(z)-\psi(\zeta)|}\,d\mu(\zeta)\,d\mu(z).$$

If μ_ψ is the measure on I' given by the formula

$$\mu_\psi(A) \;=\; \mu(\psi^{-1}(A)), \quad A \subseteq I',$$

the integral on the right can be rewritten

$$\int_{I'}\int_{I'} \log\frac{1}{|z'-\zeta'|}\,d\mu_\psi(z')d\mu_\psi(\zeta');$$

it is, moreover clear that *any* positive measure of total mass 1 on I' can be obtained as a μ_ψ for proper choice of the measure μ on E_+ with $\mu(E_+) = 1$. Choose, then, μ so as to make μ_ψ the *equilibrium charge distribution* for the arc I' (article 1). The integral just written is then equal to γ', the *conductor potential* for I', so, by the previous relation,

$$\gamma' \;\geqslant\; \int_{E_+}\int_{E_+} \log\frac{1}{|z-\zeta|}\,d\mu(\zeta)\,d\mu(z).$$

The expression on the right is, however, $\geqslant \gamma$ according to the observation made above. Thus.

$$\gamma' \geqslant \gamma,$$

or, in terms of $\operatorname{Cap} I' = e^{-\gamma'}$ and $\operatorname{Cap} E_+$,

$$\operatorname{Cap} I' \leqslant \operatorname{Cap} E_+.$$

The logarithmic capacity of the *single arc* I' may be expressed directly in terms of $|I'| = |E_+|$ after going through an elementary but somewhat tedious computation. We, however, do not require any great precision, and content ourselves with a simple lower bound on $\operatorname{Cap} I'$. Suppose, wlog, that I' is the (counterclockwise) arc from $e^{-i\alpha}$ to $e^{i\alpha}$, where $0 < \alpha \leqslant \pi/2$, so that $|I'| = 2\alpha$, and denote by φ the horizontal projection from I' onto the vertical line through the point 1:

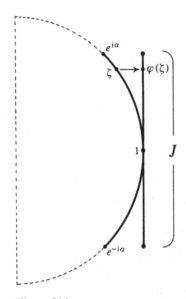

Figure 216

$\varphi(I')$ is thus the straight segment

$$J = [1 - i \sin \alpha, \ 1 + i \sin \alpha].$$

For z and $\zeta \in I'$, we clearly have

$$|\varphi(z) - \varphi(\zeta)| \leqslant |z - \zeta|,$$

whence, by an argument like the one made above,

$$\operatorname{Cap} J \leqslant \operatorname{Cap} I'.$$

The *left side* of this inequality is easily calculated. The Joukowski transformation

$$z \longrightarrow \frac{z-1}{i\sin\alpha} + \sqrt{\left(\left(\frac{z-1}{i\sin\alpha}\right)^2 - 1\right)} = w$$

takes the *exterior* of J conformally onto the domain $\{|w| > 1\}$, with ∞ going to ∞; so, if $\mathscr{D}$ denotes the domain

$$(\mathbb{C} \cup \{\infty\}) \sim J,$$

we have

$$G_{\mathscr{D}}(z, \infty) = \log|w| = \log\left|\frac{z-1}{\sin\alpha} + \sqrt{\left(\left(\frac{z-1}{\sin\alpha}\right)^2 + 1\right)}\right|.$$

For large values of $|z|$, the expression on the right reduces to

$$\log|z| + \log\left(\frac{2}{\sin\alpha}\right) + O\left(\frac{1}{|z|}\right),$$

whence, from article 1,

$$\gamma_{\mathscr{D}} = \lim_{z \to \infty} (G_{\mathscr{D}}(z, \infty) - \log|z|) = \log\left(\frac{2}{\sin\alpha}\right),$$

and $\operatorname{Cap} J = e^{-\gamma_{\mathscr{D}}} = \frac{1}{2}\sin\alpha$.

Since $0 < \alpha \leqslant \pi/2$, $\alpha \leqslant \frac{1}{2}\pi\sin\alpha$, so

$$|I'| = 2\alpha \leqslant 2\pi \operatorname{Cap} J$$

by the calculation just made and hence, in view of the relations established above,

$$|I'| \leqslant 2\pi \operatorname{Cap} I' \leqslant 2\pi \operatorname{Cap} E_+ \leqslant 2\pi \operatorname{Cap} E.$$

On the other hand,

$$|E| \leqslant 2|E_+| = 2|I'|,$$

so finally

$$|E| \leqslant 4\pi \operatorname{Cap} E.$$

We are done.

Remark. For sets E on the unit circumference, $4\pi \operatorname{Cap} E$ is in general a *poor upper bound* for $|E|$; it has the same order of magnitude as $|E|$ only when E is an arc or at least a set of fairly simple structure. As a rule, the comparison between $|E|$ and $\mathrm{Cap}\, E$ becomes *worse* as the set E under consideration becomes *more disconnected*. It is not hard to construct totally disconnected closed sets E on the unit circle for which $|E| = 0$ and yet $\mathrm{Cap}\, E > 0$. Descriptions of such constructions are found in many books, including the ones by Nevanlinna, Kahane & Salem, and Tsuji.

Problem 36

(a) Let E be compact and composed of a finite number of smooth arcs, and put $\mathscr{D} = (\mathbb{C} - E) \cup \{\infty\}$. If $f(z)$, analytic in $\mathscr{D}$, has modulus < 1 there and is zero at ∞, show that

$$\lim_{z \to \infty} |zf(z)| \leqslant \mathrm{Cap}\, E$$

(Hint: Look at $\log |f(z)| + G_{\mathscr{D}}(z, \infty)$.)

(b) Let E be a finite union of closed arcs on the unit circumference. Show that

$$\mathrm{Cap}\, E \geqslant \sin(|E|/4).$$

(Hint: With

$$\varphi(z) = \frac{1}{2\pi} \int_E \frac{e^{it} + z}{e^{it} - z}\, dt, \quad z \in \mathscr{D},$$

take

$$f(z) = \frac{e^{\pi i \varphi(z)/2} - e^{\pi i \varphi(0)/2}}{z(e^{\pi i \varphi(z)/2} + e^{-\pi i \varphi(0)/2})}$$

and use the result of part (a).)

Remarks. This proof of the inequality in (b) is from Lebedev's book on the area principle. The inequality becomes an equality for single arcs E. The supremum of the left-hand limits in (a) for the kind of functions f considered there is called the *analytic capacity* of E, and the f given in (b) actually realizes that supremum. For more about analytic capacity and its rôle in approximation theory the reader should consult the monographs by Garnett and by Zalcman.

Combination of the lemma just proved with Pfluger's theorem shows that if E is a finite union of arcs on the unit circumference and Γ a simple closed curve about 0 on which $|z| < \frac{1}{2}(3 - \sqrt{5})$, we have

$$\frac{|E|}{2\pi} \leqslant 2m_\Gamma^{-1/2} e^{-\pi/\Lambda(\Delta_\Gamma, G)}$$

where $m_\Gamma = \inf_{z \in \Gamma}(|z|/(1 + |z|)^2)$, and $\Lambda(\Delta_\Gamma, G)$ is related to E and Γ in the way described at the beginning of article 2. The use of this relation to estimate harmonic measure for simply connected domains comes immediately to mind on account of the conformal invariance of both harmonic measure and extremal length. Some control on the quantity m_Γ is of course needed if the results obtained are to have any practical value. We use a couple of the elementary properties of univalent functions for that; those are covered in many texts, for instance, the ones by Nehari and by Markushevich.

Lemma. *Let $\mathcal{O}$ be a simply connected domain with $z_0 \in \mathcal{O}$, and put*

$$R_0 = \text{dist}(z_0, \partial\mathcal{O}).$$

If σ is the circle $|z - z_0| = R_0/16$ and φ is a conformal mapping of $\mathcal{O}$ onto the unit disk Δ with $\varphi(z_0) = 0$, φ takes σ onto a closed curve Γ about 0, such that

$$|w| < \tfrac{1}{2}(3 - \sqrt{5}) \quad \text{for } w \in \Gamma$$

and that

$$|w|/(1 + |w|)^2 > 1/75, \quad w \in \Gamma.$$

Proof.

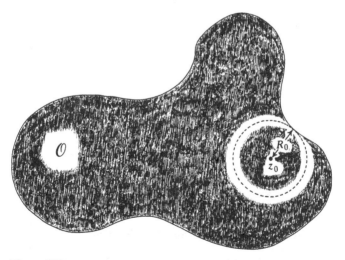

Figure 217

The function

$$f(\zeta) = \frac{\varphi(z_0 + R_0\zeta)}{R_0\varphi'(z_0)}$$

is certainly univalent for $|\zeta| < 1$, and has a Taylor expansion of the form

$$\zeta + A_2\zeta^2 + A_3\zeta^3 + \cdots$$

there. f also maps $\{|\zeta| < 1\}$ conformally onto a region *included* in the disk

$$\{|w| < 1/R_0|\varphi'(z_0)|\}.$$

Thence, by the Koebe 1/4-theorem, $1/R_0|\varphi'(z_0)| \geqslant 1/4$, or

$$R_0|\varphi'(z_0)| \leqslant 4.$$

Denoting by Φ the *inverse* to φ, we have, on the other hand,

$$\frac{\Phi(w) - z_0}{\Phi'(0)} = w + B_2w^2 + B_3w^3 + \cdots$$

for $|w| < 1$, with the left side *univalent* there. Here, $\Phi(0) = 0$ is distant by at least $R_0/|\Phi'(0)|$ units from the boundary of $\Phi(\Delta)$, so, by Koebe's 1/4-theorem, $R_0/|\Phi'(0)| \geqslant 1/4$, i.e.,

$$R_0|\varphi'(z_0)| \geqslant \frac{1}{4}.$$

According to the distortion theorem,

$$\frac{|\zeta|}{(1 + |\zeta|)^2} \leqslant |f(\zeta)| \leqslant \frac{|\zeta|}{(1 - |\zeta|)^2}$$

for $|\zeta| < 1$, so, in terms of $z = z_0 + R_0\zeta$ and φ,

$$\frac{R_0|\varphi'(z_0)||\zeta|}{(1 + |\zeta|)^2} \leqslant |\varphi(z)| \leqslant \frac{R_0|\varphi'(z_0)||\zeta|}{(1 - |\zeta|)^2}.$$

Hence, if $|z - z_0| = R_0/16$,

$$|\varphi(z)| \leqslant \frac{4/16}{(1 - 1/16)^2} < \frac{1}{3} < \frac{3 - \sqrt{5}}{2}$$

by the *first* of the above inequalities, in other words,

$$|w| < \frac{3 - \sqrt{5}}{2} \quad \text{for } w \in \Gamma = \varphi(\sigma).$$

When $|z - z_0| = R_0/16$, we also see by the second of the above inequalities and the relation involving $|\varphi(z)|$ and ζ, that

$$|\varphi(z)| \geqslant \frac{1/4 \cdot 1/16}{(1 + 1/16)^2} = \frac{4}{(17)^2} > \frac{1}{72}.$$

Thence, since $r/(1 + r)^2$ is increasing for $0 \leqslant r \leqslant 1$,

$$\frac{|w|}{(1 + |w|)^2} \; > \; \frac{1/72}{(73/72)^2} \; > \; \frac{1}{75} \;\; \text{for } w \in \Gamma.$$

The lemma is proved

Now we can give the

Theorem. *Let $\mathcal{O}$ be a simply connected domain bounded by a Jordan curve. Taking a $z_0 \in \mathcal{O}$, we denote by σ the circle of radius $\frac{1}{16}\operatorname{dist}(z_0, \partial\mathcal{O})$ about z_0, and by $\mathcal{O}_\sigma$ the ring domain bounded by $\partial\mathcal{O}$ and σ. If F is a finite union of closed arcs on $\partial\mathcal{O}$, let S be the family of curves in $\mathcal{O}_\sigma$ joining F to σ. Then*

$$\omega_{\mathcal{O}}(F, z_0) \; \leqslant \; 18 e^{-\pi/\Lambda(\mathcal{O}_\sigma, S)}.$$

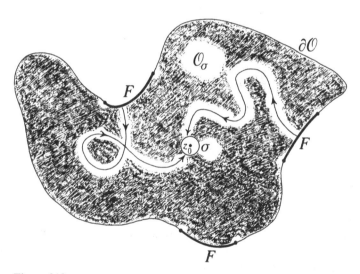

Figure 218

Proof. Let φ be a conformal mapping of $\mathcal{O}$ onto the unit disk Δ with $\varphi(z_0) = 0$; φ takes $F \subseteq \partial\mathcal{O}$ onto a finite union E of arcs on the unit circumference, and the circle σ to a simple closed curve Γ about 0, lying in Δ. The function φ thus maps the ring domain $\mathcal{O}_\sigma$ conformally onto Δ_Γ, the region bounded by Γ and the unit circumference, and takes the curves of the family S to those lying in Δ_Γ, joining E to Γ. Denoting, as usual, the collection of the latter curves by G, we have, by property 2 from article 1 (conformal invariance of extremal length),

$$\Lambda(\mathcal{O}_\sigma, S) = \Lambda(\Delta_\Gamma, G).$$

On the other hand,

$$\omega_{\mathcal{O}}(F, z_0) = \frac{|E|}{2\pi}.$$

According to the *second* of the preceding two lemmas, we can apply Pfluger's theorem with our curve Γ, so

$$\operatorname{Cap} E \leqslant m_\Gamma^{-1/2} e^{-\pi/\Lambda(\Delta_\Gamma, G)}$$

where

$$m_\Gamma = \inf_{w \in \Gamma} \frac{|w|}{(1+|w|)^2}.$$

By the same lemma, we have here,

$$m_\Gamma^{-1/2} \leqslant 9.$$

From the *first* of the above lemmas we now get

$$\frac{|E|}{2\pi} \leqslant 2 \operatorname{Cap} E \leqslant 18 e^{-\pi/\Lambda(\Delta_\Gamma, G)},$$

so, by the preceding two relations,

$$\omega_{\mathcal{O}}(F, z_0) \leqslant 18 e^{-\pi/\Lambda(\mathcal{O}_\sigma, S)}. \qquad\qquad \text{Q.E.D.}$$

Remark 1. In the theorem, one may replace $\Lambda(\mathcal{O}_\sigma, S)$ by $\Lambda(\mathcal{O}, S_\sigma)$ where S_σ is the family of *curves in $\mathcal{O}$* joining σ to F, for the two reciprocal extremal lengths are *equal*. Indeed, every curve in S_σ has on it an *arc* which, *by itself*, belongs to S. Therefore, if a weight $p(z)$ on $\mathcal{O}_\sigma$ is admissible for S, the weight $p_1(z)$ on $\mathcal{O}$ equal to $p(z)$ in $\mathcal{O}_\sigma$ and to *zero* inside the circle σ is admissible for S_σ, and hence $\Lambda(\mathcal{O}, S_\sigma) \leqslant \Lambda(\mathcal{O}_\sigma, S)$. On the other hand, $S \subseteq S_\sigma$, so if a weight $p_2(z)$ on $\mathcal{O}$ is admissible for S_σ, its *restriction*, $p(z)$, to $\mathcal{O}_\sigma$ must be admissible for S. This makes $\Lambda(\mathcal{O}_\sigma, S) \leqslant \Lambda(\mathcal{O}, S_\sigma)$.

Remark 2. As stated at the beginning of this §, Beurling's thesis already contained practically the same result.

Remark 3. Note that *here* we have the exponential

$$e^{-\pi/\Lambda(\mathcal{O}_\sigma, S)}$$

where the corresponding theorem of §E.1 (for *single* arcs) involves the expression

$$e^{-\pi/4\Lambda(\mathscr{D},G)}.$$

There is, however, *no real discrepancy* between the two results. The *extra factor of* 4 figuring in the second exponential is due to the fact that the curves of the family G involved there *start* from an arc on $\partial\mathscr{D}$ (the one whose harmonic measure at z_0 is in question), *loop around* z_0, and then *go back out* to that arc. The curves of our *present* family S *just go in* from the set F on $\partial\mathscr{O}$ to the circle σ about z_0; *they don't go back out* to F again.

Remark 4. The present result, unlike the corresponding one of §E.1, furnishes an inequality whose two sides are generally *not* of the same order of magnitude. According to the remark following the first of the above lemmas, the estimate given will usually tend to get *less and less precise* when applied to sets $F \subseteq \partial\mathscr{O}$ having *more and more components*. The *right* side of the inequality is *really* a measure (roughly speaking) of *logarithmic capacity* rather than of *harmonic measure*. Our result *does* nevertheless *have its uses*.

One application of the last theorem is to the derivation of a generalization, usually known as *Tsuji's inequality*, of the Ahlfors–Carleman estimate given at the end of §E.1. Suppose we have a simply connected domain $\mathscr{O}$ (bounded or not), with $\partial\mathscr{O}$ consisting of a Jordan curve, or of Jordan arcs going out to ∞. Let $z_0 \in \mathscr{O}$.

If $r > \text{dist}(z_0, \partial\mathscr{O})$, we denote by $\mathscr{O}_r$ the *component* of

$$\mathscr{O} \cap \{|z - z_0| < r\}$$

containing the point z_0. $\mathscr{O}_r$ is bounded by all or part of $\partial\mathscr{O}$, and, in the second case, *by certain arcs on the circle* $|z - z_0| = r$ as well. We call the union of these $\Sigma(r)$ (understanding that $\Sigma(r)$ may be *empty*), and write

$$\theta(r) = |\Sigma(r)|/r.$$

For $0 < r < \text{dist}(z_0, \partial\mathscr{O})$, $\Sigma(r)$ consists of the *entire circumference* $|z - z_0| = r$, and *then we put*

$$\theta(r) = \infty \ (sic!)$$

as in §E.1.

Figure 219

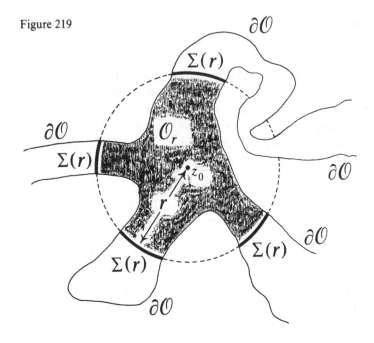

Fixing any $R > \text{dist}(z_0, \partial \mathcal{O})$, we look at harmonic measure $\omega_{\mathcal{O}_R}(\ , z)$ for the domain $\mathcal{O}_R$. Concerning the latter, we have the

Theorem. (Tsuji's inequality)

$$\omega_{\mathcal{O}_R}(\Sigma(R), z_0) \;\leqslant\; 18 \mathrm{e}^{-\pi \int_0^R \frac{dr}{r \, \theta(r)}}.$$

Proof. Consider first the case where $\Sigma(R)$ consists of a *finite* number of arcs which we may just as well look on as *closed*. Taking the circle σ of radius $\frac{1}{16}\text{dist}(z_0, \partial \mathcal{O})$ about z_0, we proceed to obtain an *upper bound* for the reciprocal extremal length $\Lambda(\mathcal{O}_\sigma, S)$, where S is the family of curves in $\mathcal{O}_\sigma$, the ring domain bounded by $\partial \mathcal{O}_R$ (*sic!*) and σ, joining $\Sigma(R)$ to σ. For this purpose, it is enough to exhibit a *single weight* $p(z) \geqslant 0$ on $\mathcal{O}_\sigma$, *admissible for the family* S. Then, by definition (§E.1), we'll have

$$\Lambda(\mathcal{O}_\sigma, S) \;\leqslant\; \iint_{\mathcal{O}_\sigma} (p(z))^2 \, dx \, dy.$$

Figure 220

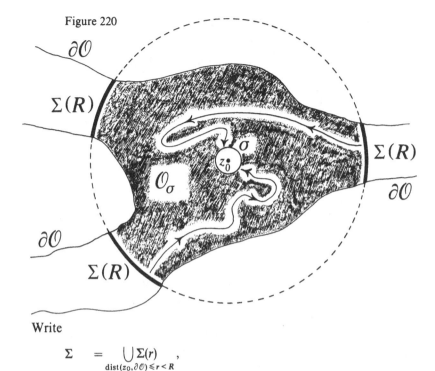

Write

$$\Sigma \;\;=\;\; \bigcup_{\mathrm{dist}(z_0,\partial\mathcal{O})\leqslant r<R}\Sigma(r) \;\;,$$

and, for $z \in \Sigma$, put

$$p(z) \;\;=\;\; \frac{k}{|z-z_0|\,\theta(|z-z_0|)},$$

where k is a constant yet to be determined. For $z \notin \Sigma$, we take $p(z) \;=\; 0$.

If λ is *any curve* in $\mathcal{O}_\sigma$ joining $\Sigma(R)$ to σ, we have (cf. proof of Ahlfors–Carleman inequality, end of §E.1),

$$\int_\lambda p(z)|\,\mathrm{d}z| \;\;\geqslant\;\; \int_\lambda p(z)|\,\mathrm{d}|z-z_0||.$$

Putting $\mathrm{dist}(z_0,\partial\mathcal{O}) \;=\; R_0$ and writing

$$z-z_0 \;=\; r\mathrm{e}^{\mathrm{i}\vartheta},$$

we see that the integral on the right is

$$\geqslant\;\; \int_{R_0}^{R} p(z)\,\mathrm{d}r \;\;=\;\; k\int_{R_0}^{R}\frac{\mathrm{d}r}{r\theta(r)}.$$

For $k \;=\; 1/\int_{R_0}^{R}\mathrm{d}r/r\theta(r)$, the last expression is equal to *unity, making $p(z)$ admissible for the family S of curves λ.*

Using the indicated value of k, we get

$$\iint_{\mathcal{O}_\sigma} (p(z))^2 \, dx \, dy \;=\; \iint_\Sigma (p(z))^2 \, dx \, dy \;=\; \int_{R_0}^R \int_{\Sigma(r)} \left(\frac{k}{r\theta(r)}\right)^2 r \, d\vartheta \, dr$$

$$=\; \int_{R_0}^R \left(\frac{k}{r\theta(r)}\right)^2 r\theta(r) \, dr \;=\; k^2 \int_{R_0}^R \frac{dr}{r\theta(r)} \;=\; k,$$

whence

$$\Lambda(\mathcal{O}_\sigma, S) \;\leqslant\; k \;=\; 1 \Big/ \int_0^R \frac{dr}{r\theta(r)}$$

(since $\theta(r) = \infty$ for $0 < r < R_0$).

Substitution of the last relation into the inequality furnished by the preceding theorem now yields the desired result when $\Sigma(R)$ is made up of finitely many arcs.

The general case (with $\Sigma(R) \subseteq \mathcal{O} \cap \{|z - z_0| = R\}$ consisting of *countably many* open arcs) is easily reduced to the one just handled, which certainly obtains whenever $\partial\mathcal{O}$ is an *analytic* Jordan curve. One simply takes limits, working with an exhaustion of $\mathcal{O}$ by domains having analytic Jordan curve boundaries. To get the latter, map $\mathcal{O}$ conformally onto the unit disk and then take the preimages of smaller concentric disks. The details of this procedure are left to the reader.

Remark. Tsuji himself did not quite arrive at the inequality found here. In its place he got

$$\omega_{\mathcal{O}_R}(\Sigma(R), \, z_0) \;\leqslant\; \frac{3}{\sqrt{(1 - \kappa)}} \, e^{-\pi \int_0^{\kappa R} \frac{dr}{r\theta(r)}},$$

with a parameter κ of arbitrary value < 1. The estimate provided by our result is better, because in it the integration goes all the way out to R.

Instead of using extremal length to derive his formula, Tsuji worked with a differential inequality due to Carleman. That procedure is known as *Carleman's method*, and its application is not limited to simply connected domains. Therefore, since our definition of the function $\theta(r)$ makes sense when $\mathcal{O}$ is *multiply connected* (i.e., has 'islands' in it), it is *very likely true* that Tsuji's original estimate (with the parameter κ) *holds* for that more general situation. Regarding this possibility and most of the other material in the present § and in §E.1, the reader should consult the *Arkiv* paper by K. Haliste; I have not checked thoroughly to make sure that all the details of Tsuji's argument go through for multiply connected domains.

It would be very good if the result proved above (*without* the parameter

κ) could be shown to hold when $\mathcal{O}$ is multiply connected.* In at least *one special situation*, this is known to be so. That is when $\Sigma(R)$ is the *whole circle* $|z - z_0| = R$ and the rest of $\partial\mathcal{O}_R$ consists of *radial segments inside that circle*; the desired inequality then follows by a celebrated theorem of Beurling which may, for instance, be found in Nevanlinna's book. For Carleman's method, the reader should first consult the little book by Heins, going afterwards to those of Nevanlinna and of Tsuji. Haliste's paper, already referred to, is recommended to anyone who wishes to become more familiar with the whole circle of ideas just discussed.

For certain geometric configurations, *Tsuji's inequality can be deduced from the one of Ahlfors and Carleman* proved in §E.1. Without attempting to describe the most general circumstances whereunder this is possible (which would oblige us to enter into all kinds of fussy geometric considerations), let us restrict our attention to the simple situation where $\mathcal{O}$ consists of *a disk together with n* (unbranched) *arms extending outward therefrom*:

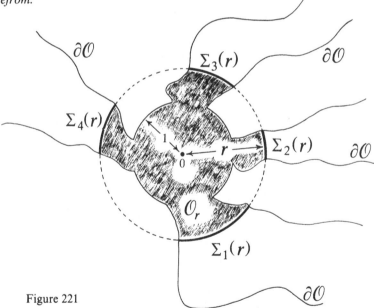

Figure 221

* Beurling's notes on extremal distance and harmonic measure have appeared in volume I of his *Collected Works*, published almost 5 years after the above lines were written. A version for *multiply connected* domains of the theorem given above on p. 149 is found in those notes on p. 372 (Theorem 3), and from it the analogue of Tsuji's inequality (without the κ) for branching channels with islands in them is readily derived (see pp 374–376 of the notes). Beurling's Theorem 3, and especially its elegant proof, were closely guarded secrets until his collected works came out.

Consider the case where $\mathcal{O}$'s central disk is just Δ. In order to keep things really simple, let us also suppose that any circle of radius $r > 1$ about the origin *intersects each of the arms along a single arc*, although this assumption is not really necessary (see the discussion of the arcs $S(R)$ and $\sigma(r)$ near the end of §E.1). *Say that* $\{|z| = r\}$ *intersects the jth arm of* $\mathcal{O}$ *along the arc* $\Sigma_j(r)$. In terms of the notation used with the previous theorem, we then have

$$\Sigma(r) = \Sigma_1(r) \cup \Sigma_2(r) \cup \cdots \cup \Sigma_n(r), \quad r > 1,$$

and if we put

$$\theta_j(r) = |\Sigma_j(r)|/r \quad \text{for} \quad r > 1,$$

we have

$$\theta(r) = \theta_1(r) + \theta_2(r) + \cdots + \theta_n(r).$$

When $0 < r < 1$, *we put each of the* $\theta_j(r)$ *equal to* ∞.

Under the present circumstances, we clearly have

$$\omega_{\mathcal{O}_R}(\Sigma(R), 0) = \sum_{j=1}^{n} \omega_{\mathcal{O}_R}(\Sigma_j(R), 0),$$

and, for each j, *by the Ahlfors–Carleman inequality*,

$$\omega_{\mathcal{O}_R}(\Sigma_j(R), 0) \leqslant C \exp\left\{-\pi \int_0^R \frac{dr}{r\theta_j(r)}\right\}$$

with an absolute constant C. Therefore, wishing to show that

$$\omega_{\mathcal{O}_R}(\Sigma(R), 0) \leqslant K \exp\left\{-\pi \int_0^R \frac{dr}{r\theta(r)}\right\}$$

with a numerical constant K, it is *more than sufficient* to verify that

$$\sum_{j=1}^{n} \exp\left\{-\pi \int_0^R \frac{dr}{r\theta_j(r)}\right\} \leqslant \exp\left\{-\pi \int_0^R \frac{dr}{r\theta(r)}\right\}$$

for $\theta(r) = \sum_{j=1}^n \theta_j(r)$, when the $\theta_j(r)$ are $\geqslant 0$ and

$$\pi \int_0^R \frac{dr}{r\theta(r)} \geqslant 2.$$

(When this last condition is *violated*, the desired inequality for $\omega_{\mathcal{O}_R}(\Sigma(R), 0)$ is *true anyway* with $K = e^2$.)

Problem 37

Prove that under the given condition the boxed inequality holds. (Hint. One may, in the first place, assume all the quantities $\int_0^R dr/r\theta_j(r)$ to be *finite*. Otherwise, if, say, the one with $j = n$ were *infinite*, we could *drop* the term corresponding to it in the sum

$$S(\theta_1, \theta_2, \ldots, \theta_n) = \sum_{j=1}^{n} \exp\left\{ -\pi \int_0^R \frac{dr}{r\theta_j(r)} \right\}$$

and then set out to prove an inequality like the *boxed* one with $n-1$ terms *on the left* and a *smaller* function $\theta(r) \geq 0$ on the *right*. The stated assumption now being *granted*, we *consider the function* $\theta(r)$ *to be fixed*. Let, wlog, the *largest among the quantities* $(\int_0^R dr/r\theta_j(r))^2 \exp\{ -\pi\int_0^R dr/r\theta_j(r)\}$ *be the one corresponding to* $j = 1$. Taking any $k > 1$, make the variations $\delta\theta_1(r) = \eta\theta_k(r)$, $\delta\theta_k(r) = -\eta\theta_k(r)$ and $\delta\theta_j(r) = 0$ for $j \neq 1$, k in the functions $\theta_j(r)$; here η is an infinitesimal > 0. These variations are *allowable* because the functions $\theta_j(r) + \delta\theta_j(r)$ are *still all* ≥ 0 and *still* add up to $\theta(r)$.

Show that, under the variations just described,

$$\delta S(\theta_1, \theta_2, \ldots, \theta_n) =$$
$$\pi\eta\left\{ \int_0^R \frac{\theta_k(r)\,dr}{r(\theta_1(r))^2} \exp\left(-\pi \int_0^R \frac{dr}{r\theta_1(r)} \right) \right.$$
$$\left. - \int_0^R \frac{dr}{r\theta_k(r)} \exp\left(-\pi \int_0^R \frac{dr}{r\theta_k(r)} \right) \right\},$$

and that the quantity in $\{\ \}$ *is certainly* > 0 *unless* $\theta_k(r) = \text{const}\,\theta_1(r)$. The *maximum value of* $S(\theta_1, \ldots, \theta_n)$ for $\theta_1 + \theta_2 + \cdots + \theta_n = \theta$ *is therefore attained when*, for each j, $\theta_j(r) = \lambda_j\theta(r)$ with a constant $\lambda_j \geq 0$. Here, $\lambda_1 + \lambda_2 + \cdots + \lambda_n = 1$. Observe finally that when $M \geq 2$, $e^{-M/\lambda}$ is an *increasing and convex function* of λ for $0 \leq \lambda \leq 1$.)

X

Why we want to have multiplier theorems

A.   Meaning of term 'multiplier theorem' in this book

Suppose we have a function $W(x) \geqslant 1$ defined on $\mathbb{R}$. We will be interested in the *question of existence* of non-zero *entire functions* $\varphi(z)$ *of arbitrarily small exponential type* for which

$$W(x)\varphi(x)$$

is *bounded*, or *belongs to some L_p class, on the real axis*. The purpose of the present chapter is to discuss some of the *reasons* for this interest.

Existence of the entire functions φ is of course not guaranteed for *arbitrary* weights $W(x) \geqslant 1$. For us, *a multiplier theorem is any result describing conditions on W from which that existence must follow*. When such conditions are *realized*, we think of W as a weight that can be 'multiplied down' by the 'multipliers' φ. In those circumstances, we also say that $W(x)$ *admits multipliers*.

▶ *Warning.* The term 'multiplier' is used *here* with *meaning entirely different* from that accepted in harmonic analysis and in the study of singular integrals.

The first restriction on a function $W(x) \geqslant 1$ which is to admit multipliers concerns its *size*. If $\varphi(z)$ is a non-zero entire function of exponential type with $W(x)|\varphi(x)| \leqslant C$ on $\mathbb{R}$, we have

$$\log|\varphi(x)| \leqslant \log C \ - \ \log W(x),$$

so, since

$$\int_{-\infty}^{\infty} \frac{\log^-|\varphi(x)|}{1+x^2}\,\mathrm{d}x$$

must be *finite* by §G.2 of Chapter III, we *have to have*

$$\int_{-\infty}^{\infty} \frac{\log W(x)}{1 + x^2}\,dx \;\; < \;\; \infty.$$

The same condition on W must hold if we require, for example, that $W(x)\varphi(x) \in L_p(\mathbb{R})$ with $p > 0$.

Problem 38
Prove the assertion just made. (Hint: Insertion of the factor $1/(1 + x^2)$ into the integrand makes the L_p integral smaller. Use the relation between arithmetic and geometric means.)

Things would be very simple if the boxed condition on W, *necessary* for that weight to admit multipliers, were *also sufficient*. That, unfortunately, is just not true, and *some additional restrictions on W's behaviour are needed.* A really adequate description of the minimal additional requirements to be imposed on a weight in order that it admit multipliers is not yet available; one has, on the one hand, some fairly straightforward *sufficient* conditions which are *more than* necessary, and, on the other, a criterion which is *both necessary and sufficient* for a very extensive class of weights, but at the same time *quite unwieldy.*

These matters will be taken up in the next chapter, the last one of this book. What we do in the present one is mainly to *show some applications* of multiplier theorems to various questions in analysis. In the following two articles we first review an elementary but quite useful such result already established in Chapter IV and then *state* a much deeper one, whose *proof* is deferred until Chapter XI.

1.   **The weight is even and increasing on the positive real axis**

As we saw in §D of Chapter IV (see especially the corollary at the end of that §), a construction used in the study of quasi-analyticity also yields the following

Theorem. (Paley and Wiener) *If $W(x) \geqslant 1$ is even, and increasing for $x > 0$, W admits multipliers if and only if*

$$\int_{-\infty}^{\infty} \frac{\log W(x)}{1 + x^2}\,dx \;\; < \;\; \infty.$$

Under the specified circumstances, then, convergence of the integral is both necessary and sufficient. This multiplier theorem is of considerable value in applications in spite of its elementary character. Levinson and Mandelbrojt have used it extensively, as did Paley and Wiener, and it will render considerable service in the construction of some important examples to be given in the next chapter. At that time, it will be helpful to refer to a *different derivation* of the result, independent of the special properties of the function $(\sin z)/z$. We proceed to give one now.

The basis for our argument here is the formula

$$\int_0^\infty \log\left|1 - \frac{x^2}{t^2}\right| dv(t) \;=\; -x \int_0^\infty \log\left|\frac{x+t}{x-t}\right| d\left(\frac{v(t)}{t}\right), \qquad x > 0,$$

valid for functions $v(t)$ positive and increasing on $[0, \infty)$ which are $O(t)$ for both $t \to 0$ and $t \longrightarrow \infty$. This is just the first lemma of §B.4 in Chapter VIII. Starting with a weight $W(x) \geqslant 1$ whose logarithmic integral *converges*, we take a suitable constant multiple $m(t)$ of $\log W(t)$ (the constant will be specified later) and then, *fixing* an arbitrary $a > 0$, put

$$v(t) \;=\; at \,-\, t\int_{\max(t,A)}^\infty \frac{m(\tau)}{\tau^2} d\tau, \qquad t \geqslant 0,$$

where A is a large number depending on a in a manner to be described immediately.

We have

$$v'(t) \;=\; a \,-\, \int_A^\infty \frac{m(\tau)}{\tau^2} d\tau, \qquad 0 < t < A,$$

and

$$v'(t) \;=\; a \,-\, \int_t^\infty \frac{m(\tau)}{\tau^2} d\tau \,+\, \frac{m(t)}{t}, \qquad t > A.$$

Therefore, if we choose A so as to make

$$\int_A^\infty \frac{m(\tau)}{\tau^2} d\tau \;<\; a,$$

which is certainly *possible* thanks to our assumption on W, $v(t)$ will be an *increasing function* of t, with $0 < v(t) < at$ for $t > 0$. It is also clear that

$$\frac{v(t)}{t} \longrightarrow a \qquad \text{for } t \longrightarrow \infty.$$

Using the function v in the above formula gives us

$$\int_0^\infty \log\left|1 - \frac{x^2}{t^2}\right| dv(t) = -x \int_A^\infty \log\left|\frac{x+t}{x-t}\right| \frac{m(t)}{t^2} dt.$$

Since, however, $m(t)$ is *increasing* according to our assumption on W, the *right side* of the last relation is

$$\leqslant -m(x) \int_x^\infty \log\left|\frac{x+t}{x-t}\right| \frac{x\,dt}{t^2}$$

for $x \geqslant A$. The substitution $\tau = t/x$ takes the integral figuring herein over to

$$\int_1^\infty \log\left|\frac{1+\tau}{1-\tau}\right| \frac{d\tau}{\tau^2},$$

a certain (finite) *numerical quantity* – call it C – whose exact value we do not need to know. Thus,

$$\int_0^\infty \log\left|1 - \frac{x^2}{t^2}\right| dv(t) \leqslant -Cm(x) \qquad \text{for } x \geqslant A.$$

Write, for $\Im z \geqslant 0$,

$$U(z) = \int_0^\infty \log\left|1 - \frac{z^2}{t^2}\right| dv(t).$$

$U(z)$ is *harmonic* in the upper half plane and, for our present function $v(t)$, *continuous up to the real axis*, where it is certainly *bounded above* in view of the previous relation. Moreover,

$$U(z) \leqslant \int_0^\infty \log\left(1 + \frac{|z|^2}{t^2}\right) dv(t),$$

and, after integrating by parts, the right side is easily seen to be $\sim \pi a|z|$ for $|z| \longrightarrow \infty$, keeping in mind that $v(t)/t \longrightarrow a$ as $t \longrightarrow \infty$. When $z = \mathrm{i}y$ with $y > 0$, the inequality just written becomes an *equality*, showing that $U(\mathrm{i}y)/y \longrightarrow \pi a$ for $y \longrightarrow \infty$.

These facts imply that

$$U(z) = \pi a \Im z + \frac{1}{\pi} \int_{-\infty}^\infty \frac{\Im z\, U(t)}{|z-t|^2} dt \qquad \text{for } \Im z > 0$$

by an argument exactly like the one used to prove the theorem of §G.1, Chapter III. Plugging the previous relation into the integral on the right then gives

$$U(x+i) \leqslant O(1) - \frac{1}{\pi} \int_{|t| \geqslant A} \frac{C m(|t|)\,dt}{(x-t)^2 + 1},$$

whence, since $m(|t|)$ increases with $|t|$,

$$U(x+i) \leqslant O(1) - \frac{C}{2} m(|x|) \quad \text{for } |x| \geqslant A.$$

Taking a larger $O(1)$ term of course ensures this estimate's validity for *all* real x.

The idea now is to observe that the integral

$$\int_0^\infty \log\left|1 - \frac{z^2}{t^2}\right| dv(t)$$

would represent the logarithm of the modulus of an entire function of exponential type, *if the increasing function $v(t)$ were integer-valued. Our $v(t)$*, of course, is *not* (it is absolutely continuous!), but one expects that

$$\int_0^\infty \log\left|1 - \frac{z^2}{t^2}\right| d[v(t)]$$

(with $[v(t)]$ designating the *greatest integer* $\leqslant v(t)$) should be *close* to

$$U(z) = \int_0^\infty \log\left|1 - \frac{z^2}{t^2}\right| dv(t).$$

This is indeed *true as soon as z gets away from the real axis*, and we have the following simple

Lemma. *If $v(t)$ is increasing and $O(t)$ on the positive real axis,*

$$\int_0^\infty \log\left|1 - \frac{z^2}{t^2}\right| (d[v(t)] - dv(t))$$

$$\leqslant \log\left\{\frac{\max(|x|, |y|)}{2|y|} + \frac{|y|}{2\max(|x|, |y|)}\right\}$$

for $\Im z = y \neq 0$.

Proof. Assuming that $\Im z \neq 0$, integrate the left-hand member by parts. Because $v(t) = O(t)$, the integrated term vanishes, and we obtain

$$\int_0^\infty (v(t) - [v(t)]) \frac{\partial}{\partial t} \log\left|1 - \frac{z^2}{t^2}\right| dt.$$

Fixing z, let us introduce the new variable $\zeta = z^2/t^2$. As t runs through $(0, \infty)$, ζ *moves in* along a certain *ray* $\mathscr{L}$ coming out from the origin. When $\Re z^2 \leqslant 0$, the distance $|1 - \zeta|$ *decreases* as t *increases*, so, since $v(t) - [v(t)] \geqslant 0$, the expression just written is $\leqslant 0$.

If, however, $\Re z^2 > 0$, $|1 - \zeta|$, for increasing t, *first decreases* to a *minimum value* $|\Im z^2|/|z|^2$ and *then increases*, *tending to* 1 *as* $t \to \infty$:

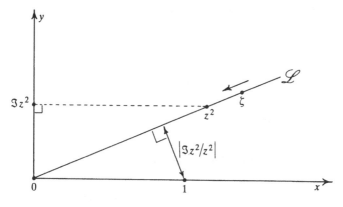

Figure 222

Hence, since $0 \leqslant v(t) - [v(t)] \leqslant 1$, the expression under consideration is

$$\leqslant \ \log(|z|^2/|\Im z^2|).$$

We see that $\int_0^\infty \log|1 - (z^2/t^2)|(\mathrm{d}[v(t)] - \mathrm{d}v(t))$ is

$$\leqslant \begin{cases} 0, & |x| \leqslant |y|, \\[2mm] \log\left(\dfrac{|x|}{2|y|} + \dfrac{|y|}{2|x|}\right), & |x| > |y|. \end{cases}$$

The right-hand side can be represented by the single expression

$$\log\left(\frac{\max(|x|, |y|)}{2|y|} + \frac{|y|}{2\max(|x|, |y|)}\right).$$

The lemma is proved

Using the lemma with our function $U(z)$, we get

$$\int_0^\infty \log\left|1 - \frac{(x + \mathrm{i})^2}{t^2}\right| \mathrm{d}[v(t)] \ \leqslant \ \log^+|x| + U(x + \mathrm{i}),$$

so, by the relation established above,

$$\int_0^\infty \log\left|1 - \frac{(x+i)^2}{t^2}\right| d[v(t)] \leq O(1) + \log^+|x| - \frac{C}{2}m(|x|), \quad x \in \mathbb{R}.$$

For each integer $k \geq 1$, denote by λ_k the positive value of t for which $v(t) = k$. Then, noting that $v(0) = 0$, we have

$$\int_0^\infty \log\left|1 - \frac{z^2}{t^2}\right| d[v(t)] = \log\left|\prod_{k=1}^\infty \left(1 - \frac{z^2}{\lambda_k^2}\right)\right|,$$

so, putting

$$\varphi(z) = \prod_{k=2}^\infty \left(1 - \frac{z^2}{\lambda_k^2}\right) \quad (sic!),$$

the preceding inequality yields

$$\log|\varphi(x+i)| \leq O(1) - \frac{C}{2}m(|x|), \quad x \in \mathbb{R}.$$

Arguing as we did above for $U(z)$, we find without trouble that

$$\log|\varphi(z)| \leq O(1) + \pi a|z|;$$

in other words, $\varphi(z)$ *is entire and of exponential type* $\leq \pi a$.

The relation involving $\varphi(x+i)$ and $m(|x|)$ can be rewritten

$$|\varphi(x+i)|e^{Cm(|x|)/2} \leq \text{const.}, \quad x \in \mathbb{R},$$

where C is a numerical constant independent of a and of $m(t)$. Going back to our even weight W, we now take

$$m(t) = \frac{2}{C}\log W(t), \quad t \geq 0,$$

and the entire function φ of exponential type $\leq \pi a$ furnished by the construction just made satisfies

$$W(x)|\varphi(x+i)| \leq \text{const.}, \quad x \in \mathbb{R}.$$

This φ is of course non-zero, so, since $a > 0$ is arbitrary, we have again arrived at the theorem stated at the beginning of the present article.

2.   Statement of the Beurling–Malliavin multiplier theorem

The result just discussed implies in particular that if an entire function $F(z)$ of exponential type has, on the real axis, a *majorant*

$M(x) \geqslant 1$, *increasing* when $x \geqslant 0$ and *decreasing* for $x < 0$, such that

$$\int_{-\infty}^{\infty} \frac{\log M(x)}{1 + x^2} \, dx \;\; < \;\; \infty,$$

then there are non-zero entire functions φ of arbitrarily small exponential type for which $F(x)\varphi(x)$ is bounded on $\mathbb{R}$. It suffices indeed to apply the theorem with the weight $W(x) \;=\; M(x)M(-x)$. In such circumstances, the function $F(z)$ is thus a *factor* of *other* non-zero entire functions, *having exponential type arbitrarily close to that of F* and *bounded* on $\mathbb{R}$.

It is very remarkable that *the monotoneity requirements on the majorant can be dispensed with here.* The *mere condition* that F be *entire and of exponential type* somehow implies *enough regularity* for the weight $1 + |F(x)|$ so that *convergence* of the *logarithmic integral* associated with the latter *already ensures its admitting of multipliers.*

Theorem (Beurling and Malliavin, 1961 – called the **theorem on the multiplier**). *Let $F(z)$ be entire and of exponential type. In order that the weight $|F(x)| + 1$ admit multipliers, it is necessary and sufficient that*

$$\int_{-\infty}^{\infty} \frac{\log^+ |F(x)|}{1 + x^2} \, dx \;\; < \;\; \infty.$$

This result is much deeper than the one of the preceding article, and there is so far no really simple way of arriving at it. A proof based on material from §C of Chapter VIII will be given in the next chapter. For the time being, the reader is only asked to take account of the theorem's *statement.* Some of its important consequences will be deduced in the following §§.

B.   Completeness of sets of exponentials on finite intervals

We return to the study of completeness of collections of functions $e^{i\lambda_n t}$, begun in §D of the last chapter. There, we obtained a *lower bound* for the *completeness radius* associated with an arbitrary *real* sequence Λ of distinct frequencies λ_n; we wish now to show that that lower bound is *also* an *upper bound*, thus arriving at a full determination of the completeness radius. The reader should perhaps again look through the beginning of §D, Chapter IX, before continuing with the present discussion.

The lower bound just referred to is most conveniently expressed in terms of the Beurling–Malliavin *effective density* $\tilde{D}_\Lambda$ for the sequence Λ, defined in §D.2 of the previous chapter. According to a theorem in that

§, the completeness radius associated with Λ is $\geqslant \pi \tilde{D}_\Lambda$; the exponentials $e^{i\lambda_n t}$, $\lambda_n \in \Lambda$, are, in other words, *complete* (in any of the usual norms) on each interval of length $< 2\pi\tilde{D}_\Lambda$.

Showing this completeness radius to be *equal* to $\pi\tilde{D}_\Lambda$ was the first use made of the multiplier theorem stated in §A.2, which was indeed elaborated for that specific purpose. This application is given in the present §. Our task here is thus to prove that the completeness radius for Λ cannot be *larger* than $\pi\tilde{D}_\Lambda$; this amounts to establishing *incompleteness* of the $e^{i\lambda_n t}$, $\lambda_n \in \Lambda$, on any interval $[-L, L]$ with $L > \pi\tilde{D}_\Lambda$.

The known procedures for doing this are all based on the *duality argument* described at the beginning of §D, Chapter IX. Desiring, for instance, to prove that linear combinations of the $e^{i\lambda_n t}$ are *not dense* in $L_1(-L, L)$, one tries to obtain a *non-zero g* in the *dual* of that space – in this case, an element of $L_\infty(-L, L)$ – for which

$$\int_{-L}^{L} g(t)e^{i\lambda_n t}\,dt = 0, \quad \lambda_n \in \Lambda.$$

Establishing incompleteness in this way thus involves proof of an existence theorem. That is why the determination of *upper bounds* on the completeness radius has always given much more difficulty than the search for *lower bounds*, which essentially depend on uniqueness theorems (based on various forms of Jensen's formula).

The idea is to arrive at the function g by constructing its *Fourier transform*

$$\int_{-L}^{L} e^{izt}g(t)\,dt.$$

Suppose, for instance, that we are able to construct a non-zero entire function $G(z)$ of exponential type $\leqslant L$, vanishing at the points of Λ, for which

$$\int_{-\infty}^{\infty} |G(x)|\,dx < \infty.$$

The function

$$g(t) = \frac{1}{2\pi}\int_{-\infty}^{\infty} e^{-itx}G(x)\,dx$$

is then continuous, and an argument just like the one used in proving the Paley–Wiener theorem shows that $g(t) \equiv 0$ for $|t| > L$ (Chapter III, §D). We therefore certainly have $g \in L_1(\mathbb{R})$, and the Fourier inversion theorem for L_1 gives

$$G(x) = \int_{-L}^{L} e^{ixt} g(t) \, dt.$$

For each $\lambda_n \in \Lambda$, we then have

$$\int_{-L}^{L} e^{i\lambda_n t} g(t) \, dt = G(\lambda_n) = 0.$$

And $g(t) \not\equiv 0$ since $G \not\equiv 0$. Our aim can thus be accomplished by showing how to get such a function G when the sequence Λ and any $L > \pi \tilde{D}_\Lambda$ are given.

Beurling and Malliavin obtained a complete solution of this problem around 1961. Considerable effort had previously been expended on it by others who had succeeded in finding various constructions of entire functions G, subject always, however, to restrictive assumptions on the sequence Λ. This was done by Paley and Wiener and then by Levinson; later on, Redheffer obtained a number of results. I have worked on the question myself. Many of the methods devised for these investigations are still of interest even though they were not powerful enough to yield the final definitive conclusion; some of them indeed find service in the present book. The reader who wants to find out more about these matters should consult Redheffer's survey article (in *Advances in Math.*), which gives a very clear exposition of most of what has been done. There, the delicate question of completeness of the $e^{i\lambda_n t}$ on intervals of length *exactly equal* to $2\pi \tilde{D}_\Lambda$ is also discussed.

Before going on to article 1, let us indicate how the work will proceed. We are given a sequence $\Lambda \subseteq \mathbb{R}$ with $\tilde{D}_\Lambda < \infty$.[*] Picking any $\eta > 0$, we wish to construct a non-zero entire function $G(z)$ of exponential type $\leqslant \pi(\tilde{D}_\Lambda + 3\eta)$, say, such that

$$G(\lambda) = 0 \quad \text{for } \lambda \in \Lambda^{[†]}$$

and

$$\int_{-\infty}^{\infty} |G(x)| \, dx < \infty.$$

Because the distribution of the $\lambda \in \Lambda$ may be very irregular, it is not

---

[*] It is best to allow Λ to have repeated points; that makes no difference for the constructions to follow.

[†] with, of course, appropriate multiplicity at the repeated points of Λ

advisable to start with the Hadamard product

$$\prod_{\substack{\lambda \in \Lambda \\ \lambda \neq 0}} \left(1 - \frac{z}{\lambda}\right) e^{z/\lambda}.$$

Instead, we first turn to the *second* lemma of §D.2; Chapter IX, and to its corollary. Given $D > \tilde{D}_\Lambda$, these provide us with a real sequence $\Sigma \supset \Lambda$ for which

$$\int_{-\infty}^{\infty} \frac{|n_\Sigma(t) - Dt|}{1 + t^2} \, dt \quad < \quad \infty.$$

Here, $n_\Sigma(t)$ denotes the number of points* of Σ in $[0, \ t]$ if $t \geqslant 0$, and *minus* the number of such points in $[t, \ 0)$ if $t < 0$. For our purposes, we take

$$D \;=\; \tilde{D}_\Lambda + \eta.$$

The points of Σ are already quite regularly distributed. Assuming, wlog, that $0 \notin \Sigma$, we form the function

$$F(z) \;=\; \prod_{\lambda \in \Sigma} \left(1 - \frac{z}{\lambda}\right) e^{z/\lambda},$$

which turns out to be of exponential type. Its behaviour is worked out in article 1.

The next step (in article 2) is to prove what is called the *little multiplier theorem*. This result (which, strictly speaking, is *not* a multiplier theorem in the sense adopted for that term at the beginning of the present chapter) gives us a non-zero entire function $\varphi(z)$ of exponential type $\pi\eta$ such that

$$\int_{-\infty}^{\infty} \frac{\log^+ |F(x)\varphi(x)|}{1 + x^2} \, dx \quad < \quad \infty.$$

The theorem stated in §A.2 is finally applied to the product $F\varphi$ in order to obtain the function G. In this way, the completeness radius associated with Λ is seen to be $\leqslant \pi(\tilde{D}_\Lambda + 3\eta),^\dagger$ and the exact determination of the former quantity thus carried out for *real* sequences Λ (article 3).

It is somewhat remarkable that all the difficulties involved in the completeness problem for sets of exponentials $e^{i\lambda_n t}$ having *complex frequencies* λ_n already occur in the one about exponentials with *real frequencies*. The more general problem is rather easily reduced to the special one, and our solution of the latter made to yield one for the former.

---

* taking multiplicities of repeated points into account

† when the points of Λ are distinct

The completeness radius associated with arbitrary *complex* sequences Λ can thus be worked out. This result also is given in article 3.

1.   **The Hadamard product over** Σ

Having fixed $\eta > 0$ and put $D = \tilde{D}_\Lambda + \eta$, we take a real sequence $\Sigma \supset \Lambda$ having, perhaps, repetitions, such that

$$\int_{-\infty}^{\infty} \frac{|n_\Sigma(t) - Dt|}{1 + t^2}\, dt \quad < \quad \infty,$$

being assured by §D.2 of Chapter IX that such Σ exist. During this article and the next one, we will assume that $n_\Sigma(t) \equiv 0$ for $|t| < 1$, i.e., that Σ (and hence surely our original Λ) has no points in $(-1, 1)$. Doing so simplifies some details in the work, but does not make the results obtained less applicable.

Let

$$F(z) \quad = \quad \prod_{\lambda \in \Sigma}\left(1 - \frac{z}{\lambda}\right) e^{z/\lambda}.$$

Then we have the

Theorem. *If* $n_\Sigma(t) = 0$ *for* $-1 < t < 1$ *and*

$$\int_{-\infty}^{\infty} \frac{|n_\Sigma(t) - Dt|}{1 + t^2}\, dt \quad < \quad \infty,$$

$F(z)$ *is of exponential type, and*

$$\limsup_{r \to \infty} \frac{\log|F(re^{i\vartheta})|}{r} \quad \leqslant \quad \pi D |\sin \vartheta| + c \cos \vartheta,$$

where c is a certain constant. When $\vartheta = \pm \pi/2$ *the limit superior on the left is an actual limit, and equality holds.*

Proof. The last lemma of §D.2, Chapter IX, tells us that

$$\frac{n_\Sigma(t)}{t} \longrightarrow D \quad \text{for } t \longrightarrow \pm\infty.$$

This, however, is not in itself enough to make $F(z)$ of exponential type; for that, boundedness of $|\int_{-R}^{R}(1/t)\,dn_\Sigma(t)|$ as $R \longrightarrow \infty$ is also necessary – and sufficient – according to the Lindelöf theorems of §B, Chapter III. In the present circumstances, we have, however,

$$\int_{-R}^{R} \frac{dn_\Sigma(t)}{t} = \frac{n_\Sigma(R)+n_\Sigma(-R)}{R} + \int_{1\le|t|\le R} \frac{n_\Sigma(t)}{t^2}\,dt$$

$$= o(1) + \int_{1\le|t|\le R} \frac{n_\Sigma(t)-Dt}{t^2}\,dt = O(1)$$

since

$$\int_{|t|\ge 1} \frac{|n_\Sigma(t)-Dt|}{t^2}\,dt < \infty,$$

so F is of exponential type.

For y real, we have

$$\log|F(iy)| = \frac{1}{2}\int_{-\infty}^{\infty} \log\left(1+\frac{y^2}{t^2}\right)dn_\Sigma(t).$$

After integrating by parts and then using the asymptotic behaviour of $n_\Sigma(t)$, we easily find that

$$\log|F(iy)| \sim \pi D|y| \quad \text{for } y \longrightarrow \pm\infty.$$

To study the behaviour of $F(z)$ on the real axis, we take

$$c = \int_{|t|\ge 1} \frac{n_\Sigma(t)-Dt}{t^2}\,dt$$

(the integral being absolutely convergent), and then look at

$$e^{-cx}F(x)$$

for real x. Here, we are able to fall back on work already done for parts (a) – (d) of problem 29 (§B.1, Chapter IX). Denote

$$\Sigma \cap (0,\infty) \quad \text{by} \quad \Sigma_+$$

and

$$(-\Sigma)\cap(0,\infty) \ (sic!) \quad \text{by} \quad \Sigma_-.$$

Then, if $x > 0$, we can write

$$\log|F(x)| = \sum_{\lambda\in\Sigma_+} \log\left|1-\frac{x^2}{\lambda^2}\right| + \left(\sum_{\lambda\in\Sigma_-} - \sum_{\lambda\in\Sigma_+}\right)\left(\log\left|1+\frac{x}{\lambda}\right| - \frac{x}{\lambda}\right).$$

Since $\lim_{t\to\infty}(n_{\Sigma_+}(t)/t)$ exists, the *first* sum on the right is $\le o(x)$ for $x\longrightarrow\infty$ by problem 29(d). (For the solution of parts (a) – (e) of that

problem, it is not necessary that the zeros of the function $C(z)$ considered
there be *integers* – they need only be *real and positive*.)
  What is left on the right side of the previous relation can be rewritten as

$$\int_0^\infty \left(\log\left(1 + \frac{x}{t}\right) - \frac{x}{t}\right)(dn_{\Sigma_-}(t) - dn_{\Sigma_+}(t)).$$

This is integrated by parts, upon which all the integrated terms vanish
($n_{\Sigma_+}(t)$ and $n_{\Sigma_-}(t)$ are zero for $0 < t < 1$!), and we end with

$$\int_0^\infty \frac{x^2}{x+t} \frac{n_{\Sigma_+}(t) - n_{\Sigma_-}(t)}{t^2} \, dt.$$

Since

$$\int_0^\infty \frac{n_{\Sigma_+}(t) - n_{\Sigma_-}(t)}{t^2} \, dt \;=\; \int_{|t| \geqslant 1} \frac{n_\Sigma(t) - Dt}{t^2} \, dt$$

with the right-hand integral absolutely convergent and equal to c, the
left-hand integral is also absolutely convergent and equal to c, so the
previous expression is $\sim cx$ for $x \longrightarrow \infty$.
  We see that

$$\log|F(x)| \;\leqslant\; cx + o(|x|) \qquad \text{for } x \longrightarrow \infty.$$

In like manner, the same is seen to hold for $x \longrightarrow -\infty$. The function
$e^{-cz}F(z)$ is thus in modulus $\leqslant e^{o(|x|)}$ on the *real* axis when x is large, and
has the *same* growth as $F(z)$ on the *imaginary* axis; it is, moreover, of
exponential type. Our desired result now follows by application of a
Phragmén–Lindelöf theorem, as in part (e) of problem 29.

  We shall have to look more closely at the behaviour of $|F(x)|$ on the real
axis. Of course,

$$\log|F(x)| \;=\; \int_{-\infty}^\infty \left(\log\left|1 - \frac{x}{t}\right| + \frac{x}{t}\right) dn_\Sigma(t), \qquad x \in \mathbb{R}.$$

Regarding integrals like the one on the right, one has the following generalization of the formula derived in problem 29(b):

Lemma. *Let* $v(t)$, *zero on a neighborhood of* 0 *(N.B.!), be increasing on*
$(-\infty, \infty)$ *and* $O(t)$ *there. Then*

$$\int_{-\infty}^\infty \left(\log\left|1 - \frac{x}{t}\right| + \frac{x}{t}\right) dv(t) \;=\; \int_{-\infty}^\infty \frac{x^2}{x-t} \frac{v(t)}{t^2} \, dt.$$

at the $x \in \mathbb{R}$ where $v'(x)$ exists and is finite, and also at those where $v(t)$ has a jump discontinuity.

Remark. The expression on the right is a *Cauchy principal value*, viz.,

$$\lim_{\varepsilon \to 0} \int_{|t-x| > \varepsilon} \frac{x^2}{x-t} \frac{v(t)}{t^2} \, dt.$$

See the end of §C.1 in Chapter VIII.

Proof. Taking an $\varepsilon > 0$, integrate

$$\int_{|t-x| \geqslant \varepsilon} \left(\log \left| 1 - \frac{x}{t} \right| + \frac{x}{t} \right) dv(t)$$

by parts. Under the given conditions, the integrated terms corresponding to $t = \pm \infty$ vanish, and, if $v'(x)$ exists and is finite, the *sum* of the ones corresponding to $t = x \pm \varepsilon$ tends to zero as $\varepsilon \longrightarrow 0$, leaving us with the right side of the identity in question. When v has a jump discontinuity at x, that identity is valid because each of its sides is then equal to $-\infty$.

Application of the lemma to our function F (formed from $n_\Sigma(t)$ which vanishes for $|t| < 1$) yields

$$\log |F(x)| \quad = \quad \int_{|t| \geqslant 1} \frac{x^2}{x-t} \frac{n_\Sigma(t)}{t^2} \, dt.$$

In using this relation, we will want to take advantage of the condition

$$\int_{|t| \geqslant 1} \frac{|n_\Sigma(t) - Dt|}{1 + t^2} \, dt \quad < \quad \infty,$$

and for that we will be helped by the formula

$$\int_{|t| \geqslant 1} \frac{x^2}{x-t} \frac{dt}{t} \quad = \quad -x \log \left| \frac{x+1}{x-1} \right|,$$

which is easily verified by direct calculation. From this and the previous, we get

$$\log |F(x)| \quad = \quad \int_{|t| \geqslant 1} \frac{x^2}{x-t} \frac{n_\Sigma(t) - Dt}{t^2} \, dt \quad - \quad Dx \log \left| \frac{x+1}{x-1} \right|.$$

The second term on the right is $\geqslant 0$, and tends to $2D$ as $x \longrightarrow \pm \infty$; for it, we certainly have

$$\int_{-\infty}^{\infty} \frac{1}{1 + x^2} Dx \log \left| \frac{x+1}{x-1} \right| dx \quad < \quad \infty.$$

As far as we are concerned, then, the behaviour of $\log|F(x)|$ is governed by that of the Cauchy principal value on the right, involving the *integrable* function $(n_\Sigma(t) - Dt)/t^2$. It will be convenient in the next article to denote that principal value by $U(x)$, i.e.,

$$U(x) \;=\; \int_{|t| \geqslant 1} \frac{x^2}{x-t} \frac{n_\Sigma(t) - Dt}{t^2}\, dt.$$

2. The little multiplier theorem

We proceed to construct a non-zero entire function $\varphi(z)$ of exponential type $\pi\eta$ which will make

$$\int_{-\infty}^{\infty} \frac{\log^+|F(x)\varphi(x)|}{1+x^2}\, dx \;<\; \infty,$$

F being the Hadamard product formed above. According to what was observed at the end of the last article, the relation just written will certainly hold if

$$\int_{-\infty}^{\infty} \frac{|U(x) + \log|\varphi(x)||}{1+x^2}\, dx \;<\; \infty$$

with the function $U(x)$ defined there. Let us write

$$\Delta(t) \;=\; \begin{cases} n_\Sigma(t) - Dt, & |t| \geqslant 1, \\ 0, & -1 < t < 1. \end{cases}$$

Then, as we have seen,

$$U(x) \;=\; \log|F(x)| \;+\; Dx \log\left|\frac{x+1}{x-1}\right|$$

is equal to

$$\int_{-\infty}^{\infty} \frac{x^2}{x-t} \frac{\Delta(t)}{t^2}\, dt.$$

For the function φ we are seeking, $\log|\varphi(x)|$ will be related to a similar expression,

$$\int_{-\infty}^{\infty} \frac{x^2}{x-t} \frac{\delta(t)}{t^2}\, dt,$$

involving a $\delta(t)$ obtained in a certain way from $\Delta(t)$. The property

$$\int_{-\infty}^{\infty} (|\Delta(t)|/t^2)\, dt \ < \ \infty$$

and the fact that $\Delta(t) + Dt$ increases both play important rôles in that construction. Here is how it goes:

Lemma. *Given the function $\Delta(t)$, zero on $(-1, 1)$, fulfilling the conditions just mentioned, and a number $\eta > 0$, there are two sequences*

$$1 = x_1 < x_2 < x_3 < \cdots < x_k \xrightarrow[k]{} \infty,$$
$$-1 = x_{-1} > x_{-2} > x_{-3} > \cdots > x_{-k} \xrightarrow[k]{} -\infty,$$

and a function $\delta(t)$, zero on $(-1, 1)$, with the following properties:

(i) $\displaystyle\sum_{k=1}^{\infty} \left(\frac{x_{k+1} - x_k}{x_k}\right)^2 + \sum_{k=-1}^{-\infty} \left(\frac{x_k - x_{k-1}}{x_k}\right)^2 \ < \ \infty;$

(ii) $\eta t + \delta(t)$ *is increasing on* $(-\infty, \infty);$

(iii) $\delta(t)$ *is* o(t) *for* $t \longrightarrow \pm\infty;$

(iv) $\displaystyle\int_{-\infty}^{\infty} \frac{|\delta(t)|}{t^2}\, dt \ < \ \infty;$

(v) *for each* $k \geqslant 1$ *or* < -1,

$$\int_{x_k}^{x_{k+1}} \frac{\Delta(t) + \delta(t)}{t^2}\, dt \ = \ 0;$$

(vi) *for* $x_k \leqslant t \leqslant x_{k+1}$,

$$|\Delta(t) + \delta(t)| \ \leqslant \ (D + \eta)(x_{k+2} - x_{k-1}),$$

where, to cover the cases $k = -2$ and $k = 1$, we put $x_0 = 0$.

Remark. Property (i) certainly implies that $x_k/x_{k-1} \longrightarrow 1$ for $k \longrightarrow \pm\infty$. Keeping this in mind, the reader familiar with the modern theory of H_1 and BMO will recognize in properties (v) and (vi) a stipulation that the functions

$$s_k(t) \ = \ \begin{cases} \dfrac{x_k^2(\Delta(t) + \delta(t))}{(D + \eta)(x_{k+2} - x_{k-1})(x_{k+1} - x_k)t^2}, & x_k \leqslant t \leqslant x_{k+1}, \\[3ex] 0, & t \notin [x_k, x_{k+1}] \end{cases}$$

be *atoms* for $\Re H_1$ (to within constant factors tending to 1 for $k \longrightarrow \pm \infty$). About this, more later on.

Proof of lemma. It suffices to show how to get the x_k with $k \geqslant 1$ and the function $\delta(t)$ when $t \geqslant 1$, the constructions on $(-\infty, -1]$ being exactly the same.

We start by putting $x_1 = 1$. Then, assuming that x_k has already been determined (and $\delta(t)$ specified on $[1, x_k)$ if $k > 1$), let us see how to find x_{k+1}, and how to define $\delta(t)$ for $x_k \leqslant t < x_{k+1}$.

As $x > x_k$ increases, the integral

$$\int_{x_k}^{x} \frac{t - x_k}{t^2} \, dt \;=\; \log \frac{x}{x_k} \;-\; \frac{x \;-\; x_k}{x}$$

tends to ∞, while

$$\int_{x_k}^{x} \frac{|\Delta(t)|}{t^2} \, dt$$

remains bounded, by hypothesis. Hence, unless the ratio

$$\int_{x_k}^{x} \frac{|\Delta(t)|}{t^2} \, dt \Big/ \int_{x_k}^{x} \frac{t - x_k}{t^2} \, dt$$

remains always $< \eta$ for $x > x_k$, there is a value of x for which it is *equal* to our given number η. If equality last obtains for a value $x > x_k + 1$ we call that value x_{k+1}; in any other case we put $x_{k+1} = x_k + 1$. We thus have $x_{k+1} \geqslant x_k + 1$ and also

$$\int_{x_k}^{x_{k+1}} \frac{|\Delta(t)|}{t^2} \, dt \;\leqslant\; \eta \int_{x_k}^{x_{k+1}} \frac{t - x_k}{t^2} \, dt,$$

with equality holding when $x_{k+1} > x_k + 1$.

We have

$$\int_{x_k}^{x_{k+1}} \frac{x_{k+1} - t}{t^2} \, dt \;-\; \int_{x_k}^{x_{k+1}} \frac{t - x_k}{t^2} \, dt \;>\; 0,$$

for the difference on the left can be rewritten as

$$-2 \int_{-l}^{l} \frac{\tau}{(\tau + c)^2} \, d\tau \;=\; 2 \int_{0}^{l} \left(\frac{1}{(c - \tau)^2} - \frac{1}{(c + \tau)^2} \right) \tau \, d\tau$$

with $c = (x_k + x_{k+1})/2$, $l = (x_{k+1} - x_k)/2$, and the new variable

$\tau = t - c$. Therefore, as x' *increases* from x_k to x_{k+1},

$$\eta \int_{x_k}^{x_{k+1}} \frac{x'-t}{t^2} \, dt$$

increases from $-\eta \int_{x_k}^{x_{k+1}} ((t-x_k)/t^2)\,dt$ to $\eta \int_{x_k}^{x_{k+1}} ((x_{k+1}-t)/t^2)\,dt >$ $\eta \int_{x_k}^{x_{k+1}} ((t-x_k)/t^2)\,dt$, and, by the previous relation, there must be an $x' \in [x_k, x_{k+1}]$ for which

$$\int_{x_k}^{x_{k+1}} \frac{\Delta(t)}{t^2} \, dt + \eta \int_{x_k}^{x_{k+1}} \frac{x'-t}{t^2} \, dt = 0.$$

We denote that value of x' by x_k', and put

$$\delta(t) = \eta(x_k' - t) \quad \text{for } x_k \leqslant t < x_{k+1}.$$

In this way, the function $\delta(t)$ is defined piece by piece on the successive intervals $[x_k, x_{k+1})$ and thus on all of $[1, \infty)$, since our requirement that $x_{k+1} \geqslant x_k + 1$ ensures that $x_k \xrightarrow{k} \infty$.

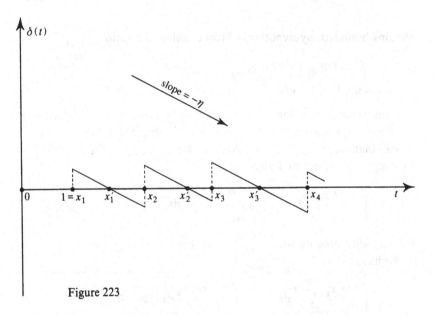

Figure 223

We must verify properties (i) – (vi) for this $\delta(t)$ and the sequence $\{x_k\}$.

Property (ii) is obvious, and (v) guaranteed by our choice of the x_k'. To check (i), observe that when $x_{k+1} > x_k + 1$,

$$\frac{\eta}{2} \frac{(x_{k+1}-x_k)^2}{x_{k+1}^2} < \eta \int_{x_k}^{x_{k+1}} \frac{t-x_k}{t^2} \, dt = \int_{x_k}^{x_{k+1}} \frac{|\Delta(t)|}{t^2} \, dt,$$

whence

$$\sum_{x_{k+1} > x_k + 1} \left(\frac{x_{k+1} - x_k}{x_{k+1}} \right)^2 < \frac{2}{\eta} \cdot \int_1^\infty \frac{|\Delta(t)|}{t^2} \, dt < \infty.$$

The sum of the $(x_{k+1} - x_k)^2 / x_{k+1}^2$ with $x_{k+1} = x_k + 1$ is, on the other hand, obviously convergent, so we have

$$\sum_{k=1}^\infty \left(\frac{x_{k+1} - x_k}{x_{k+1}} \right)^2 < \infty.$$

This certainly implies that

$$\frac{x_{k+1}}{x_k} \longrightarrow 1, \qquad k \longrightarrow \infty,$$

and we must also have

$$\sum_{k=1}^\infty \left(\frac{x_{k+1} - x_k}{x_k} \right)^2 < \infty.$$

For (iii) and (iv), we use the fact that

$$|\delta(t)| = \eta |x'_k - t| \leqslant \eta(x_{k+1} - x_k) \quad \text{for } x_k \leqslant t < x_{k+1}.$$

Thence $|\delta(t)/t| \leqslant \eta(x_{k+1} - x_k)/x_k$ on $[x_k, x_{k+1})$, but, by what we have just seen, the right-hand quantity tends to zero for $k \longrightarrow \infty$. Again,

$$\int_{x_k}^{x_{k+1}} \frac{|\delta(t)|}{t^2} \, dt < \eta \frac{(x_{k+1} - x_k)^2}{x_k^2},$$

and the convergence of $\int_1^\infty (|\delta(t)|/t^2) \, dt$ follows from property (i), already verified.

We are left with property (vi). Given $k \geqslant 1$, we have

$$\int_{x_{k-1}}^{x_k} \frac{\Delta(t) + \delta(t)}{t^2} \, dt = \int_{x_{k+1}}^{x_{k+2}} \frac{\Delta(t) + \delta(t)}{t^2} \, dt = 0$$

by (v) and (for $k = 1$) the fact that $\Delta(t) = \delta(t) = 0$ for $x_0 = 0 \leqslant t < 1 = x_1$. There are thus points t' and t'', in $[x_{k-1}, x_k)$ and $[x_{k+1}, x_{k+2})$ respectively, for which

$$\Delta(t') + \delta(t') \geqslant 0$$
$$\Delta(t'') + \delta(t'') \leqslant 0.$$

According to (ii), $\delta(t) + \eta t$ increases, and $\Delta(t) + Dt$ is increasing by hypothesis. Therefore, if $x_k \leqslant t \leqslant x_{k+1}$,

$$\Delta(t) + \delta(t) \geqslant \Delta(t') + \delta(t') - (D + \eta)(t - t') \geqslant -(D + \eta)(x_{k+1} - x_{k-1}),$$

and

$$\Delta(t) + \delta(t) \leqslant \Delta(t'') + \delta(t'') + (D + \eta)(t'' - t) \leqslant (D + \eta)(x_{k+2} - x_k):$$

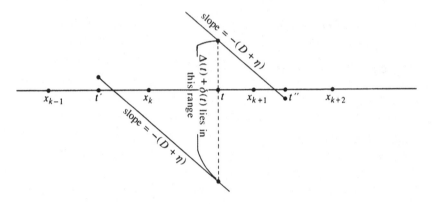

Figure 224

We see that for $x_k \leqslant t \leqslant x_{k+1}$,

$$|\Delta(t) + \delta(t)| \leqslant (D + \eta) \max\{(x_{k+2} - x_k), (x_{k+1} - x_{k-1})\},$$

more than what is asserted by (vi).

The lemma is proved.

Theorem. *If $\Delta(t)$, zero on $(-1, 1)$ and with $\Delta(t) + Dt$ increasing on $\mathbb{R}$, satisfies*

$$\int_{-\infty}^{\infty} \frac{|\Delta(t)|}{t^2} \, dt \;\; < \;\; \infty$$

and if $\delta(t)$ is the function furnished by the lemma, we have

$$\int_{-\infty}^{\infty} \frac{|U(x) + v(x)|}{1 + x^2} \, dx \;\; < \;\; \infty,$$

where

$$U(x) \;\; = \;\; \int_{-\infty}^{\infty} \frac{x^2}{x - t} \frac{\Delta(t)}{t^2} \, dt$$

and

$$v(x) \;=\; \int_{-\infty}^{\infty} \frac{x^2}{x-t} \frac{\delta(t)}{t^2} \, dt.$$

Proof. Taking the sequences $x_1, x_2, x_3, \ldots$ and $x_{-1}, x_{-2}, x_{-3}, \ldots$ provided by the lemma, we can write

$$U(x) + v(x) \;=\; \sum_{k \neq -1, 0} \int_{x_k}^{x_{k+1}} \frac{x^2}{x-t} \frac{\Delta(t) + \delta(t)}{t^2} \, dt.$$

Putting, therefore,

$$V_k(x) \;=\; \int_{x_k}^{x_{k+1}} \frac{x^2}{x-t} \frac{\Delta(t) + \delta(t)}{t^2} \, dt,$$

it is, in order to prove the theorem, more than sufficient to verify that

$$\sum_{k \neq -1, 0} \int_{-\infty}^{\infty} \frac{|V_k(x)|}{x^2} \, dt \;<\; \infty.$$

For this purpose, we use what has become a standard tool of singular integral theory.

Denote by I_k the interval $[x_k, \; x_{k+1})$ and by I_k^* the one having *the same midpoint as* I_k, *but twice its length*:

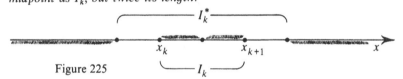

Figure 225

We then break up each separate expression

$$\int_{-\infty}^{\infty} \frac{|V_k(x)|}{x^2} \, dx$$

as

$$\left(\int_{\mathbb{R} \sim I_k^*} + \int_{I_k^*} \right) \frac{|V_k(x)|}{x^2} \, dx,$$

and estimate the last two integrals with the help of different techniques.

In considering the *first* one, we call on property (v) from the lemma. According to it, if $x \notin I_k^*$,

$$\frac{V_k(x)}{x^2} \;=\; \int_{I_k} \frac{1}{x-t} \frac{\Delta(t) + \delta(t)}{t^2} \, dt$$

$$=\; \int_{I_k} \left(\frac{1}{x-t} - \frac{1}{x-c} \right) \frac{\Delta(t) + \delta(t)}{t^2} \, dt,$$

where the constant c is arbitrary. We take c equal to the *abscissa* of I_k's *midpoint*, and thus find that

$$\int_{c+|I_k|}^{\infty} \frac{|V_k(x)|}{x^2}\,dx \;\leqslant\; \int_{c+|I_k|}^{\infty} \int_{I_k} \frac{|t-c|}{(x-t)(x-c)}\,\frac{|\Delta(t)+\delta(t)|}{t^2}\,dt\,dx$$

$$\leqslant\; \int_{I_k} \frac{|\Delta(t)+\delta(t)|}{t^2}\,dt \int_{c+|I_k|}^{\infty} \frac{|I_k|/2}{(x-x_{k+1})^2}\,dx,$$

which boils down to just

$$\int_{I_k} \frac{|\Delta(t)+\delta(t)|}{t^2}\,dt \quad (!).$$

In exactly the same way, we get

$$\int_{-\infty}^{c-|I_k|} \frac{|V_k(x)|}{x^2}\,dx \;\leqslant\; \int_{I_k} \frac{|\Delta(t)+\delta(t)|}{t^2}\,dt,$$

so finally

$$\int_{\mathbb{R}\sim I_k^*} \frac{|V_k(x)|}{x^2}\,dx \;\leqslant\; 2\int_{I_k} \frac{|\Delta(t)+\delta(t)|}{t^2}\,dt.$$

To estimate the integral of $|V_k(x)|/x^2$ over I_k^*, we begin by using Schwarz:

$$\int_{I_k^*} \frac{|V_k(x)|}{x^2}\,dx \;\leqslant\; \sqrt{|I_k^*|}\left(\int_{I_k^*} \left(\frac{V_k(x)}{x^2}\right)^2 dx\right)^{1/2}.$$

At this point one must recognize that

$$\frac{V_k(x)}{x^2} \;=\; \pi\tilde{\psi}_k(x),$$

where

$$\tilde{\psi}_k(x) \;=\; \frac{1}{\pi}\int_{-\infty}^{\infty} \frac{1}{x-t}\,\psi_k(t)\,dt$$

is the *Hilbert transform* of the function

$$\psi_k(t) \;=\; \begin{cases} (\Delta(t)+\delta(t))/t^2, & t\in I_k, \\ 0, & t\notin I_k. \end{cases}$$

The L_2 theory of Hilbert transforms was discussed in the scholium at the end of §C.1, Chapter VIII, and according to that theory we have

$$\int_{-\infty}^{\infty} (\tilde{\psi}_k(x))^2\,dx \;=\; \int_{-\infty}^{\infty} (\psi_k(t))^2\,dt.$$

Thus,

$$\int_{-\infty}^{\infty} \left(\frac{V_k(x)}{x^2}\right)^2 dx = \pi^2 \int_{I_k} \left(\frac{\Delta(t) + \delta(t)}{t^2}\right)^2 dt,$$

which, used in the previous relation, yields

$$\int_{I_k^*} \frac{|V_k(x)|}{x^2} dx \leqslant \pi(2|I_k|)^{1/2} \left(\int_{I_k} \left(\frac{\Delta(t) + \delta(t)}{t^2}\right)^2 dt\right)^{1/2}.$$

Combining this inequality with the one for the integral over $\mathbb{R} \sim I_k^*$, we get

$$\int_{-\infty}^{\infty} \frac{|V_k(x)|}{x^2} dx \leqslant 2\int_{I_k} \frac{|\Delta(t) + \delta(t)|}{t^2} dt$$

$$+ \pi(2|I_k|)^{1/2} \left(\int_{I_k} \left(\frac{\Delta(t) + \delta(t)}{t^2}\right)^2 dt\right)^{1/2}.$$

We can now obtain an estimate good enough for our purpose by using a very crude procedure on the right-hand integrals. We simply plug the inequality

$$|\Delta(t) + \delta(t)| \leqslant (D + \eta)(|I_{k-1}| + |I_k| + |I_{k+1}|), \qquad t \in I_k,$$

(property (vi) of the lemma) into each of them and find, for $k \geqslant 1$, the right side of the previous relation to be

$$\leqslant (2 + \sqrt{2}\pi)(D + \eta)\frac{(|I_{k-1}| + |I_k| + |I_{k+1}|)|I_k|}{x_k^2}$$

We therefore certainly have

$$\int_{-\infty}^{\infty} \frac{|V_k(x)|}{x^2} dx \leqslant 7(D + \eta)\left(\frac{|I_{k-1}| + |I_k| + |I_{k+1}|}{x_k}\right)^2$$

for $k \geqslant 1$.

From this we see that

$$\sum_{k=1}^{\infty} \int_{-\infty}^{\infty} \frac{|V_k(x)|}{x^2} dx \leqslant 7(D + \eta) \sum_{k=1}^{\infty} \left(\frac{x_{k+2} - x_{k-1}}{x_k}\right)^2.$$

We have

$$(x_{k+2} - x_{k-1})^2 \leqslant 3\{(x_k - x_{k-1})^2 + (x_{k+1} - x_k)^2 + (x_{k+2} - x_{k+1})^2\},$$

and, by property (i) of the lemma,

$$\sum_{k=1}^{\infty} \left(\frac{x_{k+1} - x_k}{x_k} \right)^2 \; < \; \infty,$$

so that $x_{k+1}/x_k \longrightarrow 1$ as $k \to \infty$. The right-hand sum in the previous inequality is hence certainly convergent, and we conclude that

$$\sum_{k=1}^{\infty} \int_{-\infty}^{\infty} \frac{|V_k(x)|}{x^2} \, dx \; < \; \infty.$$

By the same reasoning, it is also shown that

$$\sum_{-\infty}^{-1} \int_{-\infty}^{\infty} \frac{|V_k(x)|}{x^2} \, dx \; < \; \infty.$$

Combination of this with the previous relation finally yields

$$\int_{-\infty}^{\infty} \frac{|U(x) + v(x)|}{1 + x^2} \, dx \; < \; \infty$$

by the observation at the beginning of this proof.   Q.E.D.

Remark. Thinking back to the remark immediately following the statement of the previous lemma, the reader acquainted with the modern theory of H_1 should recognize that in the argument just given, *exhibition of a specific atomic decomposition* for $(\Delta(t) + \delta(t))/t^2$ was used to show that that function belonged to $\Re H_1$. This reasoning was employed by Beurling and Malliavin some 13 years before Coifman brought atomic decomposition into H_p space theory* as a systematic tool in 1974. True, the work of Beurling and Malliavin was never widely circulated, and the version of it finally published by them in 1967 is very hard to understand.

It turns out that a function belongs to $\Re H_1$ *if and only if* it has an atomic decomposition like the one figuring in our proof (in general, with atoms having non-disjoint supporting intervals). Fefferman's celebrated theorem on the duality of $\Re H_1$ and BMO is easily seen to be equivalent to this result, whose *if* part was essentially verified in the course of the above reasoning. The *only if* part, which guarantees the *existence* of atomic decompositions for arbitrary functions in $\Re H_1$, is deeper. What Coifman did was to obtain a *direct proof* of that existence, and thus arrive at a new proof of the Fefferman duality theorem.

The reader who wishes to go into these matters should first look at

---

* For *martingale* H_1, he was preceded in this by Herz.

problem 11 on page 274 of Garnett's book* and then consult the articles referred to there.

We have now done most of the work needed to establish the

Little Multiplier Theorem (Beurling and Malliavin, 1961). *Let* Σ *be a sequence of (perhaps repeated) real numbers lying outside* $(-1, 1)$, *with*

$$\int_{-\infty}^{\infty} \frac{|n_\Sigma(t) - Dt|}{1 + t^2} \, dt \;\; < \;\; \infty$$

for some $D \geqslant 0$, *and put*

$$F(z) \;\; = \;\; \prod_{\lambda \in \Sigma} \left(1 - \frac{z}{\lambda} \right) e^{z/\lambda}.$$

Given $\eta > 0$, *there is a sequence* S *of real numbers lying outside* $(-1, 1)$ *such that*

$$\frac{n_S(t)}{t} \;\; \longrightarrow \;\; \eta \quad \text{for } t \longrightarrow \pm \infty$$

and, for a suitable real number γ, *the function*

$$\varphi(z) \;\; = \;\; e^{-\gamma z} \prod_{\lambda \in S} \left(1 - \frac{z}{\lambda} \right) e^{z/\lambda}$$

satisfies

$$\int_{-\infty}^{\infty} \frac{\log^+ |F(x)\varphi(x)|}{1 + x^2} \, dx \;\; < \;\; \infty.$$

$\varphi(z)$ *is entire and of exponential type.*

Remark. According to the theorem in article 1, $F(z)$ is of exponential type – so, then, is the product $F(z)\varphi(z)$. The limits of

$$\frac{\log |F(iy)\varphi(iy)|}{|y|}$$

for $y \longrightarrow \pm \infty$ both exist, and are equal to $\pi(D + \eta)$, *a quantity as close as we like to* πD. Because the product has a convergent logarithmic integral, one can show by the method of §B.2, Chapter VI, that in fact

$$|F(z)\varphi(z)| \;\; \leqslant \;\; C_\varepsilon \exp\!\big(\pi(D + \eta)|\Im z| + \varepsilon|z|\big)$$

---

* The one on bounded analytic functions – the recent publication by Garcia-Cuerva and Rubio de Francia is also (and especially!) called to the reader's attention.

for each $\varepsilon > 0$. At the same time, $F(z)\varphi(z)$ vanishes (with appropriate multiplicity) at each point of the given sequence Σ.

Proof of theorem. Fixing the number $\eta > 0$, we take the function $\delta(t)$ corresponding to it furnished by the lemma and $v(x)$, related to $\delta(t)$ as in the statement of the preceding theorem. The function $U(x)$ figuring in that result is, as we know, related to our F by the formula

$$U(x) \;\; = \;\; \log|F(x)| \;\; + \;\; Dx\log\left|\frac{x+1}{x-1}\right|.$$

Let us obtain a similar representation for $v(x)$.
  Write

$$v(t) \;\; = \;\; \begin{cases} \eta t + \delta(t), & |t| \geq 1, \\ 0, & |t| < 1. \end{cases}$$

By property (ii) of the lemma, $v(t)$ is *increasing* on $(-\infty, \infty)$, and, by property (iii),

$$\frac{v(t)}{t} \;\; \longrightarrow \;\; \eta \quad\text{as } t \longrightarrow \pm\infty.$$

$v(t)$ is in fact *piecewise constant*, with jump discontinuities at (and *only* at) the points x_k, $k = \pm 1, \pm 2, \ldots$ mentioned in the lemma's statement:

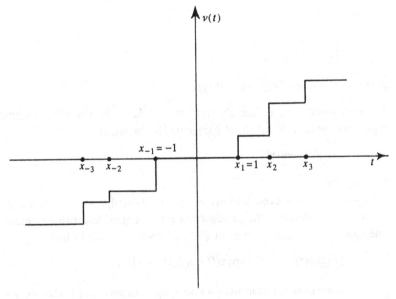

Figure 226

According to the lemma and discussion at the end of the last article, we have

$$
v(x) \;=\; \int_{-\infty}^{\infty} \frac{x^2}{x-t}\frac{\delta(t)}{t^2}\, \mathrm{d}t \;=\; \int_{|t|\geqslant 1} \frac{x^2}{x-t}\frac{v(t)}{t^2}\, \mathrm{d}t \;+\; \eta x \log\left|\frac{x+1}{x-1}\right|
$$

$$
=\; \int_{-\infty}^{\infty}\left(\log\left|1-\frac{x}{t}\right| + \frac{x}{t}\right)\mathrm{d}v(t) \;+\; \eta x \log\left|\frac{x+1}{x-1}\right|.
$$

In terms of

$$
H(z) \;=\; \int_{-\infty}^{\infty}\left(\log\left|1-\frac{z}{t}\right| + \frac{\Re z}{t}\right)\mathrm{d}v(t),
$$

we thus have our desired representation:

$$
v(x) \;=\; H(x) \;+\; \eta x \log\left|\frac{x+1}{x-1}\right|.
$$

Using this formula with the previous one for $U(x)$, we can reformulate the conclusion of the last theorem to get

$$
\int_{-\infty}^{\infty} \frac{|\log|F(x)| + H(x)|}{1+x^2}\, \mathrm{d}x \;<\; \infty.
$$

The use of

$$
\int_{-\infty}^{\infty}\left(\log\left|1-\frac{z}{t}\right| + \frac{\Re z}{t}\right)\mathrm{d}[v(t)],
$$

obviously the logarithm of the modulus of an entire function, in place of $H(z)$ comes now immediately to mind – one recalls the lemma of §A.1. ▶(N.B. For $p \geqslant 0$, $[p]$ denotes, as usual, the greatest integer $\leqslant p$, *but when* $p < 0$, *we take* $[p]$ *as the least integer* $\geqslant p$, *so as to have* $[-p] = -[p]$.)

Here, one must be somewhat careful. The expression

$$
\int_{-\infty}^{\infty}\left(\log\left|1-\frac{z}{t}\right| + \frac{\Re z}{t}\right)\mathrm{d}v(t)
$$

is very sensitive to small changes in v because of the term $\Re z/t$ in the integrand; replacement of $v(t)$ by $[v(t)]$ usually produces a new term linear in $\Re z = x$ which spoils the convergence of the integral involving $\log|F|$

and *H*. What we do have is a relation

$$\int_{-\infty}^{\infty}\left(\log\left|1-\frac{x+i}{t}\right|+\frac{x}{t}\right)(d[v(t)]-dv(t))$$

$$\leqslant \ \gamma x \ + \ 2\log^+|x| \ + \ O(1), \qquad x\in\mathbb{R},$$

valid with a certain real constant γ.

To show this, a device from the proof of the theorem in article 1 is used.*
Assuming, wlog, that $x \geqslant 0$, we observe that the *left-hand member* of the
relation in question can be rewritten as

$$\int_0^{\infty}\log\left|1-\left(\frac{x+i}{t}\right)^2\right|(d[v(t)]-dv(t))$$

$$+ \ \int_0^{\infty}\left(\log\left|1+\frac{x+i}{t}\right|-\frac{x}{t}\right)(d[-v(-t)]+dv(-t)-d[v(t)]+dv(t)).$$

To estimate the *first* integral we fall back on the lemma from §A.1,
according to which it is

$$\leqslant \ \log^+|x| \ + \ O(1).$$

The *second* one we integrate by parts, remembering that $v(t) = 0$ for
$|t| < 1$. When that is done, the integrated terms (involving the *differences*
$(-v(-t))-[-v(-t)]$ and $v(t)-[v(t)]$) all vanish, leaving

$$x\int_1^{\infty}\frac{[v(t)]-v(t)+(-v(-t))-[-v(-t)]}{t^2}\,dt$$

$$+ \ \int_1^{\infty}\left(\frac{\partial}{\partial t}\log\left|1+\frac{x+i}{t}\right|\right)\{(-v(-t)-[-v(-t)]) \ - \ (v(t)-[v(t)])\}\,dt.$$

The first term here is just γx, where

$$\gamma \ = \ \int_1^{\infty}\frac{[v(t)]-v(t)+(-v(-t))-[-v(-t)]}{t^2}\,dt$$

(a quantity between -1 and 1). In the second term, the expression in { }
lies between -1 and 1, while

$$\frac{\partial}{\partial t}\log\left|1+\frac{x+i}{t}\right| \ < \ 0$$

---

* One may also work directly with the expression
$\int_{|t|\geqslant 1}\{\log|1 - (x+i)/t|| + x/t\}d([v(t)]-v(t))$, adapting the proof of the lemma
in §A.1 to the (improper) integral involving the first term in { } and using partial
integration on what remains.

for $1 \leqslant t < \infty$ since $x \geqslant 0$. The second term is therefore $\leqslant \log|1 + x + i|$ (cf. proof of the lemma in §A.1).

Putting these results together, we see that

$$\int_{-\infty}^{\infty} \left(\log\left| 1 - \frac{x+i}{t} \right| + \frac{x}{t} \right)(\mathrm{d}[v(t)] - \mathrm{d}v(t))$$

$$\leqslant \log^+|x| + O(1) + \gamma x + \log|1 + x + i|$$

for $x \geqslant 0$, proving the desired inequality when that is the case. A similar argument may be used when x is negative.

Having established our relation, we proceed to construct the entire function $\varphi(z)$. Take S as the sequence of points where $[v(t)]$ *jumps*, each of those being *repeated* a number of times equal to the *magnitude* of the *jump* corresponding to it; S simply consists of some of the points x_k, $k = \pm 1, \pm 2, \ldots$, with certain repetitions. We then put

$$\varphi(z) = \mathrm{e}^{-\gamma z} \prod_{\lambda \in S} \left(1 - \frac{z}{\lambda} \right) \mathrm{e}^{z/\lambda},$$

taking care to repeat each of the factors on the right as many times as the λ corresponding to it is repeated in S. This function $\varphi(z)$ is entire, and clearly

$$\log|\varphi(z)| = -\gamma \Re z + \int_{-\infty}^{\infty} \left(\log\left| 1 - \frac{z}{t} \right| + \frac{\Re z}{t} \right) \mathrm{d}[v(t)].$$

The inequality proved above now yields

$$\log|\varphi(x + i)| \leqslant H(x + i) + 2\log^+|x| + O(1), \qquad x \in \mathbb{R}.$$

Obviously, $n_S(t) = [v(t)]$, so

$$\frac{n_S(t)}{t} \longrightarrow \eta \quad \text{as } t \longrightarrow \pm\infty.$$

Also,

$$\int_{-\infty}^{\infty} \frac{|n_S(t) - \eta t|}{1 + t^2} \mathrm{d}t \leqslant \int_{-\infty}^{\infty} \frac{|v(t) - \eta t| + 1}{1 + t^2} \mathrm{d}t = \int_{-\infty}^{\infty} \frac{|\delta(t)| + 1}{1 + t^2} \mathrm{d}t < \infty$$

by property (iv) of the lemma. The hypothesis of the theorem in article 1 therefore holds for the function $\mathrm{e}^{\gamma z}\varphi(z)$, so it – and hence $\varphi(z)$ – is of *exponential type*. That result (as well as the second Lindelöf theorem of §B, Chapter III, on which it depends) is also easily adapted so as to apply to functions like $H(z)$, and we thus find, reasoning as for $\mathrm{e}^{\gamma z}\varphi(z)$, that

$$H(z) \leqslant \text{const.}|z| + O(1).$$

$H(z)$ is, of course, *harmonic* in $\Im z > 0$. These properties of H are used to get a grip on $\log|F(x+i)| + H(x+i)$, the idea being to then make use of our relation between $\log|\varphi(x+i)|$ and $H(x+i)$.

Because $F(z)$ is of exponential type, we certainly have

$$\log|F(z)| + H(z) \leqslant \text{const.}|z| + O(1),$$

with the left side harmonic in $\Im z > 0$. The functions $|F(z)|$ and $e^{H(z)}$ are actually continuous right up to the real axis, as long as we take the value of the latter one to be *zero* at the points of S; moreover,

$$\int_{-\infty}^{\infty} \frac{(\log|F(x)| + H(x))_+}{1+x^2}\, dx \; < \; \infty$$

by the observations made at the beginning of this proof ($(a)_+$ denotes $\max(a,0)$ for real a). We can therefore use the theorem of §E, Chapter III (actually, a variant of it having, however, exactly the same proof) so as to conclude that (with an appropriate constant A)

$$\log|F(x+i)| \; + \; H(x+i) \; \leqslant \; A \; + \; \frac{1}{\pi}\int_{-\infty}^{\infty} \frac{(\log|F(t)| + H(t))_+}{(x-t)^2 + 1}\, dt.$$

Now we bring in the above relation involving $|\varphi(x+i)|$ and $H(x+i)$, and get

$$\log|F(x+i)| \; + \; \log|\varphi(x+i)| \; \leqslant \; O(1) \; + \; 2\log^+|x|$$

$$+ \; \frac{1}{\pi}\int_{-\infty}^{\infty} \frac{|\log|F(t)| + H(t)|}{(x-t)^2 + 1}\, dt,$$

from which we easily see that

$$\int_{-\infty}^{\infty} \frac{\log^+|F(x+i)\varphi(x+i)|}{x^2+1}\, dx \; < \; \infty,$$

following the procedure so often used in Chapter VI and elsewhere. Having arrived at this point, we may use the theorem from §E, Chapter III once more, this time in the half plane $\{\Im z \leqslant 1\}$ – the function $F(z)\varphi(z)$ *is* entire, and of exponential type. Doing that and then repeating the argument just referred to, we find that

$$\int_{-\infty}^{\infty} \frac{\log^+|F(x)\varphi(x)|}{1+x^2}\, dx \; < \; \infty,$$

which is what we wanted to prove.

We are done.

Remark. If the function $v(t) = \eta t + \delta(t)$ were known to be *integral valued*, $\log|\varphi(z)|$ could have been taken *equal* to $H(z)$ in the above proof, and the discussion about the effect of replacing $v(t)$ by $[v(t)]$ avoided. To realize this simplification, we would have had to modify the lemma's construction so as to make it yield a function $\delta(t)$ with $\delta(t) + \eta t$ integral valued. As a matter of fact, that can be done without too much difficulty, and one thus arrives at an alternative derivation of the preceding result. Such is the procedure followed by Redheffer in his survey article.

3. **Determination of the completeness radius for real and complex sequences Λ**

We are finally ready to apply the result stated in §A.2.

Theorem (Beurling and Malliavin, 1961). *Let Λ be a sequence of distinct real numbers having effective density $\tilde{D}_\Lambda < \infty$. Then the completeness radius associated with Λ is equal to $\pi\tilde{D}_\Lambda$.*

Proof. According to the discussion at the beginning of this §, it is enough to show that the $e^{i\lambda t}$, $\lambda\in\Lambda$, are not complete on any interval of length $> 2\pi\tilde{D}_\Lambda$, and for that purpose it suffices, as explained there, to establish, for arbitrary $\eta > 0$, the existence of a non-zero entire function $G(z)$ of exponential type $\leqslant \pi(\tilde{D}_\Lambda + 3\eta)$ with

$$\int_{-\infty}^{\infty} |G(x)|\,dx < \infty$$

and

$$G(\lambda) = 0 \quad \text{for } \lambda\in\Lambda.$$

Writing $D = \tilde{D}_\Lambda + \eta$, we take the real sequence $\Sigma \supseteq \Lambda$ such that

$$\int_{-\infty}^{\infty} \frac{|n_\Sigma(t) - Dt|}{1 + t^2}\,dt < \infty,$$

used in article 1 at the start of our constructions. We then *throw away any points that Σ may have in* $(-1, 1)$, *so as to ensure that $n_\Sigma(t) = 0$ there.** This perhaps leaves us with a certain finite number of $\mu\in\Lambda$ *not belonging to* Σ (the points of Λ in $(-1, 1)$); those will be taken care of in a moment. We next form

$$F(z) = \prod_{\substack{\lambda\in\Sigma\\\lambda\notin(-1,1)}} \left(1 - \frac{z}{\lambda}\right)e^{z/\lambda}$$

as in article 1, and use the little multiplier theorem from article 2 to get

---

* That does not affect the preceding relation!

the non-zero entire function $\varphi(z)$ of exponential type described there, such that

$$\int_{-\infty}^{\infty} \frac{\log^+ |F(x)\varphi(x)|}{1 + x^2}\, dx \;\; < \;\; \infty.$$

As remarked just after the statement of the little multiplier theorem (and as one checks immediately), when $y \longrightarrow \pm\infty$,

$$\frac{\log|F(iy)\varphi(iy)|}{|y|} \;\longrightarrow\; \pi(D + \eta) \;\;=\;\; \pi(\tilde{D}_\Lambda + 2\eta).$$

Put now

$$F_0(z) \;\;=\;\; F(z) \prod_{\substack{\mu \in \Lambda \\ -1 < \mu < 1}} (z - \mu).$$

The relations just written obviously still hold with F_0 standing in place of F. The *theorem on the multiplier* enunciated in §A.2 therefore gives us a non-zero entire function $\psi(z)$ of exponential type $\leqslant \pi\eta$ with

$$|F_0(x)\varphi(x)\psi(x)| \;\leqslant\; \text{const.}, \qquad x \in \mathbb{R}.$$

In view of the previous relation, we clearly have

$$\limsup_{y \to \pm\infty} \frac{\log|F_0(iy)\varphi(iy)\psi(iy)|}{|y|} \;\leqslant\; \pi(\tilde{D}_\Lambda + 3\eta),$$

so the *boundedness* of $F_0\varphi\psi$ on the real axis implies that that product is *of exponential type* $\leqslant \pi(\tilde{D}_\Lambda + 3\eta)$ by the *third* Phragmén–Lindelöf theorem from §C of Chapter III.

The function $\varphi(z)$ furnished by the little multiplier theorem has a zero at each point of the real sequence S, with

$$\frac{n_S(t)}{t} \;\longrightarrow\; \eta \;\; \text{for } t \longrightarrow \pm\infty.$$

We can thus certainly take *two different points* $s_1, s_2 \in S$ (!), and $\varphi(z)/(z - s_1)(z - s_2)$ will still be entire. Let, finally,

$$G(z) \;\;=\;\; \frac{F_0(z)\varphi(z)\psi(z)}{(z - s_1)(z - s_2)}.$$

This function is entire and of exponential type $\leqslant \pi(\tilde{D}_\Lambda + 3\eta)$, and

$$\int_{-\infty}^{\infty} |G(x)|\, dx \;\; < \;\; \infty.$$

Also, $G(z)$ vanishes at each point of Λ since $F_0(z)$ does. We are done.

Remark. If Λ has repeated points, the argument just made goes through without change. The entire function $G(z)$ thus obtained then vanishes with appropriate multiplicitly at each point of Λ.

In order to now obtain the completeness radius for sequences of *complex* numbers Λ, we proceed somewhat as in §H.3 of Chapter III.

Notation. For complex λ *with non-zero real part*, we write

$$\lambda' = 1/\Re(1/\lambda).$$

If Λ is any sequence of complex numbers, we let Λ' be the real sequence consisting of the λ' corresponding to the $\lambda \in \Lambda$ having non-zero real part. Should several members of Λ correspond to the *same* value for λ', we look on that value as *repeated an appropriate number of times* in Λ'.

We then have the

Theorem (Beurling and Malliavin, 1967). *Let Λ be any sequence of distinct complex numbers. If*

$$\sum_{\substack{\lambda \in \Lambda \\ \lambda \neq 0}} \frac{|\Im\lambda|}{|\lambda|^2} = \infty,$$

the exponentials $e^{i\lambda t}$, $\lambda \in \Lambda$, are complete on any interval of finite length.

Otherwise, they are complete on any interval of length $< 2\pi\tilde{D}_{\Lambda'}$, and, if that quantity is finite, incomplete on any interval of length $> 2\pi\tilde{D}_{\Lambda'}$.

Proof. If the $e^{i\lambda t}$ are incomplete on (say) the interval $[-L, \ L]$, there is (as at the beginning of §D, Chapter IX) a non-zero measure μ on $[-L, \ L]$ with

$$\int_{-L}^{L} e^{i\lambda t} d\mu(t) = 0, \qquad \lambda \in \Lambda.$$

The non-zero entire function

$$G(z) = \int_{-L}^{L} e^{izt} d\mu(t)$$

of exponential type $\leqslant L$ thus vanishes at each point of Λ, so, since G is bounded on the real axis, we have

$$\sum_{\substack{\lambda \in \Lambda \\ \lambda \neq 0}} \frac{|\Im\lambda|}{|\lambda|^2} < \infty$$

by §G.3 of Chapter III.

Let Σ denote the complete sequence of zeros of G (with repetitions according to multiplicities, as usual). The Hadamard representation for G is then

$$G(z) \;=\; Az^p e^{cz} \prod_{\substack{\mu \in \Sigma \\ \mu \neq 0}} \left(1 - \frac{z}{\mu}\right) e^{z/\mu}.$$

Denote by S the difference set

$$\Sigma \;\sim\; \{\lambda \in \Lambda : \, \Re\lambda \neq 0\};$$

then

$$G(z) \;=\; Az^p e^{cz} \prod_{\substack{\mu \in S \\ \mu \neq 0}} \left(1 - \frac{z}{\mu}\right) e^{z/\mu} \cdot \prod_{\substack{\lambda \in \Lambda \\ \Re\lambda \neq 0}} \left(1 - \frac{z}{\lambda}\right) e^{z/\lambda}.$$

Take now the function

$$G_0(z) \;=\; Az^p e^{cz} \prod_{\substack{\mu \in S \\ \mu \neq 0}} \left(1 - \frac{z}{\mu}\right) e^{z/\mu} \cdot \prod_{\lambda' \in \Lambda'} \left(1 - \frac{z}{\lambda'}\right) e^{z/\lambda}$$

(with exponentials $e^{z/\lambda}$ and *not* $e^{z/\lambda'}$ in the second product!). By work done in §H.3 of Chapter III, we see that $G_0(z)$ is of exponential type, and that

$$|G_0(x)| \;\leqslant\; |G(x)|, \qquad x \in \mathbb{R},$$

so that $G_0(x)$ is bounded on the real axis (like $G(x)$).

Write

$$B \;=\; \limsup_{y \to \infty} \frac{\log|G_0(iy)|}{y}$$

and

$$B' \;=\; \limsup_{y \to -\infty} \frac{\log|G_0(iy)|}{|y|}.$$

Observe also that

$$\log|G(z)| \;\leqslant\; L|\Im z| + O(1)$$

by the third Phragmén–Lindelöf theorem of §C, Chapter III, since G is of exponential type $\leqslant L$ and bounded on the real axis.

Apply now Levinson's theorem (Chapter III, §H.3) to the zero distribution for $G_0(z)$, and then use Jensen's formula on the one for $G(z)$ together with the estimate just written for the latter function. The two zero distributions are the same asymptotically,* so, by an argument just

---

* Refer to volume I, pp. 74–5.

like the one at the end of §H.3, Chapter III, it is found that

$$\frac{B + B'}{2} \leqslant L.$$

Once this is known, we have by §D of Chapter IX,

$$\pi \tilde{D}_{\Lambda'} \leqslant \frac{B + B'}{2} \leqslant L,$$

since $G_0(z)$ vanishes* at the points of $\Lambda' \subseteq \mathbb{R}$. *Incompleteness* of the $e^{i\lambda t}$, $\lambda \in \Lambda$, on $[-L, L]$ thus implies that $L \geqslant \pi \tilde{D}_{\Lambda'}$, and those exponentials must therefore be complete on any interval of length $< 2\pi \tilde{D}_{\Lambda'}$.

We must now show that if

$$\sum_{\substack{\lambda \in \Lambda \\ \lambda \neq 0}} \frac{|\Im \lambda|}{|\lambda|^2} < \infty$$

and $\tilde{D}_{\Lambda'} < \infty$, the $e^{i\lambda x}$, $\lambda \in \Lambda$, are *incomplete* on any interval of length $> 2\pi \tilde{D}_{\Lambda'}$. Fix any $\eta > 0$. The previous theorem and remark then give us an entire function $f(z) \not\equiv 0$ of exponential type $\leqslant \pi(\tilde{D}_{\Lambda'} + 3\eta)$, vanishing[†] at each point of Λ', and such that (wlog)

$$|f(x)| \leqslant 1, \qquad x \in \mathbb{R}.$$

Denote by Ξ the set of *non-zero, purely imaginary* $\mu \in \Lambda$, and then put

$$g(z) = f(z) \cdot \prod_{\lambda' \in \Lambda'} \left(\frac{1 - z/\lambda}{1 - z/\lambda'} \right) \cdot \prod_{\mu \in \Xi} \left(1 - \frac{z}{\mu} \right).$$

Using the two Lindelöf theorems of §B, Chapter III we easily see that $g(z)$ is of exponential type, thanks to the convergence of the above sum of the $|\Im \lambda|/|\lambda|^2$. $g(z)$ vanishes at each $\lambda \in \Lambda$ (save that at the origin, in case $0 \in \Lambda$). By calculations like one made in §H.3, Chapter III, we also verify without difficulty that

$$\log \prod_{\lambda' \in \Lambda'} \left| \frac{1 - iy/\lambda}{1 - iy/\lambda'} \right| \leqslant o(|y|)$$

and

$$\log \prod_{\mu \in \Xi} \left| 1 - \frac{iy}{\mu} \right| \leqslant o(|y|)$$

---

* with the appropriate multiplicity at repeated points of Λ', which may well *have* some, although Λ does not

† with the appropriate multiplicity

as $y \longrightarrow \pm \infty$, the second on account of the convergence of $\sum_{\mu \in \Xi} 1/|\mu|$.
Therefore, since $f(z)$ is of exponential type $\leqslant \pi(\tilde{D}_{\Lambda'} + 3\eta)$, we have

$$\limsup_{y \to \pm \infty} \frac{\log|g(iy)|}{|y|} \leqslant \pi(\tilde{D}_{\Lambda'} + 3\eta).$$

It is now claimed that

$$\int_{-\infty}^{\infty} \frac{\log^+|g(x)|}{1 + x^2} \, dx \; < \; \infty.$$

Because $|f(x)| \leqslant 1$, this will follow from the convergence of

$$\int_{-\infty}^{\infty} \frac{1}{1 + x^2} \log^+ \left| \prod_{\mu \in \Xi} \left(1 - \frac{x}{\mu} \right) \right| dx$$

and of

$$\int_{-\infty}^{\infty} \frac{1}{1 + x^2} \log^+ \left| \prod_{\lambda' \in \Lambda'} \left(\frac{1 - x/\lambda}{1 - x/\lambda'} \right) \right| dx.$$

Using the pure imaginary character of the $\mu \in \Xi$ and the relation
$1/\lambda' = \Re(1/\lambda)$, we see at once that the factors $1 - x/\mu$, $\mu \in \Xi$,
$(1 - x/\lambda)/(1 - x/\lambda')$, $\lambda' \in \Lambda'$, are all in modulus $\geqslant 1$ for real x. *The* $\log^+$
may therefore be replaced by log *in these integrals.*

Once this is done, the resulting expressions are easily worked out
explicitly using Poisson's formula for a half plane. In that way we find
the *first* integral to be equal to

$$\pi \sum_{\mu \in \Xi} \log \left(1 + \frac{1}{|\mu|} \right)$$

and the *second* to be

$$\leqslant \; \pi \sum_{\lambda' \in \Lambda'} \log \left(1 + \frac{|\Im\lambda|}{|\lambda|^2} \right).$$

Both of these sums, however, are *finite* since $\sum_{\substack{\lambda \in \Lambda \\ \lambda \neq 0}} |\Im\lambda|/|\lambda|^2 < \infty$.
Convergence of the logarithmic integral involving g is thus established.

Thanks to that convergence, *the theorem from §A.2 applies to our entire
function* g. There is, in other words, a non-zero entire function $\psi(z)$ of
exponential type $\leqslant \pi\eta$ with $g(x)\psi(x)$ bounded on $\mathbb{R}$; reasoning as at the
end of the preceding theorem's proof we can even choose ψ so as to have

$$\int_{-\infty}^{\infty} |xg(x)\psi(x)| \, dx \; < \; \infty.$$

Referring to the above estimate on $\log|g(iy)|/|y|$ and applying the usual
Phragmén–Lindelöf theorem, we see that $zg(z)\psi(z)$ is of exponential type

$\leqslant \pi(\tilde{D}_{\Lambda'} + 4\eta)$. This function certainly vanishes at each $\lambda \in \Lambda$, so the $e^{i\lambda t}$, $\lambda \in \Lambda$, are *not* complete on $[-\pi(\tilde{D}_{\Lambda'} + 4\eta), \pi(\tilde{D}_{\Lambda'} + 4\eta)]$ according to the discussion at the beginning of this §.

The theorem is completely proved.

Problem 39

Let Λ be any sequence of distinct complex numbers. Show that the completeness radius associated with Λ is equal to π times the *infimum* of the numbers $c > 0$ with the following property: there exist *distinct integers* n_λ corresponding to the different non-zero λ in Λ such that

$$\sum_{\substack{\lambda \in \Lambda \\ \lambda \neq 0}} \left| \frac{1}{\lambda} - \frac{c}{n_\lambda} \right| \; \leqslant \; \infty.$$

This criterion is due to Redheffer. (Hint: Look again at the constructions in §D.2 of Chapter IX.)

C. The multiplier theorem for weights with uniformly continuous logarithms

The result of Beurling and Malliavin enunciated in §A.2 is broader in scope than may appear at first sight. One can, for instance, deduce from it another multiplier theorem for weights fulfilling a simple descriptive regularity condition. This is done in article 1 below; the work depends on some elementary material from Chapter VI and the first part of Chapter VII.

In article 2, the theorem of article 1 is used to extend a result obtained in problem 11 (Chapter VII, §A.2) to certain unbounded measures on $\mathbb{R}$.

1   The multiplier theorem

Theorem (Beurling and Malliavin, 1961). *Let* $W(x) \geqslant 1$, *and let* $\log W(x)$ *be uniformly continuous* on* $\mathbb{R}$. *Then* W *admits multipliers iff*

---

* Beurling and Malliavin require only that
$$\omega(s) \;\; = \;\; \text{ess sup}_{x \in \mathbb{R}} |\log W(x + s) \; - \; \log W(x)| \text{ be } finite \text{ for a set of } s \in \mathbb{R} \text{ having}$$
positive Lebesgue measure. To reduce the treatment under this less stringent assumption to that of the uniform Lip 1 case handled below, they observe that there must be some $M < \infty$ with $\omega(s) \leqslant M$ on a Lebesgue measurable set E with $|E| > 0$. But then $E - E$ includes a whole interval $(-h, h)$, $h > 0$, so $\omega(s) \leqslant 2M$ for $|s| < h$, ω being clearly *even* and *subadditive*. From this point, one proceeds as in the text, passing from W to W_h; the only changes are in the constants.

This argument is valid as long as $W(x) \geqslant 1$ is Lebesgue measurable, for then
$$\omega(s) \;\; = \;\; \lim_{p \to \infty} \left(\int_{-\infty}^{\infty} |\log W(x + s) \; - \; \log W(x)|^p e^{-2|x|} \, dx \right)^{1/p}$$
is also Lebesgue measurable (the *integrals* are by Tonelli's theorem, $\log W(x + s) - \log W(x)$ being Lebesgue measurable on $\mathbb{R}^2$).

$$\int_{-\infty}^{\infty} \frac{\log W(x)}{1+x^2}\,dx \;\; < \;\; \infty.$$

Remark. The weaker assumption that $\log\log W(x)$ is uniformly continuous on $\mathbb{R}$ *does not imply* that W admits multipliers when the above integral is convergent. For an example, see the following chapter.

Proof of theorem. As explained at the beginning of §A, convergence of the integral in question is certainly *necessary if* W is to admit multipliers; we therefore need only concern ourselves with the *sufficiency* of that convergence in the present circumstances.

We may, to begin with, replace the hypothesis of *uniform continuity* for $\log W(x)$ by the stronger one that

$$|\log W(x) - \log W(x')| \;\leqslant\; C|x - x'| \qquad \text{for } x, x' \in \mathbb{R},$$

i.e., that $\log W$ be *uniformly* Lip 1 on $\mathbb{R}$. Indeed, the former property gives us a fixed $h > 0$ such that

$$|\log W(x) - \log W(x')| \;\leqslant\; 1 \quad \text{whenever } |x - x'| \;\leqslant\; h.$$

Take any smooth positive function φ supported on $[-h, h]$ with

$$\int_{-h}^{h} \varphi(t)\,dt \;=\; 1,$$

and define a new weight $W_h(x)$ by putting

$$\log W_h(x) \;=\; \int_{-h}^{h} (\log W(x-t))\varphi(t)\,dt \;=\; \int_{-\infty}^{\infty} \varphi(x-s)\log W(s)\,ds.$$

Then, by our choice of h,

$$\log W(x) \;-\; 1 \;\leqslant\; \log W_h(x) \;\leqslant\; \log W(x) \;+\; 1.$$

Again,

$$\frac{d\log W_h(x)}{dx} \;=\; \int_{-\infty}^{\infty} \varphi'(x-s)\log W(s)\,ds \;=\; \int_{-h}^{h} \varphi'(t)\log W(x-t)\,dt.$$

Here, since $\varphi(-h) = \varphi(h) = 0$,

$$\int_{-h}^{h} \varphi'(t)\log W(x)\,dt \;=\; 0,$$

so

$$\frac{d\log W_h(x)}{dx} \;=\; \int_{-h}^{h} \varphi'(t)(\log W(x-t) - \log W(x))\,dt.$$

By the choice of h, the integral on the right is in absolute value

$$\leqslant \int_{-h}^{h} |\varphi'(t)|\, dt \;=\; C, \quad \text{say},$$

so

$$\left| \frac{d \log W_h(x)}{dx} \right| \;\leqslant\; C, \qquad x \in \mathbb{R},$$

and $\log W_h(x)$ satisfies the Lipschitz condition written above.
We also have

$$e^{-1} W(x) \;\leqslant\; W_h(x) \;\leqslant\; e\, W(x)$$

by the previous estimate, so

$$\int_{-\infty}^{\infty} \frac{\log W_h(x)}{1+x^2}\, dx \;<\; \infty$$

provided that the corresponding integral with W is finite. If, then, we can conclude that $W_h(x)$ admits multipliers, the left-hand half of the preceding double inequality shows that $W(x)$ also does so, and it is enough to establish the theorem for weights W with $\log W$ uniformly Lip 1 on $\mathbb{R}$.

Assuming henceforth this Lipschitz condition on $\log W(x)$ and the convergence of the corresponding logarithmic integral, we set out to show that W admits multipliers. Our idea is to produce an entire function $K(z)$ of exponential type such that

$$\int_{-\infty}^{\infty} \frac{\log^+ |K(x)|}{1+x^2}\, dx \;<\; \infty,$$

while

$$4K(x) \;\geqslant\; (W(x))^{\alpha} \qquad \text{for } x \in \mathbb{R}$$

with a certain constant $\alpha > 0$. Application of the theorem from §A.2 to $K(z)$ will then yield multipliers for W.

Following a procedure of Akhiezer used in Chapters VI and VII, we form the new weight

$$W_1(x) \;=\; \sup\{|f(x)|:\, f \text{ entire, of exponential type } \leqslant 1,$$
$$\text{bounded on } \mathbb{R}, \text{ and } |f(t)/W(t)| \leqslant 1 \text{ on } \mathbb{R}\}.$$

If $\log W(x)$ satisfies the Lipschitz condition written above (with Lipschitz constant C) we have, by the *first* theorem of §A.1, Chapter VII,

$$W_1(x) \;\geqslant\; \tfrac{1}{2}(W(x))^{1/\sqrt{(C^2+1)}} \qquad \text{for } x \in \mathbb{R}.$$

What we want, then, is an entire function $K(z)$ of exponential type with convergent logarithmic integral, such that

$$K(x) \geqslant (W_1(x))^2,$$

say, for $x \in \mathbb{R}$.

In order to obtain $K(z)$, we use Akhiezer's theory of weighted approximation by sums of exponentials, presented in Chapter VI. We work, however, with a weighted L_2 norm instead of the weighted uniform one used there. Taking

$$\Omega(x) = (1 + x^2)^{1/2} W(x),$$

let us consider approximation by finite linear combinations of the $e^{i\lambda x}$, $-1 \leqslant \lambda \leqslant 1$, in the norm $\| \quad \|_{\Omega,2}$ defined by

$$\| g \|_{\Omega,2} = \sqrt{\left(\frac{1}{\pi} \int_{-\infty}^{\infty} \left| \frac{g(t)}{\Omega(t)} \right|^2 dt \right)}.$$

According to our assumed convergence of the logarithmic integral involving W, we have

$$\int_{-\infty}^{\infty} \frac{\log \Omega(x)}{1 + x^2} dx < \infty.$$

Hence, by a version of T. Hall's theorem (the *first* one of §D, Chapter VI) appropriate to approximation in the norm $\| \quad \|_{\Omega,2}$ (see §§E.2 and G of Chapter VI), linear combinations of the $e^{i\lambda x}$, $-1 \leqslant \lambda \leqslant 1$, are *not* $\| \quad \|_{\Omega,2}$ dense in the space of functions for which that norm is finite. This, and the Akhiezer theorem (Chapter VI, §E.2) corresponding to the norm $\| \quad \|_{\Omega,2}$ (see again §G, Chapter VI) imply that

$$\int_{-\infty}^{\infty} \frac{\log \Omega_1(x)}{1 + x^2} dx < \infty,$$

where

$$\Omega_1(x) = \sup \{ |f(x)| : f \text{ entire, of exponential type} \leqslant 1,$$
$$\text{bounded on } \mathbb{R}, \text{ and } \| f \|_{\Omega,2} \leqslant 1 \}.$$

Observe now that for any function f with $|f(t)/W(t)| \leqslant 1$ on $\mathbb{R}$, we certainly have $\| f \|_{\Omega,2} \leqslant 1$. Clearly, then,

$$\Omega_1(x) \geqslant W_1(x),$$

and, if we can show that $(\Omega_1(x))^2$ *coincides with an entire function of exponential type on the real axis*, we can simply take the latter as the function K we are seeking.

For that purpose we resort to a simple general argument. The space of Lebesgue measurable functions with finite $\| \ \|_{\Omega,2}$ norm is certainly *separable*, so, since the entire functions of exponential type $\leqslant 1$ bounded on $\mathbb{R}$ *belong* to that space, we may choose a (countable!) *sequence* of those which is $\| \ \|_{\Omega,2}$ *dense in the collection of all of them.* Using the inner product

$$\langle f, g \rangle_\Omega \ = \ \frac{1}{\pi} \int_{-\infty}^{\infty} \frac{f(x)\overline{g(x)}}{(\Omega(x))^2} \, dx,$$

one then applies *Schmidt's orthogonalization procedure* to that dense sequence, obtaining, after normalization, a sequence of entire functions $\varphi_n(z)$ of exponential type $\leqslant 1$, bounded on $\mathbb{R}$, with $\| \varphi_n \|_{\Omega,2} = 1$ and

$$\langle \varphi_n, \varphi_m \rangle_\Omega \ = \ 0 \quad \text{if } n \neq m.$$

Finite linear combinations of these φ_n are also $\| \ \|_{\Omega,2}$ dense in the collection of all such entire functions.

Fixing any $x_0 \in \mathbb{R}$ and any N, we look at the finite linear combinations

$$S(x) \ = \ \sum_{k=1}^{N} a_k \varphi_k(x)$$

such that $\| S \|_{\Omega,2} \leqslant 1$, seeking the one which makes $|S(x_0)|$ *a maximum*. Since the φ_k are orthonormal with respect to $\langle \ , \ \rangle_\Omega$, the condition on $\| S \|_{\Omega,2}$ is equivalent to

$$\sum_{k=1}^{N} |a_k|^2 \ \leqslant \ 1,$$

so, by Schwarz' inequality,

$$|S(x_0)| \ \leqslant \ \sqrt{\left(\sum_{k=1}^{N} |\varphi_k(x_0)|^2 \right)}.$$

For proper choice of the coefficients a_k, the two sides are the same; the *maximum value* of $|S(x_0)|$ for the sums S is *thus equal to the quantity on the right.*

Put

$$K_N(z) \ = \ \sum_{k=1}^{N} \varphi_k(z) \overline{\varphi_k(\bar{z})};$$

this function is entire, of exponential type $\leqslant 2$ (*sic* !), and bounded on the real axis, where it is also $\geqslant 0$. As we have just seen, for each given $x \in \mathbb{R}$, the maximum value of $|S(x)|$ for sums S of the kind just specified is equal to $\sqrt{(K_N(x))}$. It is now claimed that when $N \longrightarrow \infty$, the $K_N(z)$ converge u.c.c to a certain entire function $K(z)$ of exponential type $\leqslant 2$, and that

$$K(x) = (\Omega_1(x))^2, \qquad x \in \mathbb{R}.$$

Given any particular N, we have, for each of the sums S,

$$|S(x)| \leqslant \Omega_1(x)$$

by *definition* of the function on the right, so

$$0 \leqslant K_N(x) \leqslant (\Omega_1(x))^2.$$

As we already know,

$$\int_{-\infty}^{\infty} \frac{\log(\Omega_1(x))^2}{1+x^2} dx \; < \; \infty.$$

Therefore, since the $K_N(z)$ are of exponential type $\leqslant 2$, the last relation implies that they all satisfy a *uniform estimate* of the form

$$|K_N(z)| \leqslant C_\varepsilon \exp(2|\Im z| + \varepsilon|z|), \qquad z \in \mathbb{C}.$$

Here $\varepsilon > 0$ is arbitrary, and C_ε depends on it, but is *completely independent* of N. The statement just made is nothing other than an *adaptation*, to approximation in the norm $\| \quad \|_{\Omega,2}$, of the *fourth* theorem in §E.2, Chapter VI, proved by the familiar Akhiezer argument of §B.2 in that chapter.

By the estimate just found, the K_N form a normal family in the complex plane, and any convergent sequence of them tends to an entire function for which the same estimate holds. However, $K_{N+1}(x) \geqslant K_N(x)$ on $\mathbb{R}$, so the entire sequence of the K_N is *already* convergent, and

$$K(z) = \lim_{N \to \infty} K_N(z)$$

is an entire function, obviously of exponential type $\leqslant 2$.

We still have to prove that

$$K(x) = (\Omega_1(x))^2$$

on $\mathbb{R}$. Of course, $0 \leqslant K(x) \leqslant (\Omega_1(x))^2$ since each $K_N(x)$ has that property, and it suffices to show the reverse inequality. Take any $x_0 \in \mathbb{R}$, and choose an entire function $f(z)$ of exponential type $\leqslant 1$, bounded on $\mathbb{R}$, with $\| f \|_{\Omega,2} \leqslant 1$ and at the same time $|f(x_0)|$ *close to* $\Omega_1(x_0)$. By our choice of the φ_n, the orthogonal series development

$$\sum_k \langle f, \varphi_k \rangle_\Omega \varphi_k(x)$$

converges in norm $\| \quad \|_{\Omega,2}$ to $f(x)$. For the partial sums

$$P_N(x) = \sum_{k=1}^{N} \langle f, \varphi_k \rangle_\Omega \, \varphi_k(x)$$

we have, however,

$$\|P_N\|_{\Omega,2} \leqslant \|f\|_{\Omega,2} \leqslant 1,$$

so by definition,

$$|P_N(x)| \leqslant \Omega_1(x), \quad x \in \mathbb{R}.$$

Hence, since the P_N are of exponential type $\leqslant 1$, another application of our version of the fourth theorem from §E.2, Chapter VI, gives us the uniform estimate

$$|P_N(z)| \leqslant \tilde{C}_\varepsilon \exp(|\Im z| + \varepsilon|z|), \quad z \in \mathbb{C},$$

on them. (Again, $\varepsilon > 0$ is arbitrary and $\tilde{C}_\varepsilon$ depends on it, but is independent of N.) The function $f(z)$ of course satisfies the same kind of estimate, and u.c.c convergence of the $P_N(z)$ to $f(z)$ now follows from the relation

$$\|f - P_N\|_{\Omega,2} \xrightarrow[N]{} 0$$

by a simple normal family argument. We see in particular that

$$P_N(x_0) \xrightarrow[N]{} f(x_0).$$

Since $\|f\|_{\Omega,2} \leqslant 1$, however,

$$|P_N(x_0)| \leqslant \sqrt{\left(\sum_{k=1}^{N} |\langle f, \varphi_k \rangle_\Omega|^2\right)} \sqrt{(K_N(x_0))} \leqslant \sqrt{(K_N(x_0))},$$

so

$$\sqrt{(K_N(x_0))} \geqslant |f(x_0)| - \varepsilon$$

with arbitrary $\varepsilon > 0$ for large enough N. Thence,

$$\sqrt{(K(x_0))} \geqslant |f(x_0)|.$$

But we chose f with $|f(x_0)|$ *close* to $\Omega_1(x_0)$ – indeed, *as close as we like*. Finally, then,

$$\sqrt{(K(x_0))} \geqslant \Omega_1(x_0),$$

whence

$$K(x) = (\Omega_1(x))^2, \quad x \in \mathbb{R},$$

the reverse inequality having already been noted.

We are at this point essentially done. The entire function $K(z)$ of exponential type $\leqslant 2$ satisfies the relation just written. We have

$$\Omega_1(x) \;\geqslant\; W_1(x) \;\geqslant\; \tfrac{1}{2}(W(x))^{1/\sqrt{(C^2+1)}}$$

on $\mathbb{R}$, where $W(x) \geqslant 1$, so $4K(x) \geqslant 1$, $\;x \in \mathbb{R}$, and it follows from

$$\int_{-\infty}^{\infty} \frac{\log \Omega_1(x)}{1+x^2}\,dx \;<\; \infty$$

that

$$\int_{-\infty}^{\infty} \frac{\log^+ K(x)}{1+x^2}\,dx \;<\; \infty.$$

The theorem of Beurling and Malliavin from §A.2 now gives us, for any $\eta > 0$, an entire function $\psi(z) \not\equiv 0$ of exponential type $\leqslant \eta$ with

$$4K(x)|\psi(x)| \;\leqslant\; 1, \qquad x \in \mathbb{R},$$

i.e.,

$$|\psi(x)|(W(x))^{2/\sqrt{(C^2+1)}} \;\leqslant\; 1, \qquad x \in \mathbb{R}.$$

Taking any fixed integer m with

$$\frac{1}{m} \;<\; \frac{2}{\sqrt{(C^2+1)}},$$

we get

$$W(x)|(\psi(x))^m| \;\leqslant\; 1 \quad \text{on } \mathbb{R}.$$

Here $(\psi(z))^m$ is entire, of exponential type $\leqslant m\eta$, and not identically zero. Hence $W(x)$ admits multipliers, $\eta > 0$ being arbitrary. The theorem is proved.

Remark. Beurling and Malliavin did not derive this result from their theorem stated in §A.2. Instead, they gave an independent proof similar to the one furnished by them for the latter result. See the end of §C.5 in Chapter XI.

2.   A theorem of Beurling

In order to indicate the location and extent of the intervals on $\mathbb{R}$ where a complex measure μ has little or no mass, Beurling, in his Stanford lectures, used a function $\sigma(x)$ related to μ by the formula

$$e^{-\sigma(x)} \;=\; \int_{-\infty}^{\infty} e^{-|x-t|}\,|d\mu(t)|.$$

(Truth to tell, Beurling wrote $\sigma(x)$ where we write $-\sigma(x)$. Some of the formulas used in working with this function look a little simpler if the minus sign is taken in the exponent as we do here.)

For *finite* measures μ, $\sigma(x)$ is bounded below $-\sigma(x)$ is *positive* if $\int_{\mathbb{R}} |d\mu(t)| \leq 1$. *Large* values of $\sigma(x)$ then correspond to the abscissae near which μ has *very little mass*. In problem 11 (§A.2, Chapter VII) the reader was asked to show that if the function $\sigma(x)$ associated with a finite complex measure μ is *so large that*

$$\int_{-\infty}^{\infty} \frac{\sigma(x)}{1+x^2}\,dx = \infty,$$

then the Fourier–Stieltjes transform

$$\hat{\mu}(\lambda) = \int_{-\infty}^{\infty} e^{i\lambda t}\,d\mu(t)$$

cannot vanish over any interval of positive length without μ's vanishing identically. Beurling originally established the gap theorem in §A.2, Chapter VII, with the help of this result, which is also due to him.

The multiplier theorem from the preceding article may be used to show that the result quoted is, in a certain sense, *best possible*. This application, set as problem 40, may be found at the end of the present article. Right now, we have in mind another application of that multiplier theorem, namely, *Beurling's extension of his result to certain unbounded complex measures μ.* This is also from his Stanford lectures.

When μ is *unbounded*, we can still define $\sigma(x)$ by means of the formula

$$e^{-\sigma(x)} = \int_{-\infty}^{\infty} e^{-|x-t|}\,|d\mu(t)|$$

as long as we admit the possibility that $\sigma(x) = -\infty$. In the case, however, that $\sigma(x) > -\infty$ for *any value* of x, *it is* $> -\infty$ *for all*. The reason for this is that $\sigma(x)$, if it is $> -\infty$ anywhere on $\mathbb{R}$, *is uniformly* Lip 1 there. To see that, we need only note that

$$|x'-t| \geq |x-t| - |x-x'|, \qquad t\in\mathbb{R},$$

whence

$$\int_{-\infty}^{\infty} e^{-|x'-t|}|d\mu(t)| \leq e^{|x'-x|}\int_{-\infty}^{\infty} e^{-|x-t|}|d\mu(t)|,$$

and

$$\sigma(x') \geq \sigma(x) - |x'-x|.$$

Interchanging x and x', we find that

$$|\sigma(x') - \sigma(x)| \leqslant |x' - x|,$$

a relation used several times in the following discussion.

Let us consider an unbounded μ for which $\sigma(x) > -\infty$. In this more general situation, $\sigma(x)$ is usually *not* bounded below (as it was for finite μ), and we need to look separately at

$$\sigma^+(x) \quad = \quad \max(\sigma(x),\ 0)$$

and

$$\sigma^-(x) \quad = \quad -\min(\sigma(x),\ 0) \qquad (sic!).$$

Since $\sigma(x)$ is uniformly Lip 1, *so are* $\sigma^+(x)$ *and* $\sigma^-(x)$.

The functions σ^+ and σ^- serve different purposes. *Large values of* $\sigma^+(x)$ correspond (as in the case of $\sigma(x)$ when dealing with finite measures) to the abscissae near which μ *has very little mass.* $\sigma^-(x)$, on the other hand, is *large near the places where* μ *has a great deal of mass.* With unbounded μ, one expects to come upon more and more such places (where $\sigma^-(x)$ assumes ever larger values) as x goes out to $+\infty$ or $-\infty$ along the real axis.

Beurling considered unbounded measures μ having *growth limited in such a way as to make*

$$\int_{-\infty}^{\infty} \frac{\sigma^-(x)}{1+x^2}\,dx \ < \ \infty.$$

Lemma. *Under the boxed condition on* σ^-, $\sigma^-(x)$ *is* $o(|x|)$ *for* $x \longrightarrow \pm\infty$.

Proof. Let $0 < c < 1$, and suppose that for any large x_0, we have

$$\sigma^-(x_0) \geqslant 2cx_0.$$

Then, by the Lip 1 property of σ^-,

$$\sigma^-(x) \geqslant cx_0 \quad \text{for } (1-c)x_0 \leqslant x \leqslant (1+c)x_0,$$

so

$$\int_{(1-c)x_0}^{(1+c)x_0} \frac{\sigma^-(x)}{x^2}\,dx \ \geqslant \ \frac{2c^2}{(1+c)^2}.$$

If the boxed relation holds, this cannot happen for arbitrarily large values of x_0, and $\sigma^-(x)$ must be $o(|x|)$ for $x \longrightarrow \infty$. Similarly for $x \longrightarrow -\infty$.

Lemma

$$\int_{x-1}^{x+1} e^{-\sigma^-(t)} |d\mu(t)| \;\leqslant\; e^2.$$

Proof. By definition,

$$e^{-\sigma(x)} \;\geqslant\; \int_{x-1}^{x+1} e^{-|x-t|} |d\mu(t)| \;\geqslant\; e^{-1} \int_{x-1}^{x+1} |d\mu(t)|.$$

This holds *a fortiori* if $\sigma(x)$ is replaced by $-\sigma^-(x) \leqslant \sigma(x)$. The Lip 1 property of σ^- now makes

$$\sigma^-(t) \;\geqslant\; \sigma^-(x) - 1, \qquad x - 1 \leqslant t \leqslant x + 1,$$

so we have

$$\int_{x-1}^{x+1} e^{-\sigma^-(t)} |d\mu(t)| \;\leqslant\; e \int_{x-1}^{x+1} e^{-\sigma^-(x)} |d\mu(t)| \;\leqslant\; e^2.$$

Done.

From these two lemmas we see that if the function σ^- corresponding to an unbounded complex measure μ fulfills the above boxed condition,

$$\int_{-\infty}^{\infty} e^{-\delta|t|} |d\mu(t)|$$

is convergent for every $\delta > 0$. In that circumstance, *the Fourier–Stieltjes transforms*

$$\hat{\mu}_\delta(\lambda) \;=\; \int_{-\infty}^{\infty} e^{i\lambda t} e^{-\delta|t|} d\mu(t)$$

are available. The $\hat{\mu}_\delta$ are nothing but the Abel means frequently used in harmonic analysis to try to give meaning to the expression

$$\int_{-\infty}^{\infty} e^{i\lambda t} d\mu(t)$$

when the integral is not absolutely convergent. It is often possible to interpret the latter as a limit (in some sense) of the $\hat{\mu}_\delta(\lambda)$ as $\delta \longrightarrow 0$ for certain (or sometimes even all) values of $\lambda \in \mathbb{R}$. We have examples of such treatment in the first lemma and theorem of §H.1, Chapter VI.

Definition. If $\int_{-\infty}^{\infty} e^{-\delta|t|} |d\mu(t)| < \infty$ for each $\delta > 0$, we say that "$\hat{\mu}(\lambda)$" ($\hat{\mu}(\lambda)$ itself is in general not defined!) vanishes on a closed interval $[a, b]$ of $\mathbb{R}$ provided that

$$\hat{\mu}_\delta(\lambda) \longrightarrow 0 \text{ uniformly for } a \leqslant \lambda \leqslant b$$

as $\delta \longrightarrow 0$.

In terms of this notion, Beurling's extension of the result established in problem 11 takes the following form:

Theorem (Beurling) *Let μ be a complex-valued Radon measure on $\mathbb{R}$ for which*

$$\int_{-\infty}^{\infty} \frac{\sigma^-(x)}{1+x^2} dx < \infty,$$

but at the same time,

$$\int_{-\infty}^{\infty} \frac{\sigma^+(x)}{1+x^2} dx = \infty,$$

σ^- *and* σ^+ *being the functions related to* μ *in the manner described above. If also "$\hat{\mu}(\lambda)$" vanishes on an interval of positive length, μ is identically zero.*

Remark. According to the previous discussion, the condition on $\sigma^-(x)$ means that μ, although (perhaps) unbounded, *does not accumulate too much mass anywhere.* The one involving $\sigma^+(x)$ means that there are also *large parts* of $\mathbb{R}$ where μ has *very little mass.*

Proof of theorem. Let, wlog, "$\hat{\mu}(\lambda)$" vanish on $[-A, A]$, where $A > 0$. The function

$$\sigma^-(x) + \log(1+x^2)$$

is uniformly Lip 1 on $\mathbb{R}$, so, *by the theorem of the preceding article*, the integral condition on σ^- makes it possible for us to get a non-zero entire function $f(z)$ of exponential type $a < A$ such that

$$|f(x)| \leqslant \frac{e^{-\sigma^-(x)}}{1+x^2}.$$

From this, the second of the above lemmas yields

$$\int_{x-1}^{x+1} |f(t)| \, |d\mu(t)| \leqslant \frac{1}{1+(|x|-1)^2} \int_{x-1}^{x+1} e^{-\sigma^-(t)} |d\mu(t)|$$

$$\leqslant \frac{e^2}{1+(|x|-1)^2} \quad \text{for } |x| \geqslant 1,$$

so by summation over integer values of x, we get

$$\int_{-\infty}^{\infty} |f(t)|\,|d\mu(t)| \quad < \quad \infty,$$

making

$$d\nu(t) \;=\; f(t)d\mu(t)$$

a totally finite measure on $\mathbb{R}$.

To ν we now apply the result from problem 11. Write

$$e^{-\tau(x)} \;=\; \int_{-\infty}^{\infty} e^{-|x-t|}|d\nu(t)|;$$

$\tau(x)$ is just the analogue of the function $\sigma(x)$ corresponding to the finite measure ν. Without loss of generality, $\int_{-\infty}^{\infty}|d\nu(t)| \leqslant 1$, so $\tau(x) \geqslant 0$. Also, $|f(x)| \leqslant 1$, so $|d\nu(t)| \leqslant |d\mu(t)|$ and

$$e^{-\tau(x)} \;\leqslant\; e^{-\sigma(x)},$$

i.e., $\tau(x) \geqslant \sigma(x)$. Combining this with the previous inequality, we get

$$\tau(x) \;\geqslant\; \sigma^+(x),$$

whence

$$\int_{-\infty}^{\infty} \frac{\tau(x)}{1+x^2}\,dx \;=\; \infty$$

by hypothesis.

It is now claimed that

$$\int_{-\infty}^{\infty} e^{i\lambda_0 t}\,d\nu(t) \;=\; 0 \quad \text{for } |\lambda_0| \leqslant A - a.$$

By the Paley–Wiener theorem, we have

$$f(t) \;=\; \int_{-a}^{a} e^{it\lambda}\varphi(\lambda)\,d\lambda,$$

where φ is (under the present circumstances) a continuous function on $[-a, a]$. Therefore,

$$\int_{-\infty}^{\infty} e^{i\lambda_0 t}d\nu(t) \;=\; \lim_{\delta \to 0} \int_{-\infty}^{\infty} e^{-\delta|t|}e^{i\lambda_0 t}f(t)d\mu(t)$$

$$=\; \lim_{\delta \to 0} \int_{-\infty}^{\infty}\int_{-a}^{a} e^{i\lambda_0 t}e^{it\lambda}\varphi(\lambda)e^{-\delta|t|}\,d\lambda\,d\mu(t)$$

$$=\; \lim_{\delta \to 0} \int_{-a}^{a}\int_{-\infty}^{\infty} e^{-\delta|t|}e^{i(\lambda_0 + \lambda)t}\,d\mu(t)\varphi(\lambda)\,d\lambda;$$

here, for each $\delta > 0$, absolute convergence holds throughout. The last limit is just

$$\lim_{\delta \to 0} \int_{-a}^{a} \varphi(\lambda)\hat{\mu}_{\delta}(\lambda + \lambda_0)\,dx$$

which is, however, *zero* when $|\lambda_0| \leqslant A - a$ since then $\hat{\mu}_{\delta}(\lambda + \lambda_0) \to 0$ uniformly for $|\lambda| \leqslant a$ as $\delta \to 0$.

The claim just established and the integral condition on $\tau(x)$ *now make* $\nu \equiv 0$ *by problem* 11. That is,

$$f(x)\,d\mu(x) \equiv 0.$$

If the function f vanishes at all on $\mathbb{R}$, it does so only at certain points x_n isolated from each other, for f is *entire* and *not identically zero*. What we have just proved is that μ, *if not identically zero*, has all its mass distributed on the points x_n. Then there must be *one* of those points, say x_0, for which

$$\mu(\{x_0\}) \neq 0.$$

That, however, *cannot happen*. If, for instance, x_0 is a k-fold zero of $f(z)$, we may *repeat the above argument using the entire function*

$$f_0(z) \;\; = \;\; \frac{f(z)}{(z - x_0)^k}$$

instead of f; doing so, we then find that

$$f_0(t)\,d\mu(t) \equiv 0.$$

Since $f_0(x_0) \neq 0$, we thus have

$$\mu(\{x_0\}) \;=\; 0,$$

a contradiction.

The measure μ must hence vanish identically. We are done.

Problem 40

Show that the result from problem 11 is best possible in the following sense:

If μ is a finite complex measure and

$$e^{-\sigma(x)} \;\; = \;\; \int_{-\infty}^{\infty} e^{-|x-t|}|d\mu(t)|,$$

then, in the case that

$$\int_{-\infty}^{\infty} \frac{\sigma(x)}{1+x^2}\,dx \; < \; \infty,$$

there *is* a finite non-zero complex measure v on $\mathbb{R}$ such that

$$\int_{-\infty}^{\infty} e^{-|x-t|}|dv(t)| \; \leqslant \; e^{-\sigma(x)}$$

but $\hat{v}(\lambda) \equiv 0$ outside some finite interval.

(Hint. One takes $dv(t) = f(t)dt$ where $f(t)$ is an entire function of exponential type chosen to satisfy $|f(x)| \leqslant e^{-\sigma(x)}/\pi(1+x^2)$.)

Yet another application of the result from article 1 is found near the end of Louis de Branges' book.

D. Poisson integrals of certain functions having given weighted quadratic norms

A condition involving the existence of multipliers is encountered when one desires to estimate certain harmonic functions whose boundary data are controlled by weighted norms. As a very simple example, let us consider the problem of estimating

$$U(z) \;\; = \;\; \frac{1}{\pi}\int_{-\infty}^{\infty} \frac{\Im z}{|z-t|^2}\,U(t)\,dt, \qquad \Im z \, > \, 0,$$

when it is known that

$$\int_{-\infty}^{\infty} |U(t)|^2 w(t)\,dt \;\; \leqslant \;\; 1$$

with some given function $w(t) \geqslant 0$ belonging to $L_1(\mathbb{R})$. We may, if we like, require that

$$|U(t)| \;\; \leqslant \;\; \text{some } M \quad \text{for } t \in \mathbb{R},$$

where M is *unknown and beyond our control*. Is it possible, in these circumstances, to say anything about the magnitude of $|U(z)|$?

If the bounded function $U(t)$ is permitted to be *arbitrary*, a simple condition on w is both necessary and sufficient for the existence of an estimate on $U(z)$. Then, we may wlog take $z = i$, and rewrite $\pi U(i)$ as

$$\int_{-\infty}^{\infty} \frac{1}{w(t)(t^2+1)}\,U(t)w(t)\,dt.$$

The very rudiments of analysis now tell us that this integral is bounded for

$$\int_{-\infty}^{\infty} |U(t)|^2 w(t) dt \;\leqslant\; 1$$

if and only if

$$\int_{-\infty}^{\infty} \frac{1}{w(t)(t^2+1)^2} dt \;<\; \infty.$$

It is for such w, then, and only for them, that the estimate in question (with arbitrary z having $\Im z > 0$) is available.

The situation alters when we restrict the *spectrum* of the functions $U(t)$ under consideration. In order not to get bogged down here in questions of harmonic analysis not really germane to the matter at hand, let us simply say that we look at *arbitrary finite sums*

$$S(t) \;=\; \sum_{\lambda \in \Sigma} A_\lambda e^{i\lambda t}$$

with some prescribed closed $\Sigma \subseteq \mathbb{R}$ — the *spectrum* for those sums. When $\Sigma = \mathbb{R}$, such sums are of course w^* dense in $L_\infty(\mathbb{R})$, because an L_1 function whose Fourier transform is everywhere zero must vanish identically. In that case we may think crudely of the collection of sums S as filling out the set of bounded functions U 'for all practical purposes', and our problem boils down to the simple one with the solution just described. It is thus natural to ask what happens when $\Sigma \neq \mathbb{R}$, and the simplest situation in which this occurs is the one where $\mathbb{R} \sim \Sigma$ *consists of one finite interval.* Then, we may take the complementary interval to be symmetric about 0, and it is possible to describe completely the functions w for which estimates of the above kind on the sums $S(t)$ exist. The description is in terms of multipliers.

Theorem. *Let* $w \geqslant 0$ *belong to* $L_1(\mathbb{R})$ *and let* $a > 0$. *A necessary and sufficient condition for the existence of a* β, $\Im\beta > 0$, *and corresponding constant* C_β, *such that*

$$\left| \int_{-\infty}^{\infty} \frac{\Im\beta}{|t-\beta|^2} S(t)\,dt \right| \;\leqslant\; C_\beta \sqrt{\left(\int_{-\infty}^{\infty} |S(t)|^2 w(t)\,dt \right)}$$

for all finite sums

$$S(t) \;=\; \sum_{|\lambda| \geqslant a} A_\lambda e^{i\lambda t},$$

is that there exist a non-zero entire function $\varphi(t)$ *of exponential type* $\leqslant a$

which makes

$$\int_{-\infty}^{\infty} \frac{|\varphi(t)|^2}{w(t)(t^2+1)^2}\,\mathrm{d}t \quad < \quad \infty.$$

Proof: Necessity. It is convenient to work with the Hilbert space norm

$$\|f\| \;=\; \sqrt{\left(\int_{-\infty}^{\infty} |f(t)|^2 w(t)\,\mathrm{d}t\right)}$$

and corresponding inner product

$$\langle f, g\rangle \;=\; \int_{-\infty}^{\infty} f(t)\overline{g(t)}w(t)\mathrm{d}t.$$

Assuming, then, that for some β, $\Im\beta > 0$, we have

$$\left|\int_{-\infty}^{\infty} \frac{\Im\beta}{|\beta - t|^2} S(t)\,\mathrm{d}t\right| \;\leqslant\; C_\beta \|S\|$$

for all sums S of the given form, there must, by the Hahn–Banach theorem*, be a measurable $k(t)$ for which

$$\|k\| \;<\; \infty$$

and

$$\int_{-\infty}^{\infty} \frac{\Im\beta}{|\beta - t|^2} S(t)\,\mathrm{d}t \;=\; \langle S, k\rangle$$

for such S. Taking just $S(t) = \mathrm{e}^{\mathrm{i}\lambda t}$ with $|\lambda| \geqslant a$, we find that

$$\int_{-\infty}^{\infty} \left(w(t)\overline{k(t)} - \frac{\Im\beta}{|\beta - t|^2}\right)\mathrm{e}^{\mathrm{i}\lambda t}\,\mathrm{d}t \;=\; 0$$

for such λ.

This shows, to begin with, that $w(t)\overline{k(t)}$ *is certainly not a.e. zero.* Since $\|k\| < \infty$ and $w\in L_1(\mathbb{R})$, we see by Schwarz' inequality that $w(t)\overline{k(t)} \in L_1(\mathbb{R})$. According to the last relation, then, the Fourier transform of the integrable function

$$w(t)\overline{k(t)} \;-\; \frac{\Im\beta}{|\beta - t|^2}$$

vanishes outside $[-a,\, a]$, so the latter must coincide a.e. on the real axis with $\psi(t)$, where ψ is *an entire function of exponential type* $\leqslant a$.

* or rather that theorem's special and elementary version for *Hilbert space*

We have

$$w(t)\overline{k(t)} \;=\; \psi(t) + \frac{\Im\beta}{(t-\beta)(t-\bar\beta)} \qquad \text{a.e., } t \in \mathbb{R}.$$

Here,

$$\varphi(t) \;=\; \psi(t)(t-\beta)(t-\bar\beta) + \Im\beta$$

is *also* entire and of exponential type $\leqslant a$, and

$$\varphi(t) \;=\; w(t)\overline{k(t)}|t-\beta|^2 \qquad \text{a.e., } t \in \mathbb{R},$$

so $\varphi \not\equiv 0$. Also,

$$\int_{-\infty}^{\infty} \frac{|\varphi(t)|^2}{w(t)|t-\beta|^4}\,dt \;=\; \|k\|^2 \;<\; \infty,$$

whence

$$\int_{-\infty}^{\infty} \frac{|\varphi(t)|^2}{w(t)(t^2+1)^2}\,dt \;<\; \infty.$$

Sufficiency. We continue to use the norm symbol $\|\ \ \|$ introduced above with the same meaning as before. Suppose there *is* a non-zero entire function φ of exponential type $\leqslant a$ such that

$$\int_{-\infty}^{\infty} \frac{|\varphi(t)|^2}{w(t)(t^2+1)^2}\,dt \;<\; \infty.$$

One may, to begin with, *exclude* the case where $\varphi(t)$ is *constant*, for then we would have

$$\int_{-\infty}^{\infty} \frac{dt}{w(t)(t^2+1)^2} \;<\; \infty,$$

making

$$\left| \int_{-\infty}^{\infty} \frac{\Im\beta}{|t-\beta|^2}\,U(t)dt \right| \;\leqslant\; C_\beta \|U\|$$

for *any* β, $\Im\beta > 0$, and *all* bounded U, by the discussion at the beginning of this §.

One may also take $\varphi(t)$ to be *real-valued* on $\mathbb{R}$. Indeed,

$$\varphi(z) \;=\; \frac{\varphi(z) + \overline{\varphi(\bar z)}}{2} + \frac{\varphi(z) - \overline{\varphi(\bar z)}}{2},$$

and *one* of the two functions on the right must be $\not\equiv$ 0. *Both* are entire and of exponential type $\leqslant a$, and, on the *real axis*, the *first* coincides with $\Re\varphi(t)$ and the *second* with $i\Im\varphi(t)$. The *first* one, or else the *second* one divided by i, will thus do the job.

Let $\Im\beta > 0$. Then

$$\frac{\varphi(t)}{(t-\beta)(t-\bar{\beta})} = \frac{\varphi(t) - \dfrac{\varphi(\beta)}{\beta-\bar{\beta}}(t-\bar{\beta}) - \dfrac{\varphi(\bar{\beta})}{\bar{\beta}-\beta}(t-\beta)}{(t-\beta)(t-\bar{\beta})}$$

$$+ \frac{\varphi(\beta)}{(\beta-\bar{\beta})(t-\beta)} + \frac{\varphi(\bar{\beta})}{(\bar{\beta}-\beta)(t-\bar{\beta})}.$$

The first term on the right, $\psi(t)$, is a certain *entire function of exponential type* $\leqslant a$, and, after multiplying by $\beta - \bar{\beta}$ and collecting terms, we get

$$\frac{(\Im\beta)\varphi(t)}{|t-\beta|^2} = (\Im\beta)\psi(t) + \frac{1}{2i}\left(\frac{\varphi(\beta)}{t-\beta} - \frac{\varphi(\bar{\beta})}{t-\bar{\beta}}\right),$$

that is,

$$\frac{(\Im\beta)\varphi(t)}{|t-\beta|^2} = (\Im\beta)\psi(t) + (\Im\varphi(\beta))\frac{t-\Re\beta}{|t-\beta|^2} + (\Re\varphi(\beta))\frac{\Im\beta}{|t-\beta|^2},$$

since $\varphi(\bar{\beta}) = \overline{\varphi(\beta)}$, φ being real on $\mathbb{R}$.

Suppose we can choose β in such a way that $\Im\varphi(\beta) = 0$ but $\Re\varphi(\beta) \neq 0$. Then

$$\left|\int_{-\infty}^{\infty} \frac{\Im\beta}{|t-\beta|^2} S(t)\,dt\right| \leqslant C_\beta \|S\|$$

for the sums

$$S(t) = \sum_{|\lambda|\geqslant a} A_\lambda e^{i\lambda t}.$$

Indeed,

$$\int_{-\infty}^{\infty} \frac{|\varphi(t)|}{|t-\beta|^2}\,dt \leqslant \sqrt{\left(\int_{-\infty}^{\infty} \frac{|\varphi(t)|^2}{w(t)|t-\beta|^4}\,dt \int_{-\infty}^{\infty} w(t)\,dt\right)},$$

a finite quantity, so the *left side* of the last identity is in $L_1(\mathbb{R})$. On the *right side*, the term

$$(\Im\varphi(\beta))\frac{t-\Re\beta}{|t-\beta|^2}$$

is *absent*, so, since $(\Re\varphi(\beta))\cdot\Im\beta/|t-\beta|^2$ is in $L_1(\mathbb{R})$, *so is* $\psi(t)$. ψ being *entire and of exponential type* $\leqslant a$, we have, however,

$$\int_{-\infty}^{\infty} \psi(t)S(t)dt = 0$$

for each of the sums S. Therefore, keeping in mind that $\Im\varphi(\beta) = 0$, we have for the latter

$$(\Re\varphi(\beta))\int_{-\infty}^{\infty}\frac{\Im\beta}{|t-\beta|^2}S(t)dt = (\Im\beta)\int_{-\infty}^{\infty}\frac{\varphi(t)}{|t-\beta|^2}S(t)\,dt.$$

The right side is in modulus

$$\leqslant \Im\beta\sqrt{\left(\int_{-\infty}^{\infty}\frac{|\varphi(t)|^2}{w(t)|t-\beta|^4}dt\right)}\cdot\|S\|,$$

so we are *done* as long as $\Re\varphi(\beta) \neq 0$.

We need therefore only show that there are β, $\Im\beta > 0$, with $\Im\varphi(\beta) = 0$ but $\Re\varphi(\beta) \neq 0$. It is claimed in the first place that there are β, $\Im\beta > 0$, with $\Im\varphi(\beta) = 0$. *Otherwise*, the harmonic function $\Im\varphi(z)$ *would be of one sign*, say $\Im\varphi(z) \geqslant 0$, for $\Im z > 0$. In such case, the *second* theorem of §F.1, Chapter III, gives us a number $\alpha \geqslant 0$ and a positive measure μ on $\mathbb{R}$ with

$$\Im\varphi(z) = \alpha\Im z + \frac{1}{\pi}\int_{-\infty}^{\infty}\frac{\Im z}{|z-t|^2}d\mu(t), \quad \Im z > 0.$$

Our function $\Im\varphi(z)$ is continuous right up to $\mathbb{R}$, φ being entire, so we readily see in the usual way that

$$d\mu(t) = (\Im\varphi(t))dt.$$

In our circumstances, however, $\Im\varphi(t) \equiv 0$, since we took $\varphi(t)$ to be *real on* $\mathbb{R}$. Hence $\Im\varphi(z) = \alpha\Im z$ for $\Im z > 0$ and finally

$$\varphi(z) = \alpha z + C.$$

Here, *we cannot have* $\alpha > 0$. For, if that were so, we would get

$$\int_{-\infty}^{\infty}\frac{|\alpha t+C|^2}{w(t)(t^2+1)^2}dt = \int_{-\infty}^{\infty}\frac{|\varphi(t)|^2}{w(t)(t^2+1)^2}dt < \infty,$$

whence, for A larger than $|C/\alpha|$,

$$\int_A^{\infty}\frac{dt}{w(t)(t^2+1)} < \infty.$$

From this, however, it would follow that

$$\int_A^\infty \frac{dt}{\sqrt{(t^2+1)}} \leqslant \sqrt{\left(\int_A^\infty \frac{dt}{w(t)(t^2+1)} \int_A^\infty w(t)\,dt\right)} < \infty,$$

which is nonsense.

Thus, $\alpha = 0$ and $\varphi(z)$ reduces to a constant C. This possibility was, however, excluded at the very beginning of the present argument – *our φ is not constant*. The function $\Im\varphi(z)$, then, *cannot be of one sign* in $\Im z > 0$, and we *have* points β in that half plane for which $\Im\varphi(\beta) = 0$.

Take any one of those – call it β_0. Since $\Im\varphi(z)$ is harmonic (everywhere!), we have

$$\int_{-\pi}^\pi \Im\varphi(\beta_0 + \rho e^{i\vartheta})\,d\vartheta = 2\pi\Im\varphi(\beta_0) = 0$$

for $\rho > 0$, and there must be a point on *each circle about* β_0 where $\Im\varphi$ also vanishes. There is thus a sequence of points $\beta_n \neq \beta_0$ in the upper half plane with $\beta_n \xrightarrow[n]{} \beta_0$ and $\Im\varphi(\beta_n) = 0$ for each n. If now $\Re\varphi(\beta_n)$ also vanished for each n, we would have $\varphi(z) \equiv 0$. But $\varphi \not\equiv 0$. Hence,

$$\Im\varphi(\beta_n) = 0 \quad \text{but} \quad \Re\varphi(\beta_n) \neq 0$$

for some n, and, taking that β_n as our β, we have what was needed. The *sufficiency* of our condition on w is thus established.

We are done.

When dealing with the harmonic functions

$$\frac{1}{2\pi}\int_{-\infty}^\infty \frac{\Im z}{|z-t|^2}\,S(t)\,dt,$$

one usually needs estimates on them in the *whole* upper half plane, and not just for *certain values* of z therein. The availability of these for sums $S(t)$ like those figuring in the last theorem with

$$\int_{-\infty}^\infty |S(t)|^2 w(t)\,dt \leqslant 1$$

is governed by a different condition on w.

Theorem. *Let $w(t) \geqslant 0$ belong to $L_1(\mathbb{R})$, and let $a > 0$. In order that the finite sums*

$$S(t) = \sum_{|\lambda| \geqslant a} A_\lambda e^{i\lambda t}$$

satisfy a relation

$$\left| \int_{-\infty}^{\infty} \frac{\Im z}{|z - t|^2} S(t) dt \right| \;\leqslant\; K_z \sqrt{\left(\int_{-\infty}^{\infty} |S(t)|^2 w(t) dt \right)}$$

for every z, $\Im z > 0$, it is necessary and sufficient that there exist a non-zero entire function ψ of exponential type $\leqslant a$ such that

$$\int_{-\infty}^{\infty} \frac{|\psi(t)|^2}{w(t)(t^2 + 1)} dt \;<\; \infty.$$

When this condition is met, the numbers K_z can be taken to be bounded above on compact subsets of $\{\Im z > 0\}$.

Proof: Sufficiency. If there *is* such a function ψ, we have, for any complex β,

$$\frac{\psi(t)}{t - \beta} \;=\; f_\beta(t) + \frac{\psi(\beta)}{t - \beta},$$

with

$$f_\beta(t) \;=\; \frac{\psi(t) - \psi(\beta)}{t - \beta}$$

entire and of exponential type $\leqslant a$. As long as $\beta \notin \mathbb{R}$, we may take $\psi(\beta)$ to be $\neq 0$. Indeed, we may assume that both $|\psi(\beta)|$ and $|\psi(\bar{\beta})|$ are bounded away from zero when β ranges over any compact subset E of $\{\Im z > 0\}$. To see this, observe that ψ has at most a finite number of zeros on $E \cup E^*$, where E^* is the reflection of E in $\mathbb{R}$. Calling those $z_1, z_2, \ldots, z_n$ (repetitions according to multiplicities, as usual), we may work with

$$\psi_E(t) \;=\; \frac{\psi(t)}{(t - z_1)(t - z_2) \cdots (t - z_n)}$$

instead of ψ. This function is entire, of exponential type $\leqslant a$, and bounded away from zero in modulus on $E \cup E^*$. And

$$\int_{-\infty}^{\infty} \frac{|\psi_E(t)|^2}{w(t)(t^2 + 1)} dt \;<\; \infty$$

because none of the z_k, $1 \leqslant k \leqslant n$, are on the real axis.*

All this being granted, we fix a β with $\Im \beta \neq 0$ and look at the entire function f_β figuring in the above relation. It is claimed that $f_\beta(t)$ is in

---

* $\psi(z)$ may *need* to vanish at some points on the real axis in order to offset certain zeros that the given weight w might have there! See the scholium at the end of this §.

$L_2(\mathbb{R})$. We have

$$\int_{-\infty}^{\infty} \frac{|\psi(t)|}{|t - \beta|} \, dt \;\; \leqslant \;\; \sqrt{\left(\int_{-\infty}^{\infty} \frac{|\psi(t)|^2}{w(t)|t - \beta|^2} \, dt \int_{-\infty}^{\infty} w(t) \, dt \right)} \;\; < \;\; \infty,$$

i.e., $|\psi(t)|/|t - \beta|$ is in $L_1(\mathbb{R})$. This ratio is also *bounded* on the real axis. Indeed, if ψ had no zeros at all its Hadamard factorization would reduce to $\psi(t) = Ce^{\gamma t}$ with constants C and γ. Then, however, $|\psi(t)|/|t - \beta|$ could not be integrable over $\mathbb{R}$. Consequently, ψ *has* a zero, say at z_0, and then $\psi(t)/(t - z_0)$ is *entire, of exponential type* $\leqslant a$, and in $L_1(\mathbb{R})$ since $\psi(t)/(t - \beta)$ is. A simple version of the Paley–Wiener theorem (Chapter III, §D) now shows that

$$\frac{\psi(t)}{t - z_0} \;\; = \;\; \int_{-a}^{a} e^{it\lambda} p(\lambda) \, d\lambda,$$

with (here) $p(\lambda)$ some *continuous* function on $[-a, a]$. By this formula, we see at once that $\psi(t)/(t - z_0)$ is bounded on $\mathbb{R}$ – so, then, is $\psi(t)/(t - \beta)$.

The ratio $|\psi(t)|/|t - \beta|$ is thus both in $L_1(\mathbb{R})$ and bounded on $\mathbb{R}$. Therefore it is in $L_2(\mathbb{R})$. So, however, is $\psi(\beta)/(t - \beta)$. The difference

$$f_\beta(t) \;\; = \;\; \frac{\psi(t)}{t - \beta} - \frac{\psi(\beta)}{t - \beta}$$

must hence also be square integrable.

Because f_β is entire and of exponential type $\leqslant a$, we now have

$$\text{l.i.m.} \int_{-A}^{A} e^{i\lambda t} f_\beta(t) \, dt \;\; = \;\; 0$$

for almost all $\lambda \notin [-a, a]$ by the L_2 form of the Paley–Wiener theorem (Chapter III, §D). At the same time, when $A \longrightarrow \infty$, the integrals

$$\int_{-A}^{A} e^{i\lambda t} f_\beta(t) \, dt$$

tend, for $\lambda \neq 0$, to a certain function of λ *continuous* on $\mathbb{R} \sim \{0\}$. This, indeed, is certainly true if, in those integrals, we replace $f_\beta(t)$ by $\psi(t)/(t - \beta) \in L_1(\mathbb{R})$. *Direct verification* shows that the same holds good when $f_\beta(t)$ is replaced by $\psi(\beta)/(t - \beta)$. The statement therefore holds for the difference $f_\beta(t)$ of these functions.

The (continuous) pointwise limit of the expressions

$$\int_{-A}^{A} e^{i\lambda t} f_{\beta}(t)\, dt, \qquad \lambda \neq 0,$$

for $A \longrightarrow \infty$ must, however, coincide a.e. with their limit in mean, known to be zero a.e. for $|\lambda| \geqslant a$, as we have just seen. Hence

$$\lim_{A \to \infty} \int_{-A}^{A} e^{i\lambda t} f_{\beta}(t)\, dt \;=\; 0, \qquad |\lambda| \geqslant a.$$

That is,

$$\lim_{A \to \infty} \int_{-A}^{A} \frac{\psi(\beta)}{t - \beta} e^{i\lambda t}\, dt \;=\; \int_{-\infty}^{\infty} \frac{\psi(t) e^{i\lambda t}}{(t - \beta)}\, dt$$

for $|\lambda| \geqslant a$, and finally,

$$\lim_{A \to \infty} \int_{-A}^{A} \frac{\psi(\beta) S(t)}{t - \beta}\, dt \;=\; \int_{-\infty}^{\infty} \frac{\psi(t) S(t)}{t - \beta}\, dt$$

for each of our sums $S(t)$.

Let us continue to write $\| \ \|$ for the norm appearing in the proof of the preceding theorem. In terms of this notation, we have for the modulus of the *right-hand* member of the last relation the upper bound

$$\sqrt{\left(\int_{-\infty}^{\infty} \frac{|\psi(t)|^2}{w(t)|t - \beta|^2}\, dt \right)} \cdot \| S \|.$$

The condition that

$$\int_{-\infty}^{\infty} \frac{|\psi(t)|^2}{w(t)(t^2 + 1)}\, dt \;<\; \infty$$

clearly makes the square root $\leqslant$ a quantity C_β, *bounded above* when β ranges over *compact subsets of* $\mathbb{C} \sim \mathbb{R}$. We thus have

$$\left| \lim_{A \to \infty} \int_{-A}^{A} \frac{S(t)}{t - \beta}\, dt \right| \;\leqslant\; \frac{C_\beta}{|\psi(\beta)|} \| S \|$$

for the sums S, when $\Im\beta > 0$. In like manner,

$$\left| \lim_{A \to \infty} \int_{-A}^{A} \frac{S(t)}{t - \bar{\beta}}\, dt \right| \;\leqslant\; \frac{C_{\bar{\beta}}}{|\psi(\bar{\beta})|} \| S \|,$$

so finally, since

$$\frac{1}{t-\beta} - \frac{1}{t-\bar{\beta}} = \frac{2i\Im\beta}{|t-\beta|^2},$$

$$\left| \int_{-\infty}^{\infty} \frac{\Im\beta}{|t-\beta|^2} S(t)\,dt \right| \leqslant K_\beta \|S\|.$$

If $E \subseteq \{\Im z > 0\}$ is compact,

$$K_\beta = \frac{C_\beta}{2|\psi(\beta)|} + \frac{C_{\bar{\beta}}}{2|\psi(\bar{\beta})|}$$

may be taken to be bounded above on E, since, as explained at first, we can choose ψ with $|\psi(\beta)|$ and $|\psi(\bar{\beta})|$ bounded away from 0 on E. Sufficiency is proved.

Necessity. Suppose that for *every* β, $\Im\beta > 0$, the last inequality (at the end of the preceding discussion) *holds*, with some finite K_β. Then, by the previous theorem, we certainly have a non-zero entire function φ of exponential type $\leqslant a$, with

$$\int_{-\infty}^{\infty} \frac{|\varphi(t)|^2}{w(t)(t^2+1)^2}\,dt < \infty.$$

As we saw at the beginning of the *sufficiency* part of that theorem's proof, we may take $\varphi(t)$ to be *real* on $\mathbb{R}$.

Let $\Im\beta \neq 0$. We have an identity

$$\frac{(\Im\beta)\varphi(t)}{|t-\beta|^2} = (\Im\beta)g_\beta(t) + (\Im\varphi(\beta))\frac{t-\Re\beta}{|t-\beta|^2} + (\Re\varphi(\beta))\frac{\Im\beta}{|t-\beta|^2}$$

like the one used in establishing the preceding theorem, where g_β is an entire function of exponential type $\leqslant a$.

The relation involving $\varphi(t)$ and w implies that $\varphi(t)/(t-\beta)^2 \in L_1(\mathbb{R})$ by the usual application of Schwarz' inequality. Then, if φ is *not* a pure exponential (when it *is*, it must be *bounded* on $\mathbb{R}$), it must have *at least two zeros*, for, if it had only *one*, $\varphi(t)/(t-\beta)^2$ would, by φ's resulting Hadamard factorization, be prevented from being in $L_1(\mathbb{R})$. This being the case, an argument like the one made during the preceding sufficiency proof shows that $\varphi(t)/(t-\beta)^2$ is *bounded* on $\mathbb{R}$. The conclusion is that $\varphi(t)/(t-\beta)^2$ is also in $L_2(\mathbb{R})$, and, referring to the above formula, we see that $g_\beta(t)$ is square integrable.

The function $g_\beta(t)$ is, however, entire and of exponential type $\leqslant a$, so we may essentially *repeat* the reasoning followed above, based on the

Paley–Wiener theorem, to conclude from the previous formula that

$$\Im\varphi(\beta)\cdot \lim_{A\to\infty}\int_{-A}^{A}\frac{t-\Re\beta}{|t-\beta|^2}S(t)\,dt \;+\; \Re\varphi(\beta)\cdot\int_{-\infty}^{\infty}\frac{\Im\beta}{|t-\beta|^2}S(t)\,dt$$

$$= \; \Im\beta\cdot\int_{-\infty}^{\infty}\frac{\varphi(t)}{|t-\beta|^2}S(t)\,dt$$

for the sums $S(t)$.

Here, we have

$$\int_{-\infty}^{\infty}\frac{|\varphi(t)|^2}{w(t)|t-\beta|^4}\,dt \;<\; \infty,$$

so the *right side* of the preceding relation is in modulus

$$\leqslant \; \text{const.}\,\|S\|,$$

where $\|\;\|$ has the same meaning as before. At the same time, our *assumption* is that

$$\left|\int_{-\infty}^{\infty}\frac{\Im\beta}{|t-\beta|^2}S(t)\,dt\right| \;\leqslant\; \text{const.}\,\|S\|$$

for our sums S (with the constant depending, of course, on β). This may now be combined with the result just found to yield

$$\left|\Im\varphi(\beta)\cdot \lim_{A\to\infty}\int_{-A}^{A}\frac{t-\Re\beta+i\Im\beta}{|t-\beta|^2}S(t)\,dt\right| \;\leqslant\; \text{const.}\,\|S\|.$$

For each β, then, with $\Im\beta>0$ there is a finite L_β such that

$$\left|\Im\varphi(\beta)\cdot \lim_{A\to\infty}\int_{-A}^{A}\frac{S(t)}{t-\beta}\,dt\right| \;\leqslant\; L_\beta\|S\|$$

for the sums S.

Suppose that $\Im\varphi(\beta)\neq 0$ *for some β with $\Im\beta>0$. Then we are done.* We can, indeed, argue as at the very *start* of the previous theorem's proof to obtain, thanks to the last relation, a $k(t)$ with $\|k\|<\infty$ (and hence $w(t)k(t)\in L_1(\mathbb{R})$ by Schwarz) such that

$$\lim_{A\to\infty}\int_{-A}^{A}\left(w(t)k(t)-\frac{1}{t-\beta}\right)e^{i\lambda t}\,dt \;=\; 0$$

for $|\lambda|\geqslant a$.

Here, the integrable function $w(t)k(t)$ must also be in $L_2(\mathbb{R})$. Indeed, the (*bounded!*) Fourier transform

$$\int_{-\infty}^{\infty} e^{i\lambda t} w(t) k(t) \, dt$$

coincides with the L_2 Fourier transform of $1/(t - \beta)$ for large $|\lambda|$, and is thus itself in L_2. Then, however, $w(t)k(t) \in L_2(\mathbb{R})$ by Plancherel's theorem.

We may now apply the L_2 Paley–Wiener theorem (Chapter III, §D) to the function

$$w(t)k(t) \ - \ \frac{1}{t - \beta}$$

and conclude from the preceding relation that it coincides a.e. on $\mathbb{R}$ with an entire function $f(t)$ of exponential type $\leqslant a$. The function

$$\psi(t) \ = \ (t - \beta)f(t) \ + \ 1$$

is also entire and of exponential type a, and

$$\psi(t) \ = \ (t - \beta)w(t)k(t) \qquad \text{a.e., } t \in \mathbb{R}.$$

The above integral relation clearly implies that $w(t)k(t)$ cannot vanish a.e., so $\psi \not\equiv 0$. Finally,

$$\int_{-\infty}^{\infty} \frac{|\psi(t)|^2}{w(t)|t - \beta|^2} \, dt \ = \ \|k\|^2 \ < \ \infty,$$

so

$$\int_{-\infty}^{\infty} \frac{|\psi(t)|^2}{w(t)(t^2 + 1)} \, dt \ < \ \infty.$$

The *necessity is thus established* provided that for some β, $\ \Im\beta > 0$, the *original entire function* φ has *non-zero imaginary part at* β. If, however, there is *no such* β, we are *also* finished! *Then*, $\Im\varphi(\beta) \equiv 0$ for $\Im\beta > 0$, so $\varphi(z)$ must be *constant*, wlog, $\varphi(z) \equiv 1$. This means that

$$\int_{-\infty}^{\infty} \frac{1}{w(t)(t^2 + 1)^2} \, dt \ < \ \infty.$$

In that case,

$$\int_{-\infty}^{\infty} \frac{|\psi(t)|^2}{w(t)(t^2 + 1)} \, dt \ < \ \infty$$

with, e.g., $\psi(t) = \sin at/at$, and *this* function ψ is entire, of exponential type a, and $\not\equiv 0$.

The theorem is completely proved.

Scholium. The discrepancy between the conditions on w involved in the above two theorems is annoying. How can there be a $w \geqslant 0$ such that the sums

$$S(t) = \sum_{|\lambda| \geqslant a} A_\lambda e^{i\lambda t}$$

satisfying

$$\int_{-\infty}^{\infty} |S(t)|^2 w(t)\, dt \;\leqslant\; 1$$

yield harmonic functions

$$\frac{1}{\pi} \int_{-\infty}^{\infty} \frac{\Im z}{|z-t|^2} S(t)\, dt$$

with values bounded at *some* points z in the upper half plane, but not at *each* of those points? If there is a non-constant entire function $\varphi \not\equiv 0$ of exponential type $\leqslant a$ for which

$$\int_{-\infty}^{\infty} \frac{|\varphi(t)|^2}{w(t)(t^2+1)^2}\, dt \;<\; \infty,$$

can we not *divide out* one of the *zeros* of φ to get another such function ψ making

$$\int_{-\infty}^{\infty} \frac{|\psi(t)|^2}{w(t)(t^2+1)}\, dt \;<\; \infty \; ?$$

The present situation illustrates the care that must be taken in the investigation of such matters, straightforward though they may appear. The conditions involved in the two results are *not* equivalent, and there *really do* exist functions $w \geqslant 0$ *satisfying one, but not the other*. None of the zeros of φ can be divided out if they are all needed to cancel those of $w(t)$!

Here is a simple example. Let

$$w(t) \;=\; \frac{\sin^2 \pi t}{t^2+1}.$$

The condition

$$\int_{-\infty}^{\infty} \frac{|\varphi(t)|^2}{w(t)(t^2+1)^2}\, dt \;<\; \infty$$

is satisfied here with

$$\varphi(t) = \sin \pi t,$$

an entire function of exponential type π. *The kind of estimate furnished by the first theorem* is therefore *available* for the sums

$$S(t) = \sum_{|\lambda| \geq \pi} A_\lambda e^{i\lambda t}.$$

Here, however, the estimates provided by the second theorem are not all valid! To see this, consider the functions

$$T_\eta(t) = \frac{i}{\sin \pi(t + i\eta)},$$

where η is a small parameter > 0. We have

$$\Re T_\eta(t) = \frac{\Im \sin \pi(t + i\eta)}{|\sin \pi(t + i\eta)|^2} = \frac{\sinh \pi\eta \cos \pi t}{\sin^2 \pi t + \sinh^2 \pi\eta}.$$

Clearly $\Re T_\eta(t + 2) = \Re T_\eta(t)$ and $\Re T_\eta$ is $\mathscr{C}_\infty$ on the real axis, so

$$\Re T_\eta(t) = \sum_{-\infty}^{\infty} a_n e^{\pi int}, \qquad t \in \mathbb{R},$$

the series being absolutely convergent. Since $\Re T_\eta(t - \frac{1}{2})$ and $\Re T_\eta(t + \frac{1}{2})$ are *odd* functions of t, we have

$$a_0 = \frac{1}{2} \int_{-1}^{1} (\Re T_\eta)(t) \, dt = 0,$$

and $\Re T_\eta(t)$ is a (uniform!) limit of sums

$$\sum_{1 \leq |n| \leq N} a_n e^{\pi int},$$

each of the form

$$\sum_{|\lambda| \geq \pi} A_\lambda e^{i\lambda t}.$$

If the estimates furnished by the second theorem held for the present w and for $a = \pi$, we would now have

$$\left| \int_{-\infty}^{\infty} \frac{(\Re T_\eta)(t)}{1 + t^2} \, dt \right| \leq C \sqrt{\left(\int_{-\infty}^{\infty} (\Re T_\eta(t))^2 w(t) \, dt \right)}$$

with a constant C *independent* of $\eta > 0$. That, however, is not the case.

Because

$$(\Re T_\eta(t))^2 w(t) \;\leqslant\; |T_\eta(t)|^2 w(t) \;=\; |T_\eta(t)|^2 \frac{\sin^2 \pi t}{t^2+1} \;\leqslant\; \frac{1}{t^2+1}$$

and

$$\Re T_\eta(t) \;\longrightarrow\; 0 \qquad \text{as } \eta \longrightarrow 0$$

for $t \neq 0, \pm 1, \pm 2, \ldots$, we have

$$\int_{-\infty}^{\infty} (\Re T_\eta(t))^2 w(t)\,dt \;\longrightarrow\; 0$$

when $\eta \longrightarrow 0$, by dominated convergence.

At the same time, since each of the functions

$$T_\eta(z) \;=\; \frac{i}{\sin \pi(z+i\eta)}$$

is *analytic* and *bounded* in $\Im z > 0$,

$$\int_{-\infty}^{\infty} \frac{(\Re T_\eta)(t)}{1+t^2}\,dt \;=\; \pi \Re T_\eta(i) \;=\; \frac{\pi}{\sinh \pi(1+\eta)} \;\longrightarrow\; \frac{\pi}{\sinh \pi} \;>\; 0$$

as $\eta \longrightarrow 0$. This does it.

It is not hard to see that here, for the sums

$$S(t) \;=\; \sum_{|\lambda| \geqslant \pi} A_\lambda e^{i\lambda t},$$

the condition

$$\int_{-\infty}^{\infty} |S(t)|^2 w(t)\,dt \;\leqslant\; 1$$

gives us control on the integrals

$$\int_{-\infty}^{\infty} \frac{\Im \beta}{|t-\beta|^2} S(t)\,dt$$

when $\Re \beta = \pm\frac{1}{2}, \pm\frac{3}{2}, \pm\frac{5}{2}, \ldots$, $\Im \sin \pi\beta$ vanishing precisely for such values of β.

One may pose a problem similar to the one discussed in this §, but with the sums

$$S(t) \;=\; \sum_{|\lambda| \geqslant a} A_\lambda e^{i\lambda t}$$

replaced by others of the form

$$\sum_{|\lambda| \leqslant a} A_\lambda e^{i\lambda t}$$

(i.e., by entire functions of exponential type $\leqslant a$ bounded on $\mathbb{R}$!). That seems *harder*. Some of the material in the first part of de Branges' book is relevant to it.

E. Hilbert transforms of certain functions having given weighted quadratic norms.

We continue along the lines of the preceding §'s discussion. Taking, as we did there, some fixed $w \geqslant 0$ belonging to $L_1(\mathbb{R})$, let us suppose that we are given a certain class of functions $U(t)$, bounded on the real axis, whose harmonic extensions

$$U(z) \;=\; \frac{1}{\pi} \int_{-\infty}^{\infty} \frac{\Im z}{|z - t|^2}\, U(t)\,\mathrm{d}t$$

to the upper half plane are controlled by the weighted norm

$$\sqrt{\left(\int_{-\infty}^{\infty} |U(t)|^2 w(t)\,\mathrm{d}t \right)}.$$

A suitably defined *harmonic conjugate* $\tilde{U}(z)$ of each of our functions $U(z)$ will then also be controlled by that norm. As we have seen in Chapter III, §F.2 and in the scholium to §H.1 of that chapter, the $\tilde{U}(z)$ have well defined non-tangential boundary values a.e. on $\mathbb{R}$ and thereby give rise to Lebesgue measurable functions $\tilde{U}(t)$ of the real variable t. Each of the latter is a *Hilbert transform* of the corresponding original bounded function $U(t)$; we say *a* Hilbert transform because that object, like the harmonic conjugate, is really only defined to within an additive constant. The reader can arrive at a fairly clear idea of these transforms by referring first to the §§ mentioned above and then to the middle of §C.1, Chapter VIII, and the scholium at the end of it.

Whatever specification is adopted for the Hilbert transforms $\tilde{U}(t)$ of our functions U, one may ask whether their *size* is governed by the weighted norm in question when that is the case for the harmonic extensions $U(z)$. To be more definite, let us ask whether there is some integrable function $\omega(t) \geqslant 0$, not a.e. zero on $\mathbb{R}$, such that

$$\int_{-\infty}^{\infty} |\tilde{U}(t)|^2 \omega(t)\,\mathrm{d}t \;\leqslant\; \int_{-\infty}^{\infty} |U(t)|^2 w(t)\,\mathrm{d}t$$

for the particular class of functions U under consideration. In the present

§, we study this question for the exponential sums

$$U(t) = \sum_{|\lambda| \geqslant a} A_\lambda e^{i\lambda t}$$

worked with in §D. Although the problem, as formulated, no longer refers directly to the harmonic extensions $U(z)$, it will turn out to have a positive solution (for given w) precisely when the latter are controlled by $\int_{-\infty}^{\infty} |U(t)|^2 w(t) \, dt$ in $\{\Im z > 0\}$ (and only then). For this reason, multipliers will again be involved in our discussion.

The work will require some material from the theory of H_p spaces. In order to save the reader the trouble of digging up that material elsewhere, we give it (and no more) in the next article, starting from scratch. This is *not* a book about H_p spaces, and anyone wishing to really learn about them should refer to such a book. Several are now available, including (and *why* not!) my own.*

1.   *H_p spaces for people who don't want to really learn about them*

We will need to know some things about H_1, H_∞ and H_2, and proceed to take up those spaces in that order. Most of the real work involved here has actually been done already in various parts of the present book.

For our purposes, it is most convenient to use the

Definition. $H_1(\mathbb{R})$, or, as we usually write, H_1, is the set of f in $L_1(\mathbb{R})$ for which the Fourier transform

$$\hat{f}(\lambda) = \int_{-\infty}^{\infty} e^{i\lambda t} f(t) \, dt$$

vanishes for all $\lambda \geqslant 0$.

---

* As much as I want that book to sell, I should warn the reader that there are a fair number of misprints and also some actual mistakes in it. The *statement* of the lemma on p. 104 is inaccurate; boundedness only holds for r away from 0 when $F(0) = 0$. *Statement* of the lemma on p. 339 is wrong; ν may also contain a point mass at 0. That, however, makes no difference for the subsequent application of the lemma. The *argument* at the bottom of p. 116 is nonsense. *Instead*, one should say that if $B|B_\alpha$ and $d\sigma' \leqslant d\sigma_\alpha$ for each α, then every f_α is in ΩH_2, where Ω is given by the formula displayed there. Hence $\omega H_2 = E$ is $\subseteq \Omega H_2$, so $B|B$ and $d\sigma' \leqslant d\sigma$ by reasoning like that at the top of p. 116. There are confusing *misprints* in the proof of the first theorem on p 13; near the end of that proof, F should be replaced by G.

Lemma. *If $f \in H_1$, $e^{i\lambda t} f(t) \in H_1$ for each $\lambda \geqslant 0$.*

Proof. Clear.

Lemma. *If $f \in H_1$ and $\Im z > 0$, $f(t)/(t - \bar z) \in H_1$.*

Proof. For $\Im z > 0$ (i.e., $\Re(-i\bar z) < 0$), we have

$$\frac{i}{t - \bar z} = \int_0^\infty e^{-i\bar z \lambda} e^{i\lambda t} \, d\lambda, \qquad t \in \mathbb{R}.$$

Therefore, if $f \in H_1$,

$$i \int_{-\infty}^\infty \frac{f(t)}{t - \bar z} \, dt = \int_{-\infty}^\infty \int_0^\infty e^{-i\bar z \lambda} e^{i\lambda t} f(t) \, d\lambda \, dt.$$

The double integral on the right is absolutely convergent, and hence can be rewritten as

$$\int_0^\infty \int_{-\infty}^\infty e^{-i\bar z \lambda} e^{i\lambda t} f(t) \, dt \, d\lambda = \int_0^\infty e^{-i\bar z \lambda} \hat{f}(\lambda) \, d\lambda = 0.$$

If $\alpha \geqslant 0$ and $f \in H_1$, $e^{i\alpha t} f(t)$ is also in H_1 by the preceding lemma, so, using it in place of $f(t)$ in the computation just made, we get

$$\int_{-\infty}^\infty e^{i\alpha t} \frac{f(t)}{t - \bar z} \, dt = 0.$$

$f(t)/(t - \bar z)$ is thus in H_1 by definition.

Theorem. *If, for $f \in H_1$, we write*

$$f(z) = \frac{1}{\pi} \int_{-\infty}^\infty \frac{\Im z}{|z - t|^2} f(t) \, dt$$

for $\Im z > 0$, the function $f(z)$ is analytic in the upper half plane.

Proof. We have

$$f(z) = \frac{1}{2\pi i} \int_{-\infty}^\infty \left(\frac{1}{t - z} - \frac{1}{t - \bar z} \right) f(t) \, dt.$$

By the last lemma, the right side equals

$$\frac{1}{2\pi i} \int_{-\infty}^\infty \frac{f(t)}{t - z} \, dt$$

for $\Im z > 0$, and this expression is clearly analytic in the upper half plane. We are done.

Theorem. *The function* $f(z)$ *defined in the statement of the preceding result has the following properties:*

(i) $f(z)$ *is continuous and bounded in each half plane* $\{\Im z \geqslant h\}$, $h > 0$, *and tends to* 0 *as* $z \longrightarrow \infty$ *in any one of those;*

(ii) $\displaystyle\int_{-\infty}^{\infty} |f(x + iy)|\,dx \leqslant \|f\|_1$ *for* $y > 0$;

(iii) $\displaystyle\int_{-\infty}^{\infty} |f(t + iy) - f(t)|\,dt \longrightarrow 0$ *as* $y \longrightarrow 0$;

(iv) $f(t + iy) \longrightarrow f(t)$ a.e. *as* $y \longrightarrow 0$.

Remark. Properties (iii) and (iv) justify our denoting

$$\frac{1}{\pi}\int_{-\infty}^{\infty} \frac{\Im z}{|z - t|^2}\, f(t)\,dt$$

by $f(z)$.

Proof of theorem. Property (i) is verified by inspection; (ii) and (iii) hold because the Poisson kernel is a (positive) approximate identity. Property (iv) comes out of the discussion beginning in Chapter II, §B and then continuing in §F.2 of Chapter III and in the scholium to §H.1 of that chapter. These ideas have already appeared frequently in the present book.

Theorem. *If* $f(t) \in H_1$ *is not zero a.e. on* $\mathbb{R}$, *we have*

$$\int_{-\infty}^{\infty} \frac{\log^- |f(t)|}{1 + t^2}\,dt \;\; < \;\; \infty,$$

and, for each z, $\;\Im z > 0$,

$$\log|f(z)| \;\leqslant\; \frac{1}{\pi}\int_{-\infty}^{\infty} \frac{\Im z}{|z - t|^2} \log|f(t)|\,dt,$$

the integral on the right being absolutely convergent. Here, $f(z)$ *has the same meaning as in the preceding two results.*

Proof. For each $h > 0$ we can apply the results from Chapter III, §G.2 to $f(z + ih)$ in the half plane $\Im z > 0$, thanks to property (i), guaranteed by the last theorem. In this way we get

$$\log|f(z + ih)| \;\leqslant\; \frac{1}{\pi}\int_{-\infty}^{\infty} \frac{\Im z}{|z - t|^2} \log|f(t + ih)|\,dt$$

for $\Im z > 0$, with the integral on the right absolutely convergent.

Fix for the moment any z, $\Im z > 0$, for which $f(z) \neq 0$. The *left* side of the relation just written then tends to a limit $> -\infty$ as $h \longrightarrow 0$. At the same time, the *right* side is equal to

$$\frac{1}{\pi}\int_{-\infty}^{\infty}\frac{\Im z}{|z-t|^2}\log^+|f(t+ih)|\,dt \;\;-\;\; \frac{1}{\pi}\int_{-\infty}^{\infty}\frac{\Im z}{|z-t|^2}\log^-|f(t+ih)|\,dt,$$

where

$$\int_{-\infty}^{\infty}\big|\log^+|f(t+ih)| - \log^+|f(t)|\big|\,dt \;\;\leqslant\;\; \int_{-\infty}^{\infty}\big||f(t+ih)| - |f(t)|\big|\,dt,$$

which tends to zero as h does, according to property (iii) in the preceding result. Therefore

$$\frac{1}{\pi}\int_{-\infty}^{\infty}\frac{\Im z}{|z-t|^2}\log^+|f(t+ih)|\,dt \;\;\longrightarrow\;\; \frac{1}{\pi}\int_{-\infty}^{\infty}\frac{\Im z}{|z-t|^2}\log^+|f(t)|\,dt,$$

a *finite quantity* (by the inequality between arithmetic and geometric means), as $h \longrightarrow 0$.

From property (iv) in the preceding theorem and Fatou's lemma, we have, however,

$$\frac{1}{\pi}\int_{-\infty}^{\infty}\frac{\Im z}{|z-t|^2}\log^-|f(t)|\,dt \;\;\leqslant\;\; \liminf_{h\to 0}\frac{1}{\pi}\int_{-\infty}^{\infty}\frac{\Im z}{|z-t|^2}\log^-|f(t+ih)|\,dt.$$

Using this and the preceding relation we see, by making $h \longrightarrow 0$ in our initial one, that

$$-\infty \;\; < \;\; \log|f(z)| \;\; \leqslant \;\; \frac{1}{\pi}\int_{-\infty}^{\infty}\frac{\Im z}{|z-t|^2}\log^+|f(t)|\,dt$$

$$-\; \frac{1}{\pi}\int_{-\infty}^{\infty}\frac{\Im z}{|z-t|^2}\log^-|f(t)|\,dt.$$

Since the *first* integral on the right is finite, the *second* must also be so. That, however, is equivalent to the relation

$$\int_{-\infty}^{\infty}\frac{\log^-|f(t)|}{1+t^2}\,dt \;\; < \;\; \infty.$$

Putting the two right-hand integrals together, we see that

$$\frac{1}{\pi}\int_{-\infty}^{\infty}\frac{\Im z}{|z-t|^2}\log|f(t)|\,dt$$

is absolutely convergent for our particular z, and hence for any z with

$\Im z > 0$. That quantity is $\geqslant \log|f(z)|$ as we have just seen, provided that $|f(z)| > 0$. It is of course $> \log|f(z)|$ in case $f(z) = 0$. We are done.

Corollary. *If $f(t) \in H_1$ is not a.e. zero, $|f(t)|$ is necessarily > 0 a.e. .*

Proof. The theorem's boxed inequality makes $\log^-|f(t)| > -\infty$ a.e..

Definition. $H_\infty(\mathbb{R})$, or, as we frequently write, H_∞, is the collection of g in $L_\infty(\mathbb{R})$ satisfying

$$\int_{-\infty}^\infty g(t)f(t)\,dt = 0$$

for all $f \in H_1$.

H_∞ is thus the subspace of L_∞, dual of L_1, consisting of functions *orthogonal* to the closed subspace H_1 of L_1. As such, it is closed, and even w^* closed, in L_∞.

By definition of H_1 we have the

Lemma. *Each of the functions $e^{i\lambda t}$, $\lambda \geqslant 0$, belongs to H_∞.*

Corollary. *A function $f \in L_1(\mathbb{R})$ belongs to H_1 iff*

$$\int_{-\infty}^\infty g(t)f(t)\,dt = 0$$

for all $g \in H_\infty$.

Lemma. *If $f \in H_1$ and $g \in H_\infty$, $g(t)f(t) \in H_1$.*

Proof. First of all, $gf \in L_1$. Also, when $\lambda \geqslant 0$, $e^{i\lambda t}f(t) \in H_1$ by a previous lemma, so by definition of H_∞,

$$\int_{-\infty}^\infty g(t)e^{i\lambda t}f(t)\,dt = 0,$$

i.e.,

$$\int_{-\infty}^\infty e^{i\lambda t}g(t)f(t)\,dt = 0$$

for *each* $\lambda \geqslant 0$. Therefore $gf \in H_1$.

Lemma. *If g and h belong to H_∞, $g(t)h(t)$ does also.*

Proof. If f is any member of H_1, gf is also in H_1 by the previous lemma. Therefore

$$\int_{-\infty}^{\infty} h(t){\cdot}g(t)f(t)\,dt \;\; = \;\; 0.$$

This, holding for all $f \in H_1$, makes $hg \in H_\infty$ by definition.

Theorem. *Let* $g \in H_\infty$. *Then the function*

$$g(z) \;\; = \;\; \frac{1}{\pi}\int_{-\infty}^{\infty} \frac{\Im z}{|z-t|^2}\,g(t)\,dt$$

is analytic for $\Im z > 0$.

Proof. Fix z, $\Im z > 0$, and, for the moment, a large $A > 0$. The function

$$f(t) \;\; = \;\; \frac{1}{t-\bar{z}}\,\frac{iA}{t+iA}$$

belongs to H_1. This is easily verified directly by showing that

$$\int_{-\infty}^{\infty} e^{i\lambda t} f(t)\,dt \;\; = \;\; 0$$

for $\lambda \geqslant 0$ using contour integration. One takes large semi-circular contours in the upper half plane with base on the real axis; the details are left to the reader.

By definition of H_∞, we thus have

$$\int_{-\infty}^{\infty} g(t)\,\frac{1}{t-\bar{z}}\,\frac{iA}{iA+t}\,dt \;\; = \;\; 0.$$

Subtracting the left side from

$$\int_{-\infty}^{\infty} g(t){\cdot}\frac{1}{t-z}\,\frac{iA}{iA+t}\,dt$$

and then dividing by $2i$, we see that

$$\int_{-\infty}^{\infty} \frac{\Im z}{|z-t|^2}\,\frac{iA}{iA+t}\,g(t)\,dt \;\; = \;\; \frac{1}{2i}\int_{-\infty}^{\infty} \frac{1}{t-z}\,\frac{iA}{iA+t}\,g(t)\,dt.$$

For each $A > 0$, then,

$$g_A(z) \;\; = \;\; \frac{1}{\pi}\int_{-\infty}^{\infty} \frac{\Im z}{|z-t|^2}\,\frac{iA}{iA+t}\,g(t)\,dt$$

is analytic for $\Im z > 0$ (by inspection).

As $A \longrightarrow \infty$, the functions $g_A(z)$ tend u.c.c. in $\{\Im z > 0\}$ to

$$\frac{1}{\pi} \int_{-\infty}^{\infty} \frac{\Im z}{|z - t|^2} g(t) \, dt \ = \ g(z).$$

The latter is therefore also analytic there.

Remark. For the function $g(z)$ figuring in the above theorem we have, for each z, $\Im z > 0$,

$$|g(z)| \leqslant \|g\|_\infty,$$

where the L_∞ norm on the right is taken for $g(t)$ on $\mathbb{R}$. This is evident by inspection. The same reasoning which shows that

$$f(t + iy) \ \longrightarrow \ f(t) \quad \text{a.e. as } y \longrightarrow 0$$

for functions f in H_1 also applies here, yielding the result that

$$g(t + iy) \ \longrightarrow \ g(t) \quad \text{a.e. as } y \longrightarrow 0$$

when $g \in H_\infty$. Unless $g(t)$ is uniformly continuous, however, we do *not* have

$$\|g(t + iy) \ - \ g(t)\|_\infty \ \longrightarrow \ 0$$

for $y \longrightarrow 0$. Instead, we are only able to affirm that $g(t + iy)$ tends w^* to $g(t)$ (in $L_\infty(\mathbb{R})$) as $y \longrightarrow 0$.

The theorem just proved has an important *converse*:

Theorem. *Let $G(z)$ be analytic and bounded for $\Im z > 0$. Then there is a $g \in H_\infty$ such that*

$$G(z) \ = \ \frac{1}{\pi} \int_{-\infty}^{\infty} \frac{\Im z}{|z - t|^2} g(t) \, dt$$

for $\Im z > 0$, and

$$\|g\|_\infty \ = \ \sup_{\Im z > 0} |G(z)|.$$

Proof. It is claimed first of all that each of the functions $G(t + ih)$, $h > 0$, belongs to H_∞ (as a function of t). Take any $f \in H_1$, and put

$$f(z) \ = \ \frac{1}{\pi} \int_{-\infty}^{\infty} \frac{\Im z}{|z - t|^2} f(t) \, dt, \qquad \Im z > 0.$$

Our definition of H_∞ requires us to verify that

$$\int_{-\infty}^{\infty} G(t + ih) f(t) \, dt \ = \ 0.$$

Since

$$\| f(t + ib) - f(t) \|_1 \longrightarrow 0$$

as $b \longrightarrow 0$, it is enough to show that

$$\int_{-\infty}^{\infty} G(t + ih)f(t + ib)\,dt = 0$$

for each $b > 0$.

Fix any such b. According to a previous result, $f(z + ib)$ is then analytic and *bounded* for $\Im z > 0$, and continuous up to the real axis. The same is true for $G(z + ih)$. These properties make it easy for us to see by contour integration that

$$\int_{-\infty}^{\infty} \left(\frac{iA}{iA + t} \right)^2 G(t + ih)f(t + ib)\,dt = 0$$

for $A > 0$; one just integrates

$$\left(\frac{iA}{iA + z} \right)^2 G(z + ih)f(z + ib)$$

around large semi-circles in $\Im z \geqslant 0$ having their diameters on the real axis. Since $f(t + ib) \in L_1(\mathbb{R})$, we may now make $A \longrightarrow \infty$ in the relation just found to get

$$\int_{-\infty}^{\infty} G(t + ih)f(t + ib)\,dt = 0$$

and thus ensure that $G(t + ih) \in H_\infty(\mathbb{R})$.

For each $h > 0$ the first lemma of §H.1, Chapter III, makes

$$G(z + ih) = \frac{1}{\pi} \int_{-\infty}^{\infty} \frac{\Im z}{|z - t|^2} G(t + ih)\,dt$$

when $\Im z > 0$. Here,

$$|G(t + ih)| \leqslant \sup_{\Im z > 0} |G(z)| < \infty.$$

Hence, since L_∞ is the *dual* of L_1, a procedure just like the one used in establishing the first theorem of §F.1, Chapter III, gives us a sequence of numbers $h_n > 0$ tending to zero and a g in L_∞ with

$$G(t + ih_n) \longrightarrow g(t) \quad \text{w*}$$

as $n \longrightarrow \infty$. From this we see, referring to the preceding formula, that

$$G(z) = \lim_{n \to \infty} G(z + ih_n) = \frac{1}{\pi} \int_{-\infty}^{\infty} \frac{\Im z}{|z - t|^2} g(t)\,dt$$

for $\Im z > 0$. By the w^* convergence we also have

$$\|g\|_\infty \;\leqslant\; \liminf_{n\to\infty} \|G(t+ih_n)\|_\infty \;\leqslant\; \sup_{\Im z>0} |G(z)|.$$

However, the representation just found for $G(z)$ implies the reverse inquality, so

$$\|g\|_\infty \;=\; \sup_{\Im z>0} |G(z)|.$$

As we have seen, each of the functions $G(t+ih_n)$ is in H_∞. Their w^* limit $g(t)$ must then also be in H_∞.

The theorem is proved.

Remark. An analogous theorem is true about H_1. Namely, if $F(z)$, analytic for $\Im z > 0$, is such that the integrals

$$\int_{-\infty}^{\infty} |F(x+iy)|\,dx$$

are *bounded* for $y > 0$, there *is* an $f \in H_1$ for which

$$F(z) \;=\; \frac{1}{\pi}\int_{-\infty}^{\infty} \frac{\Im z}{|z-t|^2}\, f(t)\,dt, \qquad \Im z > 0.$$

This result will not be needed in the present §; it is deeper than the one just found because $L_1(\mathbb{R})$ is *not the dual* of any Banach space. The F. and M. Riesz theorem is required for its proof; see §B.4 of Chapter VII.

Problem 41

Let $g \in H_\infty$, and write

$$g(z) \;=\; \frac{1}{\pi}\int_{-\infty}^{\infty} \frac{\Im z}{|z-t|^2} g(t)\,dt$$

for $\Im z > 0$.

(a) If $\Im c > 0$, both functions

$$\frac{g(t)-g(c)}{t-\bar c} \quad\text{and}\quad \frac{g(t)-g(c)}{t-c}$$

belong to H_∞. (Hint: In considering the first function, begin by noting that $1/(t-\bar c) \in H_\infty$ according to the second lemma about H_1. To investigate the second function, look at $(g(z)-g(c))/(z-c)$ in the upper half plane.)

(b) Hence show that if $f \in H_1$ and

$$f(z) \;=\; \frac{1}{\pi}\int_{-\infty}^{\infty} \frac{\Im z}{|z-t|^2} f(t)\,dt$$

for $\Im z > 0$, one has

$$f(c)g(c) \;=\; \frac{1}{\pi}\int_{-\infty}^{\infty} \frac{\Im c}{|c-t|^2}\, f(t)g(t)\,\mathrm{d}t$$

for each c with $\Im c > 0$.

(c) If, for the $f(z)$ of part (b) one has $f(c) = 0$ for some c, $\Im c > 0$, show that $f(t)/(t-c)$ belongs to H_1. (Hint: Follow the argument of (b) using the function $g(t) = \mathrm{e}^{\mathrm{i}\lambda t}$, where $\lambda \geqslant 0$ is arbitrary.)

Theorem. *If $g(t) \in H_\infty$ is not a.e. zero on $\mathbb{R}$, we have*

$$\int_{-\infty}^{\infty} \frac{\log^-|g(t)|}{1+t^2}\,\mathrm{d}t \;<\; \infty,$$

and, for $\Im z > 0$,

$$\log|g(z)| \;\leqslant\; \frac{1}{\pi}\int_{-\infty}^{\infty} \frac{\Im z}{|z-t|^2}\log|g(t)|\,\mathrm{d}t,$$

the integral on the right being absolutely convergent. Here, $g(z)$ has its usual meaning:

$$g(z) \;=\; \frac{1}{\pi}\int_{-\infty}^{\infty} \frac{\Im z}{|z-t|^2}\, g(t)\,\mathrm{d}t.$$

Proof. By the first of the preceding two theorems, $g(z)$ is analytic (and of course bounded) for $\Im z > 0$. Therefore, by the results of §G.2 in Chapter III, for each $h > 0$,

$$\log|g(z+\mathrm{i}h)| \;\leqslant\; \frac{1}{\pi}\int_{-\infty}^{\infty} \frac{\Im z}{|z-t|^2}\log|g(t+\mathrm{i}h)|\,\mathrm{d}t$$

when $\Im z > 0$.

We may, wlog, take $\|g\|_\infty$ to be $\leqslant 1$, so that $|g(z)| \leqslant 1$ and $\log|g(z)| \leqslant 0$ for $\Im z > 0$. As $h \longrightarrow 0$, $g(t+\mathrm{i}h) \longrightarrow g(t)$ a.e. according to a previous remark, so, by Fatou's lemma,

$$\limsup_{h\to 0} \frac{1}{\pi}\int_{-\infty}^{\infty} \frac{\Im z}{|z-t|^2}\log|g(t+\mathrm{i}h)|\,\mathrm{d}t \;\leqslant\; \frac{1}{\pi}\int_{-\infty}^{\infty} \frac{\Im z}{|z-t|^2}\log|g(t)|\,\mathrm{d}t.$$

The right-hand quantity must thus be $\geqslant \log|g(z)|$ by the previous relation, proving the second inequality of our theorem.

In case $g(t)$ is not a.e. zero, there must be some z, $\Im z > 0$, with $g(z) \neq 0$, again because $g(t + ih) \longrightarrow g(t)$ a.e. for $h \longrightarrow 0$. Using this z in the inequality just proved, we see that

$$\frac{1}{\pi} \int_{-\infty}^{\infty} \frac{\Im z}{|z - t|^2} \log|g(t)| \, dt \quad > \quad -\infty,$$

whence

$$\int_{-\infty}^{\infty} \frac{\log^-|g(t)|}{1 + t^2} \, dt \quad < \quad \infty,$$

and the former integral is actually absolutely convergent for all z with $\Im z > 0$, whether $g(z) \neq 0$ or not.

We are done.

Come we now to the space H_2.

Definition. A function $f \in L_2(\mathbb{R})$ belongs to $H_2(\mathbb{R})$, usually designated as H_2, iff

$$\int_{-\infty}^{\infty} \frac{f(t)}{t - \bar{z}} \, dt \quad = \quad 0$$

for all z with $\Im z > 0$.

H_2 is clearly a closed subspace of $L_2(\mathbb{R})$.

Theorem. *If* $f \in H_2$, *the function*

$$f(z) \quad = \quad \frac{1}{\pi} \int_{-\infty}^{\infty} \frac{\Im z}{|z - t|^2} f(t) \, dt$$

is analytic for $\Im z > 0$.

Proof. Is like that of the corresponding result for H_1.

Theorem. *If* $f \in H_2$, *the function* $f(z)$ *in the preceding theorem has the following properties:*

(i) $|f(z)| \leqslant \|f\|_2 / \sqrt{(\pi \Im z)}, \qquad \Im z > 0;$

(ii) $\displaystyle\int_{-\infty}^{\infty} |f(x + iy)|^2 dx \leqslant \|f\|_2^2 \qquad \text{for } y > 0;$

(iii) $\displaystyle\int_{-\infty}^{\infty} |f(t+iy)-f(t)|^2\,dt \;\longrightarrow\; 0$ as $y\rightarrow 0$;

(iv) $f(t+iy) \;\longrightarrow\; f(t)$ a.e. as $y\rightarrow 0$.

Proof. Property (i) follows by applying Schwarz' inequality to the formula for $f(z)$. The remaining properties are verified by arguments like those used in proving the corresponding theorem about H_1, given above.

As is the case for H_∞ (and for H_1), these results have a converse:

Theorem. *Let $F(z)$ be analytic for $\Im z > 0$, and suppose that*

$$\int_{-\infty}^{\infty} |F(x+iy)|^2\,dx$$

is bounded for $y > 0$. Then there is an $f \in H_2$ with

$$F(z) \;=\; \frac{1}{\pi}\int_{-\infty}^{\infty} \frac{\Im z}{|z-t|^2}\,f(t)\,dt, \qquad \Im z > 0,$$

and

$$\|f\|_2^2 \;=\; \sup_{y>0}\int_{-\infty}^{\infty} |F(x+iy)|^2\,dx.$$

Proof. For each $h > 0$, put

$$F_h(z) \;=\; \frac{1}{2h}\int_{-h}^{h} F(z+s)\,ds, \qquad \Im z > 0.$$

By Schwarz' inequality,

$$|F_h(z)| \;\leqslant\; (2h)^{-1/2}\sqrt{\left(\int_{-\infty}^{\infty} |F(z+s)|^2\,ds\right)} \;\leqslant\; \frac{C}{\sqrt{(2h)}}$$

where C is independent of z or h; each function $F_h(z)$ is therefore *bounded* in $\Im z > 0$, besides being analytic there.

A previous theorem therefore gives us functions $f_h \in H_\infty$ such that

$$F_h(z) \;=\; \frac{1}{\pi}\int_{-\infty}^{\infty} \frac{\Im z}{|z-t|^2}\,f_h(t)\,dt, \qquad \Im z > 0,$$

and, as already remarked,

$$F_h(t+iy) \;\longrightarrow\; f_h(t) \quad \text{a.e. for } y\rightarrow 0.$$

We have, for each h and $y > 0$,

$$\int_{-\infty}^{\infty} |F_h(x+iy)|^2 \, dx \;\leqslant\; \frac{1}{2h} \int_{-\infty}^{\infty} \int_{-h}^{h} |F(x+s+iy)|^2 \, ds \, dx$$

$$= \int_{-\infty}^{\infty} |F(x+iy)|^2 \, dx$$

by Schwarz' inequality and Fubini. Since the right side is bounded by a quantity $M < \infty$ independent of y (and h), the limit relation just written guarantees that

$$\|f_h\|_2^2 \;\leqslant\; M$$

for $h > 0$, according to Fatou's lemma.

Once it is known that the norms $\|f_h\|_2$ are bounded we can, as in the proof of the corresponding theorem about H_∞, get a sequence of $h_n > 0$ tending to zero for which the f_{h_n} converge *weakly, this time* in L_2, to some $f \in L_2(\mathbb{R})$. Then, for each z, $\Im z > 0$,

$$F_{h_n}(z) \;=\; \frac{1}{\pi} \int_{-\infty}^{\infty} \frac{\Im z}{|z-t|^2} f_{h_n}(t) \, dt \;\longrightarrow\; \frac{1}{\pi} \int_{-\infty}^{\infty} \frac{\Im z}{|z-t|^2} f(t) \, dt$$

as $n \longrightarrow \infty$. At the same time,

$$F_{h_n}(z) \;\xrightarrow{n}\; F(z),$$

so we *have* our desired representation of $F(z)$ if we can show that $f \in H_2$.

For this purpose, it is enough to verify that when $\Im z > 0$,

$$\int_{-\infty}^{\infty} \frac{f_h(t)}{t-\bar{z}} \, dt \;=\; 0,$$

since the f_{h_n} tend to f weakly in L_2. However, the f_h belong to H_∞, and, when $\Im z > 0$ and $A > 0$, the function

$$\frac{1}{t-\bar{z}} \frac{iA}{iA+t}$$

belongs to H_1, as we have noted during the proof of a previous result. Hence

$$\int_{-\infty}^{\infty} \frac{iA}{iA+t} \frac{f_h(t)}{t-\bar{z}} \, dt \;=\; 0.$$

Here, $f_h(t)/(t-\bar{z})$ belongs to L_1, so we may make $A \longrightarrow \infty$ in this relation, which yields the desired one.

We still need to show that $\|f\|_2^2 = \sup_{y>0} \int_{-\infty}^{\infty} |F(x+iy)|^2 \, dx$. Here, we now know that the function $F(z)$ is nothing but the $f(z)$ figuring in the preceding theorem. The statement in question thus follows from properties (ii) and (iii) of that result.

We are done.

Remark. Using the theorems just proved, one readily verifies that H_2 consists precisely of the functions $u(t) + i\tilde{u}(t)$, with u an arbitrary real-valued member of $L_2(\mathbb{R})$ and $\tilde{u}$ its L_2 Hilbert transform – the one studied in the scholium to §C.1 of Chapter VIII. The reader should carry out this verification.

Our use of the space H_2 in the following articles of this § is based on a relation between H_2 and H_1, established by the following two results.

Theorem. *If f and g belong to H_2, $f \cdot g$ is in H_1.*

Proof. Certainly $fg \in L_1$, so the quantity

$$\int_{-\infty}^{\infty} e^{i\lambda t} f(t) g(t) \, dt$$

varies continuously with λ. It is therefore enough to show that it vanishes for $\lambda > 0$ (*sic*) in order to prove that $fg \in H_1$.

Let, as usual,

$$f(z) = \frac{1}{\pi} \int_{-\infty}^{\infty} \frac{\Im z}{|z-t|^2} f(t) \, dt$$

for $\Im z > 0$, and

$$g(z) = \frac{1}{\pi} \int_{-\infty}^{\infty} \frac{\Im z}{|z-t|^2} g(t) \, dt$$

there.

Using the facts that $\|f(t+ih) - f(t)\|_2 \longrightarrow 0$ and $\|g(t+ih) - g(t)\|_2 \longrightarrow 0$ for $h \longrightarrow 0$ (property (iii) in the first of the preceding two theorems) and applying Schwarz' inequality to the identity

$$f(t+ih) g(t+ih) - f(t) g(t)$$

$$= [f(t+ih) - f(t)] g(t) + f(t+ih)[g(t+ih) - g(t)],$$

one readily sees that

$$\|f(t+ih) g(t+ih) - f(t) g(t)\|_1 \longrightarrow 0$$

as $h \longrightarrow 0$. It is therefore sufficient to check that

$$\int_{-\infty}^{\infty} e^{i\lambda t} f(t + ih) g(t + ih) \, dt \quad = \quad 0$$

for each $h > 0$ when $\lambda > 0$.

Fix any such h. By property (i) from the result just referred to,

$$|f(z + ih)| \quad \leqslant \quad \frac{\text{const.}}{\sqrt{h}} \qquad \text{for } \Im z \geqslant 0.$$

Also, since $f(t) \in L_2(\mathbb{R})$, the function

$$f(z + ih) \quad = \quad \frac{1}{\pi} \int_{-\infty}^{\infty} \frac{\Im z + h}{|z + ih - t|^2} f(t) \, dt$$

tends uniformly to zero for z tending to ∞ in any *fixed strip* $0 \leqslant \Im z \leqslant L$. The function $g(z + ih)$ has the same behaviour.

These properties make it possible for us to now virtually *copy* the contour integral argument made in proving the Paley–Wiener theorem, Chapter III, §D, replacing the function $f_h(z)$ figuring there* by $f(z + ih) g(z + ih)$. In that way we find that

$$\int_{-\infty}^{\infty} e^{i\lambda t} f(t + ih) g(t + ih) \, dt \quad = \quad 0$$

for $\lambda > 0$, the relation we needed.

The theorem is proved.

The last result has an important converse:

Theorem. *Given $\varphi \in H_1$, there are functions f and g in H_2 with $\varphi = fg$ and* $\| f \|_2 = \| g \|_2 = \sqrt{(\| \varphi \|_1)}$.

Proof. There is no loss of generality in assuming that $\varphi(t)$ is not a.e. zero on $\mathbb{R}$, for otherwise our theorem is trivial. Putting, then,

$$\varphi(z) \quad = \quad \frac{1}{\pi} \int_{-\infty}^{\infty} \frac{\Im z}{|z - t|^2} \varphi(t) \, dt$$

for $\Im z > 0$, we know by previous results that $\varphi(z)$ is analytic in the upper

---

* Here, the condition $\lambda > 0$ plays the rôle that the relation $\lambda > A$ did in the discussion referred to.

half plane and that

$$\log|\varphi(z)| \;\leqslant\; \frac{1}{\pi}\int_{-\infty}^{\infty}\frac{\Im z}{|z-t|^2}\log|\varphi(t)|\,dt$$

there, the integral on the right being absolutely convergent.

Thanks to the absolute convergence, we can define a function $F(z)$ analytic for $\Im z > 0$ by writing

$$F(z) \;=\; \exp\left\{\frac{1}{2\pi i}\int_{-\infty}^{\infty}\left(\frac{1}{t-z}-\frac{t}{t^2+1}\right)\log|\varphi(t)|\,dt\right\};$$

the idea here is that $F(z) \neq 0$ for $\Im z > 0$, with

$$\log|F(z)| \;=\; \frac{1}{2\pi}\int_{-\infty}^{\infty}\frac{\Im z}{|z-t|^2}\log|\varphi(t)|\,dt,$$

one half the right side of the preceding inequality. The ratio

$$G(z) \;=\; \frac{\varphi(z)}{F(z)}$$

is then *analytic* for $\Im z > 0$, and we have

$$\log|G(z)| \;\leqslant\; \frac{1}{2\pi}\int_{-\infty}^{\infty}\frac{\Im z}{|z-t|^2}\log|\varphi(t)|\,dt \;=\; \log|F(z)|,$$

i.e.,

$$|G(z)| \;\leqslant\; |F(z)|, \qquad \Im z > 0.$$

By the inequality between arithmetic and geometric means,

$$|F(z)|^2 \;\leqslant\; \frac{1}{\pi}\int_{-\infty}^{\infty}\frac{\Im z}{|z-t|^2}|\varphi(t)|\,dt,$$

so, for each $y > 0$,

$$\int_{-\infty}^{\infty}|G(x+iy)|^2\,dx \;\leqslant\; \int_{-\infty}^{\infty}|F(x+iy)|^2\,dx$$

$$\leqslant\; \frac{1}{\pi}\int_{-\infty}^{\infty}\int_{-\infty}^{\infty}\frac{y\,|\varphi(t)|}{(x-t)^2+y^2}\,dt\,dx \;=\; \|\varphi\|_1.$$

According to a previous theorem, there are thus functions f and g in H_2 with

$$F(z) \;=\; \frac{1}{\pi}\int_{-\infty}^{\infty}\frac{\Im z}{|z-t|^2}f(t)\,dt, \qquad \Im z > 0,$$

$$G(z) = \frac{1}{\pi} \int_{-\infty}^{\infty} \frac{\Im z}{|z-t|^2} g(t)\, dt, \qquad \Im z > 0,$$

and

$$\|g\|_2^2 \leqslant \|f\|_2^2 \leqslant \|\varphi\|_1.$$

For $\Im z > 0$, we have

$$\varphi(z) = F(z)G(z),$$

However, when $y \longrightarrow 0$,

$$\varphi(t+iy) \longrightarrow \varphi(t) \quad \text{a.e.}$$

while at the same time

$$F(t+iy) \longrightarrow f(t) \quad \text{a.e.}$$

and

$$G(t+iy) \longrightarrow g(t) \quad \text{a.e..}$$

Therefore,

$$\varphi(t) = f(t)g(t) \qquad \text{a.e., } t \in \mathbb{R},$$

our desired factorization.

Schwarz' inequality now yields

$$\|\varphi\|_1 \leqslant \|f\|_2 \|g\|_2.$$

We already know, however, that

$$\|g\|_2 \leqslant \|f\|_2 \leqslant \sqrt{(\|\varphi\|_1)}.$$

Hence $\|g\|_2 = \|f\|_2 = \sqrt{(\|\varphi\|_1)}$.
We are done.

Remark. For the function $F(z)$ used in the above proof, we have

$$\log|F(z)| = \frac{1}{2\pi} \int_{-\infty}^{\infty} \frac{\Im z}{|z-t|^2} \log|\varphi(t)|\, dt,$$

so

$$\log|F(t+iy)| \longrightarrow \left|\tfrac{1}{2}\log|\varphi(t)|\right| \quad \text{a.e.}$$

as $y \longrightarrow 0$ by the property of the Poisson kernel already used frequently in this article. This means, however, that

$$|F(t+iy)| \longrightarrow \sqrt{(|\varphi(t)|)} \quad \text{a.e.}$$

for $y \longrightarrow 0$. At the same time,

$$F(t + iy) \longrightarrow f(t) \quad \text{a.e.,}$$

so we have

$$|f(t)| = \sqrt{(|\varphi(t)|)} \quad \text{a.e., } t \in \mathbb{R},$$

for the H_2 function f furnished by the last theorem.

Since $\varphi \in H_1$, we must have $|\varphi(t)| > 0$ a.e. by a previous corollary (unless $\varphi(t) \equiv 0$ a.e., a trivial special case which we are excluding). The H_2 function g with $fg = \varphi$ must then also satisfy

$$|g(t)| = \sqrt{(|\varphi(t)|)} \quad \text{a.e., } t \in \mathbb{R}.$$

In spite of the fact that the H_2 functions f and g involved here have a.e. the same moduli on $\mathbb{R}$, they are in general essentially different. It is usually true that their extensions F and G to the upper half plane satisfy

$$|G(z)| < |F(z)|$$

there.

Later on in this §, our work will involve the products

$$e^{i\lambda t} f(t)$$

with $\lambda \geqslant 0$, where f is a given function in H_2. Our first observation about these is the

Lemma. *If $f \in H_2$ and $\lambda \geqslant 0$, $e^{i\lambda t} f(t) \in H_2$.*

Proof. If $\Im z > 0$, the function $1/(t - \bar{z})$ belongs to H_2. This is most easily checked by referring to the definition of H_2 and doing a contour integral; such verification is left to the reader. According to a previous theorem, then, $f(t)/(t - \bar{z})$ belongs to H_1. Hence

$$\int_{-\infty}^{\infty} \frac{e^{i\lambda t} f(t)}{t - \bar{z}} \, dt = 0$$

for each $\lambda \geqslant 0$. Here, z with $\Im z > 0$ is arbitrary, so the functions $e^{i\lambda t} f(t)$ with $\lambda \geqslant 0$ belong to H_2 by definition. Done.

When $f \in H_2$, finite linear combinations of the products $e^{i\lambda t} f(t)$ with $\lambda \geqslant 0$ form, by the lemma just proved, a certain vector subspace of H_2. We want to know when the L_2 closure of that subspace is *all* of H_2. This question was answered by Beurling. His argument uses material from the proof of the preceding theorem about factorization of functions in H_1. We need first of all to note the following analogue of a result already established for functions in H_1 or in H_∞:

Theorem. *If $f \in H_2$ and $f(t)$ is not a.e. zero on $\mathbb{R}$,*

$$\int_{-\infty}^{\infty} \frac{\log^- |f(t)|}{1+t^2} \, dt \quad < \quad \infty.$$

Also, for

$$f(z) \quad = \quad \frac{1}{\pi} \int_{-\infty}^{\infty} \frac{\Im z}{|z-t|^2} f(t) \, dt,$$

one has

$$\log |f(z)| \quad \leqslant \quad \frac{1}{\pi} \int_{-\infty}^{\infty} \frac{\Im z}{|z-t|^2} \log |f(t)| \, dt$$

when $\Im z > 0$, the integral on the right converging absolutely.

Proof. Is very similar to that of the corresponding theorem in H_1.* Here, when considering the difference

$$\int_{-\infty}^{\infty} \frac{\Im z}{|z-t|^2} \log^+ |f(t+ih)| \, dt \quad - \quad \int_{-\infty}^{\infty} \frac{\Im z}{|z-t|^2} \log^+ |f(t)| \, dt,$$

one first observes that it is bounded in absolute value by

$$\int_{-\infty}^{\infty} \frac{\Im z}{|z-t|^2} \big| |f(t+ih)| - |f(t)| \big| \, dt$$

and then applies Schwarz' inequality. The rest of the argument is the same as for H_1.

Corollary. *Unless $f(t) \in H_2$ vanishes a.e., $|f(t)| > 0$ a.e. on $\mathbb{R}$.*

Definition (Beurling). A function f in H_2 which is not a.e. zero on $\mathbb{R}$ is called *outer* if, for the function $f(z)$ of the above theorem we have

$$\log |f(z)| \quad = \quad \frac{1}{\pi} \int_{-\infty}^{\infty} \frac{\Im z}{|z-t|^2} \log |f(t)| \, dt$$

whenever $\Im z > 0$.

---

* One may also appeal directly to that theorem after noting that $f^2 \in H_1$.

Theorem. *Let $f \in H_2$, not a.e. zero on $\mathbb{R}$, be outer. Then the finite linear combinations of the $e^{i\lambda t} f(t)$ with $\lambda \geqslant 0$ are $\| \quad \|_2$ dense in H_2.*

Remark. This result is due to Beurling, who also established its *converse*. The latter will not be needed in our work; it is set at the end of this article as problem 42.

Proof of theorem. In order to show that the $e^{i\lambda t} f(t)$ with $\lambda \geqslant 0$ generate H_2, it suffices to verify that if φ is any element of L_2 such that

$$\int_{-\infty}^{\infty} e^{i\lambda t} f(t)\varphi(t)\,dt = 0$$

for all $\lambda \geqslant 0$, then

$$\int_{-\infty}^{\infty} g(t)\varphi(t)\,dt = 0$$

for each $g \in H_2$. This will follow if we can show that such a φ belongs to H_2, for then the products $g\varphi$ with $g \in H_2$ will be in H_1.

Since f and $\varphi \in L_2$, $f\varphi \in L_1$, and our assumed relation makes $f\varphi$ in H_1. The function

$$F(z) = \frac{1}{\pi} \int_{-\infty}^{\infty} \frac{\Im z}{|z-t|^2} f(t)\varphi(t)\,dt$$

is thus *analytic* for $\Im z > 0$.

If $\varphi(t) \equiv 0$ a.e. there is nothing to prove, so we may assume that this is not the case. By the preceding corollary, $|f(t)| > 0$ a.e.; therefore $f(t)\varphi(t)$ is not a.e. zero on $\mathbb{R}$. Hence, by an earlier result,

$$\log|F(z)| \leqslant \frac{1}{\pi} \int_{-\infty}^{\infty} \frac{\Im z}{|z-t|^2} \log|f(t)\varphi(t)|\,dt$$

when $\Im z > 0$, with the right-hand integral *absolutely convergent*.
At the same time, for

$$f(z) = \frac{1}{\pi} \int_{-\infty}^{\infty} \frac{\Im z}{|z-t|^2} f(t)\,dt$$

we have

$$\log|f(z)| = \frac{1}{\pi} \int_{-\infty}^{\infty} \frac{\Im z}{|z-t|^2} \log|f(t)|\,dt$$

by hypothesis whenever $\Im z > 0$. The integral on the right is certainly $> -\infty$, being absolutely convergent, so $F(z)/f(z)$ is *analytic* in $\Im z > 0$.

For that ratio, the previous two relations give

$$\log\left|\frac{F(z)}{f(z)}\right| \leqslant \frac{1}{\pi}\int_{-\infty}^{\infty}\frac{\Im z}{|z-t|^2}\log|\varphi(t)|\,dt, \qquad \Im z > 0.$$

Thence, by the inequality between arithmetic and geometric means,

$$\left|\frac{F(z)}{f(z)}\right|^2 \leqslant \frac{1}{\pi}\int_{-\infty}^{\infty}\frac{\Im z}{|z-t|^2}|\varphi(t)|^2\,dt, \qquad \Im z > 0,$$

from which, by Fubini's theorem,

$$\int_{-\infty}^{\infty}\left|\frac{F(x+iy)}{f(x+iy)}\right|^2 dx \leqslant \|\varphi\|_2^2.$$

According to a previous theorem, there is hence a function $\psi \in H_2$ with

$$\frac{F(z)}{f(z)} = \frac{1}{\pi}\int_{-\infty}^{\infty}\frac{\Im z}{|z-t|^2}\psi(t)\,dt$$

for $\Im z > 0$, and

$$\frac{F(t+iy)}{f(t+iy)} \longrightarrow \psi(t) \quad \text{a.e.}$$

as $y \longrightarrow 0$.

We have, however, by the formula for $F(z)$,

$$F(t+iy) \longrightarrow f(t)\varphi(t) \quad \text{a.e. as } y \longrightarrow 0,$$

and, for $f(z)$,

$$f(t+iy) \longrightarrow f(t) \quad \text{a.e. as } y \longrightarrow 0.$$

Therefore, since $|f(t)| > 0$ a.e.,

$$\varphi(t) = \psi(t) \quad \text{a.e.},$$

i.e., $\varphi \in H_2$, as we needed to show.

The theorem is proved.

Remark. The function f in H_2 appearing near the end of the above factorization theorem for H_1 is outer. In general, given *any* function $M(t) \geqslant 0$ such that

$$\int_{-\infty}^{\infty}\frac{\log^- M(t)}{1+t^2}\,dt < \infty$$

and

$$\int_{-\infty}^{\infty} (M(t))^2 \, dt \ < \ \infty,$$

we can construct an outer function $f \in H_2$ for which

$$|f(t)| \ = \ M(t) \qquad \text{a.e. on } \mathbb{R}.$$

To do this, one first puts

$$F(z) \ = \ \exp\left\{\frac{1}{\pi i}\int_{-\infty}^{\infty}\left(\frac{1}{t-z} - \frac{t}{t^2+1}\right)\log M(t)\,dt\right\}$$

for $\Im z > 0$; the conditions on M ensure absolute convergence of the integral figuring on the right. We have

$$\log|F(z)| \ = \ \frac{1}{\pi}\int_{-\infty}^{\infty}\frac{\Im z}{|z-t|^2}\log M(t)\,dt, \qquad \Im z > 0,$$

so that, in the first place,

$$\log|F(t+iy)| \ \longrightarrow \ \log M(t) \quad \text{a.e.}$$

for $y \rightarrow 0$. In the second place, since geometric means do not exceed arithmetic means,

$$\int_{-\infty}^{\infty}|F(x+iy)|^2\,dx \ \leqslant \ \int_{-\infty}^{\infty}(M(t))^2\,dt$$

for $y > 0$, by an argument like one in the above proof. There is thus an $f \in H_2$ with

$$F(z) \ = \ \frac{1}{\pi}\int_{-\infty}^{\infty}\frac{\Im z}{|z-t|^2}f(t)\,dt, \qquad \Im z > 0,$$

and

$$F(t+iy) \ \longrightarrow \ f(t) \quad \text{a.e.}$$

as $y \rightarrow 0$.

Comparing the above two limit relations we see, first of all, that

$$|f(t)| \ = \ M(t) \qquad \text{a.e., } t \in \mathbb{R}.$$

Therefore

$$\log|F(z)| \ = \ \frac{1}{\pi}\int_{-\infty}^{\infty}\frac{\Im z}{|z-t|^2}\log|f(t)|\,dt$$

for $\Im z > 0$. Here, our function $F(z)$ is in fact the $f(z)$ figuring in the proof of the last theorem. Hence f is *outer*.

This construction works in particular whenever $M(t) = |g(t)|$ with $g(t)$ in H_2 not a.e. zero on $\mathbb{R}$. Therefore, *any such g in H_2 coincides a.e. in modulus with an outer function in H_2*.

Problem 42

Prove the converse of the preceding result. Show, in other words, that if $f \in H_2$ is *not outer*, the $e^{i\lambda t}f(t)$ with $\lambda \geqslant 0$ *do not generate H_2* (in norm $\| \ \|_2$). (Hint: One may as well assume that $f(t)$ is not a.e. zero on $\mathbb{R}$. Take then the *outer* function $g \in H_2$ with $|g(t)| = |f(t)|$ a.e., furnished by the preceding remark. Show first that the ratio $\omega(t) = f(t)/g(t)$ – it is of modulus 1 a.e. – belongs to H_∞. For this purpose, one may look at $f(z)/g(z)$ in $\Im z > 0$.

Next observe that

$$\int_{-\infty}^{\infty} e^{i\lambda t}f(t)\overline{\omega(t)}g(t)\,dt = 0$$

for all $\lambda \geqslant 0$, so that it suffices to show that

$$\int_{-\infty}^{\infty} \varphi(t)\overline{\omega(t)}g(t)\,dt$$

cannot be zero for all $\varphi \in H_2$. Assume that *were* the case. Then

$$\int_{-\infty}^{\infty} e^{i\lambda t}\psi(t)\overline{\omega(t)}g(t)\,dt = 0,$$

i.e.,

$$\int_{-\infty}^{\infty} \overline{\omega(t)}\psi(t)e^{i\lambda t}g(t)\,dt = 0$$

for all $\lambda \geqslant 0$ and every $\psi \in H_2$.

Use now the preceding theorem (!) and another result to argue that

$$\int_{-\infty}^{\infty} \overline{\omega(t)}h(t)\,dt = 0$$

for all $h \in H_1$, making $\overline{\omega(t)}$ also in H_∞, together with $\omega(t)$. This means that

$$\omega(z) = \frac{1}{\pi}\int_{-\infty}^{\infty} \frac{\Im z}{|z-t|^2}\omega(t)\,dt$$

and $\overline{\omega(z)}$ are *both analytic* in $\Im z > 0$. Since f is *not outer*, however, $|\omega(z)| = |f(z)/g(z)| < 1$ for some such z. A contradiction is now easily obtained.)

Remark. The $\omega \in H_\infty$ figuring in the argument just indicated is called an *inner function.*

2. **Statement of the problem, and simple reductions of it**

Given a function $w \geqslant 0$ belonging to $L_1(\mathbb{R})$, we want to know whether there is an $\omega \geqslant 0$ defined on $\mathbb{R}$, *not a.e. zero*, such that

$$\int_{-\infty}^{\infty} |\tilde{U}(t)|^2 \omega(t)\, \mathrm{d}t \;\leqslant\; \int_{-\infty}^{\infty} |U(t)|^2 w(t)\, \mathrm{d}t$$

for the Hilbert transforms $\tilde{U}(t)$ (specified in some definite manner) of the functions $U(t)$ belonging to a certain class. Depending on that class, the answer is different for different specifications of $\tilde{U}(t)$.

Two particular specifications are in common use in analysis. The *first* is preferred when dealing with functions U for which only the convergence of

$$\int_{-\infty}^{\infty} \frac{|U(t)|}{1 + t^2}\, \mathrm{d}t$$

is assured; in that case one takes

$$\tilde{U}(x) \;=\; \frac{1}{\pi}\int_{-\infty}^{\infty} \left(\frac{1}{x - t} + \frac{t}{t^2 + 1} \right) U(t)\, \mathrm{d}t.$$

The expression on the right – not really an integral – is a *Cauchy principal value*, defined for almost all real x. (At this point the reader should look again at §H.1, Chapter III and the second part of §C.1, Chapter VIII.)

A *second* definition of $\tilde{U}$ is adopted when, for $\delta > 0$, the integrals

$$\int_{|t-x| \geqslant \delta} \frac{U(t)}{x - t}\, \mathrm{d}t$$

are *already absolutely convergent*. In that case, one drops the term $t/(t^2 + 1)$ figuring in the previous expression and simply takes

$$\tilde{U}(x) \;=\; \frac{1}{\pi}\int_{-\infty}^{\infty} \frac{U(t)}{x - t}\, \mathrm{d}t,$$

in other words, $1/\pi$ times the limit of the preceding integral for $\delta \longrightarrow 0$. This specification of $\tilde{U}$ was employed in §C.1 of Chapter VIII (see especially the scholium to that article). It is useful even in cases where the above integrals are *not absolutely convergent for $\delta > 0$ but merely exist as limits,* viz.,

$$\lim_{A \to \infty} \left(\int_{-A}^{x-\delta} + \int_{x+\delta}^{A} \right) \frac{U(t)}{x - t} dt.$$

This happens, for instance, with certain kinds of functions $U(t)$ bounded on $\mathbb{R}$ and *not* dying away to zero as $t \to \pm \infty$. The Hilbert transforms thus obtained are the ones listed in various tables, such as those issued in the Bateman Project series.

If now our question is posed for the *first* kind of Hilbert transform, it turns out to have substance when the given class of functions U is so large as to *include all bounded ones*. In those circumstances, it is most readily treated by first making the substitution $t = \tan(\vartheta/2)$ and then working with functions $U(t)$ equal to *trigonometric polynomials in ϑ* and with certain auxiliary functions analytic in the unit disk. One finds in that way that the question has a positive answer (i.e., that a non-zero $\omega \geqslant 0$ exists) if and only if

$$\int_{-\infty}^{\infty} \frac{1}{(t^2 + 1)^2 w(t)} dt \quad < \quad \infty$$

(under the initial assumption that $w \in L_1(\mathbb{R})$); the reader will find this work set as problem 43 below, which may serve as a test of how well he or she has assimilated the procedures of the present §.

Except in problem 43, we do not consider the first kind of Hilbert transform any further. Instead, we turn to the *second* kind, taking $\int_{-\infty}^{\infty}$ in its most general sense, as $\lim_{\delta \to \infty} \lim_{A \to \infty} (\int_{-A}^{x-\delta} + \int_{x+\delta}^{A})$. Then

$$\tilde{U}(x) \quad = \quad \frac{1}{\pi} \int_{-\infty}^{\infty} \frac{U(t)}{x - t} dt$$

is defined for $U(t)$ equal to $\sin \lambda t$ and $\cos \lambda t$, and hence for finite linear combinations of such functions (the so-called *trigonometric sums*). In the following articles, we restrict our attention to trigonometric sums $U(t)$, for which the definition of $\tilde{U}$ by means of the preceding formula presents no problem. As explained at the beginning of §D, one may think crudely of the collection of trigonometric sums as 'filling out' $L_\infty(\mathbb{R})$ 'for all practical purposes'.

By elementary contour integration, one readily finds that

$$\frac{1}{\pi} \int_{-\infty}^{\infty} \frac{e^{i\lambda t}}{x - t} dt \quad = \quad \begin{cases} - ie^{i\lambda x}, & \lambda > 0, \\ ie^{i\lambda x}, & \lambda < 0. \end{cases}$$

When $\lambda = 0$, the quantity on the left is *zero*. The reader should do this computation. One of the original applications made of contour integration by Cauchy, who *invented* it, was precisely the evaluation of such principal values! In terms of real valued functions, the formula just written goes as follows:

$$\frac{1}{\pi}\int_{-\infty}^{\infty}\frac{\cos \lambda t}{x-t}\,\mathrm{d}t \;=\; \sin \lambda t, \qquad \lambda > 0;$$

$$\frac{1}{\pi}\int_{-\infty}^{\infty}\frac{\sin \lambda t}{x-t}\,\mathrm{d}t \;=\; -\cos \lambda t, \qquad \lambda > 0.$$

From this we see already that our question (about the existence of non-zero $\omega \geq 0$) is *without substance* for the *present* specification of the Hilbert transform, when posed for *all trigonometric sums U. There can never be an $\omega \geq 0$, not a.e. zero, such that*

$$\int_{-\infty}^{\infty}|\tilde{U}(t)|^2\omega(t)\,\mathrm{d}t \;\leq\; \int_{-\infty}^{\infty}|U(t)|^2 w(t)\,\mathrm{d}t$$

for all such U, when w is integrable. This follows immediately on taking

$$U(t) \;=\; \sin \lambda t, \qquad \tilde{U}(t) \;=\; -\cos \lambda t$$

in such a presumed relation and then making $\lambda \longrightarrow 0$; in that way one concludes by Fatou's lemma that

$$\int_{-\infty}^{\infty} \omega(t)\,\mathrm{d}t \;=\; 0.$$

The same state of affairs prevails whenever our given class of functions U includes pure oscillations of arbitrary phase with frequencies tending to zero. For this reason, we should require the class of trigonometric sums $U(t)$ under consideration to *only contain terms involving frequencies bounded away from zero,* as we did in §D. The simplest non-trivial version of our problem thus has the following formulation:

> *Let $a > 0$. Under what conditions on the given $w \geq 0$ belonging to $L_1(\mathbb{R})$ does there exist an $\omega \geq 0$, not a.e. zero, such that*
>
> $$\int_{-\infty}^{\infty}|\tilde{U}(t)|^2\omega(t)\,\mathrm{d}t \;\leq\; \int_{-\infty}^{\infty}|U(t)|^2 w(t)\,\mathrm{d}t$$
>
> *for all finite trigonometric sums*
>
> $$U(t) \;=\; \sum_{|\lambda| \geq a} C_\lambda \mathrm{e}^{\mathrm{i}\lambda t} \;\;?$$

Here, we are dealing with the *second* kind of Hilbert transform, so, for the sum $U(t)$ just written,

$$\tilde{U}(t) \; = \; \sum_{|\lambda| \geq a} (-iC_\lambda \operatorname{sgn} \lambda) e^{i\lambda t}.$$

Such functions $U(t)$ can, of course, also be expressed thus:

$$U(t) \; = \; \sum_{\lambda \geq a} (A_\lambda \cos \lambda t + B_\lambda \sin \lambda t).$$

Then

$$\tilde{U}(t) \; = \; \sum_{\lambda \geq a} (A_\lambda \sin \lambda t - B_\lambda \cos \lambda t).$$

This manner of writing our trigonometric sums will be preferred in the following discussion; it has the advantage of making the *real-valued* sums $U(t)$ be *precisely* the ones involving *only real coefficients* A_λ and B_λ.

We see in particular that if $U(t)$ is a complex-valued sum of the above kind, $\Re U(t)$ and $\Im U(t)$ are also sums of the same form. This means that our relation

$$\int_{-\infty}^{\infty} |\tilde{U}(t)|^2 \omega(t) \, dt \; \leqslant \; \int_{-\infty}^{\infty} |U(t)|^2 w(t) \, dt$$

holds for all complex-valued U of the above form iff it holds for the real valued ones.

Given any trigonometric sum $U(t)$ (real-valued or not) of the form in question, we have

$$U(t) \, + \, i\tilde{U}(t) \; = \; \sum_{\lambda \geq a} C_\lambda e^{i\lambda t}$$

with certain coefficients C_λ. Conversely, if $F(t)$ is *any* finite sum like the one on the right,

$$\Re F(t) \; = \; U(t)$$

is a sum of the form under consideration, and then

$$\tilde{U}(t) \; = \; \Im F(t).$$

These statements are immediately verified by simple calculation.

Lemma. *Given $w \geqslant 0$ in $L_1(\mathbb{R})$, let $a > 0$. The relation*

$$\int_{-\infty}^{\infty} |\tilde{U}(t)|^2 \omega(t) \, dt \; \leqslant \; \int_{-\infty}^{\infty} |U(t)|^2 w(t) \, dt$$

holds for all trigonometric sums

$$U(t) \;=\; \sum_{\lambda \geq a} (A_\lambda \cos \lambda t + B_\lambda \sin \lambda t)$$

with the function $\omega(t) \geq 0$ *iff*

$$\left| \int_{-\infty}^{\infty} (w(t) + \omega(t))(F(t))^2 \, dt \right| \;\leq\; \int_{-\infty}^{\infty} (w(t) - \omega(t)) |F(t)|^2 \, dt$$

for all finite sums

$$F(t) \;=\; \sum_{\lambda \geq a} C_\lambda e^{i\lambda t}.$$

Proof. As remarked above, our relation holds for trigonometric sums U of the given form iff

$$\int_{-\infty}^{\infty} (\tilde{U}(t))^2 \omega(t) \, dt \;\leq\; \int_{-\infty}^{\infty} (U(t))^2 w(t) \, dt$$

for all such *real-valued* U. Multiply this relation by 2 and then add to both sides of the result the quantity

$$\int_{-\infty}^{\infty} \{ (\tilde{U}(t))^2 w(t) \;-\; (U(t))^2 w(t) \;-\; (\tilde{U}(t))^2 \omega(t) \;-\; (U(t))^2 \omega(t) \} \, dt.$$

We obtain the relation

$$\int_{-\infty}^{\infty} (w(t) \;+\; \omega(t)) \{ (\tilde{U}(t))^2 \;-\; (U(t))^2 \} \, dt$$

$$\leq\; \int_{-\infty}^{\infty} (w(t) \;-\; \omega(t)) \{ (U(t))^2 \;+\; (\tilde{U}(t))^2 \} \, dt$$

which must thus be *equivalent* to our original one (see the remark immediately following this proof).

In terms of $F(t) = U(t) + i\tilde{U}(t)$, the last inequality becomes

$$-\Re \int_{-\infty}^{\infty} (w(t) + \omega(t))(F(t))^2 \, dt \;\leq\; \int_{-\infty}^{\infty} (w(t) - \omega(t)) |F(t)|^2 \, dt,$$

so, according to the statements preceding the lemma, our original relation holds with the trigonometric sums $U(t)$ iff the present one is valid for the finite sums

$$F(t) \;=\; \sum_{\lambda \geq a} C_\lambda e^{i\lambda t}.$$

If, however, $F(t)$ is of this form, so is $e^{i\gamma}F(t)$ for each real constant γ. The preceding condition is thus equivalent to the requirement that

$$-\Re e^{2i\gamma} \int_{-\infty}^{\infty} (w(t) + \omega(t))(F(t))^2 \, dt \;\leqslant\; \int_{-\infty}^{\infty} (w(t) - \omega(t))|F(t)|^2 \, dt$$

for each function F and all real γ, and that happens iff

$$\left| \int_{-\infty}^{\infty} (w(t) + \omega(t))(F(t))^2 \, dt \right|$$

is $\leqslant$ the integral on the right for any such F. This last condition is hence equivalent to our original one,   Q.E.D.

Remark. The argument just made tacitly assumes finiteness of $\int_{-\infty}^{\infty} (\tilde{U}(t))^2 \omega(t) \, dt$ and $\int_{-\infty}^{\infty} (U(t))^2 \omega(t) \, dt$, as well as that of $\int_{-\infty}^{\infty} (\tilde{U}(t))^2 w(t) \, dt$ and $\int_{-\infty}^{\infty} (U(t))^2 w(t) \, dt$. About the latter two quantities, there can be no question, w being assumed integrable. Then, however, the former two must also be finite, whether we suppose the *first relation* of the lemma to hold or the *second*. Indeed, if it is the *first* one that holds,

$$\int_{-\infty}^{\infty} \{(U(t))^2 + (\tilde{U}(t))^2\} \omega(t) \, dt$$

$$\leqslant \int_{-\infty}^{\infty} \{(U(t))^2 + (\tilde{U}(t))^2\} w(t) \, dt \;<\; \infty,$$

since, for $F(t) = U(t) + i\tilde{U}(t)$, $\tilde{F}(t) = -iF(t)$. And, if the *second* holds, we surely have

$$\int_{-\infty}^{\infty} (w(t) - \omega(t))((U(t))^2 + (\tilde{U}(t))^2) \, dt \;\geqslant\; 0.$$

Theorem. *Given* $w \geqslant 0$ *in* $L_1(\mathbb{R})$ *and* $a > 0$, *any* $\omega \geqslant 0$ *such that*

$$\int_{-\infty}^{\infty} |\tilde{U}(t)|^2 \omega(t) \, dt \;\leqslant\; \int_{-\infty}^{\infty} |U(t)|^2 w(t) \, dt$$

for all sums U *of the form*

$$U(t) = \sum_{\lambda \geqslant a} (A_\lambda \cos \lambda t + B_\lambda \sin \lambda t)$$

must satisfy

$$\omega(t) \leqslant w(t) \qquad \text{a.e. on } \mathbb{R}.$$

Proof. Such an ω must in the first place belong to $L_1(\mathbb{R})$. For, putting first

$U(t) = \sin at$, $\tilde{U}(t) = -\cos at$ in our relation, and then $U(t) = \sin 2at$, $\tilde{U}(t) = -\cos 2at$, we get

$$\int_{-\infty}^{\infty} (\cos^2 at + \cos^2 2at)\omega(t)\,dt < \infty.$$

Here,

$$\cos^2 at + \cos^2 2at = \tfrac{1}{2}(1 + \cos 2at + 2\cos^2 2at) \geqslant \tfrac{7}{16}$$

for $t \in \mathbb{R}$, so ω is integrable.

Knowing that w and ω are both integrable, we can prove the theorem by verifying that

$$\int_{-\infty}^{\infty} (w(t) - \omega(t))\varphi(t)\,dt \geqslant 0$$

for each continuous function $\varphi \geqslant 0$ of compact support.

Fix any such φ, and pick an $\varepsilon > 0$. Choose first an L so large that $\varphi(t)$ *vanishes identically* outside $(-L, L)$ and that

$$\|\varphi\|_\infty \cdot \int_{|t| \geqslant L} (w(t) + \omega(t))\,dt < \varepsilon;$$

since w and ω are in $L_1(\mathbb{R})$, such a choice is possible. Then expand $\sqrt{(\varphi(t))}$ in a Fourier series on $[-L, L]$:

$$\sqrt{(\varphi(t))} \sim \sum_{n=-\infty}^{\infty} a_n e^{\pi i n t / L}, \quad -L \leqslant t \leqslant L.$$

According to the rudiments of harmonic analysis, the Fejér means

$$s_N(t) = \sum_{n=-N}^{N} \left(1 - \frac{|n|}{N}\right) a_n e^{\pi i n t / L}$$

of this Fourier series tend *uniformly* to $\sqrt{(\varphi(t))}$ on $[-L, L]$ as $N \longrightarrow \infty$. Also,

$$\|s_N\|_\infty \leqslant \|\sqrt{(\varphi)}\|_\infty.$$

We thus have

$$\int_{-\infty}^{\infty} (w(t) - \omega(t))\varphi(t)\,dt = \int_{-L}^{L} (w(t) - \omega(t))\varphi(t)\,dt$$

$$= \lim_{N \to \infty} \int_{-L}^{L} (w(t) - \omega(t))(s_N(t))^2\,dt.$$

And, for each N,

$$\int_{-L}^{L} (w(t) - \omega(t))(s_N(t))^2 \, dt = \left(\int_{-\infty}^{\infty} - \int_{|t| \geq L} \right)(w(t) - \omega(t))(s_N(t))^2 \, dt$$

$$\geq \int_{-\infty}^{\infty} (w(t) - \omega(t))(s_N(t))^2 \, dt - \|s_N\|_{\infty}^2 \int_{|t| \geq L} (w(t) + \omega(t)) \, dt$$

$$\geq \int_{-\infty}^{\infty} (w(t) - \omega(t))(s_N(t))^2 \, dt - \varepsilon$$

by choice of L, since $\|s_N\|_{\infty}^2 \leq \|\varphi\|_{\infty}$.

Putting

$$F_N(t) = e^{iat} e^{\pi i N t / L} s_N(t),$$

we have $|F_N(t)|^2 = (s_N(t))^2$. $F_N(t)$, however, is of the form

$$\sum_{\lambda \geq a} C_\lambda e^{i\lambda t},$$

so

$$\int_{-\infty}^{\infty} (w(t) - \omega(t))(s_N(t))^2 \, dt = \int_{-\infty}^{\infty} (w(t) - \omega(t))|F_N(t)|^2 \, dt$$

is ≥ 0 by the lemma. Using this in the last member of the previous chain of inequalities, we see that

$$\int_{-L}^{L} (w(t) - \omega(t))(s_N(t))^2 \, dt \geq -\varepsilon$$

for each N, so, by the above limit relation,

$$\int_{-\infty}^{\infty} (w(t) - \omega(t))\varphi(t) \, dt \geq -\varepsilon.$$

Squeezing ε, we see that the integral on the left is ≥ 0, which is what we needed to show to prove the theorem. Done.

Lemma. *Given $w \geq 0$ in $L_1(\mathbb{R})$, let $a > 0$. A necessary and sufficient condition that there be an $\omega \geq 0$, not a.e. zero on $\mathbb{R}$, such that*

$$\int_{-\infty}^{\infty} |\tilde{U}(t)|^2 \omega(t) \, dt \leq \int_{-\infty}^{\infty} |U(t)|^2 w(t) \, dt$$

for the finite sums

$$U(t) = \sum_{\lambda \geq a} (A_\lambda \cos \lambda t + B_\lambda \sin \lambda t),$$

is that there exist a function $\rho(t)$ not a.e. zero, $0 \leqslant \rho(t) \leqslant w(t)$, with

$$\left| \int_{-\infty}^{\infty} w(t)(F(t))^2 \, dt \right| \leqslant \int_{-\infty}^{\infty} (w(t) - \rho(t)) |F(t)|^2 \, dt$$

for all functions F of the form

$$F(t) = \sum_{\lambda \geqslant a} C_\lambda e^{i\lambda t}.$$

When an ω fulfilling the above condition exists, ρ may be taken equal to it. When, on the other hand, a function ρ is known, the ω equal to $\frac{1}{2}\rho$ will work.

Proof. If a function ω with the stated properties exists, we know by the previous theorem that $0 \leqslant \omega(t) \leqslant w(t)$ a.e.. Therefore, if $U(t)$ is any sum of the above form,

$$\frac{1}{2} \int_{-\infty}^{\infty} |\tilde{U}(t)|^2 \omega(t) \, dt \leqslant \frac{1}{2} \int_{-\infty}^{\infty} |U(t)|^2 w(t) \, dt$$

$$\leqslant \int_{-\infty}^{\infty} |U(t)|^2 \left(w(t) - \tfrac{1}{2}\omega(t) \right) dt.$$

The first condition of the previous lemma is thus fulfilled with

$$\omega_1(t) = \tfrac{1}{2}\omega(t)$$

in place of $\omega(t)$ and

$$w_1(t) = w(t) - \tfrac{1}{2}\omega(t)$$

in place of $w(t)$. Hence, by that lemma,

$$\left| \int_{-\infty}^{\infty} (w_1(t) + \omega_1(t))(F(t))^2 \, dt \right| \leqslant \int_{-\infty}^{\infty} (w_1(t) - \omega_1(t)) |F(t)|^2 \, dt$$

for the functions F of the form described. This relation goes over into the asserted one on taking $\rho(t) = \omega(t)$.

If, conversely, the relation involving functions F holds for some ρ, $0 \leqslant \rho(t) \leqslant w(t)$, we certainly have

$$\left| \int_{-\infty}^{\infty} \left(w(t) + \tfrac{1}{2}\rho(t) \right)(F(t))^2 \, dt \right| \leqslant \int_{-\infty}^{\infty} (w(t) - \rho(t)) |F(t)|^2 \, dt$$

$$+ \frac{1}{2} \int_{-\infty}^{\infty} \rho(t) |F(t)|^2 \, dt$$

$$= \int_{-\infty}^{\infty} \left(w(t) - \tfrac{1}{2}\rho(t) \right) |F(t)|^2 \, dt$$

for such F, so, by the previous lemma,

$$\frac{1}{2}\int_{-\infty}^{\infty} |\tilde{U}(t)|^2 \rho(t)\, dt \;\leqslant\; \int_{-\infty}^{\infty} |U(t)|^2 w(t)\, dt$$

for the sums U. Our relation for the latter thus *holds* with $\omega(t) = \rho(t)/2$, and this is not a.e. zero if $\rho(t)$ is not. Done.

Theorem. *If, for given* $w \geqslant 0$ *in* $L_1(\mathbb{R})$ *and some* $a > 0$ *there is any* $\omega \geqslant 0$, *not a.e. zero, such that*

$$\int_{-\infty}^{\infty} |\tilde{U}(t)|^2 \omega(t)\, dt \;\leqslant\; \int_{-\infty}^{\infty} |U(t)|^2 w(t)\, dt$$

for the finite sums

$$U(t) \;=\; \sum_{\lambda \geqslant a} (A_\lambda \cos \lambda t + B_\lambda \sin \lambda t),$$

we have

$$\int_{-\infty}^{\infty} \frac{\log^- w(t)}{1 + t^2}\, dt \;<\; \infty.$$

Remark. Of course,

$$\int_{-\infty}^{\infty} \frac{\log^+ w(t)}{1 + t^2}\, dt \;<\; \infty$$

by the inequality between arithmetic and geometric means, w being in L_1.

Proof of theorem. If an ω having the stated properties exists, there is, by the preceding lemma, a function ρ, not a.e. zero, $0 \leqslant \rho(t) \leqslant w(t)$), such that

$$\left| \int_{-\infty}^{\infty} w(t)(F(t))^2\, dt \right| \;\leqslant\; \int_{-\infty}^{\infty} (w(t) - \rho(t))|F(t)|^2\, dt$$

for the functions

$$F(t) \;=\; \sum_{\lambda \geqslant a} C_\lambda e^{i\lambda t}.$$

Suppose now that

$$\int_{-\infty}^{\infty} \frac{\log^- w(t)}{1 + t^2}\, dt \;=\; \infty;$$

then we will show that the function $\rho(t)$ figuring in the previous relation must be zero a.e., thus obtaining a contradiction. For this purpose, we use a variant of *Szegö's theorem* which, under our assumption on $\log^- w(t)$, gives us a sequence of functions $F_N(t)$, having the form just indicated, such that

$$\int_{-\infty}^{\infty} |1 - F_N(t)|^2 \, w(t) \, dt \xrightarrow[N]{} 0.$$

The reader should refer to Chapter II and to problem 2 at the end of it. *There*, Szegö's theorem was established for the weighted L_1 norm, and problem 2 yielded functions $F_N(t)$ of the above form for which

$$\int_{-\infty}^{\infty} |1 - F_N(t)| w(t) \, dt \xrightarrow[N]{} 0.$$

However, after making a simple modification in the argument of Chapter II, §A, which should be apparent to the reader, one obtains a proof of Szegö's theorem for weighted L_2 norms – indeed, for weighted L_p ones, where $1 \leqslant p < \infty$. There is then no difficulty in carrying out the steps of problem 2 for the weighted L_2 norm.

Once we have functions F_N satisfying the above relation, we see that

$$\int_{-\infty}^{\infty} w(t)(F_N(t))^2 \, dt \xrightarrow[N]{} \int_{-\infty}^{\infty} w(t) \, dt.$$

Indeed, using Schwarz and the triangle inequality, we have

$$\int_{-\infty}^{\infty} w(t) |(F_N(t))^2 - 1| \, dt \;=\; \int_{-\infty}^{\infty} w(t) |F_N(t) - 1| \, |F_N(t) + 1| \, dt$$

$$\leqslant \sqrt{\left(\int_{-\infty}^{\infty} w(t) |F_N(t) + 1|^2 \, dt \cdot \int_{-\infty}^{\infty} w(t) |F_N(t) - 1|^2 \, dt \right)}$$

$$\leqslant \left(\sqrt{\left(\int_{-\infty}^{\infty} w(t) |F_N(t) - 1|^2 \, dt \right)} + \sqrt{\left(4 \int_{-\infty}^{\infty} w(t) \, dt \right)} \right) \times$$

$$\times \sqrt{\left(\int_{-\infty}^{\infty} w(t) |F_N(t) - 1|^2 \, dt \right)},$$

and the last expression goes to zero as $N \longrightarrow \infty$.

We also see by this computation that

$$\int_{-\infty}^{\infty} w(t) |F_N(t)|^2 \, dt \xrightarrow[N]{} \int_{-\infty}^{\infty} w(t) \, dt,$$

and again, since $0 \leqslant \rho(t) \leqslant w(t)$, that

$$\int_{-\infty}^{\infty} \rho(t)|F_N(t)|^2 \, dt \xrightarrow[N]{} \int_{-\infty}^{\infty} \rho(t) \, dt.$$

Using these relations and making $N \longrightarrow \infty$ in the inequality

$$\left| \int_{-\infty}^{\infty} w(t)(F_N(t))^2 \, dt \right| \leqslant \int_{-\infty}^{\infty} (w(t) - \rho(t))|F_N(t)|^2 \, dt,$$

we get

$$\int_{-\infty}^{\infty} w(t) \, dt \leqslant \int_{-\infty}^{\infty} (w(t) - \rho(t)) \, dt,$$

i.e.,

$$\rho(t) = 0 \quad \text{a.e.,}$$

since $\rho(t) \geqslant 0$.

We have reached our promised contradiction. This shows that the integral $\int_{-\infty}^{\infty} (\log^- w(t)/(1 + t^2)) \, dt$ must indeed be finite, as claimed. The theorem is proved.

3.   Application of H_p space theory; use of duality

The last theorem of the preceding article shows that our problem can have a positive solution only when

$$\int_{-\infty}^{\infty} \frac{\log^- w(t)}{1 + t^2} \, dt < \infty;$$

we may thus limit our further considerations to functions $w \geqslant 0$ in $L_1(\mathbb{R})$ fulfilling this condition. According to a remark at the end of article 1, there is, corresponding to any such w, an *outer function* φ in H_2 with

$$|\varphi(t)| = \sqrt{(w(t))} \quad \text{a.e., } t \in \mathbb{R}.$$

Theorem. *Let $w \geqslant 0$, belonging to $L_1(\mathbb{R})$, satisfy the above condition on its logarithm, and let $a > 0$. In order that there exist an $\omega \geqslant 0$, not a.e. zero, such that*

$$\int_{-\infty}^{\infty} |\tilde{U}(t)|^2 \omega(t) \, dt \leqslant \int_{-\infty}^{\infty} |U(t)|^2 w(t) \, dt$$

for the functions

$$U(t) \;=\; \sum_{\lambda \geqslant a} (A_\lambda \cos \lambda t + B_\lambda \sin \lambda t),$$

it is necessary and sufficient that there be a function $\sigma(t)$, *not a.e. zero, with*

$$0 \leqslant \sigma(t) \leqslant 1 \quad \text{a.e.}$$

and

$$\left| \int_{-\infty}^{\infty} e^{2iat} \frac{\overline{\varphi(t)}}{\varphi(t)} f(t)\, dt \right| \;\leqslant\; \int_{-\infty}^{\infty} (1 - \sigma(t)) |f(t)|\, dt$$

for all $f \in H_1$. *Here,* $\varphi(t)$ *is any outer function in* H_2 *with*

$$|\varphi(t)| \;=\; \sqrt{(w(t))} \quad \text{a.e.,} \quad t \in \mathbb{R}.$$

If we have a function ω *for which the above relation holds,* $\sigma(t)$ *can be taken equal to* $\omega(t)/w(t)$. *If, on the other hand, a* σ *is furnished,* $\omega(t)$ *can be taken equal to* $\sigma(t)w(t)/2$.

Remark. Any two outer functions φ in H_2 with

$$|\varphi(t)| \;=\; \sqrt{(w(t))} \quad \text{a.e.}$$

differ by a constant factor of modulus 1.

Proof of theorem. Is based on an idea from the *Bologna Annali* paper of Helson and Szegö.

According to the second lemma of the preceding article, the existence of an ω having the properties in question is *equivalent* to that of a ρ not a.e. zero, $0 \leqslant \rho(t) \leqslant w(t)$, such that

$$\left| \int_{-\infty}^{\infty} w(t)(F(t))^2\, dt \right| \;\leqslant\; \int_{-\infty}^{\infty} (w(t) - \rho(t)) |F(t)|^2\, dt$$

for the functions

$$F(t) \;=\; \sum_{\lambda \geqslant a} C_\lambda e^{i\lambda t}.$$

This relation is in turn equivalent to the requirement that

$$\left| \int_{-\infty}^{\infty} w(t) P(t) Q(t)\, dt \right| \;\leqslant\; \frac{1}{2} \int_{-\infty}^{\infty} (w(t) - \rho(t))(|P(t)|^2 + |Q(t)|^2)\, dt$$

for all *pairs P, Q* of finite sums of the form

$$\sum_{\lambda \geqslant a} C_\lambda e^{i\lambda t}.$$

To see this, one notes in the first place that the present inequality goes over into the preceding one on taking $P = Q = F$. If, on the other hand, the preceding one always holds for our functions F, it is true both with

$$F = P + Q$$

and with

$$F = P - Q$$

whenever P and Q are two sums of the given form. Therefore,

$$\left| \int_{-\infty}^{\infty} w(t)P(t)Q(t)\,dt \right| = \frac{1}{4}\left| \int_{-\infty}^{\infty} w(t)\{(P(t)+Q(t))^2 - (P(t)-Q(t))^2\}\,dt \right|$$

$$\leqslant \frac{1}{4}\int_{-\infty}^{\infty} (w(t)-\rho(t))\{|P(t)+Q(t)|^2 + |P(t)-Q(t)|^2\}\,dt$$

$$= \frac{1}{2}\int_{-\infty}^{\infty} (w(t)-\rho(t))(|P(t)|^2 + |Q(t)|^2)\,dt,$$

and we have the second of the two inequalities.

Take now an outer function $\varphi \in H_2$ such that $|\varphi(t)|^2 = w(t)$ a.e., and write

$$\sigma(t) = \rho(t)/w(t),$$

so that $0 \leqslant \sigma(t) \leqslant 1$ a.e.. Our condition on $\log w(t)$ makes

$$w(t) > 0 \quad \text{a.e.},$$

so the ratio $\sigma(t)$ will be > 0 on a set of positive measure iff $\rho(t)$ is. In terms of φ and σ, the relation involving P and Q can be rewritten

$$\left| \int_{-\infty}^{\infty} \frac{\overline{\varphi(t)}}{\varphi(t)}(\varphi(t)P(t))(\varphi(t)Q(t))\,dt \right|$$

$$\leqslant \frac{1}{2}\int_{-\infty}^{\infty} (1-\sigma(t))\{|\varphi(t)P(t)|^2 + |\varphi(t)Q(t)|^2\}\,dt.$$

Write now

$$P(t) = e^{iat}p(t), \qquad Q(t) = e^{iat}q(t),$$

so that $p(t)$ and $q(t)$ are sums of the form

$$\sum_{\lambda \geq 0} C_\lambda e^{i\lambda t}.$$

Then the last inequality becomes

$$\left| \int_{-\infty}^{\infty} e^{2iat} \frac{\overline{\varphi(t)}}{\varphi(t)} (\varphi(t)p(t))(\varphi(t)q(t)) \, dt \right|$$

$$\leq \frac{1}{2} \int_{-\infty}^{\infty} (1 - \sigma(t)) \{ |\varphi(t)p(t)|^2 + |\varphi(t)q(t)|^2 \} \, dt.$$

Since $\varphi \in H_2$ is *outer*, the products $\varphi(t)p(t)$, $\varphi(t)q(t)$ are $\| \ \|_2$ dense in H_2 when p and q range through the collection of finite sums

$$\sum_{\lambda \geq 0} C_\lambda e^{i\lambda t},$$

according to the last theorem of article 1. The preceding relation is therefore *equivalent* to the condition that

$$\left| \int_{-\infty}^{\infty} e^{2iat} \frac{\overline{\varphi(t)}}{\varphi(t)} g(t)h(t) \, dt \right| \leq \frac{1}{2} \int_{-\infty}^{\infty} (1 - \sigma(t)) \{ |g(t)|^2 + |h(t)|^2 \} \, dt$$

for *all* g and h in H_2.

Suppose $f \in H_1$. According to the factorization theorem from article 1 and the remark thereto, there are functions $g, h \in H_2$ such that

$$f(t) = g(t)h(t) \quad \text{a.e.}$$

and

$$|g(t)| = |h(t)| = \sqrt{(|f(t)|)} \quad \text{a.e..}$$

Substituting these relations into the previous one, we get

$$\left| \int_{-\infty}^{\infty} e^{2iat} \frac{\overline{\varphi(t)}}{\varphi(t)} f(t) \, dt \right| \leq \int_{-\infty}^{\infty} (1 - \sigma(t)) |f(t)| \, dt.$$

This inequality is thus a *consequence* of the preceding one. But it also *implies* the latter. Let, indeed, g and h be in H_2. Then gh is in H_1 by a result of article 1, so, if the present relation holds,

$$\left| \int_{-\infty}^{\infty} e^{2iat} \frac{\overline{\varphi(t)}}{\varphi(t)} g(t)h(t) \, dt \right| \leq \int_{-\infty}^{\infty} (1 - \sigma(t)) |g(t)h(t)| \, dt$$

$$\leq \frac{1}{2} \int_{-\infty}^{\infty} (1 - \sigma(t)) \{ |g(t)|^2 + |h(t)|^2 \} \, dt.$$

Our final inequality, involving functions $f \in H_1$ and the quantity $\sigma(t)$, is hence *fully equivalent* with the initial one for our functions $F(t)$, involving the quantity $\rho(t)$. Since, as we have observed, $\rho(t)$ is > 0 on a set of positive measure iff $\sigma(t)$ is, the first and main conclusion of our theorem now follows directly from the second lemma of the preceding article. Again, since $\rho(t) = \sigma(t)w(t)$, the second conclusion also follows by that lemma. We are done.

In order to proceed further, we use the duality between $L_1(\mathbb{R})$ and $L_\infty(\mathbb{R})$. When one says that the latter space is the *dual* of the former, one means that each (bounded) linear functional Ψ on L_1 corresponds to a unique $\psi \in L_\infty$ such that

$$\Psi(F) = \int_{-\infty}^{\infty} F(t)\psi(t)\,dt$$

for $F \in L_1$. Here, we need the linear functionals on the *closed subspace* H_1 of L_1. These can be described according to a well known recipe from functional analysis, in the following way.

Take the (w^*) closed subspace E of L_∞ consisting of the ψ therein for which

$$\int_{-\infty}^{\infty} f(t)\psi(t)\,dt = 0$$

whenever $f \in H_1$; the *quotient space* L_∞/E can then be identified with the dual of H_1. This is how the identification goes: to each bounded linear functional Λ on H_1 corresponds precisely one subset of L_∞ of the form $\psi_0 + E$ (called a *coset* of E) such that

$$\Lambda(f) = \int_{-\infty}^{\infty} f(t)\psi(t)\,dt$$

whenever $f \in H_1$ for any $\psi \in \psi_0 + E$, and *only* for those ψ.

From article 1, we know that E is H_∞. *The dual of H_1 can thus be identified with the quotient space L_∞/H_∞.* We want to use this fact to investigate the criterion furnished by the last result. For this purpose, we resort to a trick, consisting of the *introduction of new norms, equivalent to the usual ones,* for L_1 and L_∞. If the inequality in the conclusion of the last theorem *holds* with any function σ, $0 \leqslant \sigma(t) \leqslant 1$, it certainly does so when $\sigma(t)/2$ stands in place of $\sigma(t)$. According to that theorem, however, it is the *existence* of such functions σ *different from zero on a set of positive*

measure which is of interest to us here. We may therefore limit our search
for one for which the inequality is valid to those satisfying

$$0 \leqslant \sigma(t) \leqslant 1/2 \quad \text{a.e..}$$

This restriction on our functions σ we henceforth assume.

Given such a σ, we then put

$$\|f\|_1^\sigma = \int_{-\infty}^{\infty} (1 - \sigma(t))|f(t)|\,\mathrm{d}t$$

for $f \in L_1$; $\|f\|_1^\sigma$ is a norm equivalent to the usual one on L_1, because

$$\tfrac{1}{2}\|f\|_1 \leqslant \|f\|_1^\sigma \leqslant \|f\|_1.$$

On L_∞, we use the dual norm

$$\|\psi\|_\infty^\sigma = \operatorname*{ess\,sup}_{t \in \mathbb{R}} \frac{|\psi(t)|}{1 - \sigma(t)};$$

here, the $1 - \sigma(t)$ goes in the denominator although we multiply by it
when defining $\| \ \|_1^\sigma$. We have

$$\|\psi\|_\infty \leqslant \|\psi\|_\infty^\sigma \leqslant 2\|\psi\|_\infty$$

for $\psi \in L_\infty$, so $\| \ \|_\infty^\sigma$ and $\| \ \|_\infty$ are equivalent on that space.

If $\psi \in L_\infty$, we have, for the functional

$$\Psi(f) = \int_{-\infty}^{\infty} f(t)\psi(t)\,\mathrm{d}t$$

on L_1 corresponding to it,

$$|\Psi(f)| \leqslant \|\psi\|_\infty^\sigma \|f\|_1^\sigma.$$

Moreover, the *supremum* of $|\Psi(f)|$ for the $f \in L_1$ with $\|f\|_1^\sigma \leqslant 1$ is
precisely $\|\psi\|_\infty^\sigma$. These facts are easily verified by writing

$$f(t)\psi(t) \quad \text{as} \quad (1 - \sigma(t))f(t) \cdot \frac{\psi(t)}{1 - \sigma(t)}$$

in the preceding integral.

For elements of the quotient space L_∞/H_∞ – these are just the cosets
$\psi_0 + H_\infty$, $\psi_0 \in L_\infty$ – we write, following standard practice,

$$\|\psi_0 + H_\infty\|_\infty^\sigma = \inf\{\|\psi_0 + h\|_\infty^\sigma : \ h \in H_\infty\}.$$

We have already observed that to each such coset corresponds a linear
functional Λ on H_1 given by the formula

$$\Lambda(f) \;=\; \int_{-\infty}^{\infty} f(t)\psi(t)\,dt, \quad f \in H_1,$$

where ψ is *any* element of $\psi_0 + H_\infty$. By choosing the $\psi \in \psi_0 + H_\infty$ to have $\|\psi\|_\infty^\sigma$ arbitrarily close to $\|\psi_0 + H_\infty\|_\infty^\sigma$, we see that

$$|\Lambda(f)| \;\leqslant\; \|\psi_0 + H_\infty\|_\infty^\sigma \|f\|_1^\sigma, \quad f \in H_1.$$

It is important that this inequality is *sharp*. Even more than that is true:

> The infimum appearing in the above formula for $\|\psi_0 + H_\infty\|_\infty^\sigma$ is actually attained for some $h \in H_\infty$, and is equal to the supremum of $|\Lambda(f)|$ for $f \in H_1$ and $\|f\|_1^\sigma \leqslant 1$.

This statement is a straightforward consequence of results from elementary functional analysis. However, lest the reader suspect that something is being produced out of nothing here by mere juggling of notation, let us give the proof.

Denote the supremum of $|\Lambda(f)|$ for $f \in H_1$ and $\|f\|_1^\sigma \leqslant 1$ by M. *According to the Hahn–Banach theorem*, there is an extension Λ^* of the linear functional Λ *to all of* L_1, such that

$$|\Lambda^*(F)| \;\leqslant\; M\|F\|_1^\sigma$$

for $F \in L_1$. Corresponding to Λ^* there is, as observed earlier, a $\psi \in L_\infty$ with

$$\Lambda^*(F) \;=\; \int_{-\infty}^{\infty} F(t)\psi(t)\,dt, \quad F \in L_1,$$

and, according to what was also noted above,

$$\|\psi\|_\infty^\sigma \;=\; \sup\{|\Lambda^*(F)| : \; F \in L_1 \text{ and } \|F\|_1^\sigma \leqslant 1\} \;\leqslant\; M.$$

Then, for $f \in H_1$,

$$\Lambda(f) \;=\; \Lambda^*(f) \;=\; \int_{-\infty}^{\infty} f(t)\psi(t)\,dt,$$

so $\psi \in \psi_0 + H_\infty$; there is, in other words, an $h \in H_\infty$ with $\psi = \psi_0 + h$, and

$$\|\psi_0 + h\|_\infty^\sigma \;=\; \|\psi\|_\infty^\sigma \;\leqslant\; M.$$

But, as we remarked previously,

$$|\Lambda(f)| \;\leqslant\; \|\psi_0 + H_\infty\|_\infty^\sigma \|f\|_1^\sigma, \quad f \in H_1,$$

so $\|\psi_0 + h\|_\infty^\sigma \geqslant \|\psi_0 + H_\infty\|_\infty^\sigma \geqslant M$. Hence

$$\|\psi_0 + H_\infty\|_\infty^\sigma \;=\; \|\psi_0 + h\|_\infty^\sigma \;=\; M.$$

Once we are in possession of the above facts, it is easy to establish the following key result.

Theorem. *Let $w \geqslant 0$ in $L_1(\mathbb{R})$ and a number $a > 0$ be given. In order that there exist an $\omega \geqslant 0$, not a.e. zero, such that*

$$\int_{-\infty}^{\infty} |\tilde{U}(t)|^2 \omega(t)\,dt \;\; \leqslant \;\; \int_{-\infty}^{\infty} |U(t)|^2 w(t)\,dt$$

for the sums

$$U(t) \;\; = \;\; \sum_{\lambda \geqslant a} (A_\lambda \cos \lambda t + B_\lambda \sin \lambda t),$$

it is necessary and sufficient that, first of all,

$$\int_{-\infty}^{\infty} \frac{\log^- w(t)}{1 + t^2}\,dt \;\; < \;\; \infty$$

and that then, if φ is any outer function in H_2 with

$$|\varphi(t)| \;\; = \;\; \sqrt{(w(t))} \quad \text{a.e.,}$$

we have

$$\left| e^{2iat} \frac{\overline{\varphi(t)}}{\varphi(t)} - h(t) \right| \;\; \leqslant \;\; 1 \qquad \text{a.e., } t \in \mathbb{R},$$

for some h, not a.e. zero, belonging to H_∞.

A function ω equal to a constant multiple of w will satisfy our conditions iff there is an $h \in H_\infty$ for which

$$\left| e^{2iat} \frac{\overline{\varphi(t)}}{\varphi(t)} - h(t) \right| \;\; \leqslant \;\; \text{const.} \;\; < \;\; 1 \quad \text{a.e..}$$

Proof. As we saw at the end of the last article, there can be no ω with the above properties unless

$$\int_{-\infty}^{\infty} \frac{\log^- w(t)}{1 + t^2}\,dt \;\; < \;\; \infty.$$

Assuming, then, this condition, we take one of the outer functions φ specified in the statement, and see by the preceding theorem and discussion following it that the existence of an ω having the properties in question is *equivalent* to that of a σ, not a.e. zero,

$$0 \leqslant \sigma(t) \leqslant 1/2 \quad \text{a.e.},$$

such that

$$\left| \int_{-\infty}^{\infty} e^{2iat} \frac{\overline{\varphi(t)}}{\varphi(t)} f(t)\,dt \right| \leqslant \|f\|_1^{\sigma}, \quad f \in H_1.$$

According to what we observed above, however, the last relation is equivalent to the existence of an $h \in H_\infty$ for which

$$\frac{|(e^{2iat}\overline{\varphi(t)}/\varphi(t)) - h(t)|}{1 - \sigma(t)} \leqslant 1 \quad \text{a.e., } t \in \mathbb{R}.$$

Suppose in the first place that *there is no non-zero $h \in H_\infty$ for which*

$$|(e^{2iat}\overline{\varphi(t)}/\varphi(t)) - h(t)| \leqslant 1 \quad \text{a.e.}.$$

Then, since $1/2 \leqslant 1 - \sigma(t) \leqslant 1$ a.e., no $h \in H_\infty$ other than the zero one could satisfy the previous relation. The latter must therefore reduce to

$$|e^{2iat}\overline{\varphi(t)}/\varphi(t)|/(1 - \sigma(t)) \leqslant 1 \quad \text{a.e.},$$

i.e., $1 - \sigma(t) \geqslant 1$ a.e., so that

$$\sigma(t) \equiv 0 \quad \text{a.e., } t \in \mathbb{R}.$$

As has just been said, this means that *there can be no non-zero ω fulfilling our conditions,* and necessity is proved.

Consider now the situation where there *is* a non-zero $h \in H_\infty$ making

$$|(e^{2iat}\overline{\varphi(t)}/\varphi(t)) - h(t)| \leqslant 1 \quad \text{a.e.}.$$

Then; since

$$(e^{2iat}\overline{\varphi(t)}/\varphi(t)) - \tfrac{1}{2}h(t) = \tfrac{1}{2}(\{(e^{2iat}\overline{\varphi(t)}/\varphi(t)) - h(t)\}$$
$$+ \ e^{2iat}\overline{\varphi(t)}/\varphi(t)),$$

the expression on the *left* also has modulus $\leqslant 1$ a.e.. It is in fact of modulus < 1 *on a set of positive measure.* Indeed, the expression in curly brackets on the right has modulus $\leqslant 1$, and the remaining right-hand term (without the factor $1/2$) has modulus equal to 1. Therefore, since the unit circle is *strictly convex,* the whole right side cannot have modulus equal to 1 *unless* the expression in curly brackets and $e^{2iat}\overline{\varphi(t)}/\varphi(t)$ are *equal,* that is, *unless $h(t) = 0$.* We are, however, assuming that $h(t) \neq 0$

on a set of positive measure; the modulus in question must hence be < 1 on such a set.

Put

$$\sigma(t) \;=\; \min\big(1/2, \; 1 - |(e^{2iat}\overline{\varphi(t)}/\varphi(t)) - (h(t)/2)|\,\big).$$

We then have $0 \leqslant \sigma(t) \leqslant 1/2$ a.e., and $\sigma(t) > 0$ on a set of positive measure by what we have just shown. Finally,

$$\frac{|(e^{2iat}\overline{\varphi(t)}/\varphi(t)) - (h(t)/2)|}{1 - \sigma(t)} \;\leqslant\; 1 \qquad \text{a.e.,}$$

so there must, by the above equivalency statements, be a non-zero $\omega \geqslant 0$ for which our inequality on functions U is satisfied. Sufficiency is proved.

We have still to verify the last part of our theorem. It follows, however, from the last part of the preceding one that an ω *equal to a constant multiple of w will work* iff, in the relation involving H_1 functions and σ, we can take σ equal to some constant c, $0 < c < 1/2$. By the above discussion, this is *equivalent* to the existence of an $h \in H_\infty$ for which

$$\frac{|(e^{2iat}\overline{\varphi(t)}/\varphi(t)) - h(t)|}{1 - c} \;\leqslant\; 1 \qquad \text{a.e.,}$$

and we have what was needed.

The theorem is completely proved.

Establishment of the remaining results in this § is based on the criterion furnished by the one just obtained. In that way, we get, first of all, the

Theorem. *Let $w \geqslant 0$ in $L_1(\mathbb{R})$ and a number $a > 0$ be given. If there is any $\omega \geqslant 0$ at all, different from zero on a set of positive measure, such that*

$$\int_{-\infty}^{\infty} |\tilde{U}(t)|^2 \omega(t)\,dt \;\leqslant\; \int_{-\infty}^{\infty} |U(t)|^2 w(t)\,dt$$

for the sums

$$U(t) \;=\; \sum_{\lambda \geqslant a} (A_\lambda \cos \lambda t + B_\lambda \sin \lambda t),$$

there is one with

$$\int_{-\infty}^{\infty} \frac{\log^- \omega(t)}{1 + t^2}\,dt \;<\; \infty.$$

Proof. If any ω enjoying the above properties exists, we know by the preceding result that

$$\int_{-\infty}^{\infty} \frac{\log^- w(t)}{1+t^2} \, dt \;<\; \infty,$$

and, taking an outer function $\varphi \in H_2$ with $\varphi(t) = \sqrt{(w(t))}$ a.e., that

$$\left| e^{2iat} \frac{\overline{\varphi(t)}}{\varphi(t)} - h(t) \right| \;\leqslant\; 1 \qquad \text{a.e.}$$

for some *non-zero* $h \in H_\infty$.

The proof of the sufficiency part of the last theorem shows, however, that *once we have such an* h, we can put

$$\sigma(t) \;=\; \min\left\{ 1/2, \; 1 - |(e^{2iat}\overline{\varphi(t)}/\varphi(t)) - (h(t)/2)| \right\},$$

and then, with *this* function σ, the conditions of the *first* result in the present article will be satisfied, ensuring that

$$\omega(t) \;=\; \tfrac{1}{2}\sigma(t)w(t)$$

has the desired properties. It is claimed that

$$\int_{-\infty}^{\infty} \frac{\log^- \omega(t)}{1+t^2} \, dt \;<\; \infty$$

for *this* function ω. In view of the above condition on $\log^- w(t)$, it is enough to verify that

$$\int_{-\infty}^{\infty} \frac{\log \sigma(t)}{1+t^2} \, dt \;>\; -\infty$$

for our present function σ.

Write

$$\psi(t) \;=\; e^{-2iat}\varphi(t)/\overline{\varphi(t)};$$

then $|\psi(t)| = 1$ a.e., and the last inequality is implied by

$$\int_{-\infty}^{\infty} \frac{1}{1+t^2} \log\left(1 - |1 - (\psi(t)h(t)/2)| \right) dt \;>\; -\infty$$

which we proceed now to establish.

We have $|1 - \psi(t)h(t)| \leqslant 1$ a.e., so

$$1 - \tfrac{1}{2}\psi(t)h(t) \;=\; \tfrac{1}{2} + \tfrac{1}{2}(1 - \psi(t)h(t))$$

lies, for almost all t, in a circle of radius $1/2$ about the point $1/2$:

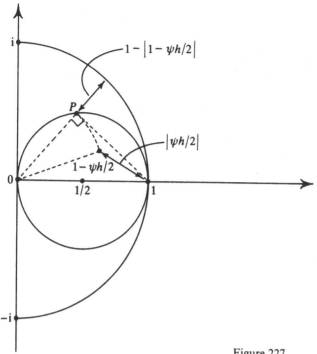

Figure 227

In this figure,

$$1 \; - \; |1 \; - \; (\psi(t)h(t)/2)| \;\; = \;\; 1 \; - \; \overline{OP},$$

where

$$\overline{OP}^2 \;\; \leqslant \;\; 1 \; - \; |\psi(t)h(t)/2|^2.$$

Therefore, for almost all t,

$$1 \; - \; |1 \; - \; (\psi(t)h(t)/2)|$$

$$\geqslant \;\; 1 \; - \; \sqrt{\left(1 \; - \; \frac{|\psi(t)h(t)|^2}{4} \right)} \;\; \geqslant \;\; \frac{|\psi(t)h(t)|^2}{8} \;\; = \;\; \frac{|h(t)|^2}{8}.$$

Thus,

$$\int_{-\infty}^{\infty} \frac{1}{1+t^2} \log(\; 1 \; - \; |1 \; - \; (\psi(t)h(t)/2)| \;)\,\mathrm{d}t \;\; \geqslant \;\; \int_{-\infty}^{\infty} \frac{\log|(h(t))^2/8|}{1+t^2}\,\mathrm{d}t.$$

However, $h \in H_\infty$ is not a.e. zero. Therefore, by a theorem from article 1,

$$\int_{-\infty}^{\infty} \frac{\log|h(t)|}{1+t^2}\,\mathrm{d}t \;\; > \;\; -\infty.$$

The preceding integral is thus also $> -\infty$ and our desired relation is established. We are done.

4.   Solution of our problem in terms of multipliers

We are now able to prove the

Theorem. *Let $w(t) \geqslant 0$ belonging to $L_1(\mathbb{R})$ and the number $a > 0$ be given. In order that there exist an $\omega \geqslant 0$, not a.e. zero, such that*

$$\int_{-\infty}^{\infty} |\tilde{U}(t)|^2 \omega(t)\,dt \;\leqslant\; \int_{-\infty}^{\infty} |U(t)|^2 w(t)\,dt$$

for the sums

$$U(t) \;=\; \sum_{\lambda \geqslant a} (A_\lambda \cos \lambda t + B_\lambda \sin \lambda t),$$

it is necessary and sufficient that there be a non-zero entire function $f(z)$ of exponential type $\leqslant a$ making

$$\int_{-\infty}^{\infty} \frac{|f(t)|^2}{(1+t^2)w(t)}\,dt \;<\; \infty.$$

Remark. We see that in the typical situation where $w(t)$ is bounded above and *very small for large values of* $|t|$, one has, under the conditions of the theorem, an entire function f of exponential type acting as a *multiplier* (in the sense adopted at the beginning of §A) for the *large* function $1/\sqrt{((1+t^2)w(t))}$.

Proof of theorem: necessity. Is based partly on a result from §D.

Suppose there *is* an ω having the properties in question. Then, by the last theorem of the preceding article, we have one for which

$$\int_{-\infty}^{\infty} \frac{\log^- \omega(t)}{1+t^2}\,dt \;<\; \infty.$$

Let $U(t)$ be any *real-valued* sum of the form indicated above, i.e., one for which the coefficients A_λ and B_λ are real. The function

$$F(t) \;=\; U(t) + i\tilde{U}(t)$$

is of the form

$$\sum_{\lambda \geqslant a} C_\lambda e^{i\lambda t}$$

and hence belongs to H_∞ according to a simple lemma in article 1. By the first theorem of article 2,

$$\omega(t) \leqslant w(t) \quad \text{a.e.,}$$

so

$$\int_{-\infty}^{\infty} |F(t)|^2 \omega(t)\,dt = \int_{-\infty}^{\infty} \{(U(t))^2 + (\tilde{U}(t))^2\}\omega(t)\,dt$$

$$\leqslant 2\int_{-\infty}^{\infty} (U(t))^2 w(t)\,dt.$$

For $\Im z > 0$, put, as in article 1,

$$F(z) = \frac{1}{\pi}\int_{-\infty}^{\infty} \frac{\Im z}{|z-t|^2} F(t)\,dt;$$

then, since $F \in H_\infty$,

$$\log|F(z)| \leqslant \frac{1}{\pi}\int_{-\infty}^{\infty} \frac{\Im z}{|z-t|^2} \log|F(t)|\,dt$$

according to a result from that article. (Here, of course, $F(z)$ is just

$$\sum_{\lambda \geqslant a} C_\lambda e^{i\lambda z},$$

a function continuous up to $\mathbb{R}$, so one may, if one prefers, apply §G.2 of Chapter III directly.) The right side of the relation just written equals

$$\frac{1}{2\pi}\int_{-\infty}^{\infty} \frac{\Im z \log(|F(t)|^2\omega(t))}{|z-t|^2}\,dt - \frac{1}{2\pi}\int_{-\infty}^{\infty} \frac{\Im z \log\omega(t)}{|z-t|^2}\,dt,$$

and this, by the inequality between arthmetic and geometric means, is

$$\leqslant \frac{1}{2}\log\left\{\frac{1}{\pi}\int_{-\infty}^{\infty} \frac{\Im z}{|z-t|^2}|F(t)|^2\omega(t)\,dt\right\} + \frac{1}{2\pi}\int_{-\infty}^{\infty} \frac{\Im z}{|z-t|^2}\log^-\omega(t)\,dt.$$

Here, the *first* term is

$$\leqslant \frac{1}{2}\log\left\{\frac{1}{\pi\Im z}\int_{-\infty}^{\infty} |F(t)|^2\omega(t)\,dt\right\}$$

which is in turn

$$\leqslant \frac{1}{2}\log\left\{\frac{2}{\pi\Im z}\int_{-\infty}^{\infty} (U(t))^2 w(t)\,dt\right\}$$

by what we have already seen. The *second* term – call it $p(z)$ – is *finite* (and harmonic!) for $\Im z > 0$, thanks to the above condition on $\log^- \omega(t)$. Our inequality for $\log|F(z)|$ thus boils down to the relation

$$|F(z)| \;\leqslant\; e^{p(z)} \sqrt{\left(\frac{2}{\pi\Im z} \int_{-\infty}^{\infty} (U(t))^2 w(t)\,dt\right)}, \qquad \Im z > 0,$$

whence, in the upper half-plane,

$$\left|\frac{1}{\pi}\int_{-\infty}^{\infty} \frac{\Im z}{|z-t|^2} U(t)\,dt\right| \;=\; |\Re F(z)| \;\leqslant\; e^{p(z)}\sqrt{\left(\frac{2}{\pi\Im z}\int_{-\infty}^{\infty} |U(t)|^2 w(t)\,dt\right)}$$

for the *real-valued* sums

$$U(t) \;=\; \sum_{\lambda \geqslant a} (A_\lambda \cos \lambda t + B_\lambda \sin \lambda t).$$

This last relation must indeed then hold for such *complex*-valued sums $U(t)$, because any of those can be written as $U_1(t) + iU_2(t)$ with *real-valued* ones U_1, U_2 of the same form.*

Invoke now the second theorem of §D! According to it, the estimate just obtained implies the existence of a non-zero entire function f of exponential type $\leqslant a$ such that

$$\int_{-\infty}^{\infty} \frac{|f(t)|^2}{(1+t^2)w(t)}\,dt \;<\; \infty.$$

Necessity is proved.

We turn to the **sufficiency**. Suppose we *have* a non-zero entire function $f(z)$ of exponential type $\leqslant a$ with

$$\int_{-\infty}^{\infty} \frac{|f(t)|^2}{(1+t^2)w(t)}\,dt \;<\; \infty.$$

As in the sufficiency proof for the first theorem in §D, f may be taken to be *real* on $\mathbb{R}$. Also, by reasoning as in the sufficiency argument for the second theorem of that §, we see that $f(t)/(t+i)$ is *bounded* on $\mathbb{R}$. The last statement certainly implies that

$$\int_{-\infty}^{\infty} \frac{\log^+ |f(t)|}{1+t^2}\,dt \;<\; \infty,$$

---

* In that circumstance, $|\int_{-\infty}^{\infty}(\Im z U(t)/|z-t|^2)dt|^2 =$
$\{\int_{-\infty}^{\infty}(\Im z U_1(t)/|z-t|^2)dt\}^2 + \{\int_{-\infty}^{\infty}(\Im z U_2(t)/|z-t|^2)dt\}^2$. Use of the inequality on each of the integrals on the right (for which it is already known to hold) yields the upper bound $(2\pi/\Im z)e^{2p(z)}\int_{-\infty}^{\infty}\{(U_1(t))^2 + (U_2(t))^2\}w(t)\,dt$.

so the first theorem of §G.2, Chapter III applies to our function f, and we have

$$\int_{-\infty}^{\infty} \frac{\log^{-}|f(t)|}{1+t^2}\,dt \quad < \quad \infty.$$

At the same time, by the inequality between arithmetic and geometric means,

$$\frac{1}{\pi}\int_{-\infty}^{\infty} \frac{1}{1+t^2}\log\left(\frac{|f(t)|^2}{w(t)}\right)dt \quad \leqslant \quad \log\left(\frac{1}{\pi}\int_{-\infty}^{\infty} \frac{|f(t)|^2}{(1+t^2)w(t)}\,dt\right) \quad < \quad \infty$$

which, with the previous relation, yields

$$\int_{-\infty}^{\infty} \frac{\log^{-}w(t)}{1+t^2}\,dt \quad < \quad \infty.$$

This condition, however, gives us an outer function $\varphi \in H_2$ for which

$$|\varphi(t)| \quad = \quad \sqrt{(w(t))} \quad > \quad 0 \quad \text{a.e.}.$$

According, then, to the second theorem of the preceding article, a function $\omega \geqslant 0$ possessing the desired properties *will exist* provided that we can find a *non-zero* $h \in H_\infty$ such that

$$\left| e^{2iat}\,\frac{\overline{\varphi(t)}}{\varphi(t)} - h(t) \right| \quad \leqslant \quad 1 \quad \text{a.e.}.$$

We proceed to exhibit such an h.

For $\Im z > 0$, write, as in article 1,

$$\varphi(z) \quad = \quad \frac{1}{\pi}\int_{-\infty}^{\infty} \frac{\Im z}{|z-t|^2}\varphi(t)\,dt;$$

by a theorem from that article, $\varphi(z)$ is *analytic* in the upper half-plane and

$$\varphi(t+iy) \quad \longrightarrow \quad \varphi(t) \quad \text{a.e.}$$

as $y \longrightarrow 0$. Saying that φ is *outer* means, as we recall, that

$$\log|\varphi(z)| \quad = \quad \frac{1}{\pi}\int_{-\infty}^{\infty} \frac{\Im z}{|z-t|^2}\log|\varphi(t)|\,dt, \qquad \Im z > 0;$$

$\varphi(z)$ has, in particular, *no zeros in the upper half plane*. The ratio

$$R(z) \quad = \quad e^{2iaz}\left(\frac{f(z)}{\varphi(z)}\right)^2$$

is thus *analytic* for $\Im z > 0$. Since $f(z)$ is *entire* and $|\varphi(t)| > 0$ a.e., $R(t + iy)$ approaches for almost every $t \in \mathbb{R}$ a *definite limit*,

$$R(t) = e^{2iat}\left(\frac{f(t)}{\varphi(t)}\right)^2,$$

as $y \longrightarrow 0$. Because our function f is real on $\mathbb{R}$,

$$R(t) \qquad \text{and} \qquad e^{2iat}\frac{\overline{\varphi(t)}}{\varphi(t)}$$

have *the same argument* there, and we see, referring to our requirement on h, that *if $R(t)$ were in H_∞*, we could take for h a suitable constant multiple of R. Usually, however, $R(t)$ is not bounded, so this will not be the case, and we have to do a supplementary construction.

We have

$$|R(t)| = \frac{(f(t))^2}{w(t)}, \qquad t \in \mathbb{R},$$

so by hypothesis,

$$\int_{-\infty}^{\infty} \frac{|R(t)|}{1 + t^2}\, dt < \infty.$$

Following an idea from a paper of Adamian, Arov and Krein we now put

$$Q(z) = \frac{1}{\pi i}\int_{-\infty}^{\infty}\left(\frac{1}{t - z} - \frac{t}{t^2 + 1}\right)|R(t)|\, dt$$

for $\Im z > 0$; the previous relation guarantees absolute convergence of the integral on the right, and $Q(z)$ is *analytic* in the upper half plane, with $\Re Q(z) > 0$ there. The quotient $R(z)/Q(z)$ is thus analytic for $\Im z > 0$.

It is now claimed that

$$\left|\frac{R(z)}{Q(z)}\right| \leqslant 1, \qquad \Im z > 0.$$

The function f is of exponential type $\leqslant a$ and fulfills the above condition involving $\log^+|f(t)|$. Hence, by §G.2, Chapter III,

$$\log|f(z)| \leqslant a\Im z + \frac{1}{\pi}\int_{-\infty}^{\infty}\frac{\Im z}{|z - t|^2}\log|f(t)|\, dt, \qquad \Im z > 0,$$

which, with the previous formula for $\log|\varphi(z)|$, yields

$$\log|R(z)| \leqslant \frac{1}{\pi}\int_{-\infty}^{\infty}\frac{\Im z}{|z - t|^2}\log|R(t)|\, dt, \qquad \Im z > 0.$$

Returning to our function $Q(z)$, we get, by the inequality between arithmetic and geometric means,

$$|Q(z)| \ge \Re Q(z) = \frac{1}{\pi} \int_{-\infty}^{\infty} \frac{\Im z}{|z-t|^2} |R(t)| \, dt$$

$$\ge \exp\left\{\frac{1}{\pi} \int_{-\infty}^{\infty} \frac{\Im z}{|z-t|^2} \log|R(t)| \, dt\right\}.$$

The preceding relation says, however, that the right-hand member is $\ge |R(z)|$. We thus have $|Q(z)| \ge |R(z)|$ for $\Im z > 0$, and the above inequality is verified.

Thanks to that inequality we see, by a result from article 1, that there is an $h \in H_\infty$ with

$$\frac{R(z)}{Q(z)} = h(z) = \frac{1}{\pi} \int_{-\infty}^{\infty} \frac{\Im z}{|z-t|^2} h(t) \, dt$$

for $\Im z > 0$, and that $\|h\|_\infty \le 1$. This function h cannot be a.e. zero on $\mathbb{R}$ because $R(z)$ is not identically zero – the entire function $f(z)$ isn't! A result from article 1 therefore implies that

$$|h(t)| > 0 \qquad \text{a.e.,} \quad t \in \mathbb{R}.$$

As $y \longrightarrow 0$,

$$\frac{R(t+iy)}{Q(t+iy)} = h(t+iy) \longrightarrow h(t) \ne 0 \quad \text{a.e..}$$

At the same time,

$$R(t+iy) \longrightarrow R(t) \quad \text{a.e.,}$$

so $Q(t+iy)$ must approach a certain definite limit, $Q(t)$, for almost all $t \in \mathbb{R}$ as $y \longrightarrow 0$. (This also follows directly from §F.2 of Chapter III.) Since

$$\Re Q(z) = \frac{1}{\pi} \int_{-\infty}^{\infty} \frac{\Im z}{|z-t|^2} |R(t)| \, dt,$$

we see, by the usual property of the Poisson kernel, that

$$\Re Q(t) = |R(t)| \quad \text{a.e..}$$

Finally, then,

$$h(t) = \frac{R(t)}{|R(t)| + i \Im Q(t)} \qquad \text{a.e.,} \quad t \in \mathbb{R}.$$

Recall that

$$R(t) = e^{2iat}\left(\frac{f(t)}{\varphi(t)}\right)^2 = e^{2iat}\frac{\overline{\varphi(t)}}{\varphi(t)}|R(t)|,$$

$f(t)$ being real. This and the preceding formula thus give

$$\left|e^{2iat}\frac{\overline{\varphi(t)}}{\varphi(t)} - h(t)\right| = \left|1 - \frac{|R(t)|}{|R(t)| + i\Im Q(t)}\right| \quad \text{a.e.,}$$

and the right side is clearly $\leqslant 1$. Our function $h \in H_\infty$ therefore has the required properties, and our proof of sufficiency is finished.

We are done.

Remark. From this theorem and the second one of §D, we see that *only* by virtue of the existence of a non-zero $\omega \geqslant 0$ making

$$\int_{-\infty}^{\infty} |\tilde{U}(t)|^2 \omega(t)\,dt \leqslant \int_{-\infty}^{\infty} |U(t)|^2 w(t)\,dt$$

can the harmonic extension of a general sum

$$U(t) = \sum_{\lambda \geqslant a} (A_\lambda \cos \lambda t + B_\lambda \sin \lambda t)$$

to the upper half plane be controlled there by the integral on the right.

In the next problem, we consider bounded functions $u(\vartheta)$ defined on $[-\pi, \pi]$, using for them a Hilbert transform given by the formula

$$\tilde{u}(\vartheta) = \frac{1}{2\pi}\int_{-\pi}^{\pi} \frac{u(\tau)}{\tan((\vartheta - \tau)/2)}\,d\tau,$$

as is customary in the study of Fourier series. (The expression on the right is a Cauchy principal value.) If one puts $\tan(\vartheta/2) = x$, $\tan(\tau/2) = t$, and then writes $u(\tau) = U(t)$, the function $\tilde{u}(\vartheta)$ goes over into the *first kind* of Hilbert transform $\tilde{U}(x)$ for functions U defined on $\mathbb{R}$, described at the beginning of article 2.

If a function $w(\vartheta) \geqslant 0$ belonging to $L_1(-\pi, \pi)$ is given, one may ask whether there exists an $\omega(\vartheta) \geqslant 0$, not a.e. zero on $[-\pi, \pi]$, such that

$$\int_{-\pi}^{\pi} |\tilde{u}(\vartheta)|^2 \omega(\vartheta)\,d\vartheta \leqslant \int_{-\pi}^{\pi} |u(\vartheta)|^2 w(\vartheta)\,d\vartheta$$

for all bounded functions u. It is clear that any given ω has this property iff, with it, the relation just written holds for all u of the *special form*

$$u(\vartheta) = \sum_{-N}^{N} a_n e^{in\theta}.$$

(Here N is *finite*, but *arbitrary*.) Such a function u is called a *trigonometric polynomial*; for it we have

$$\tilde{u}(\vartheta) = -i \sum_{-N}^{N} a_n \operatorname{sgn} n\, e^{in\theta}.$$

Problem 43

Given $w \geqslant 0$ in $L_1(-\pi, \pi)$, one is to prove that there exists an $\omega \geqslant 0$, not a.e. zero on $[-\pi, \pi]$, such that

$$\int_{-\pi}^{\pi} |\tilde{u}(\vartheta)|^2 \omega(\vartheta)\, d\vartheta \;\leqslant\; \int_{-\pi}^{\pi} |u(\vartheta)|^2 w(\vartheta)\, d\vartheta$$

for all *trigonometric polynomials* u iff

$$\int_{-\pi}^{\pi} \frac{d\vartheta}{w(\vartheta)} \;<\; \infty.$$

(a) First prove Kolmogorov's theorem, which says that there is a sequence of *trigonometric polynomials* $u_k(\vartheta)$ *without constant term* (i.e., in which $a_0 = 0$) such that

$$\int_{-\pi}^{\pi} |1 - u_k(\vartheta)|^2 w(\vartheta)\, d\vartheta \xrightarrow[k]{} 0$$

iff

$$\int_{-\pi}^{\pi} \frac{d\vartheta}{w(\vartheta)} = \infty.$$

(Hint: Work with the inner product

$$\langle u, v \rangle_w = \int_{-\pi}^{\pi} u(\vartheta)\overline{v(\vartheta)}w(\vartheta)\, d\vartheta$$

and use orthogonality.)

(b) Show that the condition $\int_{-\pi}^{\pi}(d\vartheta/w(\vartheta)) < \infty$ is *necessary* for the existence of an ω enjoying the properties in question. (Hint: Let $u_0(\vartheta) = \sum_{n \neq 0} a_n e^{in\vartheta}$ be any trigonometric polynomial *without constant term*, and put

$$u_1(\vartheta) = 1 - u_0(\vartheta); \qquad u_2(\vartheta) = e^{-i\vartheta}(1 - u_0(\vartheta)).$$

Then observe that

$$e^{i\vartheta}\tilde{u}_2(\vartheta) \;-\; \tilde{u}_1(\vartheta) \;=\; i(1 - a_1 e^{i\vartheta}),$$

so, for an ω having the above properties, we would have

$$\int_{-\pi}^{\pi} |1 - a_1 e^{i\vartheta}|^2 \omega(\vartheta)\,d\vartheta \;\leqslant\; 4\int_{-\pi}^{\pi} |1 - u_0(\vartheta)|^2 w(\vartheta)\,d\vartheta.$$

We have

$$|1 - a_1 e^{i\vartheta}|^2 \;=\; (1 - |a_1|)^2 \;+\; 4|a_1|\sin^2\frac{\vartheta - \alpha}{2}$$

with $-\pi \leqslant \alpha \leqslant \pi$, so the integral on the left has a *strictly positive minimum* for $a_1 \in \mathbb{C}$ *unless* $\omega(\vartheta)$ vanishes a.e. on $[-\pi,\ \pi]$. Apply the result from (a).)

(c) Suppose now that $w(\vartheta)$ and $1/w(\vartheta)$ both belong to $L_1(-\pi,\ \pi)$. For $|z| < 1$, put

$$\Omega(z) \;=\; \frac{1}{2\pi}\int_{-\pi}^{\pi} \frac{e^{i\tau} + z}{e^{i\tau} - z}\,\frac{d\tau}{w(\tau)}.$$

$\Omega(z)$ is analytic for $\{|z| < 1\}$ and, by Chapter III, §F.2,

$$\lim_{r \to 1}\Omega(re^{i\vartheta}) \;=\; \Omega(e^{i\vartheta})$$

exists for almost all ϑ, and

$$\Re\Omega(e^{i\vartheta}) \;=\; 1/w(\vartheta) \quad \text{a.e..}$$

This makes $|1/\Omega(e^{i\vartheta})| \leqslant w(\vartheta)$ a.e., so that $1/\Omega(e^{i\vartheta}) \in L_1(-\pi,\ \pi)$. *Show that*

$$\int_{-\pi}^{\pi} \frac{e^{in\vartheta}}{\Omega(e^{i\vartheta})}\,d\vartheta \;=\; 0 \quad \text{for } n = 1, 2, 3, \ldots.$$

(Hint: The reader familiar with the theory of H_p spaces for the unit disk may use Smirnov's theorem. Otherwise, one may start from scratch, arguing as follows. For $|z| < 1$, by the inequality between arithmetic and harmonic means,

$$\left|\frac{1}{\Omega(z)}\right| \;\leqslant\; \left(\frac{1}{2\pi}\int_{-\pi}^{\pi}\frac{1 - |z|^2}{|e^{i\tau} - z|^2}\,\frac{d\tau}{w(\tau)}\right)^{-1} \;\leqslant\; \frac{1}{2\pi}\int_{-\pi}^{\pi}\frac{1 - |z|^2}{|e^{i\tau} - z|^2}\,w(\tau)\,d\tau.$$

Use this relation to show, *firstly*, that there is a *complex measure* ν on $[-\pi,\ \pi]$ for which

$$\frac{1}{\Omega(z)} \;=\; \frac{1}{2\pi}\int_{-\pi}^{\pi}\frac{1 - |z|^2}{|e^{i\tau} - z|^2}\,d\nu(\tau), \qquad |z| < 1,$$

(see proof of *first* theorem in §F.1, Chapter III), and, *secondly*, that v must be *absolutely continuous* (without appealing to the F. and M. Riesz theorem). These facts imply that

$$\frac{1}{\Omega(z)} = \frac{1}{2\pi} \int_{-\pi}^{\pi} \frac{1-|z|^2}{|e^{i\tau} - z|^2} \frac{d\tau}{\Omega(e^{i\tau})}, \qquad |z| < 1,$$

and from this the desired relation follows on taking an $r < 1$, observing that

$$\int_{-\pi}^{\pi} (e^{in\vartheta}/\Omega(re^{i\vartheta})) \, d\vartheta = 0 \quad \text{for } n = 1, 2, 3, \ldots,$$

and then using Fubini's theorem.)

(d) Let $\Omega(e^{i\vartheta})$ be the function from (c). In analogy with what was done in article 3, put

$$\sigma(\vartheta) = 1 - \left| 1 - \frac{1}{w(\vartheta)\Omega(e^{i\vartheta})} \right|.$$

Show that $0 \leqslant \sigma(\vartheta) \leqslant 1$ a.e. and that $\sigma(\vartheta)$ is *not* a.e. zero on $[-\pi, \pi]$.

(e) If $f(\vartheta)$ is a finite sum of the form $\sum_{n \geqslant 1} c_n e^{in\vartheta}$, show that

$$\Re \int_{-\pi}^{\pi} (f(\vartheta))^2 w(\vartheta) \, d\vartheta \leqslant \int_{-\pi}^{\pi} (1 - \sigma(\vartheta)) w(\vartheta) |f(\vartheta)|^2 \, d\vartheta.$$

(Hint: By (c), the integral figuring on the left equals

$$\int_{-\pi}^{\pi} w(\vartheta)\left(1 - \frac{1}{w(\vartheta)\Omega(e^{i\vartheta})}\right)(f(\vartheta))^2 \, d\vartheta.$$

Refer to (d).)

(f) Hence show that

$$\int_{-\pi}^{\pi} \sigma(\vartheta) w(\vartheta) |\tilde{u}_0(\vartheta)|^2 \, d\vartheta \leqslant 2 \int_{-\pi}^{\pi} w(\vartheta) |u_0(\vartheta)|^2 \, d\vartheta$$

for any trigonometric polynomial $u_0(\vartheta)$ *without constant term.*

(Hint: It is enough to do this for *real-valued* $u_0(\vartheta)$. Given such a one, use

$$f(\vartheta) = \tilde{u}_0(\vartheta) - iu_0(\vartheta)$$

in result from (e).)

(g) Show that

$$\int_{-\pi}^{\pi} \sigma(\vartheta) w(\vartheta) |\tilde{u}(\vartheta)|^2 \, d\vartheta \leqslant C \int_{-\pi}^{\pi} w(\vartheta) |u(\vartheta)|^2 \, d\vartheta$$

for *general* trigonometric polynomials $u(\vartheta)$, where C is a suitable constant. (Hint: If $u_0(\vartheta)$ denotes $u(\vartheta)$ minus its constant term, $\tilde{u}(\vartheta) = \tilde{u}_0(\vartheta)$. Use result from (a) to show that

$$\int_{-\pi}^{\pi} |u_0(\vartheta)|^2 w(\vartheta) \, d\vartheta \quad \leqslant \quad \text{const.} \int_{-\pi}^{\pi} |u(\vartheta)|^2 w(\vartheta) \, d\vartheta. \;)$$

(*h) Show that

$$\int_{-\pi}^{\pi} \log(\sigma(\vartheta) w(\vartheta)) \, d\vartheta \quad > \quad -\infty.$$

(Hint: Look at the proof of the last theorem in article 3; here $1/w(\vartheta)\Omega(e^{i\vartheta})$ already lies on the circle with diameter $[0, 1]$ for almost all $\vartheta \in [-\pi, \pi]$. Argument uses some H_p space theory for the unit disk.)

The result established in this problem was generalized to the case of weighted L_p norms ($1 < p < \infty$) by Carleson and Jones and, using a method different from theirs, by Rubio de Francia*. Related investigations have been made by Arocena, Cotlar, Sadoski, and their co-workers. A general result for operators in Hilbert space is due to Treil.

F.   Relation of material in preceding § to the geometry of unit sphere in L_∞/H_∞

Combination of the second theorem in §E.3 with the one from §E.4 shows immediately that if we take any outer function $\varphi \in H_2$, the *existence* of a non-zero entire function f of exponential type $\leqslant a$ making

$$\int_{-\infty}^{\infty} \frac{|f(t)|^2}{(1 + t^2)|\varphi(t)|^2} \, dt \quad < \quad \infty$$

is *equivalent* to that of a *non-zero* $h \in H_\infty$ such that

$$\left| e^{2iat} \frac{\overline{\varphi(t)}}{\varphi(t)} - h(t) \right| \quad \leqslant \quad 1 \quad \text{a.e.}.$$

---

* Chapter VI of his recent book with Garcia-Cuerva has in it a rather general treatment of the corresponding question about singular integrals on $\mathbb{R}^n$.

The second of these two conditions has an interpretation in terms of the quotient space L_∞/H_∞ that deserves mention; we look at it briefly in the present §.

The space L_∞/H_∞ has already appeared in §E.3; as explained there, its elements are the *cosets*

$$\psi + H_\infty,$$

where ψ ranges over L_∞. Instead of the *ad hoc* norm $\|\ \|_\infty^\sigma$ employed for such cosets in §E.3, we will here use the standard one to which $\|\ \|_\infty^\sigma$ reduces when $\sigma = 0$, viz.

$$\|\psi + H_\infty\|_\infty = \inf\{\|\psi + h\|_\infty : \ h \in H_\infty\}.$$

Equipped with $\|\ \|_\infty$, L_∞/H_∞ becomes a Banach space, and we denote by Σ the *unit sphere* (unit ball) of that space; that is simply the collection of cosets $\psi + H_\infty$ for which

$$\|\psi + H_\infty\|_\infty \leqslant 1.$$

In §E.3, essential use was made of the fact that L_∞/H_∞ is the *dual* of H_1; now we observe that this makes Σ w^* *compact* by what boils down to Tychonoff's theorem.

A member P of Σ is called an *extreme point* of Σ if, whenever

$$P = \lambda Q + (1 - \lambda)R$$

with Q and R in Σ and $0 < \lambda < 1$, we *must have* $Q = R = P$. Geometrically, this means that *there cannot be any straight segment lying in Σ and passing through P* (i.e., with P *strictly between* its endpoints).

If $0 < \|P\|_\infty < 1$ it is clear that P *cannot* be an extreme point of Σ; the *zero coset* cannot be one either, for, since $L_\infty \neq H_\infty$, Σ contains cosets P and $-P$ with $P \neq 0 + H_\infty$. Any extreme points that Σ can have must thus be included among the set of P with $\|P\|_\infty = 1$ which we may refer to as the *surface* of Σ. Knowledge about Σ's extreme points can be used to gain insight into the geometrical structure of that surface. Such an approach is familiar to functional analysts, and one may get an idea of some of its possibilities by consulting the *Proceedings* of the A.M.S. symposium on convexity. Phelps' beautiful little book is also recommended.

The convexity and w^* compactness of Σ ensure that it has lots of extreme points according to the celebrated Krein–Milman theorem. So many, in fact, that Σ is their w^* closed convex hull. From a theorem of Bishop and Phelps (about which more later) we can furthermore deduce a much stronger result in the present circumstances: the extreme points of Σ are actually $\| \; \|_\infty$ dense on its surface. We may thus think of that surface as being 'filled out, for all practical purposes' by Σ's extreme points.

For this very reason, it seems of interest to have a procedure for exhibiting points on Σ's surface which are *not* extreme points of Σ. An outer function $\varphi \in H_2$ satisfying *either* (and hence *both*) of the two conditions set down above will frequently *give* us such a point, thanks to the following simple

Lemma. *Let* $|u(t)| \equiv 1$ *a.e.. Then* $u + H_\infty$ *is an extreme point of* Σ, *the unit sphere of* L_∞/H_∞, *iff there is no non-zero* $h \in H_\infty$ *for which*

$$|u(t) + h(t)| \leqslant 1 \quad \text{a.e..}$$

Proof. Since $\| u \|_\infty = 1$, $u + H_\infty$ is certainly in Σ.

Suppose in the first place that $u + H_\infty$ is *not* an extreme point of Σ, then there are two *different* cosets $v_1 + H_\infty$, $v_2 + H_\infty$, both of norm $\leqslant 1$, and a λ, $0 < \lambda < 1$, with

$$u + H_\infty = \lambda(v_1 + H_\infty) + (1 - \lambda)(v_2 + H_\infty).$$

According to a result proved in §E.3 (recall that $\| \; \|_\infty^\sigma$ is just $\| \; \|_\infty$ when $\sigma = 0$!),

$$\inf \{ \| v_1 + h \|_\infty : h \in H_\infty \} = \| v_1 + H_\infty \|_\infty$$

is actually realized for some $h \in H_\infty$. Therefore, since $v_1 + h + H_\infty = v_1 + H_\infty$, there is no loss of generality in assuming that $\| v_1 \|_\infty \leqslant 1$. Similarly, we may suppose that $\| v_2 \|_\infty \leqslant 1$.

The previous relation means that there is some $h_0 \in H_\infty$ for which

$$u + h_0 = \lambda v_1 + (1 - \lambda)v_2.$$

Here, the right side has norm $\leqslant 1$. Therefore

$$\| u + h_0 \|_\infty \leqslant 1.$$

In this relation, however, h_0 cannot be zero. Indeed, assuming it *were*, we would have

$$u(t) = \lambda v_1(t) + (1 - \lambda)v_2(t) \quad \text{a.e.,}$$

with $0 < \lambda < 1$, $|u(t)| = 1$, and $|v_1(t)| \leqslant 1$, $|v_2(t)| \leqslant 1$. *Strict convexity* of the unit circle would then make $v_1(t) = v_2(t)$ a.e., so the cosets $v_1 + H_\infty$ and $v_2 + H_\infty$ would be *equal*, contrary to our initial assumption.

We thus have a *non-zero* $h_0 \in H_\infty$ such that

$$|u(t) + h_0(t)| \leqslant 1 \quad \text{a.e.,}$$

and our lemma is proved in one direction.

Going the other way, assume that there *is* a non-zero $h \in H_\infty$ such that $\|u + h\|_\infty \leqslant 1$. Put then

$$\sigma(t) = 1 - |u(t) + \tfrac{1}{2}h(t)|;$$

we have

$$0 \leqslant \sigma(t) \leqslant 1 \quad \text{a.e.,}$$

and see, as in proving *sufficiency* for the second theorem of §E.3, that $\sigma(t) > 0$ *on a set of positive measure*.

We now have

$$|u(t) + \sigma(t) + \tfrac{1}{2}h(t)| \leqslant 1 \quad \text{a.e.}$$

and

$$|u(t) - \sigma(t) + \tfrac{1}{2}h(t)| \leqslant 1 \quad \text{a.e.,}$$

so, since

$$u + H_\infty = \tfrac{1}{2}(u + \sigma + H_\infty) + \tfrac{1}{2}(u - \sigma + H_\infty),$$

it will follow that $u + H_\infty$ is *not* an extreme point of Σ as long as the two cosets on the right are *different*, i.e., as long as $\sigma \notin H_\infty$.

If, however, $\sigma \in H_\infty$,

$$\sigma(z) = \frac{1}{\pi} \int_{-\infty}^{\infty} \frac{\Im z}{|z - t|^2} \sigma(t)\,\mathrm{d}t$$

is *analytic* in $\Im z > 0$ by §E.1; it is, at the same time, *real* there, since $\sigma(t)$ is real. This makes $\sigma(z)$ *constant* and hence $\sigma(t)$, equal a.e. to $\lim_{y \to 0}\sigma(t + \mathrm{i}y)$, *also constant.*

It thus follows from our present assumption that $u + H_\infty$ is *not* an extreme point of Σ, save perhaps in the case where $\sigma(t)$ is constant. *But then $u + H_\infty$ cannot be an extreme point either*, for the constant must be > 0, $\sigma(t)$ being > 0 on a set of positive measure. We have, in other words,

$$|u(t) + \tfrac{1}{2}h(t)| = 1 - c \quad \text{a.e.}$$

where $c > 0$, so $\|u + H_\infty\|_\infty < 1$. As already noted, such a coset $u + H_\infty$ is not an extreme point of Σ.

The lemma is proved.

This result and the equivalence noted at the beginning of the present § yield without further ado the

Theorem. *Let* φ *be an outer function in* H_2. *Given* $a > 0$, *the coset*

$$e^{2iat}\frac{\overline{\varphi(t)}}{\varphi(t)} \quad + \quad H_\infty$$

fails *to be an extreme point of* Σ, *the unit sphere in* L_∞/H_∞, *iff*

$$\int_{-\infty}^{\infty} \frac{|f(t)|^2}{(1+t^2)|\varphi(t)|^2}\,dt \quad < \quad \infty$$

for some non-zero entire function f *of exponential type* $\leqslant a$.

From the theorem we have the following recipe for obtaining points on Σ's surface that are *not* extreme points of Σ: take any outer $\varphi \in H_2$ such that, for some $a > 0$, an entire $f \not\equiv 0$ of exponential type $\leqslant a$ satisfying the relation in the statement exists. Then the point

$$P \quad = \quad e^{2\,iat}\frac{\overline{\varphi(t)}}{\varphi(t)} \quad + \quad H_\infty$$

will have the property in question as long as $\|P\|_\infty = 1$.

It will indeed *frequently happen* that $\|P\|_\infty = 1$. Should that *fail* to come about, in which case

$$\|(e^{2iat}\overline{\varphi(t)}/\varphi(t)) + H_\infty\|_\infty < 1,$$

the relation

$$\int_{-\infty}^{\infty} |\tilde{U}(t)|^2 |\varphi(t)|^2\,dt \quad \leqslant \quad \text{const.} \int_{-\infty}^{\infty} |U(t)|^2 |\varphi(t)|^2\,dt$$

will in fact hold for the sums

$$U(t) \quad = \quad \sum_{\lambda \geqslant a} (A_\lambda \cos \lambda t + B_\lambda \sin \lambda t)$$

according to the second theorem of §E.3. This can only occur for rather special φ in H_2, closely related to the entire functions of exponential type $\leqslant a$ of a particular kind, *integrable* on the real axis but at the same time *not too small* there. The possibility may be fully investigated by the method used in proving the Helson–Szegő theorem. In spite of the matter's relevance to the study of various questions, we cannot go further into it here; very similar material is taken up in the paper of Hruščev, Nikolskii and Pavlov.

Except in the circumstance just mentioned, an outer function $\varphi \in H_2$ will satisfy

$$\| (e^{2iat}\overline{\varphi(t)}/\varphi(t)) + H_\infty \|_\infty = 1.$$

Use of the above procedure with such φ can lead to interesting examples even when only the simple Paley–Wiener multiplier theorem from §A.1 is called on. The reader is encouraged to carry out one or two constructions in such fashion.

For any coset $u + H_\infty$ lying on Σ's surface but not an extreme point of Σ one has a beautiful parametric representation, due to Adamian, Arov and Krein, of the functions $h \in H_\infty$ with $\| u + h \|_\infty = 1$. This was first obtained by means of operator theory, but Garnett has since found an easier function-theoretic derivation, given in his book.*

The work in §E was originally done at the end of the 1960s, in hopes that the connection established by the last theorem would make possible a *proof* of the Beurling–Malliavin multiplier theorem (stated in §A.2) *based on* Banach space and Banach algebra techniques. That approach did not work, and I think now that it is probably not feasible. Whatever value the result may have seems rather to lie in the possibility of its helping us understand the structure of L_∞/H_∞ *when used with multiplier theorems* to construct various examples, according to the above scheme.

The quotient space L_∞/H_∞ is the foundation for the theory of Hankel and Toeplitz forms. Study of these is not really part of this book's subject matter, and the present § is included merely to show some ways of applying multiplier theorems therein. The reader interested in that study should first of all consult Sarason's Blacksburg notes. Wishing to go further, he or she should next take up the papers of that author and his co-workers, perusing, at the same time, a book on H_p spaces so as to get a good grounding in their theory. It is then essential that one become familiar with the remarkable papers of Adamian, Arov and Krein. Those make heavy use of Hilbert space operator theory.

There is, in general, much mutual interplay between operator theory and the investigation of L_∞/H_∞; so vast, indeed, is the region common to these two fields that it seems hopeless to try to furnish even sketchy references here. Let us at least mention the so-called *Nagy–Foiaş model*; the book about it by those two authors is well known. For more recent

---

* the one on bounded analytic functions

treatments, see Nikolskii's book and (especially) his survey article with Hruščev.

Before closing this § and the present chapter, let us see how the extreme points of Σ are related to its *support points*, to be defined in a moment. As we saw in §E.3, each coset $u + H_\infty$ in L_∞/H_∞ corresponds to a *linear functional* Λ on H_1 given by the formula

$$\Lambda(f) = \int_{-\infty}^{\infty} u(t)f(t)\,\mathrm{d}t, \qquad f \in H_1,$$

and the *supremum* of $|\Lambda(f)|$ for the $f \in H_1$ with $\|f\|_1 \leqslant 1$ is equal to $\|u + H_\infty\|_\infty$. (The reader is again reminded that the norm $\|\ \|_\infty^\sigma$ used in §E.3 reduces to $\|\ \|_\infty$ when $\sigma(t) \equiv 0$.) We know that there *is* some $h \in H_\infty$ for which $\|u + h\|_\infty = \|u + H_\infty\|_\infty$; *that, however, does not mean that there need be an* $f \in H_1$ *of norm* 1 *with* $\Lambda(f) = \|u + H_\infty\|_\infty$. Since the space H_1 is not reflexive there is no reason why this should be the case; it is in fact *true* for *some* cosets $u + H_\infty$ and *false* for *others*.

Definition. A coset $u + H_\infty$ with $\|u + H_\infty\|_\infty = 1$ is said to be a *support point* for Σ if there is an $f \in H_1$ with

$$\int_{-\infty}^{\infty} u(t)f(t)\,\mathrm{d}t = \|f\|_1 = 1.$$

There is then the

Theorem. *A support point of* Σ *is an extreme point of* Σ.

Proof. Let $u + H_\infty$ be a support point of Σ. There is, as we know, a $v \in u + H_\infty$ with $\|v\|_\infty = \|u + H_\infty\|_\infty = 1$; it is enough to show that such a v is of modulus 1 a.e. and *uniquely determined*, for then there can be *no* non-zero $h \in H_\infty$ with $\|v + h\|_\infty \leqslant 1$, and $u + H_\infty = v + H_\infty$ must hence be an extreme point of Σ by the previous lemma.

There is by definition an $f \in H_1$ with

$$\int_{-\infty}^{\infty} v(t)f(t)\,\mathrm{d}t = \int_{-\infty}^{\infty} u(t)f(t)\,\mathrm{d}t = \|f\|_1 = 1.$$

Here, $|v(t)| \leqslant 1$ a.e., so we must have

$$v(t)f(t) = |f(t)| \quad \text{a.e..}$$

However, $|f(t)| > 0$ a.e. by §E.1, so we get

$$v(t) \;=\; \frac{|f(t)|}{f(t)} \quad \text{a.e.,}$$

which determines v makes it of modulus 1 a.e.. Done.

The *converse* of this theorem is *not true*; the example provided by problems 44 and 46 given below shows that. It is, however, true that the *surface* of Σ is *full of support points*; those are, indeed, $\|\ \|_\infty$ dense on that surface according to a remarkable theorem, due to Bishop and Phelps, whose proof may be found in the above mentioned A.M.S. volume on convexity. Since all these support points are extreme points by the result just obtained, it follows that *the extreme points of* Σ *are* $\|\ \|_\infty$ *dense on its surface*. This is a much higher concentration of extreme points than could be surmised from the Krein–Milman theorem. Still, there are lots of points on Σ's surface that are *not* extreme, and we have seen how to find many of them.

When a particular $u \in L_\infty$ with $\|u + H_\infty\|_\infty = 1$ is given, it is hard to tell by just looking at the qualitative behaviour of the function $u(t)$ whether $u + H_\infty$ is a support point of Σ or not. About all that is known *generally* is that $u + H_\infty$ must then be a support point if $u(t)$ is *continuous* on $\mathbb{R}$ *and tends to equal limits for* $t \longrightarrow \infty$ *and* $t \longrightarrow -\infty$. Proof of this fact, which depends on the F. and M. Riesz theorem, may be found in the more recent books about H_p spaces. If we merely require *uniform continuity* of $u(t)$ on $\mathbb{R}$, the conclusion may cease to hold. An example of this will be furnished by problems 44 and 46.

To work the following problems, a generalization of the Schwarz reflection principle due to Carleman will be needed. Suppose that we have a rectangle $\mathscr{D}_0$ in the upper half plane whose *base* is a segment of the real axis:

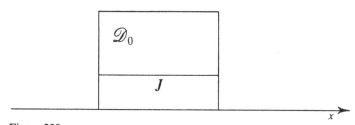

Figure 228

Carleman's result deals with functions $F(z)$ analytic in $\mathscr{D}_0$ for which

$$\int_J |F(z)||\mathrm{d}z| \;\leqslant\; \text{const.}$$

for all the *horizontal* line segments J running across the interior of $\mathscr{D}_0$. These functions are just those of the class $\mathscr{S}_1(\mathscr{D}_0)$ studied in §B.4, Chapter VII. The reader should refer again to that §. According to the first theorem proved there, when $F \in \mathscr{S}_1(\mathscr{D}_0)$, $\lim_{y\to x} F(x + \mathrm{i}y)$ exists for almost every x on the base of $\mathscr{D}_0$. Following standard practice, that limit is denoted by $F(x)$.

Lemma (Carleman). *If* $F \in \mathscr{S}_1(\mathscr{D}_0)$ *and* $F(x)$ *is real a.e. along the base of* $\mathscr{D}_0$, $F(z)$ *can be analytically continued across that base into* $\mathscr{D}_0^*$, *the reflection of* $\mathscr{D}_0$ *in the real axis, by putting* $F(z) = \overline{F(\bar z)}$ *for* $z \in \mathscr{D}_0^*$.

Proof. Let I be any segment properly included in the base of $\mathscr{D}_0$ in the manner shown in the following figure, and take any rectangle $\mathscr{D}$, entirely contained in $\mathscr{D}_0$, having I as its base.

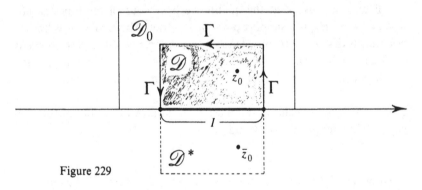

Figure 229

If $z_0 \in \mathscr{D}$, $1/(\zeta - \bar z_0)$ is analytic in ζ for ζ in $\mathscr{D}$, so, by the corollary to the *second* theorem of §B.4, Chapter VII,

$$\int_{\partial \mathscr{D}} \frac{F(\zeta)}{\zeta - \bar z_0}\,\mathrm{d}\zeta \;=\; 0;$$

here the integral is absolutely convergent according to the third lemma and first theorem* of that §.

The function $\big(F(z) - F(z_0)\big)/(z - z_0)$ clearly belongs to $\mathscr{S}_1(\mathscr{D}_0)$ if F does.

---

* In Fig. 69, accompanying the proof of that theorem (p. 287 of vol I), B_1 and B_2 should have designated the horizontal sides of $\mathscr{D}_0$ and not of $\mathscr{D}$.

Hence, by the corollary just used, we also have

$$\int_{\partial \mathscr{D}} \frac{F(\zeta) - F(z_0)}{\zeta - z_0} d\zeta = 0,$$

from which the Cauchy formula

$$F(z_0) = \frac{1}{2\pi i} \int_{\partial \mathscr{D}} \frac{F(\zeta)}{\zeta - z_0} d\zeta$$

immediately follows on using the relation

$$\int_{\partial \mathscr{D}} \frac{d\zeta}{\zeta - z_0} = 2\pi i.$$

Combining our formula for $F(z_0)$ with the one preceding it, and then dropping the subscript on z, we get

$$F(z) = \frac{1}{2\pi i} \int_{\Gamma} \left(\frac{1}{\zeta - z} - \frac{1}{\zeta - \bar{z}} \right) F(\zeta) d\zeta + \frac{1}{\pi} \int_{I} \frac{\Im z}{|\xi - z|^2} F(\xi) d\xi, \quad z \in \mathscr{D},$$

where Γ is the path consisting of the *top* of $\mathscr{D}$ together with its *two vertical sides* (see figure).

Of the two integrals on the right, the *first* certainly represents a complex-valued *harmonic* (*not* analytic!) function of z in any region disjoint from both Γ and its reflection in the real axis. That expression is, in particular, harmonic in the rectangle $\mathscr{D} \cup I \cup \mathscr{D}^*$, where $\mathscr{D}^*$ denotes the reflection of $\mathscr{D}$ in $\mathbb{R}$; its *imaginary part*, $V(z)$, is thus also harmonic in $\mathscr{D} \cup I \cup \mathscr{D}^*$.

The function $\Im F(z)$, harmonic in $\mathscr{D}$, is *equal* there to

$$V(z) + \frac{1}{\pi} \int_{I} \frac{\Im z}{|\xi - z|^2} \Im F(\xi) d\xi;$$

this, however, is just $V(z)$, since $\Im F(\xi) = 0$ a.e. on I by hypothesis. The function $\Im F(z)$ therefore has a *harmonic continuation* from $\mathscr{D}$ to the larger rectangle $\mathscr{D} \cup I \cup \mathscr{D}^*$. Its *harmonic conjugate*, $-\Re F(z)$, can thus also be continued harmonically into all of $\mathscr{D} \cup I \cup \mathscr{D}^*$, and then we obtain an *analytic continuation* of $F(z)$ into that larger rectangle by putting $F(z) = \Re F(z) + i\Im F(z)$.

This means, in particular, that $F(z)$ is *continuous* at the points of I (save perhaps at the endpoints). By hypothesis, however, $F(x)$ is *real* a.e. on I. Therefore it is *real everywhere* on I, besides being *continuous there*. Now we can apply the *classical* Schwarz reflection principle to conclude that $F(\bar{z}) = \overline{F(z)}$ for $z \in \mathscr{D}$.

Our choices of I, properly contained in $\mathscr{D}_0$'s base, and of $\mathscr{D}$, entirely included in $\mathscr{D}_0$, were arbitrary. The formula $F(\bar{z}) = \overline{F(z)}$ thus gives us an analytic continuation of F across the *whole* base of $\mathscr{D}_0$ into all of $\mathscr{D}_0^*$. Done.

Problem 44

Let Λ be any measurable sequence of distinct integers > 0, having (ordinary) density $D_\Lambda < \frac{1}{2}$ (refer to §E.3, Chapter VI).
Write

$$C(z) = \prod_{n \in \Lambda} \left(1 - \frac{z^2}{n^2} \right),$$

and then put

$$B(z) = \frac{C(z-i)}{C(z+i)};$$

$B(z)$ is just a *Blaschke product* for the upper half plane (see §G.3, Chapter III), having *zeros* at the points $\pm n + i$, $n \in \Lambda$, and *poles* at $\pm n - i$, $n \in \Lambda$. Consider the function

$$u(t) = \frac{e^{\pi i t}}{B(t)},$$

of modulus 1 on $\mathbb{R}$.

(a) Show that $u(t)$ is uniformly continuous on $\mathbb{R}$. (Hint: $B(t)$ is of the form $\exp(i\varphi(t))$ where $\varphi(t)$ is real. Express $\varphi'(t)$ in terms of the $n \in \Lambda$.)

(b) Assume that $u + H_\infty$ is a support point of Σ; this means that there *is* an $f \in H_1$ with $\int_{-\infty}^{\infty} u(t)f(t)\,dt = \|f\|_1 = 1$. For $\Im z > 0$, write, as usual,

$$f(z) = \frac{1}{\pi} \int_{-\infty}^{\infty} \frac{\Im z}{|z-t|^2} f(t)\,dt,$$

and then put

$$F(z) = e^{\pi i z}(C(z+i))^2 f(z)$$

in the upper half plane. *Show that* $F(z)$ *can be continued analytically across* $\mathbb{R}$ by putting $F(\bar{z}) = \overline{F(z)}$. (Hint: Use Carleman's lemma.)

(c) Show that the entire function $F(z)$ obtained in (b) is of *exponential type*. (Hint: See proof of the first theorem in §F.4, Chapter VI.)

(d) Hence obtain a *contradiction with the assumption made in* (b) by showing that $F(z)$ must be identically zero. (Hint: Look at the behaviour of $F(z)$ on the imaginary axis, referring to problem 29(a) from §B.1 of Chapter IX.)

By making the right choice of the sequence Λ, various interesting examples can be obtained. We need another lemma, best given as

Problem 45

(a) Let $g(w)$ be analytic in $\{|w| < 1\}$, with $\Re g(w) \geqslant 0$ there. Show that for any $p < 1$, the integrals

$$\int_{-\pi}^{\pi} |g(re^{i\vartheta})|^p \, d\vartheta$$

are bounded for $r < 1$. (Hint: By the principle of conservation of domain, $g(w)$ can never be zero for $|w| < 1$, so we can define an analytic and single valued branch of $(g(w))^p$ there. Apply Cauchy's formula to the latter to get $(g(0))^p$, then take real parts and note that $\cos(p \arg g(w))$ is bounded away from 0.)

(b) If $g(w)$ is as in (a), show that $\lim_{r \to 1} g(re^{i\tau}) = g(e^{i\tau})$ exists a.e., and that for any $p < 1$,

$$(g(w))^p = \frac{1}{2\pi} \int_{-\pi}^{\pi} \frac{1 - |w|^2}{|w - e^{i\tau}|^2} (g(e^{i\tau}))^p \, d\tau, \qquad |w| < 1,$$

the integral on the right being absolutely convergent. (Hint: Fix a p', $p < p' < 1$, and apply the result from (a) to $(g(w))^{p'}$. Then argue as in the proof of the first theorem from §F.1, Chapter III, using the duality between the spaces $L_r(-\pi, \pi)$ and $L_s(-\pi, \pi)$, where $r = p'/p$ and $(1/r) + (1/s) = 1$. This will yield a function $G(\tau)$ in $L_r(-\pi, \pi)$ such that

$$(g(w))^p = \frac{1}{2\pi} \int_{-\pi}^{\pi} \frac{1 - |w|^2}{|w - e^{i\tau}|^2} G(\tau) \, d\tau, \qquad |w| < 1.$$

Appeal to standard results about the Poisson integral to describe the boundary behaviour of $(g(w))^p$ and relate $G(\tau)$ thereto.)

(c) Given $v(t)$ defined on $\mathbb{R}$ with $|v(t)| \leqslant \pi/2$ there, consider the function

$$\psi(z) = \frac{1}{\pi} \int_{-\infty}^{\infty} \left(\frac{1}{t - z} - \frac{t}{t^2 + 1} \right) v(t) \, dt,$$

analytic in $\Im z > 0$. As we know,

$$\lim_{y \to 0} \psi(t + iy) = \psi(t)$$

exists a.e. on $\mathbb{R}$, with $\psi(t) = -\tilde{v}(t) + iv(t)$ a.e. there, $\tilde{v}(t)$ being the *first* kind of Hilbert transform described at the beginning of §E.2. *Show that, when $p < 1$,*

$$\frac{e^{p\psi(t)}}{(t + i)^2}$$

belongs to H_1. (Hint: By mapping the upper half plane conformally onto the unit disk and using the result from (b), show first of all that when $p < 1$,

$$e^{p\psi(z)} = \frac{1}{\pi} \int_{-\infty}^{\infty} \frac{\Im z}{|z-t|^2} e^{p\psi(t)} dt$$

for $\Im z > 0$, the integral on the right being absolutely convergent. This representation gives us fairly good control on the *size* of $\exp(p\psi(z))$ in $\Im z > 0$ – cf. Chapter VI, §A.2 – A.3. Knowing this, show by integrating around suitable contours that if λ and $\delta > 0$,

$$\int_{-\infty}^{\infty} \frac{e^{i\lambda x}}{(x+i)^{2+\delta}} \exp(p\psi(x+ih)) dx = 0$$

for any $h > 0$ – cf. proof of theorem that the product of two H_2 functions is in H_1, §E.1. Now one may make $\delta \longrightarrow 0$ and use dominated convergence (*guaranteed* by our representation for $\exp(p\psi(z))$!) to get

$$\int_{-\infty}^{\infty} \frac{e^{i\lambda x}}{(x+i)^2} \exp(p\psi(x+ih)) dx = 0.$$

Plug the representation for $\exp(p\psi(z))$ into this result and use Fubini's theorem, noting that $e^{i\lambda t}/(t+i)^2$ is in H_∞. Finally, make $h \longrightarrow 0$.)

Let us now take a measurable sequence Λ of integers > 0 *having* (ordinary) *density zero*, whose Beurling–Malliavin *effective density* $\tilde{D}_\Lambda$ is *equal* to 1 (see §D.2 of Chapter IX). It is easy to construct such sequences. We need merely pick intervals $[a_k, b_k]$ with integral endpoints,

$$0 < a_1 < b_1 < a_2 < b_2 < a_3 < \cdots,$$

such that b_{k-1}/b_k and $(b_k - a_k)/b_k$ both tend to zero as $k \longrightarrow \infty$, while

$$\sum_{1}^{\infty} \left(\frac{b_k - a_k}{b_k} \right)^2 = \infty,$$

and then have Λ consist of the integers in the $[a_k, b_k]$.

Using such a sequence Λ, let us form the functions $C(z)$, $B(z)$ and $u(t)$ considered in problem 44. Then,

> $u + H_\infty$ is an extreme point of Σ even though it is not a support point thereof.

This will follow from problem 44, the first lemma of the present §, and

Problem 46

To show that there can be *no* non-zero $h \in H_\infty$ with

$$|u(t) - h(t)| \leq 1 \qquad \text{a.e..}$$

(a) Assuming that there *is* such an h, show how to construct a function $v(t)$ defined a.e. on $\mathbb{R}$, with $|v(t)| \leq \pi/2$ and

$$e^{-\pi i t} B(t) h(t)^{iv(t)} \geq 0 \qquad \text{a.e..}$$

(b) Using the v found in (a), form $\psi(t) = -\tilde{v}(t) + iv(t)$ as in problem 45(c). Show that in the present circumstances,

$$\int_{-\infty}^{\infty} \frac{|h(t) \exp \psi(t)|}{1 + t^2} \, dt \; < \; \infty.$$

(Hint: From the solution of problem 45(c),

$$\frac{1}{\pi} \int_{-\infty}^{\infty} \frac{\exp(p\psi(t))}{1 + t^2} \, dt \; = \; \exp(p\psi(i))$$

whenever $p < 1$. Take real parts and make $p \longrightarrow 1$; then what we have on the right tends to the *finite* value $\Re \exp \psi(i)$, so that

$$\int_{-\infty}^{\infty} \frac{e^{-\tilde{v}(t)} \cos v(t)}{1 + t^2} \, dt \; \leq \; \pi \Re e^{\psi(i)} \; < \; \infty$$

by Fatou's lemma. If, however, we write

$$g(t) \; = \; e^{-\pi i t} B(t) h(t),$$

we have the following diagram:

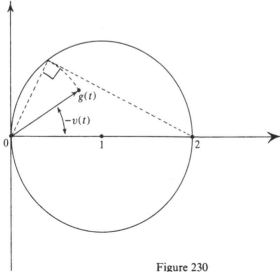

Figure 230

From this it is clear that $|g(t)| \leqslant 2\cos v(t)$.)

(c) Continuing with the notation used in (b), show that

$$\frac{B(t)h(t)\exp\psi(t)}{(t+\mathrm{i})^2}$$

belongs to H_1. (Hint: $Bh \in H_\infty$ and, for each $p < 1$, $\exp(p\psi(t)/(t+\mathrm{i})^2$ is in H_1 by problem 45(c). Thus, if $\lambda \geqslant 0$, we have

$$\int_{-\infty}^{\infty} \mathrm{e}^{\mathrm{i}\lambda t} \frac{B(t)h(t)\exp(p\psi(t))}{(t+\mathrm{i})^2}\,\mathrm{d}t \;=\; 0$$

by §E.1, whenever $p < 1$. In this relation, we may let $p \longrightarrow 1$ and use dominated convergence, referring to the result from (b).)

(d) Show that the function

$$F(z) \;=\; \mathrm{e}^{-\pi\mathrm{i}z}B(z)h(z)\mathrm{e}^{\psi(z)},$$

analytic in the upper half-plane, can be continued across the real axis, *yielding an entire function of exponential type* $\leqslant \pi$, by putting

$$F(\bar{z}) \;=\; \overline{F(z)}.$$

Here, as usual,

$$h(z) \;=\; \frac{1}{\pi}\int_{-\infty}^{\infty} \frac{\Im z}{|z-t|^2}h(t)\,\mathrm{d}t, \qquad \Im z > 0.$$

(Hint: See problem 44 again.)

(e) Hence show that the function $F(z)$ from (d) *is identically zero*, so that *in fact* $h(z) \equiv 0$, proving that the assumption made in (a) is *untenable*. (Hint: First apply the Riesz–Fejér theorem from §G.3 of Chapter III to get an entire function $f(z)$ of exponential type $\leqslant \pi/2$, *vanishing* at each of the points $\pm n + \mathrm{i}$, $n \in \Lambda$, such that

$$F(z) \;=\; f(z)\overline{f(\bar{z})}.$$

$f(z+\mathrm{i})$ then certainly vanishes at each point of Λ. Use the fact that $\tilde{D}_\Lambda = 1$, applying a suitable variant of the theorem quoted at the very beginning of §E, Chapter IX.)

Problem 47

Let $u(t)$ be as in problem 46. Show that $\bar{u} + H_\infty$ (*sic!*) is an extreme point, but not a support point, of Σ.

Remark. When Carleman's lemma is used in the above problems, one is actually dealing with functions $F \in \mathscr{S}_1(\mathscr{D}_0)$ whose boundary values are *positive* (and not merely real) along the base of $\mathscr{D}_0$. In this circumstance, analytic continuation across the base of $\mathscr{D}_0$ is possible under a *weaker* condition on F than that of membership in $\mathscr{S}_1(\mathscr{D}_0)$. It is enough that F belong to the space $H_{1/2}$ associated with each of the smaller rectangles $\mathscr{D}$ used in the proof of the lemma. That fact follows easily from an argument due to Neuwirth and Newman, and, independently, to Helson and Sarason. The reader is referred to the discussion accompanying problem 13 at the end of Chapter II in Garnett's book.*

---

* *Bounded Analytic Functions*

Multiplier Theorems

It is time to prove the multiplier theorem stated in §A.2 of the preceding chapter and then applied there, in §§B and C. We desire also to establish another result of the same kind and finally to start working towards a *description* of the weights $W(x) \geqslant 1$ that *admit multipliers* (in the sense explained at the beginning of Chapter X). All this will require the use of some elementary material from potential theory.

There is a dearth of modern expositions of that theory accessible to readers having only a general background in analysis. Moreover, the books on it that do exist* are not so readily available. It therefore seems advisable to first explain the basic results we will use from the subject without, however, getting involved in any attempt at a systematic treatment of it. That is the purpose of the first § in this chapter. Other more special potential-theoretic results called for later on will be formulated and proved as they are needed.

A   Some rudimentary potential theory

1.   Superharmonic functions; their basic properties

A function $U(z)$ harmonic in a domain $\mathscr{D}$ enjoys the *mean value property* there: for $z \in \mathscr{D}$,

$$U(z) = \frac{1}{2\pi} \int_0^{2\pi} U(z + \rho e^{i\vartheta}) \, d\vartheta, \qquad 0 < \rho < \text{dist}(z, \partial\mathscr{D}).$$

---

* The books by Carleson, Tsuji, Kellogg, Helms and Landkof are in my possession, together with a copy of Frostman's thesis; most of the time I have been able to make do with just the *first three* of these.

To Gauss is due the important *converse* of this statement: among the functions $U(z)$ continuous in $\mathscr{D}$, the mean value property *characterizes* the ones harmonic there. The proof of this contains a key to the understanding of much of the work with superharmonic functions (defined presently) to concern us here; let us therefore recall how that proof goes.

An (apparently) more general result can in fact be established by the same reasoning. Suppose that a function $U(z)$, continuous in a domain $\mathscr{D}$, enjoys a *local mean value property there*; in other words, that to each $z \in \mathscr{D}$ corresponds an r_z, $0 < r_z \leqslant \operatorname{dist}(z, \partial\mathscr{D})$ (with, *a priori*, $r_z < \operatorname{dist}(z, \partial\mathscr{D})$) such that

$$U(z) \;\; = \;\; \frac{1}{2\pi}\int_0^{2\pi} U(z + \rho e^{i\vartheta})\,\mathrm{d}\vartheta \qquad \text{for } 0 \, < \, \rho \, < \, r_z.$$

It is claimed that $U(z)$ is then *harmonic in $\mathscr{D}$*.

The main part of the argument consists in showing that the local mean value property implies the *strong maximum principle* for U on (connected) domains with compact closures lying in $\mathscr{D}$. Letting Ω be any such domain, we have to verify that for $z \in \Omega$,

$$U(z) \;\; < \;\; \sup_{\zeta \in \partial\Omega} U(\zeta)$$

unless $U(z) \equiv \operatorname{const}$ on $\bar{\Omega}$. Here, $U(z)$ has on $\bar{\Omega}$ a *maximum* – call it M – and the statement in question amounts to the assertion that $U(z) \equiv M$ on $\bar{\Omega}$ if, for any $z_0 \in \Omega$, $U(z_0) = M$.

Suppose there is such a z_0. Then, for each sufficiently small $\rho > 0$,

$$\frac{1}{2\pi}\int_0^{2\pi} U(z_0 + \rho e^{i\vartheta})\,\mathrm{d}\vartheta \;\; = \;\; U(z_0) \;\; = \;\; M$$

with $U(z_0 + \rho e^{i\theta})$ continuous in ϑ and $\leqslant M$. This makes $U(z_0 + \rho e^{i\vartheta}) \equiv M$ for such ρ, so that $U(z) \equiv M$ in a *small disk* centered at z_0. The set

$$E \;\; = \;\; \{z_0 \in \Omega\colon U(z_0) = M\}$$

is thus *open*. That set is, however, *closed* in Ω's relative topology on account of the continuity of U. Hence $E = \Omega$ since Ω is connected, and $U(z) \equiv M$ in Ω – thus finally on $\bar{\Omega}$, thanks again to the continuity of U.

To complete the proof of Gauss' result, let us take any $z_0 \in \mathscr{D}$ and an $R < \operatorname{dist}(z_0, \partial\mathscr{D})$; it is enough to establish that

$$U(z_0 + \rho e^{i\vartheta}) \;\; = \;\; \frac{1}{2\pi}\int_0^{2\pi} \frac{R^2 - \rho^2}{R^2 + \rho^2 - 2R\rho\cos(\vartheta - \tau)} U(z_0 + Re^{i\tau})\,\mathrm{d}\tau$$

for $0 \leqslant \rho < R$. Calling the expression on the right $V(z_0 + \rho e^{i\vartheta})$, we proceed to show first that

$$U(z) \leqslant V(z)$$

for $|z - z_0| < R$.

Fix any $\varepsilon > 0$. By continuity of U and the elementary properties of the Poisson kernel we know that

$$V(z_0 + re^{i\vartheta}) \longrightarrow U(z_0 + Re^{i\vartheta})$$

uniformly in ϑ for $r < R$ tending to R; the same is of course true if we replace V by U on the left. On the circles $|z - z_0| = r$ with radii $r < R$ sufficiently close to R we therefore have

$$U(z) - V(z) \leqslant \varepsilon.$$

Figure 231

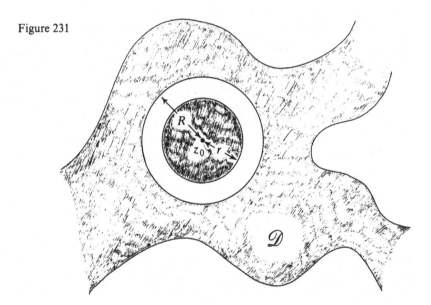

Here, both $U(z)$ and the *harmonic* function $V(z)$ enjoy the local mean value property in the open disk $\{|z - z_0| < R\}$. Hence, by what has just been shown, we have the strong maximum principle for the *difference* $U(z) - V(z)$ on the smaller disks $\{|z - z_0| < r\}$. The preceding inequality thus implies that $U(z) - V(z) \leqslant \varepsilon$ on each of those disks, and finally that $U(z) - V(z) \leqslant \varepsilon$ for $|z - z_0| < R$. Squeezing ε, we see that

$$U(z) - V(z) \leqslant 0 \quad \text{for } |z - z_0| < R.$$

By working with the difference $V(z) - U(z)$ we can, however, prove the *reverse* inequality in the same fashion. This means that one must have $U(z) = V(z)$ for $|z - z_0| < R$, and our proof is finished. It is this *argument* that the reader will find helpful to keep in mind during the following development.

Next in importance to the harmonic functions as objects of interest in potential theory come those that are *subharmonic* or *superharmonic*. One can actually work exclusively with harmonic functions and the ones belonging to *either* of the last two categories; which of the latter is singled out makes very little difference. Logarithms of the moduli of analytic functions are subharmonic, but most writers on potential theory prefer (probably on account of the customary formulation of Riesz' theorem, to be given in article 2) to deal with *superharmonic functions*, and we follow their example here. The difference between the two kinds of functions is purely one of *sign*: a given $F(z)$ is *subharmonic* if and only if $-F(z)$ is *superharmonic*.

Definition. A function $U(z)$ defined in a domain $\mathscr{D}$ with $-\infty < U(z) \leqslant \infty$ there is said to be superharmonic in $\mathscr{D}$ provided that

(i) $\liminf\limits_{z \to z_0} U(z) \geqslant U(z_0)$ for $z_0 \in \mathscr{D}$;

(ii) to each $z \in \mathscr{D}$ corresponds an r_z, $0 < r_z \leqslant \operatorname{dist}(z, \partial\mathscr{D})$, such that

$$\frac{1}{2\pi} \int_0^{2\pi} U(z + \rho e^{i\vartheta}) \, \mathrm{d}\vartheta \leqslant U(z) \quad \text{for } 0 < \rho < r_z.$$

Superharmonic functions are thus permitted to assume the value $+\infty$ at certain points. Although authors on potential theory do not generally agree to call the function *identically equal* to $+\infty$ superharmonic, we will sometimes find it convenient to do so.

Assumption of the value $-\infty$, on the other hand, is not allowed. This restriction plays a serious rôle in the subject. By it, functions like

$$U(z) = \begin{cases} \Im z, & \Im z > 0, \\ -\infty, & \Im z \leqslant 0, \end{cases}$$

are excluded from consideration.

It may seem at first sight that an extensive theory could hardly be based on the definition just given. On thinking back, however, to the proof of

Gauss' result, one begins to suspect that the simple conditions figuring in the definition involve more structure than is immediately apparent. One notices, to begin with, that (i) and (ii) signify *opposite kinds* of local behaviour. The *first* guarantees that $U(z)$ *stays almost as large* as $U(z_0)$ on small neighborhoods of z_0, and the *second* gives us lots of points z in such neighborhoods at which $U(z) \leqslant U(z_0)$. Considerable use of the interplay between these two contrary effects will be made presently; for the moment, let us simply remark that together, they entail *equality* of $\liminf_{z \to z_0} U(z)$ and $U(z_0)$ at the $z_0 \in \mathscr{D}$.

It is probably best to start our work with superharmonic functions by seeing what can be deduced from the requirement that $U(z) > -\infty$ and condition (i), *taken by themselves*. The latter is nothing other than a prescription for *lower semicontinuity* in $\mathscr{D}$; as is well known, and easily verified by the reader, it implies that $U(z)$ has an *assumed minimum* on each compact subset of $\mathscr{D}$. Together with the requirement, that means that $U(z)$ *has a finite lower bound on every compact subset of $\mathscr{D}$*. This property will be used repeatedly. (I can never remember *which* of the two kinds of semicontinuity is *upper*, and which is *lower*, and suspect that some readers of this book may have the same trouble. That is why I systematically avoid using the *terms* here, and prefer instead to specify explicitly each time which behaviour is meant.)

A *monotonically increasing* sequence of functions continuous on a domain $\mathscr{D}$ tends to a limit $U(z) > -\infty$ satisfying (i) there. This is immediate; what is less apparent is a kind of *converse*:

Lemma. *If $U(z) > -\infty$ has property* (i) *in $\mathscr{D}$ there is, for any compact subset K of $\mathscr{D}$, a monotonically increasing sequence of functions $\varphi_n(z)$ continuous on K and tending to $U(z)$ there.*

Proof. For each $n \geqslant 1$ put, for $z \in K$,

$$\varphi_n(z) = \inf_{\zeta \in K} (U(\zeta) + n|z - \zeta|).$$

Since $U(\zeta)$ is bounded below on K by the above observation, the functions $\varphi_n(z)$ are all $> -\infty$. It is evident that $\varphi_n(z) \leqslant \varphi_{n+1}(z) \leqslant U(z)$ for $z \in K$ and each n.

To show continuity of φ_n at $z_0 \in K$, we remark that the function of ζ equal to $U(\zeta) + n|\zeta - z_0|$ enjoys, like $U(\zeta)$, property (i) and thus *assumes its minimum* on K. There is hence a $\zeta_0 \in K$ such that

$$\varphi_n(z_0) = n|\zeta_0 - z_0| + U(\zeta_0),$$

so, if $z \in K$,

$$\varphi_n(z) \;\leqslant\; n|z - \zeta_0| \;+\; U(\zeta_0) \;\leqslant\; n|z - z_0| \;+\; \varphi_n(z_0).$$

In the same way, we see that

$$\varphi_n(z_0) \;\leqslant\; n|z_0 - z| \;+\; \varphi_n(z)$$

which, combined with the previous, yields

$$|\varphi_n(z) - \varphi_n(z_0)| \;\leqslant\; n|z - z_0| \quad \text{for } z_0 \text{ and } z \in K.$$

We proceed to verify that $\varphi_n(z_0) \xrightarrow[n]{} U(z_0)$ at each $z_0 \in K$. Given such a z_0, take any number $V < U(z_0)$. By property (i) there is an $\eta > 0$ such that $U(\zeta) > V$ for $|\zeta - z_0| < \eta$. $U(\zeta)$ has, as just recalled, a finite lower bound, say $-M$, on K. Then, for $n > (V + M)/\eta$, we have $n|\zeta - z_0| + U(\zeta) > V$ for $\zeta \in K$ with $|\zeta - z_0| \geqslant \eta$. But when $|\zeta - z_0| < \eta$ we also have $n|\zeta - z_0| + U(\zeta) > V$. Therefore

$$\varphi_n(z_0) \;\geqslant\; V \quad \text{for } n > (M + V)/\eta$$

Since, on the other hand, $\varphi_n(z_0) \leqslant U(z_0)$, we see that the convergence in question holds, $V < U(z_0)$ being arbitrary.

The lemma is proved.

Remark. This result figures in some introductory treatments of the Lebesque integral.

Let us give some examples of superharmonic functions. The class of these includes, to begin with, all the *harmonic* functions. Gauss' result implies indeed that a function $U(z)$ defined on a domain $\mathscr{D}$ is *harmonic* there *if and only if* both $U(z)$ and $-U(z)$ are *superharmonic* in $\mathscr{D}$. The simplest kind of functions $U(z)$ superharmonic, but not harmonic, in $\mathscr{D}$ are those of the form

$$U(z) \;=\; \log \frac{1}{|z - z_0|} \quad \text{with } z_0 \in \mathscr{D}.$$

Positive linear combinations of these are also superharmonic, and so, finally, are the expressions

$$U(z) \;=\; \int_K \log \frac{1}{|z - \zeta|} \, \mathrm{d}\mu(\zeta)$$

formed from positive measures μ supported on compact sets K. *The reader should not proceed further without verifying the last statement.* This involves the use of Fatou's lemma for property (i), and of the handy relation

$$\frac{1}{2\pi}\int_0^{2\pi} \log\frac{1}{|z + \rho e^{i\vartheta} - \zeta|}\, d\vartheta \;=\; \min\!\left(\log\frac{1}{|z-\zeta|},\; \log\frac{1}{\rho}\right)$$

(essentially the same as one appearing in the derivation of Jensen's formula, Chapter I!) for property (ii).

Integrals like the above one actually turn out to be practically capable of representing all superharmonic functions. In a sense made precise by Riesz' theorem, to be proved in article 2, the most general superharmonic function is equal to such an integral plus a harmonic function.

By such examples, one sees that superharmonic functions are far from being 'well behaved'. Consider, for instance

$$U(z) \;=\; \sum_n a_n \log\frac{1}{|z - z_n|},$$

formed with the z_n of modulus $< 1/2$ tending to 0 and numbers $a_n > 0$ chosen so as to make

$$\sum_n a_n \log\frac{1}{|z_n|} \;<\; \infty.$$

Here, $U(0) < \infty$ although U is infinite at each of the z_n. In more sophisticated versions of this construction, the z_n are *dense* in $\{|z| < 1/2\}$ and various sequences of $a_n > 0$ with $\sum_n a_n < \infty$ are used.

We now allow both properties from our definition to play their parts, (ii) as well as (i). In that way, we obtain the first general results pertaining specifically to superharmonic functions, among which the following *strong minimum principle* is probably the most important:

Lemma. *Let $U(z)$ be superharmonic in a domain $\mathcal{D}$. Then, if Ω is a (connected) domain with compact closure contained in $\mathcal{D}$,*

$$U(z) \;>\; \inf_{\zeta\in\partial\Omega} U(\zeta) \qquad \text{for } z\in\Omega$$

unless $U(z)$ is constant on $\bar{\Omega}$.

Proof. As we know, $U(z)$ attains its (finite) minimum, M, on $\bar{\Omega}$, and it is enough to show that if $U(z_0) = M$ at some $z_0\in\Omega$, we have $U(z) \equiv M$ on $\bar{\Omega}$. The reasoning here is like that followed in establishing the strong maximum principle for harmonic functions.

Assuming that there is such a z_0, we have, by property (ii),

$$M = U(z_0) \geqslant \frac{1}{2\pi} \int_0^{2\pi} U(z_0 + \rho e^{i\vartheta}) \, d\vartheta$$

whenever $\rho > 0$ is sufficiently small. Here, $U(z_0 + \rho e^{i\vartheta}) \geqslant M$ and if, at any ϑ_0, we had $U(z_0 + \rho e^{i\vartheta_0}) > M$, $U(z_0 + \rho e^{i\vartheta})$ would be $> M$ for all ϑ belonging to some *open interval including* ϑ_0, by property (i). In that event, the above right-hand integral would also be $> M$, yielding a contradiction. We must therefore have $U(z_0 + \rho e^{i\vartheta}) \equiv M$ for small enough values of $\rho > 0$.

The rest of the proof is like that of the result for harmonic functions, with $E = \{z \in \Omega : U(z) = M\}$ closed in Ω's relative topology thanks to property (i). We are done.

Corollary. *Let $U(z)$ be superharmonic in a domain $\mathscr{D}$, and let $\mathscr{O}$ be an open set with compact closure lying in $\mathscr{D}$. Then, for $z \in \mathscr{O}$,*

$$U(z) \geqslant \inf_{\zeta \in \partial \mathscr{O}} U(\zeta).$$

Proof. Apply the lemma in each component of $\mathscr{O}$.

Corollary. *Let $U(z)$ be superharmonic in $\mathscr{D}$, a domain with compact closure. If $\liminf_{z \to \zeta} U(z) \geqslant M$ at each $\zeta \in \partial \mathscr{D}$, one has $U(z) \geqslant M$ in $\mathscr{D}$.*

Proof. Fix any $\varepsilon > 0$. Then, corresponding to each $\zeta \in \partial \mathscr{D}$ there is an r_ζ, $0 < r_\zeta < \varepsilon$, such that

$$U(z) \geqslant M - \varepsilon \quad \text{for } z \in \mathscr{D} \text{ and } |z - \zeta| \leqslant r_\zeta.$$

Here, $\partial \mathscr{D}$ is compact, so it can be covered by a *finite number* of the open disks

$$\{|z - \zeta| < r_\zeta\}.$$

Let $\mathscr{O}$ be the open set equal to the *complement*, in $\mathscr{D}$, of the *union of the closures* of *those particular disks.*

The closure $\bar{\mathscr{O}}$ is compact and contained in $\mathscr{D}$. If $z \in \partial \mathscr{O}$, we have $|z - \zeta| = r_\zeta$ for some $\zeta \in \partial \mathscr{D}$, so $U(z) \geqslant M - \varepsilon$. $U(z)$ is hence $\geqslant M - \varepsilon$ in $\mathscr{O}$ by the previous corollary. $\mathscr{O}$, however, certainly includes all points of $\mathscr{D}$ distant by more than ε from $\partial \mathscr{D}$. Our result thus follows on making $\varepsilon \longrightarrow 0$.

From these results we can deduce a useful *characterization* of superharmonic functions.

Theorem. *If $U(z)$ is $> -\infty$ and enjoys property* (i) *in a domain $\mathscr{D}$, it is superharmonic there provided that for each $z_0 \in \mathscr{D}$ and every disk Δ of sufficiently small radius with centre at z_0, one has*

$$U(z_0) \geqslant h(z_0)$$

for every function $h(z)$ harmonic in Δ and continuous up to $\partial\Delta$, satisfying

$$h(\zeta) \leqslant U(\zeta)$$

on $\partial\Delta$.

Conversely, *if $U(z)$ is superharmonic in $\mathscr{D}$ and Ω is any domain having compact closure $\subseteq \mathscr{D}$, every function $h(z)$ harmonic in Ω and continuous up to $\partial\Omega$ is $\leqslant U(z)$ in Ω provided that $h(\zeta) \leqslant U(\zeta)$ on $\partial\Omega$.*

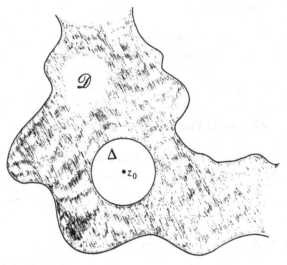

Figure 232

Proof. For the first part, we take any $z_0 \in \mathscr{D}$ and verify property (ii) for U there, assuming the hypothesis concerning disks Δ about z_0.

Let then $0 < r < \mathrm{dist}(z_0, \partial\mathscr{D})$. By the first lemma of this article, there is an *increasing sequence* of functions $u_n(\vartheta)$, *continuous* and of period 2π, such that

$$u_n(\vartheta) \xrightarrow[n]{} U(z_0 + re^{i\vartheta}), \qquad 0 \leqslant \vartheta \leqslant 2\pi.$$

Put $h_n(z_0 + re^{i\vartheta}) = u_n(\vartheta)$, and, for $0 \leqslant \rho < r$, take

$$h_n(z_0 + \rho e^{i\vartheta}) = \frac{1}{2\pi} \int_0^{2\pi} \frac{r^2 - \rho^2}{r^2 + \rho^2 - 2r\rho\cos(\vartheta - \tau)} u_n(\tau)\, d\tau.$$

Then each function $h_n(z)$ is harmonic in the disk Δ of radius r about z_0 and continuous up to $\partial\Delta$, where we of course have

$$h_n(\zeta) \;\leqslant\; U(\zeta).$$

If $r > 0$ is small enough, our assumption thus tells us that

$$h_n(z_0) \;\leqslant\; U(z_0)$$

for every n. Now Lebesgue's monotone convergence theorem ensures that

$$h_n(z_0) \;\;\xrightarrow[n]{}\;\; \frac{1}{2\pi}\int_0^{2\pi} U(z_0 + re^{i\tau})\,d\tau$$

as $n \longrightarrow \infty$. Hence

$$\frac{1}{2\pi}\int_0^{2\pi} U(z_0 + re^{i\tau})\,d\tau \;\;\leqslant\;\; U(z_0)$$

for all sufficiently small $r > 0$, and property (ii) holds.

The other part of the theorem is practically a restatement of the *second* of the above corollaries. Indeed, if $h(z)$, harmonic in Ω and continuous up to $\bar\Omega \subseteq \mathscr{D}$ satisfies $h(\zeta) \leqslant U(\zeta)$ on $\partial\Omega$, we certainly have

$$\liminf_{\substack{z\to\zeta \\ z\in\Omega}} (U(z) \;-\; h(z)) \;\;\geqslant\;\; 0$$

at each $\zeta \in \partial\Omega$ on account of property (i). At the same time, $U(z) - h(z)$ is superharmonic in Ω; it must therefore be $\geqslant 0$ there by the corollary in question.

This does it.

By combining the two arguments followed in the last proof, we immediately obtain the following inequality:

For $U(z)$ superharmonic in $\mathscr{D}$, $z_0 \in \mathscr{D}$, and $0 < r < \operatorname{dist}(z_0,\ \partial\mathscr{D})$,

$$\boxed{\; U(z_0 + \rho e^{i\vartheta}) \;\geqslant\; \frac{1}{2\pi}\int_0^{2\pi} \frac{r^2 - \rho^2}{r^2 + \rho^2 - 2r\rho\cos(\vartheta - \tau)}\, U(z_0 + re^{i\tau})\,d\tau, \;\; 0\leqslant\rho<r. \;}$$

This in turn gives us a result needed in article 2:

Lemma. *If $U(z)$ is superharmonic in a domain $\mathscr{D}$ and $z_0 \in \mathscr{D}$,*

$$\frac{1}{2\pi}\int_0^{2\pi} U(z_0 + re^{i\vartheta})\,d\vartheta$$

is a decreasing function of r for $0 < r < \operatorname{dist}(z_0,\ \partial\mathscr{D})$.

Proof. Integrate both sides of the boxed inequality with respect to ϑ and then use Fubini's theorem on the right.

Along these same lines, we have, finally, the

Theorem. *Let $U(z)$ be superharmonic in a domain $\mathscr{D}$, and suppose that $z_0 \in \mathscr{D}$ and that $0 < R < \operatorname{dist}(z_0, \partial\mathscr{D})$. Denoting by Δ the disk $\{|z - z_0| < R\}$, put $V(z) = U(z)$ for $z \in \mathscr{D} \sim \Delta$. In Δ, take*

$$V(z_0 + re^{i\vartheta}) = \frac{1}{2\pi}\int_0^{2\pi} \frac{R^2 - r^2}{R^2 + r^2 - 2rR\cos(\vartheta - \tau)} U(z_0 + Re^{i\tau})\,d\tau$$

(for $0 \leqslant r < R$). Then $V(z) \leqslant U(z)$ and $V(z)$ is superharmonic in $\mathscr{D}$.

Proof. For $z \in \mathscr{D} \sim \Delta$, the relation $V(z) \leqslant U(z)$ is manifest, and for $z \in \Delta$ it is a consequence of the above boxed inequality.

To verify property (ii) for V, suppose first of all that $z \in \mathscr{D} \sim \Delta$. Then, for sufficiently small $\rho > 0$,

$$V(z) = U(z) \geqslant \frac{1}{2\pi}\int_0^{2\pi} U(z + \rho e^{i\vartheta})\,d\vartheta.$$

By the relation just considered, the right-hand integral is in turn

$$\geqslant \frac{1}{2\pi}\int_0^{2\pi} V(z + \rho e^{i\vartheta})\,d\vartheta;$$

V thus enjoys property (ii) at z.

We must also look at the points $z \in \Delta$. On $\partial\Delta$, the function $U(\zeta)$ is *bounded below*, according to an early observation in this article. The Poisson integral used above to define $V(z)$ in Δ is therefore *either infinite for every r, $0 \leqslant r < R$, or else convergent for each such r.* In the former case, $V(z) \equiv \infty$ for $z \in \Delta$, and V (trivially) possesses property (ii) at those z. In the latter case, $V(z)$ is actually *harmonic* in Δ and hence, for any given z therein, *equal to the previous mean value when $\rho < \operatorname{dist}(z, \partial\Delta)$.* Here also, V has property (ii) at z.

Verifications of the relation $V(z) > -\infty$ and of property (i) remain. The first of these is clear; it is certainly true in $\mathscr{D} \sim \Delta$ where V coincides with U, and also true in Δ where, as a Poisson integral,

$$V(z) \geqslant \inf_{|\zeta - z_0| = R} U(\zeta)$$

with the right side $> -\infty$, as we know.

We have, then, to check property (i). The only points at which this can present any difficulty must lie on $\partial\Delta$, for, *inside Δ, V is either harmonic* and thus *continuous* or else *everywhere infinite*, and *outside* $\bar{\Delta}$, V coincides (in $\mathscr{D}$) with U, a function *having* the semicontinuity in question. Let therefore $|z - z_0| = R$. Then we surely have

$$\liminf_{\substack{\zeta \to z \\ \zeta \notin \Delta}} V(\zeta) = \liminf_{\substack{\zeta \to z \\ \zeta \notin \Delta}} U(\zeta) \geqslant U(z) = V(z),$$

so we need only examine the behaviour of $V(\zeta)$ for ζ tending to z *from within* Δ. The relation just written holds in particular, however, for $\zeta = z_0 + Re^{i\tau}$ tending to z on $\partial \Delta$. Since $U(z_0 + Re^{i\tau})$ is also bounded below on $\partial \Delta$, we see by the elementary properties of the Poisson kernel that

$$\liminf_{\substack{\zeta \to z \\ \zeta \in \Delta}} V(\zeta) \geqslant U(z) = V(z).$$

We thus have

$$\liminf_{\zeta \to z} V(z) \geqslant V(z)$$

for the points z on $\partial \Delta$, as well as at the other $z \in \mathscr{D}$, and V has property (i). The theorem is proved.

Our work will involve the consideration of certain *families* of superharmonic functions. Concerning these, one has two main results.

Theorem. *Let the $U_n(z)$ be superharmonic in a domain $\mathscr{D}$, with*

$$U_1(z) \leqslant U_2(z) \leqslant U_3(z) \leqslant \cdots \leqslant U_n(z) \leqslant \cdots$$

there. Then

$$U(z) = \lim_{n \to \infty} U_n(z)$$

is superharmonic (perhaps $\equiv \infty$) in $\mathscr{D}$.

Proof. Since $U_1(z) > -\infty$ in $\mathscr{D}$, the same is true for $U(z)$.

Verification of property (i) is almost automatic. Given $z_0 \in \mathscr{D}$, let M be any number $< U(z_0)$. Then, for some particular n, $U_n(z_0) > M$, so, since U_n enjoys property (i), $U_n(z) > M$ in a neighborhood of z_0. A fortiori, $U(z) > M$ in that same neighborhood, and $\liminf_{z \to z_0} U(z) \geqslant U(z_0)$ on account of the arbitrariness of M.

Property (ii) is a consequence of Lebesgue's monotone convergence theorem. Let $z_0 \in \mathscr{D}$ and fix any $\rho < \text{dist}(z_0, \partial \mathscr{D})$. Then, by the above boxed inequality,

$$U_n(z_0) \geqslant \frac{1}{2\pi} \int_0^{2\pi} U_n(z_0 + \rho e^{i\vartheta}) \, d\vartheta$$

for each n. Here $U_1(z_0 + \rho e^{i\vartheta})$ is bounded below for $0 \leqslant \vartheta \leqslant 2\pi$, so the right-hand integral tends to

$$\frac{1}{2\pi} \int_0^{2\pi} U(z_0 + \rho e^{i\vartheta}) \, d\vartheta$$

as $n \longrightarrow \infty$ by the monotone convergence. At the same time, $U_n(z_0) \xrightarrow[n]{} U(z_0)$, so property (ii) holds.

We are done.

A statement of opposite character is valid for *finite* collections of superharmonic functions. If, namely, $U_1(z)$, $U_2(z)$, ..., $U_N(z)$ *are superharmonic in a domain $\mathscr{D}$, so is* $\min_{1 \leqslant k \leqslant N} U_k(z)$. This observation, especially useful when the functions $U_k(z)$ involved are *harmonic*, is easily verified directly.

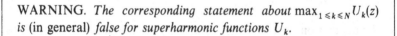

> WARNING. *The corresponding statement about* $\max_{1 \leqslant k \leqslant N} U_k(z)$
> *is* (in general) *false for superharmonic functions* U_k.

One has a version of the observation for *infinite* collections of superharmonic functions:

Theorem. *Let $\mathscr{F}$ be any family of functions superharmonic in a domain $\mathscr{D}$. For $z \in \mathscr{D}$, put*

$$W(z) = \inf\{U(z) : U \in \mathscr{F}\},$$

and then let

$$V(z) = \liminf_{\zeta \to z} W(\zeta), \quad z \in \mathscr{D}.$$

Then $V(z) \leqslant U(z)$ in $\mathscr{D}$ for every $U \in \mathscr{F}$, and (especially) *if $V(z) > -\infty$ in $\mathscr{D}$, it is superharmonic there.*

Remark. Something like the last condition is needed in order to avoid situations like the one where $\mathscr{D} = \mathbb{C}$ and $\mathscr{F}$ consists of the functions $n\Im z$, $n = 1, 2, 3, \ldots$. There, $V(z) = \Im z$ for $\Im z > 0$ but $V(z) = -\infty$ for $\Im z \leqslant 0$. Such functions V are not superharmonic.

Proof of theorem. First of all, $V(z) \leqslant W(z)$ in $\mathscr{D}$, i.e., $V(z) \leqslant U(z)$ there for each $U \in \mathscr{F}$. Indeed, since any such U is superharmonic in $\mathscr{D}$, $\liminf_{\zeta \to z} U(\zeta)$ is actually *equal* to $U(z)$ there, as observed earlier in this article (a result of playing properties (i) and (ii) against each other).

Therefore, whenever $U \in \mathcal{F}$,

$$V(z) \;=\; \liminf_{\zeta \to z} W(\zeta) \;\leqslant\; \liminf_{\zeta \to z} U(\zeta) \;=\; U(z), \qquad z \in \mathcal{D}.$$

Secondly, V has property (i) in $\mathcal{D}$. To see this, fix any $z_0 \in \mathcal{D}$ and pick* any $M < V(z_0)$; according to our definition of V, $W(z)$ is then $> M$ in some punctured open neighborhood of z_0 (i.e., an open neighborhood of z_0 with z_0 *deleted*). But this certainly makes $V(z) = \liminf_{\zeta \to z} W(\zeta) \geqslant M$ in that punctured neighborhood, so, since $M < V(z_0)$ was arbitrary, we have $\liminf_{z \to z_0} V(z) \geqslant V(z_0)$.

To complete verification of $V(z)$'s superharmonicity in $\mathcal{D}$ when that function is $> -\infty$ there, one may resort to the criterion provided by the *first* of the preceding theorems. According to the latter, it is enough to show that if $z_0 \in \mathcal{D}$ and Δ is any disk centred at z_0 with radius $< \operatorname{dist}(z_0, \partial \mathcal{D})$, we have $V(z_0) \geqslant h(z_0)$ for each function $h(z)$ *continuous* on $\bar{\Delta}$, *harmonic* in Δ, and satisfying $h(\zeta) \leqslant V(\zeta)$ on $\partial \Delta$. But for any such function h we certainly have $h(\zeta) \leqslant U(\zeta)$ on $\partial \Delta$ for every $U \in \mathcal{F}$, so, by the second part of the theorem referred to, $h(z) \leqslant U(z)$ in Δ for those U. Hence

$$h(z) \;\leqslant\; \inf_{U \in \mathcal{F}} U(z) \;=\; W(z)$$

in Δ, and finally, h being continuous at z_0 (the centre of Δ !),

$$h(z_0) \;=\; \lim_{z \to z_0} h(z) \;\leqslant\; \liminf_{z \to z_0} W(z) \;=\; V(z_0),$$

as required. We are done.

Remark. This theorem, together with the second of those preceding it, forms the basis for what is known as *Perron's method* of solution of the Dirichlet problem.

2. The Riesz representation of superharmonic functions

A superharmonic function can be approximated from below by others which are also infinitely differentiable. This is obvious for the function $U(z)$ identically infinite in a domain $\mathcal{D}$, that one being just the limit, as $n \longrightarrow \infty$, of the *constant* functions $U_n(z) = n$. We therefore turn to the construction of such approximations to functions $U(z)$ superharmonic and $\not\equiv \infty$ in $\mathcal{D}$.

Given such a U, one starts by forming the means

$$U_\rho(z) \;=\; \frac{1}{2\pi} \int_0^{2\pi} U(z + \rho e^{i\vartheta}) \, d\vartheta;$$

---

* in the case where $V(z_0) > -\infty$; otherwise property (i) clearly does hold at z_0

when $\rho > 0$ is given, these are defined for the z in $\mathscr{D}$ with dist$(z, \partial\mathscr{D}) > \rho$. According to property (ii) from our definition,

$$U_\rho(z) \leqslant U(z)$$

for all sufficiently small $\rho > 0$ (and in fact for *all* such $\rho <$ dist$(z, \partial\mathscr{D})$ by the boxed inequality near the end of the preceding article); on the other hand, $\liminf_{\rho \to 0} U_\rho(z) \geqslant U(z)$ by property (i). Thus, for each $z \in \mathscr{D}$,

$$U_\rho(z) \longrightarrow U(z) \quad \text{as } \rho \longrightarrow 0.$$

A lemma from the last article shows that this convergence is actually *monotone*; the $U_\rho(z)$ *increase* as ρ diminishes towards 0.

Concerning the U_ρ, we have the useful

Lemma. *If $U(z)$ is superharmonic in a (connected) domain $\mathscr{D}$ and not identically infinite there, the $U_\rho(z)$ are finite for $z \in \mathscr{D}$ and $0 < \rho < $ dist$(z, \partial\mathscr{D})$.*

Proof. Suppose that

$$U_r(z_0) = \frac{1}{2\pi} \int_0^{2\pi} U(z_0 + re^{i\tau}) \, d\tau = \infty$$

for some $z_0 \in \mathscr{D}$ and an r with $0 < r < $ dist$(z_0, \partial\mathscr{D})$. It is claimed that then $U(z) \equiv \infty$ in $\mathscr{D}$.

By one of our first observations about superharmonic functions in the preceding article, $U(z_0 + re^{i\tau})$ is *bounded below* for $0 \leqslant \tau \leqslant 2\pi$. The above relation therefore makes the Poisson integrals occurring in the boxed inequality near the end of that article *infinite*, and we must have $U(z) \equiv \infty$ for $|z - z_0| < r$.

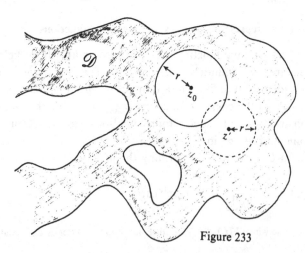

Figure 233

Let now z' be any point in $\mathscr{D}$ about which one can draw a circle – of radius r', say – *lying entirely* in $\mathscr{D}$ and also intersecting the open disk of radius r centred at z_0.* Since $U(z)$ is bounded below on that circle, we have $U_{r'}(z') = \infty$ so, by the argument just made, $U(z) \equiv \infty$ for $|z - z'| < r'$.

The process may evidently be continued indefinitely so as to gradually fill out the connected open region $\mathscr{D}$. In that way, one sees that $U(z) \equiv \infty$ therein, and the lemma is proved.

Corollary. *If $U(z)$ is superharmonic and not identically infinite in a (connected) domain $\mathscr{D}$, it is locally L_1 there (with respect to Lebesgue measure for $\mathbb{R}^2$).*

Proof. It is enough to verify that if $z_0 \in \mathscr{D}$ and $0 < r < \frac{1}{2}\mathrm{dist}(z_0, \partial\mathscr{D})$, we have

$$\iint_{r \leqslant |z - z_0| \leqslant 2r} |U(z)|\, \mathrm{d}x\, \mathrm{d}y < \infty,$$

for, since each point of $\mathscr{D}$ lies in the *interior* of some annulus like the one over which the integral is taken, any compact subset of $\mathscr{D}$ can be covered by a finite number of such annuli.

By the lower bound property already used so often, there is an $M < \infty$ such that $U(z) \geqslant -M$ when $|z - z_0| \leqslant 2r$. The preceding integral is therefore

$$\leqslant \iint_{r \leqslant |z - z_0| \leqslant 2r} (U(z) + 2M)\, \mathrm{d}x\, \mathrm{d}y$$

$$= 6\pi r^2 M + \int_r^{2r} \int_0^{2\pi} U(z_0 + \rho e^{i\vartheta})\rho\, \mathrm{d}\vartheta\, \mathrm{d}\rho$$

$$= 6\pi r^2 M + 2\pi \int_r^{2r} U_\rho(z_0)\rho\, \mathrm{d}\rho.$$

$U_\rho(z_0)$ is, as noted above, a *decreasing* function of ρ; the last expression is thus

$$\leqslant 6\pi r^2 M + 3\pi r^2 U_r(z_0).$$

This, however, is finite by the lemma.

We are done.

With the means $U_\rho(z)$ at hand, we continue our construction of superharmonic $\mathscr{C}_\infty$ approximations to a given $U(z)$ superharmonic and $\not\equiv \infty$ in a domain $\mathscr{D}$. For this purpose, one chooses any function $\varphi(\rho)$ infinitely

---

* with its interior

differentiable on $(0, \infty)$, *identically zero outside* $(1, 2)$ *and* > 0 *on that interval,* normalized so as to make

$$\int_1^2 \varphi(\rho)\rho \, d\rho \;=\; 1 :$$

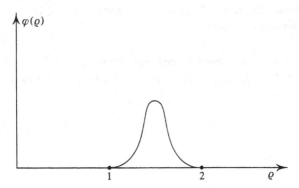

Figure 234

(As we shall see, it turns out to be convenient to work with a function $\varphi(\rho)$ *vanishing for the values of* ρ *near* 0 as well as for the large ones.) Using φ, one then forms *averages of the means* U_ρ :

$$(\Phi_r U)(z) \;=\; \frac{1}{r^2}\int_0^\infty U_\rho(z)\,\varphi(\rho/r)\rho \, d\rho;$$

for given $z \in \mathcal{D}$, these are defined when

$$0 \;<\; r \;<\; \tfrac{1}{2}\,\mathrm{dist}(z,\, \partial\mathcal{D}).$$

The $\Phi_r U$ are the approximations we set out to obtain; one has, namely, the

Theorem. *Given* $r > 0$, *denote by* $\mathcal{D}_r$ *the set of* $z \in \mathcal{D}$ *with* $\mathrm{dist}(z,\, \partial\mathcal{D}) > 2r$. *Let* U *be superharmonic in* $\mathcal{D}$, *then*:

$$(\Phi_r U)(z) \;\leqslant\; U(z) \quad \text{for } z \in \mathcal{D}_r;$$

$$(\Phi_r U)(z) \;\longrightarrow\; U(z) \quad \text{as } r \to 0 \text{ for each } z \in \mathcal{D};$$

$$(\Phi_{2r} U)(z) \;\leqslant\; (\Phi_r U)(z) \quad \text{for } z \in \mathcal{D}_{2r}.$$

If also $U(z) \not\equiv \infty$ *in the (connected) domain* $\mathcal{D}$, *each* $(\Phi_r U)(z)$ *is infinitely differentiable in the corresponding* $\mathcal{D}_r$, *and superharmonic in each connected component thereof.*

Proof. The *first two* properties of the $\Phi_r U$ follow as direct consequences of the behaviour, noted above, of the $U_\rho(z)$ together with φ's normalization. The *third* is then assured by $\varphi(\rho)$'s being supported on the interval $(1, 2)$.

Passing to the *superharmonicity* of $\Phi_r U$, we first check property (ii) for that function in $\mathscr{D}_r$. This does not depend on the condition that $U(z) \not\equiv \infty$. Fix any $z \in \mathscr{D}_r$. For $0 < \sigma < \text{dist}(z, \partial\mathscr{D}) - 2r$ we then have

$$\frac{1}{2\pi} \int_0^{2\pi} (\Phi_r U)(z + \sigma e^{i\psi}) \, d\psi$$

$$= \frac{1}{4\pi^2 r^2} \int_0^{2\pi} \int_0^\infty \int_0^{2\pi} U(z + \rho e^{i\vartheta} + \sigma e^{i\psi}) \, \varphi(\rho/r)\rho \, d\vartheta d\rho \, d\psi.$$

Since $\varphi(\rho/r)$ vanishes for $\rho \geqslant 2r$, $z + \rho e^{i\vartheta}$ lies in $\mathscr{D}$ and has distance $> \sigma$ from $\partial\mathscr{D}$ for all the values of ρ actually involved in the second expression. The argument $z + \rho e^{i\vartheta} + \sigma e^{i\psi}$ of U thus ranges over a *compact subset* of $\mathscr{D}$ in that triple integral, and on such a subset U is *bounded below*, as we know. This makes it permissible for us to *perform first the integration with respect to* ψ. Doing that, and using the boxed inequality from the preceding article, we obtain a value

$$\leqslant \frac{1}{2\pi r^2} \int_0^\infty \int_0^{2\pi} U(z + \rho e^{i\vartheta}) \varphi(\rho/r)\rho \, d\vartheta d\rho \quad = \quad (\Phi_r U)(z),$$

showing that $\Phi_r U$ has property (ii) at z. Superharmonicity of $\Phi_r U$ in the components of $\mathscr{D}_r$ thus follows if it meets our definition's other two requirements there.

Satisfaction of the latter is, however, obviously guaranteed by the *infinite differentiability* of $\Phi_r U$ in $\mathscr{D}_r$, which we now proceed to verify for functions $U(z) \not\equiv \infty$ in $\mathscr{D}$.

The left-hand member of the last relation can be rewritten as

$$\frac{1}{2\pi r^2} \int_{-\infty}^\infty \int_{-\infty}^\infty U(z + \zeta) \, \varphi(|\zeta|/r) \, d\xi \, d\eta,$$

where, as usual, $\zeta = \xi + i\eta$. Putting $z + \zeta = \zeta' = \xi' + i\eta'$, this becomes

$$\frac{1}{2\pi r^2} \int_{-\infty}^\infty \int_{-\infty}^\infty U(\zeta') \, \varphi(|\zeta' - z|/r) \, d\xi' \, d\eta'.$$

Here, $\varphi(|\zeta' - z|/r)$ vanishes for $|\zeta' - z| \leqslant r$ and $|\zeta' - z| \geqslant 2r$. Looking, then, at values of z near some *fixed* $z_0 \in \mathscr{D}_r$ – to be definite, at those, say, with

$$|z - z_0| < \delta = \frac{1}{2}\min(r, \text{dist}(z_0, \partial\mathscr{D}) - 2r),$$

we have

$$(\Phi_r U)(z) \;=\; \frac{1}{2\pi r^2} \iint_{r-\delta \,\leqslant\, |\zeta' - z_0| \,\leqslant\, 2r+\delta} U(\zeta')\,\varphi(|z - \zeta'|/r)\,\mathrm{d}\xi'\,\mathrm{d}\eta'.$$

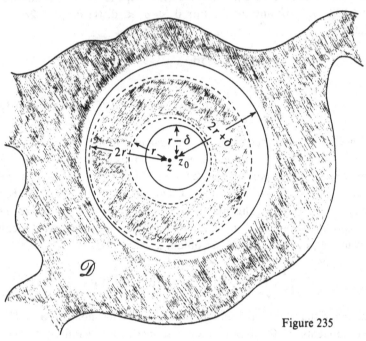

Figure 235

The annulus $K = \{\zeta': \; r-\delta \leqslant |\zeta' - z_0| \leqslant 2r+\delta\}$ over which the integration is carried out here is a *compact* subset of $\mathscr{D}$ (independent of z !), so, for the function U under consideration we have

$$\iint_K |U(\zeta')|\,\mathrm{d}\xi'\,\mathrm{d}\eta' \;<\; \infty$$

by the above corollary. Infinite differentiability of $(\Phi_r U)(z)$ at z_0 can therefore be *read off by inspection* from the last formula, provided that $\varphi(|z - \zeta'|/r)$ enjoys the same property for *each* $\zeta' \in K$ (differentiation inside the integral signs). That, however, is indeed the case, as follows by the chain rule from infinite differentiability of φ and the fact that $|z_0 - \zeta'| \geqslant \delta > 0$ for each $\zeta' \in K$. (Here we have been helped by $\varphi(\rho)$'s vanishing for $0 < \rho < 1$.) $\Phi_r U$ is thus $\mathscr{C}_\infty$ in $\mathscr{D}_r$.
The theorem is proved.

The approximations $\Phi_r U$ to a given superharmonic function U are used in establishing the *Riesz representation* for the latter. That says essentially that a function $U(z)$ superharmonic and $\not\equiv \infty$ *in and on* a bounded domain $\mathscr{D}$ (i.e., in a domain *including* $\bar{\mathscr{D}}$) is given there by a *formula*

$$U(z) \;=\; \int_{\bar{\mathscr{D}}} \log\frac{1}{|z-\zeta|}\,d\mu(\zeta) \;+\; H(z),$$

with μ a (finite) positive measure on $\bar{\mathscr{D}}$ and $H(z)$ *harmonic* in $\mathscr{D}$. (Conversely, expressions like the one on the right *are* always superharmonic in $\mathscr{D}$, according to the remarks following the first lemma of the preceding article.)

The representation is really of *local character*, for the restriction of the measure μ figuring in it to *any open disk* $\Delta \subseteq \mathscr{D}$ is *completely determined by the behaviour of U in* Δ (see problem 48 below), and at the same time, the function of z equal to

$$\int_{\mathscr{D} \sim \Delta} \log\frac{1}{|z-\zeta|}\;d\mu(\zeta)$$

is certainly *harmonic* in Δ. The general form of the result can thus be obtained from *a special version of it for disks* by simply pasting some of those together so as to cover the given domain $\mathscr{D}$! In fact, *only the version for disks* will be required in the present chapter, so that is what we prove here. Passage from it to the more general form is left as an exercise to the reader (problem 49).

We proceed, then, to the derivation of the Riesz representation formula for disks. The idea is to first get it for $\mathscr{C}_\infty$ superharmonic functions by simple application of Green's theorem and then pass from those to the general ones with the help of the $\Phi_r U$. In this, an essential rôle is played by the classical

Lemma. *A function $V(z)$ infinitely differentiable in a domain $\mathscr{D}$ is superharmonic there if and only if*

$$\frac{\partial^2 V(z)}{\partial x^2} \;+\; \frac{\partial^2 V(z)}{\partial y^2} \;\leqslant\; 0 \quad \text{for } z \in \mathscr{D}.$$

Notation. The Laplacian $\partial^2/\partial x^2 \;+\; \partial^2/\partial y^2$ is denoted by ∇^2 (following earlier usage in this book).

Proof of lemma. Supposing that V is superharmonic in $\mathscr{D}$, we take any point z_0 therein. Then, by the third lemma of the preceding article,

$$\int_0^{2\pi} V(z_0 + \rho e^{i\vartheta})\,d\vartheta$$

is a *decreasing* function of ρ for $0 < \rho < \mathrm{dist}(z_0,\, \partial\mathscr{D})$. The $\mathscr{C}_\infty$ character of V makes it possible for us to differentiate this expression under the integral sign with respect to ρ, so we have

$$\int_0^{2\pi} \frac{\partial V(z_0 + \rho e^{i\vartheta})}{\partial \rho} \, d\vartheta \;\; \leqslant \;\; 0$$

for small positive values of that parameter.

By Green's theorem, however,

$$\iint_{|z-z_0|<\rho} (\nabla^2 V)(z) \, dx \, dy \;\; = \;\; \int_0^{2\pi} \frac{\partial V(z_0 + \rho e^{i\vartheta})}{\partial \rho} \rho \, d\vartheta;$$

the left-hand integral is thus *negative*. Finally,

$$(\nabla^2 V)(z_0) \;\; = \;\; \lim_{\rho \to 0} \frac{1}{\pi \rho^2} \iint_{|z-z_0|<\rho} (\nabla^2 V)(z) \, dx \, dy,$$

showing that $\nabla^2 V \leqslant 0$ at z_0.

Assuming, on the other hand, that $\nabla^2 V \leqslant 0$ in $\mathscr{D}$, we see by the second of the above relations that

$$\int_0^{2\pi} V(z_0 + \rho e^{i\vartheta}) \, d\vartheta$$

is a decreasing function of ρ for $0 < \rho < \mathrm{dist}(z_0, \partial\mathscr{D})$ when $z_0 \in \mathscr{D}$. At the same time,

$$V(z_0) \;\; = \;\; \lim_{\rho \to 0} \frac{1}{2\pi} \int_0^{2\pi} V(z_0 + \rho e^{i\vartheta}) \, d\vartheta$$

in the present circumstances, so $V(z_0)$ must be $\geqslant$ each of the means figuring on the right for the values of ρ just indicated. This establishes property (ii) for V at z_0 and hence the superharmonicity of V in $\mathscr{D}$.

The lemma is proved.

Here is the version of Riesz' result that we will be using. It is most convenient to obtain a representation differing slightly in appearance from the one written above, but equivalent to the latter. About this, more in the remark following the proof.

Theorem (F. Riesz). *Let $U(z)$ be superharmonic and $\not\equiv \infty$ in a domain $\mathscr{D}$, and suppose that $z_0 \in \mathscr{D}$ and $0 < r < \mathrm{dist}(z_0, \partial\mathscr{D})$. Then, for $|z - z_0| < r$, one has*

$$U(z) \;\; = \;\; \int_{|\zeta - z_0| \leqslant r} \log \left| \frac{r^2 - (\overline{\zeta - z_0})(z - z_0)}{r(z - \zeta)} \right| d\mu(\zeta) \;\; + \;\; h(z),$$

where μ is a finite positive measure on the closed disk $\{|z - z_0| \leqslant r\}$, and $h(z)$ a function harmonic for $|z - z_0| < r$.

Remark. In the integrand we simply have the *Green's function* associated with the disk $\{|z - z_0| < r\}$. The integral is therefore frequently referred to as a *pure Green potential* for that disk – 'pure' because the *measure* μ is *positive*.

Proof of theorem. To simplify the writing, we take $z_0 = 0$ and $r = 1$ – *that also frees the letter r for another use during this proof*! For some $R > 1$, the closed disk

$$\bar{\Delta} = \{|z| \leqslant R\}$$

lies in $\mathscr{D}$, and the averages $\Phi_r U$ introduced previously are hence *defined*, *infinitely differentiable* and *superharmonic in and on $\bar{\Delta}$* when the parameter r (not to be confounded with the radius of the disk for which our representation is being derived!) is small enough. We *fix* such an r, and denote $\Phi_r U$ *by V for the time being* (again to help keep the notation clear).

Fix also any z, $|z| < R$, for the moment. The Green's function

$$\log \left| \frac{R^2 - \zeta \bar{z}}{R(\zeta - z)} \right|$$

is *harmonic* in ζ for $|\zeta| < R$ and $\zeta \neq z$; it is, besides, *zero* when $|\zeta| = R$. From this we see by applying Green's theorem in the region $\{\zeta : |\zeta - z| > \rho \text{ and } |\zeta| < R\}$ and afterwards causing ρ to tend to zero (cf. beginning of the proof of symmetry of the Green's function, end of

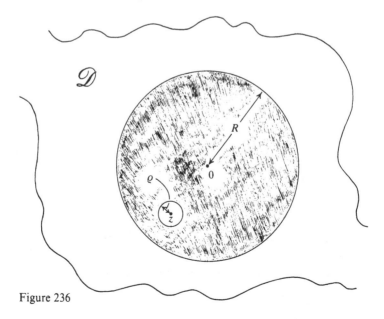

Figure 236

§A.2, Chapter VIII), that

$$V(z) \quad = \quad -\frac{1}{2\pi} \int_0^{2\pi} \left(\frac{\partial}{\partial \sigma} \log \left| \frac{R^2 - \sigma e^{i\vartheta} \bar{z}}{R(\sigma e^{i\vartheta} - z)} \right| \right)_{\sigma = R} V(Re^{i\vartheta}) R \, \mathrm{d}\vartheta$$

$$- \frac{1}{2\pi} \iint_{|\zeta| < R} \log \left| \frac{R^2 - \zeta \bar{z}}{R(\zeta - z)} \right| (\nabla^2 V)(\zeta) \, \mathrm{d}\xi \, \mathrm{d}\eta.$$

Working out the partial derivative in the first integral on the right, we get

$$V(z) \quad = \quad \frac{1}{2\pi} \int_0^{2\pi} \frac{R^2 - |z|^2}{|z - Re^{i\vartheta}|^2} V(Re^{i\vartheta}) \mathrm{d}\vartheta$$

$$- \frac{1}{2\pi} \iint_{|\zeta| < R} \log \left| \frac{R^2 - \zeta \bar{z}}{R(\zeta - z)} \right| (\nabla^2 V)(\zeta) \mathrm{d}\xi \, \mathrm{d}\eta ;$$

this, then, holds for each z of modulus $< R$.

Here, in the first integrand, we recognize the *Poisson kernel* for the disk $\{|z| < R\}$ (that's where the kernel *comes* from!); the *first* right-hand term is hence equal to a function *harmonic in that disk*. In the *second* term on the right, $(\nabla^2 V)(\zeta)$ is *negative according to the preceding lemma*, V being superharmonic in and on $\{|z| \leq R\}$. The last relation is therefore a *formula of the kind we are seeking to establish*, representing, however, the $\mathscr{C}_\infty$ superharmonic approximaton $V = \Phi_r U$ to our original superharmonic function U instead of U itself. We wish now to arrive at the desired formula for U by making $r \longrightarrow 0$.

With that in mind, we rearrange the preceding relation, writing $\Phi_r U$ in place of V:

$$-\frac{1}{2\pi} \iint_{|\zeta| < R} (\nabla^2 \Phi_r U)(\zeta) \log \left| \frac{R^2 - \zeta \bar{z}}{R(z - \zeta)} \right| \mathrm{d}\xi \, \mathrm{d}\eta$$

$$= \quad (\Phi_r U)(z) \quad - \quad \frac{1}{2\pi} \int_0^{2\pi} \frac{R^2 - |z|^2}{|z - Re^{i\vartheta}|^2} (\Phi_r U)(Re^{i\vartheta}) \mathrm{d}\vartheta, \qquad |z| < R.$$

From this, we proceed to deduce the *boundedness of*

$$- \iint_{|\zeta| \leq 1} (\nabla^2 \Phi_r U)(\zeta) \, \mathrm{d}\xi \, \mathrm{d}\eta$$

for r tending to zero.

Here we use the *first* lemma of this article, according to which there are points z *arbitrarily close to* 0 *for which* $U(z) < \infty$, it having been given that $U \not\equiv \infty$ in $\mathscr{D}$. Fixing such a z, of modulus $< 1/2$, say, and denoting it by the letter c, we have, from the preceding theorem,

$$(\Phi_r U)(c) \leq U(c),$$

so, by the last formula,

$$\frac{1}{2\pi}\iint_{|\zeta|<R} (-\nabla^2\Phi_r U)(\zeta) \log\left|\frac{R^2-\zeta\bar{c}}{R(\zeta-c)}\right| d\xi\,d\eta$$

$$\leqslant\ U(c)\ -\ \frac{1}{2\pi}\int_0^{2\pi}\frac{R^2-|c|^2}{|Re^{i\vartheta}-c|^2}(\Phi_r U)(Re^{i\vartheta})\,d\vartheta$$

for sufficiently small $r > 0$.

Now $(\Phi_r U)(Re^{i\vartheta})$ is by our construction an *average* of $U(\zeta)$ over the little annulus $r < |\zeta - Re^{i\vartheta}| < 2r$, and the union of these annuli for $0 \leqslant \vartheta \leqslant 2\pi$ is contained in the disk $\{|\zeta| \leqslant R + 2r\}$. The latter, in turn, is contained in a *fixed* disk $\{|\zeta| \leqslant R'\}$ slightly larger than $\{|\zeta| \leqslant R\}$ for values of $r < (R' - R)/2$. R, however, was chosen so as to make the disk $\{|\zeta| \leqslant R\}$ lie in $\mathscr{D}$; we may thus take $R' > R$ close enough to R to ensure that $\{|\zeta| \leqslant R'\}$ is also in $\mathscr{D}$. Once this is done, we know there is a finite M with $U(\zeta) \geqslant -M$ for $|\zeta| \leqslant R'$; this, then, holds in particular on the little annuli first mentioned when $2r < R' - R$. For the averages $\Phi_r U$ corresponding to those values of r we therefore have

$$(\Phi_r U)(Re^{i\vartheta})\ \geqslant\ -M, \quad 0 \leqslant \vartheta \leqslant 2\pi.$$

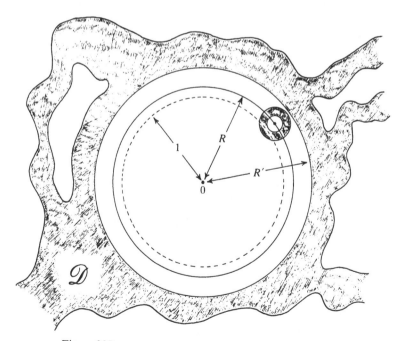

Figure 237

At the same time $R > 1$, so the expression

$$\log\left|\frac{R^2 - \zeta\bar{c}}{R(\zeta - c)}\right|,$$

positive for $|\zeta| < R$, is actually $\geqslant$ some $k > 0$ for $|\zeta| \leqslant 1$; meanwhile, $(-\nabla^2\Phi_r U)(\zeta) \geqslant 0$ for $|\zeta| \leqslant R$ as we know, when $r > 0$ is sufficiently small. Use these relations in the *left side* of the above inequality, and plug the previous one into the *right-hand* integral figuring in the latter. It is found that

$$\frac{k}{2\pi}\iint_{|\zeta| \leqslant 1} (-\nabla^2\Phi_r U)(\zeta)\,d\xi\,d\eta \quad \leqslant \quad U(c) + M,$$

a finite quantity, for $r > 0$ small enough. The integral on the left thus does remain bounded as $r \longrightarrow 0$.

By this boundedness we see, keeping positivity of the functions $-\nabla^2\Phi_r U$ in mind, that there is a certain *positive measure* μ on $\{|\zeta| \leqslant 1\}$ such that, *on the closed unit disk*,

$$-\frac{1}{2\pi}(\nabla^2\Phi_r U)(\zeta)\,d\xi\,d\eta \quad \longrightarrow \quad d\mu(\zeta) \qquad w^*$$

as $r \longrightarrow 0$ through a certain sequence of values r_n (cf. §F.1 of Chapter III, where the same kind of argument is used). There is no loss of generality in our taking $r_{n+1} < r_n/2$; this will permit us to take advantage of the relation $\Phi_{2r} U \leqslant \Phi_r U$.

Let us now rewrite *for the unit disk* the representation of the $\Phi_r U$ derived above for $\{|z| < R\}$. That takes the form

$$(\Phi_r U)(z) \quad = \quad -\frac{1}{2\pi}\iint_{|\zeta| \leqslant 1} \log\left|\frac{1 - z\bar{\zeta}}{z - \zeta}\right|(\nabla^2\Phi_r U)(\zeta)\,d\xi\,d\eta$$

$$+ \frac{1}{2\pi}\int_0^{2\pi}\frac{1 - |z|^2}{|z - e^{i\vartheta}|^2}(\Phi_r U)(e^{i\vartheta})\,d\vartheta, \qquad |z| < 1$$

(assuming, of course, as always that $r > 0$ is sufficiently small). Fixing any z of modulus < 1, we let r tend to 0 through the sequence $\{r_n\}$. According to the preceding theorem, $(\Phi_r U)(z)$ will then tend to $U(z)$, and, since $r_n > 2r_{n+1}$, $(\Phi_r U)(e^{i\vartheta})$ will, for each ϑ, *increase monotonically*, tending to $U(e^{i\vartheta})$. The *second* integral on the right will thus tend to

$$\frac{1}{2\pi}\int_0^{2\pi}\frac{1 - |z|^2}{|z - e^{i\vartheta}|^2} U(e^{i\vartheta})\,d\vartheta$$

by the monotone convergence theorem. We desire at this point to deduce simultaneous convergence of the *first* term on the right to

$$\int_{|\zeta| \leqslant 1} \log \left| \frac{1 - z\bar{\zeta}}{z - \zeta} \right| d\mu(\zeta)$$

from the w^* convergence just described, since that would complete the proof.

That, however, involves a slight difficulty, for, as a function of ζ,

$$\log \left| \frac{1 - \bar{\zeta}z}{\zeta - z} \right|$$

is discontinuous at $\zeta = z$. To deal with this, we first break up the preceding formula for $\Phi_r U$ in the following way:

$$\begin{aligned}(\Phi_r U)(z') &= -\frac{1}{2\pi} \iint_{|\zeta| \leqslant 1} \log \frac{1}{|z' - \zeta|} (\nabla^2 \Phi_r U)(\zeta) \, d\xi \, d\eta \\ &\quad - \frac{1}{2\pi} \iint_{|\zeta| \leqslant 1} \log|1 - z'\bar{\zeta}| \, (\nabla^2 \Phi_r U)(\zeta) \, d\xi \, d\eta \\ &\quad + \frac{1}{2\pi} \int_0^{2\pi} \frac{1 - |z'|^2}{|z' - e^{i\vartheta}|^2} (\Phi_r U)(e^{i\vartheta}) \, d\vartheta, \qquad |z'| < 1.\end{aligned}$$

Keeping z, of modulus < 1, fixed, we take z' in this relation equal to $z + \rho e^{i\psi}$, with $0 < \rho < 1 - |z|$, and then integrate with respect to ψ on both sides, from 0 to 2π.

When $|\zeta| \leqslant 1$, $\log|1 - z'\bar{\zeta}|$ is *harmonic* in z' for $|z'| < 1$; we thus have

$$\frac{1}{2\pi} \int_0^{2\pi} \log|1 - (z + \rho e^{i\psi})\bar{\zeta}| \, d\psi = \log|1 - z\bar{\zeta}|$$

for each such ζ and $0 < \rho < 1 - |z|$. In like manner,

$$\frac{1}{2\pi} \int_0^{2\pi} \frac{1 - |z + \rho e^{i\psi}|^2}{|z + \rho e^{i\psi} - e^{i\vartheta}|^2} \, d\psi = \frac{1 - |z|^2}{|z - e^{i\vartheta}|^2}$$

for the indicated values of ρ. There is, finally, the elementary formula

$$\frac{1}{2\pi} \int_0^{2\pi} \log \frac{1}{|z + \rho e^{i\psi} - \zeta|} \, d\psi = \min \left(\log \frac{1}{|z - \zeta|}, \log \frac{1}{\rho} \right)$$

already mentioned in the last article.

With the help of these relations we find by integration of the parameter

ψ that

$$\frac{1}{2\pi}\int_0^{2\pi}(\Phi_rU)(z+\rho e^{i\psi})\,\mathrm{d}\psi$$

$$= \quad -\frac{1}{2\pi}\iint_{|\zeta|\leqslant 1}\min\left(\log\frac{1}{|z-\zeta|},\ \log\frac{1}{\rho}\right)(\nabla^2\Phi_rU)(\zeta)\,\mathrm{d}\xi\,\mathrm{d}\eta$$

$$\quad -\frac{1}{2\pi}\iint_{|\zeta|\leqslant 1}\log|1-z\bar\zeta|\,(\nabla^2\Phi_rU)(\zeta)\,\mathrm{d}\xi\,\mathrm{d}\eta$$

$$\quad +\frac{1}{2\pi}\int_0^{2\pi}\frac{1-|z|^2}{|z-e^{i\vartheta}|^2}(\Phi_rU)(e^{i\vartheta})\,\mathrm{d}\vartheta$$

for $|z| < 1$, $0 < \rho < 1-|z|$, and r sufficiently small.

Fix now such values of z and ρ, and make $r \longrightarrow 0$ through the sequence of values r_n. The *third* integral on the right in the present relation then tends to

$$\frac{1}{2\pi}\int_0^{2\pi}\frac{1-|z|^2}{|z-e^{i\vartheta}|^2}U(e^{i\vartheta})\,\mathrm{d}\vartheta$$

as observed above, and the *left side* tends to

$$\frac{1}{2\pi}\int_0^{2\pi}U(z+\rho e^{i\psi})\,\mathrm{d}\psi \quad = \quad U_\rho(z)$$

for the same reason (monotone convergence). Here, the functions of ζ involved in the *first two integrals* on the right *are* continuous on the closed unit disk. This allows us to conclude from the w^* convergence described above that those integrals tend respectively to

$$\iint_{|\zeta|\leqslant 1}\min\left(\log\frac{1}{|z-\zeta|},\ \log\frac{1}{\rho}\right)\mathrm{d}\mu(\zeta)$$

and to

$$\iint_{|\zeta|\leqslant 1}\log|1-z\bar\zeta|\,\mathrm{d}\mu(\zeta).$$

Putting these observations together, we see that

$$U_\rho(z) \quad = \quad \iint_{|\zeta|\leqslant 1}\min\left(\log\left|\frac{1-z\bar\zeta}{z-\zeta}\right|,\ \log\frac{|1-z\bar\zeta|}{\rho}\right)\mathrm{d}\mu(\zeta)$$

$$\quad +\frac{1}{2\pi}\int_0^{2\pi}\frac{1-|z|^2}{|z-e^{i\vartheta}|^2}U(e^{i\vartheta})\,\mathrm{d}\vartheta$$

for $|z| < 1$ and $0 < \rho < 1-|z|$.

We finally let $\rho \longrightarrow 0$, continuing to hold z fixed. Then, as noted at the very beginning of this article, $U_\rho(z) \longrightarrow U(z)$. At the same time, the first right-hand integral in the formula just written tends to

$$\iint_{|\zeta| \leqslant 1} \log \left| \frac{1 - z\bar{\zeta}}{z - \zeta} \right| \, d\mu(\zeta)$$

by monotone convergence! We therefore have

$$U(z) \;=\; \iint_{|\zeta| \leqslant 1} \log \left| \frac{1 - z\bar{\zeta}}{z - \zeta} \right| d\mu(\zeta) \;+\; \frac{1}{2\pi} \int_0^{2\pi} \frac{1 - |z|^2}{|z - e^{i\vartheta}|^2} U(e^{i\vartheta}) \, d\vartheta$$

for $|z| < 1$. The function

$$h(z) \;=\; \frac{1}{2\pi} \int_0^{2\pi} \frac{1 - |z|^2}{|z - e^{i\vartheta}|^2} U(e^{i\vartheta}) \, d\vartheta$$

is *harmonic* in the open unit disk. Our theorem is thus proved.

Remark. The representation just obtained is frequently written differently. Taking, to simplify the notation, $z_0 = 0$, what we have so far reads

$$U(z) \;=\; \int_{|\zeta| \leqslant r} \log \left| \frac{r^2 - z\bar{\zeta}}{r(z - \zeta)} \right| d\mu(\zeta) \;+\; h(z), \qquad |z| \leqslant r,$$

with $h(z)$ a certain function *harmonic* in $\{|z| < r\}$. Under the circumstances of the theorem (U superharmonic in a *slightly larger* disk), we even have an explicit formula for h,

$$h(z) \;=\; \frac{1}{2\pi} \int_0^{2\pi} \frac{r^2 - |z|^2}{|z - re^{i\vartheta}|^2} U(re^{i\vartheta}) \, d\vartheta, \qquad |z| < r,$$

found at the end of the above proof.
   The integral

$$\int_{|\zeta| \leqslant r} \log |r^2 - z\bar{\zeta}| \, d\mu(\zeta),$$

however, is itself a harmonic function of z for $|z| < r$. The preceding relation can be thus rewritten as

$$U(z) \;=\; \int_{|\zeta| \leqslant r} \log \frac{1}{|z - \zeta|} \, d\mu(\zeta) \;+\; H(z), \qquad |z| < r,$$

with

$$H(z) \quad = \quad h(z) \quad + \quad \int_{|\zeta| \leqslant r} \left(\log |r^2 - z\bar{\zeta}| \quad + \quad \log \frac{1}{r} \right) d\mu(\zeta)$$

also harmonic in the disk $\{|z| < r\}$. Here we recognize on the right the familiar *logarithmic potential* (corresponding to the (here finite) positive measure μ) which has already played a rôle in §F of Chapter IX. The simplicity of this version in comparison with the original one is somewhat offset by a drawback: $H(z)$, unlike $h(z)$, is no longer determined by the boundary values $U(re^{i\vartheta})$ alone. It is often easier, nevertheless, to work with the former rather than the latter.

As they stand, the two forms of the representation are *equivalent*, with the above relation between the harmonic functions h and H serving to pass from one to the other. As long as the (finite) measure μ is *positive*, and the function $H(z)$ *harmonic* in $\{|z| < r\}$, the boxed formula does give us a function $U(z)$ superharmonic there according to observations in the preceding article; this is also true of the other formula under the same circumstances regarding μ and h. Concerning, however, such a function U, with $H(z)$, say, known *only* to be harmonic for $|z| < r$, we can say nothing about the boundary values $U(re^{i\vartheta})$ (not even as regards their existence), and thus lose the above representation for the function $h(z)$ corresponding to H as a Poisson integral in $\{|z| < r\}$. In order to have the latter, some additional information about U is necessary, its superharmonicity in a *larger* disk, for instance (this in turn implied by harmonicity of H in such a disk).

Regarding the measure μ appearing in either version of Riesz' result one has the important

Theorem. *In the representation*

$$U(z) \quad = \quad \int_{|\zeta - z_0| \leqslant r} \log \frac{1}{|z - \zeta|} \, d\mu(\zeta) \quad + \quad H(z), \qquad |z - z_0| < r,$$

of a function $U(z)$ superharmonic in and on $|z - z_0| \leqslant r$ (with μ positive on that disk and $H(z)$ harmonic in its interior), the measure μ has no mass in any open subset of the disk where $U(z)$ is harmonic.

Proof. Let $|z - z_0| < r$ and suppose that $U(z')$ is harmonic in and on the closed disk $|z' - z| \leqslant \rho$, where $0 < \rho < r - |z - z_0|$. By the mean value property we then have

$$\frac{1}{2\pi} \int_0^{2\pi} U(z + \rho e^{i\psi}) \, d\psi \;\; = \;\; U(z).$$

$H(z)$, however, has also the mean value property. Hence, using once again the formula

$$\frac{1}{2\pi} \int_0^{2\pi} \log \frac{1}{|z + \rho e^{i\psi} - \zeta|} \, d\psi \;\; = \;\; \min \left(\log \frac{1}{|z - \zeta|}, \; \log \frac{1}{\rho} \right)$$

together with the given representation for U, we see that the left-hand integral in the preceding relation equals

$$\int_{|\zeta - z_0| \leqslant r} \min \left(\log \frac{1}{|z - \zeta|}, \; \log \frac{1}{\rho} \right) d\mu(\zeta) \;\; + \;\; H(z).$$

Subtracting $U(z)$ from this, we get

$$\int_{|\zeta - z_0| \leqslant r} \log^+ \frac{\rho}{|\zeta - z|} \, d\mu(\zeta) \;\; = \;\; 0.$$

Therefore $\mu(\{|\zeta - z| < \rho\}) = 0$, μ being positive. This does it.

Problem 48

(a) In the Riesz representation

$$U(z) \;\; = \;\; \int_{|\zeta - z_0| \leqslant r} \log \frac{1}{|z - \zeta|} \, d\mu(\zeta) \;\; + \;\; H(z), \qquad |z - z_0| \;\; < \;\; r,$$

of a function $U(z)$ superharmonic in and on $|\zeta - z_0| \leqslant r$ (with the measure μ positive and $H(z)$ harmonic for $|\zeta - z_0| < r$), the *restriction of μ to the open disk $|\zeta - z_0| < r$ is unique*. (Hint: If $F(z)$ is any continuous function supported on a *compact subset* of the open disk in question, we have

$$F(\zeta) \;\; = \;\; \lim_{\rho \to 0} \frac{2}{\pi \rho^2} \int \int_{|z - z_0| < r} F(z) \log^+ \frac{\rho}{|\zeta - z|} \, dx \, dy$$

uniformly for $|\zeta - z_0| < r$.)

(b) In the representation for U written in (a), the function $H(z)$ harmonic for $|z - z_0| < r$ is *not unique* – it can be *altered* by letting μ have *more mass on the circle $|\zeta - z_0| = r$*. Show uniqueness for the function $h(z)$, harmonic in $\{|z - z_0| < r\}$, figuring in the *original form* of the Riesz representation of U in that disk:

$$U(z) = \int_{|\zeta - z_0| \leq r} \log \left| \frac{r^2 - (z - z_0)\overline{(\zeta - z_0)}}{r(z - \zeta)} \right| d\mu(\zeta) + h(z).$$

(c) Let $U(z)$, superharmonic in a domain $\mathcal{D}$, have the Riesz representation

$$U(z) = \int_{|\zeta - z_0| \leq r_0} \log \frac{1}{|z - \zeta|} d\mu_0(\zeta) + H_0(z),$$

$$U(z) = \int_{|\zeta - z_1| \leq r_1} \log \frac{1}{|z - \zeta|} d\mu_1(\zeta) + H_1(z),$$

with $H_0(z)$ and $H_1(z)$ harmonic, in the respective disks $\{|z - z_0| < r_0\}$, $\{|z - z_1| < r_1\}$, whose *closures* lie in $\mathcal{D}$. Show that the positive measures μ_0 and μ_1 *agree* on the *intersection* of those *open disks* as long as it is non-empty. (Hint: The method followed in part (a) may be used.)

The last part of this problem gives us a procedure for extending the Riesz representation from disks to more general domains – the pasting argument referred to earlier.

Problem 49

Let $U(z)$ be superharmonic in a domain $\mathcal{D}$, and let Ω be any smaller domain with *compact closure* lying in $\mathcal{D}$. Corresponding to each open disk Δ whose closure lies in $\mathcal{D}$ we have, by the Riesz representation, a positive measure μ_Δ on $\bar{\Delta}$ and a function H_Δ harmonic in Δ such that

$$U(z) = \int_{\bar{\Delta}} \log \frac{1}{|z - \zeta|} d\mu_\Delta(\zeta) + H_\Delta(z)$$

for $z \in \Delta$.

(a) Show that there is a (finite) positive measure μ on $\bar{\Omega}$ agreeing in each intersection $\bar{\Omega} \cap \Delta$ with the corresponding measure μ_Δ. (Hint: Use a finite covering of $\bar{\Omega}$ by some of the disks Δ and then refer to the result from problem 48(c).)

(b) Hence show that

$$U(z) = \int_{\bar{\Omega}} \log \frac{1}{|z - \zeta|} d\mu(\zeta) + H(z)$$

for $z \in \Omega$, where $H(z)$ is harmonic in that domain and μ is the measure obtained in (a). (Hint: It suffices to show that for each $z_0 \in \Omega$,

$$U(z) - \int_{\Delta_0} \log \frac{1}{|z - \zeta|} d\mu(\zeta)$$

is harmonic in Δ_0, a disk contained in Ω with centre at z_0.)

3. **A maximum principle for pure logarithmic potentials. Continuity of such a potential when its restriction to generating measure's support has that property**

Consider a function $U(z)$ superharmonic in and on $\{|z| \leqslant 1\}$ and thus having a Riesz representation

$$U(z) \;=\; \int_{|\zeta| \leqslant 1} \log \frac{1}{|z - \zeta|}\, d\mu(\zeta) \;+\; H(z)$$

(with μ positive and $H(z)$ harmonic) in the *open* unit disk. If U is actually *harmonic* in an open subset $\mathcal{O}$ of the latter, μ is in fact supported on the compact set

$$K \;=\; \{|\zeta| \leqslant 1\} \sim \mathcal{O}$$

according to the last theorem of the preceding article.

One is frequently interested in the *continuity* of $U(z)$ for $|z| < 1$. Because $H(z)$ is even harmonic for such z, the property in question is governed by the continuity of

$$\int_K \log \frac{1}{|z - \zeta|}\, d\mu(\zeta)$$

there. An important result of Evans and Vasilesco given in the present article guarantees the continuity of such a logarithmic potential (everywhere!), provided that *its restriction to the support K of μ enjoys that property*. This enables one to *exclude from consideration the open set $\mathcal{O}$ in which U is known to be harmonic* when checking for that function's continuity in the open unit disk.

The result referred to is based on a *version of the maximum principle*, of considerable interest in its own right.

Maria's theorem. *Let the (finite) positive measure μ be supported on a compact set K, and suppose that*

$$V(z) \;=\; \int_K \log \frac{1}{|z - \zeta|}\, d\mu(\zeta).$$

Then, if $V(z) \leqslant M$ at each $z \in K$, one has $V(z) \leqslant M$ in $\mathbb{C}$.

Remark. $V(z)$ is, of course *harmonic* in $\Omega = \mathbb{C} \sim K$ (and tends to $-\infty$ as $z \longrightarrow \infty$, unless $\mu \equiv 0$), but the theorem *does not follow without further work* from the ordinary maximum principle for harmonic functions. For $\zeta \in \partial\Omega \subseteq K$, all that the *elementary properties* of superharmonic functions tell us *directly* is that

$$\liminf_{z \to \zeta} V(z) = V(\zeta) \leqslant M.$$

If we only had limsup *on the left instead of* liminf, there would be no problem, but that's not what stands there! Such pitfalls abound in this subject.

Proof of theorem. We need only consider the situation where $M < \infty$, since otherwise the result is trivial. In that event, the quantities

$$V_\rho(z) = \int_K \min\left(\log\frac{1}{|z - \zeta|}, \log\frac{1}{\rho} \right) d\mu(\zeta)$$

increase, for *each* $z \in K$, to the *finite* limit $V(z)$ as $\rho \longrightarrow 0$. Given $\varepsilon > 0$, there is thus by *Egorov's theorem* a compact $E \subseteq K$ such that

$$V_\rho(z) \longrightarrow V(z) \quad \textit{uniformly for } z \in E$$

as $\rho \longrightarrow 0$, and

$$\mu(K \sim E) < \varepsilon.$$

Since $|z - \zeta| \leqslant \operatorname{diam} K$ for z and ζ in K, the *second* condition makes

$$\int_E \log\frac{1}{|z - \zeta|} d\mu(\zeta) \leqslant \int_K \log\frac{1}{|z - \zeta|} d\mu(\zeta) + (\log \operatorname{diam} K)\,\mu(K \sim E)$$

$$\leqslant V(z) + \varepsilon \log \operatorname{diam} K \leqslant M + \varepsilon \log \operatorname{diam} K$$

for $z \in K$, hence certainly for $z \in E$. By choosing $\varepsilon > 0$ small enough, we can ensure that the last quantity on the right, $M + \varepsilon \log \operatorname{diam} K$, *denoted henceforth by M'*, is as close as we like to M.

At the same time, when $z \notin K$,

$$\int_{K \sim E} \log\frac{1}{|z - \zeta|} d\mu(\zeta)$$

lies between

$$\varepsilon \log \frac{1}{\operatorname{dist}(z, K) + \operatorname{diam} K} \quad \text{and} \quad \varepsilon \log \frac{1}{\operatorname{dist}(z, K)}.$$

For any such fixed z, then,

$$\int_E \log\frac{1}{|z - \zeta|} d\mu(\zeta)$$

will be arbitrarily close to $V(z)$ when $\varepsilon > 0$ is sufficiently small (depending on z). We see, $\varepsilon > 0$ being arbitrary, that we will have $V(z) \leqslant M$ at each

$z \notin K$ (thus proving the theorem) *if we can deduce that*

$$\int_E \log \frac{1}{|z-\zeta|} \, d\mu(\zeta) \;\leqslant\; M'$$

outside E knowing that this holds everywhere on E.

The last implication looks just like the one affirmed by the theorem, so it may seem as though nothing has been gained. We nevertheless have more of a toehold here on account of the *first* condition on our set E, according to which

$$\int_K \log^+ \frac{\rho}{|z-\zeta|} \, d\mu(\zeta) \;=\; \int_K \log \frac{1}{|z-\zeta|} \, d\mu(\zeta)$$

$$- \int_K \min\left(\log \frac{1}{|z-\zeta|}, \; \log \frac{1}{\rho} \right) d\mu(\zeta)$$

tends to zero *uniformly for $z \in E$* as $\rho \to 0$. Thence, *a fortiori* (!),

$$\int_E \log^+ \frac{\rho}{|z-\zeta|} \, d\mu(\zeta) \;\longrightarrow\; 0 \quad \textit{uniformly for } z \in E$$

as $\rho \to 0$. This uniformity plays an essential rôle in the following argument.

It will be convenient to write

$$U(z) \;=\; \int_E \log \frac{1}{|z-\zeta|} \, d\mu(\zeta).$$

The proof of our theorem has boiled down to showing that if

$$U(z) \;\leqslant\; M' \quad \text{for } z \in E,$$

then $U(z)$ is *also* $\leqslant M'$ at each $z \notin E$.

This is where we use the maximum principle for harmonic functions. In $\mathbb{C} \sim E$, U is harmonic; also,

$$U(z) \;\longrightarrow\; -\infty \quad \text{as } z \to \infty$$

unless $\mu(E) = 0$, in which case the desired conclusion is obviously true. The principle of maximum will therefore make $U(z) \leqslant M'$ in $\mathbb{C} \sim E$ provided that $\limsup_{z \to z_0} U(z) \leqslant M'$ for each $z_0 \in E$ (cf. second corollary to the second lemma in article 1).

Take any $\delta > 0$; we wish to show that at each $z_0 \in E$,

$$U(z) \;<\; M' + 7\delta$$

for the points z in a neighborhood of z_0. Thanks to the uniformity arrived at in the preceding construction, we can *fix* a $\rho > 0$ such that

$$\int_E \log^+ \frac{\rho}{|z-\zeta|}\, d\mu(\zeta) \;\; < \;\; \delta$$

whenever $z \in E$. With such a ρ, which we can also take to be < 1, we have

$$U(z) \;\; = \;\; \int_E \min\left(\log\frac{1}{|z-\zeta|},\ \log\frac{1}{\rho}\right) d\mu(\zeta) \;\; + \;\; \int_E \log^+ \frac{\rho}{|z-\zeta|}\, d\mu(\zeta).$$

The first integral on the right is $\leqslant U(z)$ and hence $\leqslant M'$ for $z \in E$; it is, moreover, *continuous* in z. *That integral is therefore* $< M' + \delta$ *whenever z is sufficiently close to any $z_0 \in E$*; our task thus reduces to verifying that

$$\int_E \log^+ \frac{\rho}{|z-\zeta|}\, d\mu(\zeta) \;\; < \;\; 6\delta$$

for z close enough to such a z_0. The last relation holds in fact at *all* points z, as we now proceed to show with the help of an ingenious device used in Carleson's little book. The latter has the advantage of being applicable when the logarithmic potential kernel $\log(1/|z-\zeta|)$ is replaced by fairly general ones of the form $k(|z-\zeta|)$, and it can be used in $\mathbb{R}^n$ for $n > 2$ as well as in $\mathbb{R}^2$.

Fix any z. If $z \in E$, the integral in question is even $< \delta$ by choice of ρ, so we may suppose that $z \notin E$. Then, using z as vertex, we partition the complex plane into six sectors, each of $60°$ opening, and denote by $E_1, E_2, \ldots, E_6$ the respective intersections of E with those sectors (so as to have $E_1 \cup E_2 \cup \cdots \cup E_6 = E$).

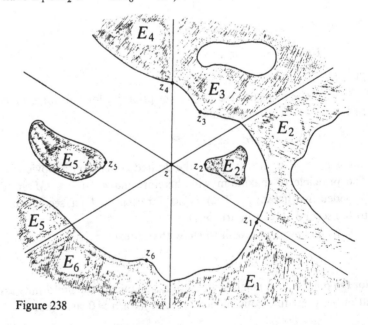

Figure 238

In each non-empty closure $\bar{E}_k$, $k = 1, 2, \ldots, 6$, pick a point z_k for which

$$|z_k - z| \;=\; \mathrm{dist}\,(z, E_k).$$

We have

$$\int_E \log^+ \frac{\rho}{|z - \zeta|}\, \mathrm{d}\mu(\zeta) \;\leqslant\; \sum_{k=1}^{6} \int_{E_k} \log^+ \frac{\rho}{|z - \zeta|}\, \mathrm{d}\mu(\zeta)$$

(with $\leqslant$ here and not $=$, because the E_k may intersect along the edges of the sectors*). However, for each k,

$$\int_{E_k} \log^+ \frac{\rho}{|z - \zeta|}\, \mathrm{d}\mu(\zeta) \;\leqslant\; \int_{E_k} \log^+ \frac{\rho}{|z_k - \zeta|}\, \mathrm{d}\mu(\zeta),$$

since

$$|z - \zeta| \;\geqslant\; |z_k - \zeta| \quad \text{when } \zeta \in E_k,$$

as one sees from the following diagram, drawn for $k = 6$:

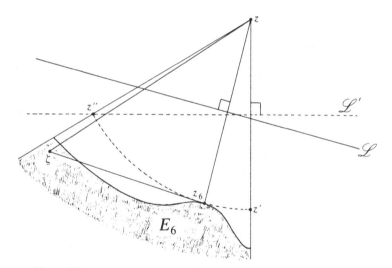

Figure 239

Here, $\mathscr{L}$ is the perpendicular bisector of the segment $[z, z_6]$ and $\mathscr{L}'$ that of $[z, z']$. Any point ζ in E_6 lies *on the same side* of $\mathscr{L}$ as z_6 and *on the opposite side thereof* from z, so the last inequality must hold. (By imagining z_6 to coincide with z' – in which case $\mathscr{L}$, coinciding with $\mathscr{L}'$, would pass

---

* if, for instance, we work with *closed* 60° sectors (which we may just as well do), in which case the sets E_k are *already* closed.

through z'' – and then allowing ζ to approach z'', we see that the $60°$ opening of the sector $\widehat{z''zz'}$ *cannot be made larger* if the inequality in question is to hold for all $\zeta \in E_6$.)

Again, for $1 \leqslant k \leqslant 6$,

$$\int_{E_k} \log^+ \frac{\rho}{|z_k - \zeta|}\, d\mu(\zeta) \;\leqslant\; \int_E \log^+ \frac{\rho}{|z_k - \zeta|}\, d\mu(\zeta).$$

But the right-hand integral is $< \delta$ by choice of ρ since $z_k \in \bar{E}_k \subseteq E$! Thence, going back to the previous relations, we find that

$$\int_E \log^+ \frac{\rho}{|z - \zeta|}\, d\mu(\zeta) \;<\; 6\delta$$

as we set out to show.

As explained above, this implies that $U(z) < M' + 7\delta$ in a suitably small neighborhood of any $z_0 \in E$ and thus finally, that $U(z) \leqslant M'$ in $\mathbb{C} \sim E$, after squeezing δ. From that, however, our result follows as we saw at the beginning of this proof. We are done.

Problem 50

With K a *compact* subset of the open (*sic!*) unit disk Δ and μ a positive measure supported on K, put

$$V(z) \;=\; \int_K \log \left| \frac{1 - \bar{\zeta}z}{z - \zeta} \right| d\mu(\zeta), \qquad |z| \leqslant 1.$$

Suppose that V is *finite* at each point of K. Show then that if $W(z)$ is superharmonic and $\geqslant 0$ in Δ, and satisfies

$$W(z) \;\geqslant\; V(z)$$

for $z \in K$, we have $V(z) \leqslant W(z)$ in Δ.

Remark. The finiteness of V at the points of μ's support cannot be dispensed with here. Consider, for example,

$$V(z) \;=\; \log \frac{1}{|z|}$$

and

$$W(z) \;=\; \frac{1}{2}\log \frac{1}{|z|} \quad !$$

(Hint: Argue first as in the above proof to get, for any given $\varepsilon > 0$, a compact subset E of K with

$$\mu(K \sim E) \;<\; \varepsilon$$

and

$$\int_E \log^+ \frac{\rho}{|z - \zeta|} \, d\mu(\zeta) \;\longrightarrow\; 0$$

uniformly for $z \in E$ as $\rho \longrightarrow 0$.

Put

$$U(z) \;=\; \int_E \log\left|\frac{1 - z\bar{\zeta}}{z - \zeta}\right| d\mu(\zeta);$$

here, $U(z) \leqslant V(z)$, so in particular $U(z) \leqslant W(z)$ on $E \subseteq K$. For any fixed $z \in \Delta \sim K$, $\; V(z) - U(z)$ is *small* if ε is, so it is enough to show that $U(z) \leqslant W(z)$ at each $z \in \Delta \sim E$.

The difference $W(z) - U(z)$ is superharmonic in $\Delta \sim E$; the last relation therefore holds (by a corollary from article 1) provided that

$$\liminf_{z \to z_0} (W(z) - U(z)) \;\geqslant\; 0$$

for each $z_0 \in \partial(\Delta \sim E)$.

When $|z_0| = 1$, this is manifest, W being $\geqslant 0$ in Δ with (here) $V(z)$ and $U(z)$ *continuous* and *zero* at z_0. It is hence only necessary to look at the behaviour near points $z_0 \in E$.

Fix any such z_0, and take any $\delta > 0$. Reasoning as in the above proof, show that

$$U(z) \;<\; U(z_0) \,+\, 7\delta$$

in a sufficiently small neighborhood of z_0. Since $W(z_0) \geqslant U(z_0)$, we therefore have

$$W(z) - U(z) \;>\; -8\delta$$

in such a neighborhood.)

We come now to the result about continuity spoken of at the beginning of this article.

Theorem (due independently to Evans and to Vasilesco). *Given a positive measure μ supported on a compact set K, put*

$$V(z) \;=\; \int_K \log\frac{1}{|z - \zeta|} \, d\mu(\zeta).$$

If the restriction of V to K is continuous at a point $z_0 \in K$, $\; V(z)$ (as a function defined in $\mathbb{C}$) is continuous at z_0.

Proof. Given $\varepsilon > 0$, there is an $\eta > 0$ (which we *fix*) such that

$$|V(z) - V(z_0)| \;\leqslant\; \varepsilon \quad \text{for } z \in K \text{ with } |z - z_0| \leqslant \eta.$$

Consider, on the *compact* set

$$K_\eta \;=\; K \cap \{|z - z_0| \leqslant \eta\}$$

the *continuous* functions

$$F_\rho(z) \;=\; \min\left\{ V(z_0) \;-\; 2\varepsilon, \;\; \int_K \min\left(\log\frac{1}{|z - \zeta|}, \; \log\frac{1}{\rho} \right) d\mu(\zeta) \right\},$$

defined for each $\rho > 0$. When ρ *diminishes* towards 0, $F_\rho(z)$ *increases* for each fixed z, tending, moreover, to $\min(V(z_0) - 2\varepsilon, V(z))$, equal to the constant $V(z_0) - 2\varepsilon$ for $z \in K_\eta$. According to *Dini's theorem*, the convergence must then be *uniform* on K_η, so, for all sufficiently small $\rho > 0$, we have

$$\int_K \min\left(\log\frac{1}{|z - \zeta|}, \; \log\frac{1}{\rho} \right) d\mu(\zeta) \;>\; V(z_0) \;-\; 3\varepsilon, \qquad z \in K_\eta.$$

The integral on the left is, however,

$$\leqslant \;\; \int_K \log\frac{1}{|z - \zeta|} \, d\mu(\zeta),$$

which is in turn $\leqslant V(z_0) + \varepsilon$ for $z \in K_\eta$; subtraction thus yields

$$\int_K \log^+ \frac{\rho}{|z - \zeta|} \, d\mu(\zeta) \;<\; 4\varepsilon, \qquad z \in K_\eta,$$

for $\rho > 0$ sufficiently small.

Fix any such $\rho < \eta/2$. We desire to use Maria's theorem so as to take advantage of the relation just found, but the appearance of $\log^+$ in the integrand instead of the logarithm gives rise to a slight difficulty.

Taking a new parameter λ with $1 < \lambda < 2$, we bring in the set

$$K_{\lambda\rho} \;=\; K \cap \{|z - z_0| \leqslant \lambda\rho\}.$$

Since $\lambda\rho < 2\rho < \eta$, we have $K_{\lambda\rho} \subseteq K_\eta$, so surely

$$\int_K \log^+ \frac{\rho}{|z - \zeta|} \, d\mu(\zeta) \;<\; 4\varepsilon$$

for $z \in K_{\lambda\rho}$, whence, *a fortiori*,

$$\int_{K_{\lambda\rho}} \log\frac{\rho}{|z - \zeta|} \, d\mu(\zeta) \;\leqslant\; \int_{K_{\lambda\rho}} \log^+ \frac{\rho}{|z - \zeta|} \, d\mu(\zeta) \;<\; 4\varepsilon$$

when z is in $K_{\lambda\rho}$.

Thence, applying Maria's result to the integral

$$\int_{K_{\lambda\rho}} \log\frac{\rho}{|z-\zeta|}\,d\mu(\zeta)$$

(which differs by but an additive constant from

$$\int_{K_{\lambda\rho}} \log\frac{1}{|z-\zeta|}\,d\mu(\zeta)\ \Big),$$

we see that it is in fact $\leqslant 4\varepsilon$ for *all z*. From this we will now deduce that

$$\int_{K_{\lambda\rho}} \log^{+}\frac{\rho}{|z-\zeta|}\,d\mu(\zeta)\ <\ 5\varepsilon$$

(with $\log^{+}$ again and not log!) whenever z is sufficiently close to z_0, provided that $\lambda > 1$ is taken near enough to 1.

We have

$$\log^{+}\frac{\rho}{|z-\zeta|}\ =\ \log\frac{\rho}{|z-\zeta|}\ +\ \log^{-}\frac{\rho}{|z-\zeta|}.$$

Here, when $\zeta \in K_{\lambda\rho}$ and $|z-z_0| \leqslant (\lambda-1)\rho$, we are assured that $|z-\zeta| \leqslant (2\lambda-1)\rho$, making

$$\log^{-}\frac{\rho}{|z-\zeta|}\ \leqslant\ \log(2\lambda-1).$$

Figure 240

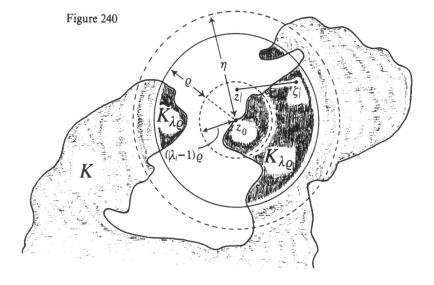

Therefore, for $|z - z_0| \leqslant (\lambda - 1)\rho$,

$$\int_{K_{\lambda\rho}} \log^+ \frac{\rho}{|z - \zeta|} \, d\mu(\zeta) \; \leqslant \; \int_{K_{\lambda\rho}} \log \frac{\rho}{|z - \zeta|} \, d\mu(\zeta) \; + \; \mu(K_{\lambda\rho}) \log(2\lambda - 1).$$

By choosing (and then fixing) $\lambda > 1$ *close enough to* 1, we ensure that the second term on the right is $< \varepsilon$; since, then, the first is $\leqslant 4\varepsilon$ as we have seen, we get

$$\int_{K_{\lambda\rho}} \log^+ \frac{\rho}{|z - \zeta|} \, d\mu(\zeta) \; < \; 5\varepsilon \quad \text{for } |z - z_0| \leqslant (\lambda - 1)\rho.$$

Now, when $|z - z_0| \leqslant (\lambda - 1)\rho$ and $\zeta \in K \sim K_{\lambda\rho}$, making $|\zeta - z_0| > \lambda\rho$, we have (see the preceding picture)

$$|z - \zeta| \; > \; \rho,$$

so

$$\log^+ \frac{\rho}{|z - \zeta|} \; = \; 0.$$

For $|z - z_0| \leqslant (\lambda - 1)\rho$, the integral in the last relation is thus equal to

$$\int_K \log^+ \frac{\rho}{|z - \zeta|} \, d\mu(\zeta),$$

which is hence $< 5\varepsilon$ then!

Let us return to $V(z)$, which can be expressed as

$$\int_K \min\left(\log \frac{1}{|z - \zeta|}, \; \log \frac{1}{\rho} \right) d\mu(\zeta) \; + \; \int_K \log^+ \frac{\rho}{|z - \zeta|} \, d\mu(\zeta).$$

When z is close enough to z_0 the *second* term is $< 5\varepsilon$ as we have just shown; the *first* term, however, is *continuous* in z, and hence tends to

$$\int_K \min\left(\log \frac{1}{|z_0 - \zeta|}, \; \log \frac{1}{\rho} \right) d\mu(\zeta) \; \leqslant \; V(z_0)$$

as $z \longrightarrow z_0$. Therefore,

$$V(z) \; < \; V(z_0) \; + \; 6\varepsilon$$

for z sufficiently close to z_0.

At the same time, V is superharmonic, so by property (i) (!),

$$\liminf_{z \to z_0} V(z) \; \geqslant \; V(z_0).$$

Thus,

$$V(z) \longrightarrow V(z_0) \quad \text{as} \quad z \longrightarrow z_0$$

since $\varepsilon > 0$ was arbitrary; the function V is thus continuous at z_0.

Q.E.D.

Corollary. *Let $U(z)$ be superharmonic in the unit disk, Δ, and harmonic in the open subset Ω thereof. If $z_0 \in \Delta \sim \Omega$ and the restriction of U to $\Delta \sim \Omega$ is continuous at z_0, $U(z)$ is continuous at z_0.*

Proof. Pick any r with $|z_0| < r < 1$; then, by the Riesz representation theorem from the preceding article,

$$U(z) \;\; = \;\; \int_{|\zeta| \leqslant r} \log \frac{1}{|z - \zeta|} \, \mathrm{d}\mu(\zeta) \;\; + \;\; H(z)$$

for $|z| < r$, where $H(z)$ is harmonic for such z and μ is a positive measure. We know also from the *last* theorem of that article that

$$\mu(\Omega \cap \{|\zeta| < r\}) \;\; = \;\; 0;$$

taking, then, the compact set

$$K \;\; = \;\; (\{|\zeta| \leqslant r\} \cap \sim \Omega) \cup \{|\zeta| = r\},$$

we can write

$$U(z) \;\; = \;\; \int_K \log \frac{1}{|z - \zeta|} \, \mathrm{d}\mu(\zeta) \;\; + \;\; H(z), \qquad |z| < r.$$

Here, since $|z_0| < r$, H is continuous at z_0, and the *restriction* of $U(z) - H(z)$ to K is also, by hypothesis. We thus arrive at the desired result by applying the theorem to

$$\int_K \log \frac{1}{|z - \zeta|} \, \mathrm{d}\mu(\zeta).$$

Done.

Problem 51

Let μ be a positive measure supported on K, a compact subset of the open unit disk, and suppose that

$$V(z) \;\; = \;\; \int_K \log \left| \frac{1 - \bar{\zeta}z}{z - \zeta} \right| \, \mathrm{d}\mu(\zeta)$$

is finite at each point of K. Show that there is a sequence of positive measures μ_n supported on K for which:

(i) $d\mu_n(\zeta) \leqslant d\mu_{n+1}(\zeta) \leqslant d\mu(\zeta)$ for each n;

(ii) $\mu(K) - \mu_n(K) \xrightarrow[n]{} 0$;

(iii) each of the functions

$$V_n(z) = \int_K \log\left|\frac{1-\bar{\zeta}z}{z-\zeta}\right| d\mu_n(\zeta)$$

is continuous on $\bar{\Delta}$;

(iv) $V_n(z) \xrightarrow[n]{} V(z)$ for each $z \in \bar{\Delta}$.

(Hint: Start by arguing as in the proof of Maria's theorem, getting compact subsets K_n of K with $\mu(K \sim K_n) < 1/n$, on each of which the convergence of

$$\int_K \min\left(\log\left|\frac{1-\bar{\zeta}z}{z-\zeta}\right|, \log\frac{|1-\bar{\zeta}z|}{\rho}\right) d\mu(\zeta)$$

to $V(z)$ for ρ tending to zero is *uniform*. This makes the restriction of V to each K_n continuous thereon.

Arranging matters so as to have $K_n \subseteq K_{n+1}$ for each n, define μ_n by putting $\mu_n(E) = \mu(E \cap K_n)$ for $E \subseteq K$. Each of the differences $\sigma_n = \mu - \mu_n$ is also a positive measure on K.

We have (with V_n as in (iii)),

$$V_n(z) = V(z) - \int_K \log\left|\frac{1-\bar{\zeta}z}{z-\zeta}\right| d\sigma_n(\zeta),$$

where the integral on the right (*without* the $-$ sign) is superharmonic in Δ. Hence, since V, restricted to any of the K_n, is continuous thereon, we see that

$$\limsup_{\substack{z \to z_0 \\ z \in K_n}} V_n(z) \leqslant V_n(z_0) \quad \text{for } z_0 \in K_n.$$

On the other hand,

$$\liminf_{z \to z_0} V_n(z) \geqslant V_n(z_0).$$

The restriction of V_n to K_n, the support of μ_n, is thus continuous. Now apply the preceding theorem.

Observe finally that

$$d\mu_n(\zeta) = \chi_{K_n}(\zeta) d\mu(\zeta)$$

with, for each fixed $z_0 \in \bar{\Delta}$,

$$\log\left|\frac{1-\bar{\zeta}z_0}{z_0-\zeta}\right| \chi_{K_n}(\zeta) \longrightarrow \log\left|\frac{1-\bar{\zeta}z_0}{z_0-\zeta}\right| \quad \text{a.e. } (\mu)$$

on K as $n \longrightarrow \infty$. This makes $V_n(z_0) \xrightarrow[n]{} V(z_0)$ by monotone convergence.)

B.   Relation of the existence of multipliers to the finiteness of a superharmonic majorant

1.   Discussion of a certain regularity condition on weights

We return to the question formulated somewhat loosely at the beginning of §A in the last chapter, which was to find the conditions a weight $W(x) \geqslant 1$ must fulfill beyond the necessary one that

$$\int_{-\infty}^{\infty} \frac{\log W(x)}{1 + x^2}\, dx \;<\; \infty,$$

in order to ensure the existence of entire functions $\varphi(z) \not\equiv 0$ of arbitrarily small exponential type making $W(x)\varphi(x)$ bounded (for instance) on $\mathbb{R}$. These must be conditions pertaining to the regularity of $W(x)$. Although an explicit minimal description of the needed regularity is not available as I write this, it seems likely that two separate requirements are involved.

One, not particularly bound up with the matter now under discussion, would serve to rule out the purely local idiosyncrasies in W's behaviour that could spoil existence of the above mentioned functions φ making $W(x)\varphi(x)$ bounded on $\mathbb{R}$ when such φ with, for example,

$$\int_{-\infty}^{\infty} |W(x)\varphi(x)|^p\, dx \;<\; \infty \qquad \text{(for some } p > 0\text{)}$$

were forthcoming. A very simple illustration helps to clarify this idea.

Consider any weight $W_0(x) \geqslant 1$ for which a non-zero entire function φ_0 of exponential type $A < \pi$ with $W_0(x)|\varphi_0(x)| \leqslant 1$ on $\mathbb{R}$ is known to exist.

Unless $W_0(x)$ is already bounded (a case without interest for us here!), $\varphi_0(z)$ must differ from a pure exponential and hence have infinitely many zeros (coming from its Hadamard factorization). Dividing out any two of those then gives us a new non-zero entire function φ, also of exponential type A, for which $W_0(x)|\varphi(x)| \leqslant \text{const.}/(1 + x^2)$, $x \in \mathbb{R}$, so that

$$\int_{-\infty}^{\infty} W_0(x)|\varphi(x)|\, dx \;<\; \infty.$$

Taking the new weight

$$W(x) \;=\; |\sin \pi x|^{-1/2} W_0(x)$$

which becomes infinite at each integer, we still have

$$\int_{-\infty}^{\infty} W(x)|\varphi(x)|\,\mathrm{d}x \;\; < \;\; \infty.$$

There is, however, no longer any entire $\psi(z) \not\equiv 0$ of exponential type $< \pi$ for which $W(x)\psi(x)$ is bounded on $\mathbb{R}$. Indeed, such a function ψ would have to be bounded on $\mathbb{R}$ and hence satisfy $|\psi(z)| \leqslant \mathrm{const.\,exp}\,(B|\Im z|)$, with $B < \pi$, by the third Phragmén–Lindelöf theorem in §C of Chapter III. In the present circumstances, ψ would also have to vanish at each integer, but then the usual application of Jensen's formula would show that it must vanish identically. Starting with a weight W_0 for which entire $\varphi_0 \not\equiv 0$ of *arbitrarily small exponential type* $A > 0$ with $\varphi_0(x)W_0(x)$ bounded on $\mathbb{R}$ are available, we thus find ourselves in a situation where – adopting the language of §A, Chapter X – the related weight $W(x)$ *admits multipliers in* $L_1(\mathbb{R})$ *but not in* $L_\infty(\mathbb{R})$.

By such rather artificial and almost trivial constructions one obtains various weights W from the original W_0 that admit multipliers in some spaces $L_p(\mathbb{R})$ but not in others. This seems to have nothing to do with the real reason (whatever it may be) for W_0 to have admitted multipliers (in $L_\infty(\mathbb{R})$) to begin with. That must also be the reason why the weights W admit multipliers in certain of the $L_p(\mathbb{R})$, and thus probably involves *some property of behaviour common to* $W_0(x)$ *and all of the* $W(x)$, independent of the special irregularities introduced in passing from the former to the latter. If this is so, it is natural to think of that behaviour property as *the essential one* governing admittance of multipliers, and the *second regularity condition* for weights would be *that they possess it*. By the *first regularity condition*, weights like $|\sin \pi x|^{-1/2}W_0(x)$ would be ruled out.

From this point of view, a search for the presumed essential second condition appears to be of primary importance. In order to be unhindered in that search, one is motivated to start by imposing on the weights W some imperfect version of the first condition, stronger than needed*, rather than seeking to express the latter in minimal form. That is how we will proceed here.

Such a version of the first condition should be both simple and sufficiently general. One, given in Beurling and Malliavin's 1962 paper, is very mild but rather elaborate. Discussion of it is postponed to the

---

* even at the cost of then arriving at a less than fully general version of the second condition

scholium at the end of this article. The following simpler variant seems adequate for most purposes; it is easy to work with and still applicable to a broad class of weights.

▶ **Regularity requirement.** *There are three strictly positive constants, L, C and α such that, for each $x \in \mathbb{R}$, one has a real interval J_x of length L containing x with*

$$W(t) \;\geqslant\; C(W(x))^{\alpha} \qquad \text{for } t \in J_x.$$

(Unless $W(x)$ is bounded – a case without interest for us here – the parameter α figuring in the condition must obviously be $\leqslant 1$.) *Much of the work in the present chapter will be limited to the weights W that meet this requirement.*[*]

What our condition does is impose a weak kind of *uniform semicontinuity* on $\log^{+}\log W(x)$. It implies, for instance, a certain boundedness property on finite intervals.

Lemma. *A weight $W(x)$ meeting the regularity requirement is either identically infinite on some interval of length L or else bounded above on every finite interval.*

Proof. Suppose that $-M \leqslant x_n \leqslant M$ and $W(x_n) \xrightarrow[n]{} \infty$. Wlog, let

$$x_n \xrightarrow[n]{} x_0.$$

To each x_n is associated an interval J_{x_n} of length L containing it, on which

$$W(x) \;\geqslant\; C(W(x_n))^{\alpha}.$$

For infinitely many values of n, J_{x_n} must extend *to the same side of x_n* (either to the right or to the left) by a distance $\geqslant L/2$. Assuming, wlog, that we have infinitely many such intervals extending by that amount *to the right* of the corresponding points x_n, we see that

$$x_0' \;=\; x_0 + \frac{L}{4}$$

lies in *infinitely many of them*. The preceding relation therefore makes $W(x_0') = \infty$. Then, however, $W(x) = \infty$ for the x belonging to the interval $J_{x_0'}$ of length L.

---

[*] Regarding its partial elimination, see Remark 5 near the end of §E.2.

Here are some of the ways in which weights fulfilling the regularity requirement arise.

Lemma. *If* $\Omega(t) \geqslant 0$, *the average*

$$W(x) \;=\; \frac{1}{2L}\int_{-L}^{L}\Omega(x+t)\,\mathrm{d}t$$

satisfies the requirement with parameters L, $C = 1/2$ *and* $\alpha = 1$.

Proof. Given any x, we have

$$\frac{1}{2L}\int_{J}\Omega(t)\,\mathrm{d}t \;\geqslant\; \frac{1}{2}W(x)$$

for an interval J equal to *one* of the two segments $[x-L,\ x]$, $[x,\ x+L]$. Taking that interval J as J_x, we then have

$$W(\xi) \;=\; \frac{1}{2L}\int_{-L}^{L}\Omega(\xi+t)\,\mathrm{d}t \;\geqslant\; \frac{1}{2L}\int_{J_x}\Omega(s)\,\mathrm{d}s \;\geqslant\; \frac{1}{2}W(x)$$

for each $\xi \in J_x$.

In like manner, one verifies:

Lemma. *If* $\Omega(t) \geqslant 0$ *and* $p > 0$ *(sic!)*,

$$W(x) \;=\; \left(\frac{1}{2L}\int_{-L}^{L}(\Omega(x+t))^p\,\mathrm{d}t\right)^{1/p}$$

satisfies the requirement with parameters L, $C = 2^{-1/p}$ *and* $\alpha = 1$.

Lemma. *If* $\Omega(t) \geqslant 1$, *the weight*

$$W(x) \;=\; \exp\left\{\frac{1}{2L}\int_{-L}^{L}\log\Omega(x+t)\,\mathrm{d}t\right\}$$

satisfies the requirement with parameters L, $C = 1$ *and* $\alpha = 1/2$.

Weights meeting the requirement are also obtained by use of the Poisson kernel:

Lemma. *Let* $\Omega(t) \geqslant 1$ *be such that*

$$\int_{-\infty}^{\infty}\frac{\log\Omega(t)}{1+t^2}\,\mathrm{d}t \;<\; \infty.$$

Then, for fixed y > 0, the weight

$$W(x) = \exp\left\{\frac{1}{\pi}\int_{-\infty}^{\infty}\frac{y\log\Omega(t)}{(x-t)^2+y^2}\,dt\right\}$$

fulfills the requirement with parameters L, C = 1 and $\alpha = e^{-L/2y}$.

Proof. Since $\log\Omega(t) \geqslant 0$, we have (Harnack!)

$$\left|\frac{d\log W(x)}{dx}\right| \leqslant \frac{1}{y}\log W(x),$$

so that

$$\log W(\xi) \geqslant (\log W(x))e^{-|\xi-x|/y}.$$

Take $J_x = [x-L/2,\ x+L/2]$.

A weight meeting the regularity requirement and also admitting multipliers has a $\mathscr{C}_\infty$ *majorant* with the same properties.

Theorem. *Let* $W(x) \geqslant 1$ *fulfill the requirement with parameters L, C, and* α, *and suppose that*

$$\int_{-\infty}^{\infty}\frac{\log W(t)}{1+t^2}\,dt < \infty.$$

There is then an infinitely differentiable weight $W_1(x) \geqslant W(x)$ *also meeting the requirement such that, corresponding to any entire function* $\varphi(z) \not\equiv 0$ *of exponential type* $\leqslant A$ *making* $W(x)|\varphi(x)| \leqslant 1$ *on* $\mathbb{R}$, *one has an entire* $\psi(z) \not\equiv 0$ *of exponential type* $\leqslant mA$ *with* $W_1(x)|\psi(x)| \leqslant$ *const.,* $x \in \mathbb{R}$. *Here, for m we can take any integer* $\geqslant 4/\alpha$.

Remark. As we know, the integral condition on $\log W$ follows from the existence of just *one* entire function φ having the properties in question.

Proof of theorem. Any entire function φ satisfying the conditions of the hypothesis must in particular have modulus $\leqslant 1$ on the real axis, so, by the second theorem of §G.2, Chapter III,

$$\log|\varphi(z)| \leqslant A\Im z + \frac{1}{\pi}\int_{-\infty}^{\infty}\frac{\Im z\log|\varphi(t)|}{|z-t|^2}\,dt$$

for $\Im z > 0$. Adding to both sides the finite quantity

$$\frac{1}{\pi}\int_{-\infty}^{\infty}\frac{\Im z\log W(t)}{|z-t|^2}\,dt$$

we see, remembering the given relation

$$\log|\varphi(t)| + \log W(t) \leqslant 0, \qquad t \in \mathbb{R},$$

that

$$\log|\varphi(z)| + \frac{1}{\pi}\int_{-\infty}^{\infty}\frac{\Im z \log W(t)}{|z-t|^2}\,dt \leqslant A\Im z, \qquad \Im z > 0.$$

Put now $z = x + iL$, and use the fact that

$$\log W(t) \geqslant \alpha \log W(x) + \log C$$

for t belonging to an interval of length L containing the point x. Since $\log W(t) \geqslant 0$, the integral on the left comes out

$$\geqslant \frac{1}{4}(\alpha \log W(x) + \log C),$$

and we find that

$$\frac{4}{\alpha}\log|\varphi(x+iL)| + \log W_1(x) \leqslant \text{const.}, \qquad x \in \mathbb{R},$$

where

$$W_1(x) = C^{-1/\alpha}\exp\left\{\frac{4}{\pi\alpha}\int_{-\infty}^{\infty}\frac{L\log W(t)}{(x-t)^2+L^2}\,dt\right\}$$

is certainly $\geqslant W(x)$. This function is, on the other hand, infinitely differentiable, and it satisfies the regularity requirement by the last lemma.

At the same time,

$$W_1(x)|\varphi(x+iL)|^{4/\alpha} \leqslant \text{const.}, \qquad x \in \mathbb{R}.$$

Because φ is bounded on the real axis, we know by the third Phragmén–Lindelöf theorem of §C, Chapter III that $\varphi(x+iL)$ is also bounded for $x \in \mathbb{R}$. Hence, taking any integer $m \geqslant 4/\alpha$, we have

$$W_1(x)|\psi(x)| \leqslant \text{const.}, \qquad x \in \mathbb{R},$$

with the entire function

$$\psi(z) = (\varphi(z+iL))^m,$$

obviously of exponential type $\leqslant mA$.

Done.

The elementary result just proved permits us to restrict our attention to *infinitely differentiable weights* when searching for the form of the 'essential' second condition that those meeting the regularity requirement must satisfy in order to admit multipliers. This observation will play a rôle in the last two §§ of the present chapter. But the main service rendered by the requirement is to make the property of admitting multipliers reduce to a more general one, easier to work with, for weights fulfilling it.

In order to explain what is meant by this, let us first consider the situation where an entire function $\varphi(z) \not\equiv 0$ of exponential type $\leqslant A$ with $W(x)|\varphi(x)| \leqslant$ const. on $\mathbb{R}$ is *known to exist*. If the weight $W(x)$ is *even*, some details of the following discussion may be skipped, making it *shorter* (although not really *easier*). One can in fact stick to just even weights (and even functions $\varphi(z)$) and still *get by* – see the remark following the last theorem in this article – and the reader is invited to make this simplification if he or she wants to. We treat the general case here in order to show that such investigations do not become *that much harder* when *evenness* is *abandoned*.

Assume that $W(x) \geqslant 1$ is either *continous*, or fulfills the *regularity requirement* (of course, one property does not imply the other). Then, since $\varphi(z) \not\equiv 0$, $W(x)$ cannot be identically infinite on any interval of length > 0. By the first of the above lemmas, this means that $W(x)$ is *bounded on finite intervals* under the second assumption. The same is of course true in the event of the first assumption.

The function $W(x)$ is, in particular, bounded near the origin, so if $\varphi(z)$ has a zero there – of order k, say – the product $W(x)\varphi(x)/x^k$ will still be bounded on $\mathbb{R}$. We can, in other words, assume wlog that $\varphi(0) \neq 0$, and hence that φ has a Hadamard factorization of the form

$$\varphi(z) = Ce^{\gamma z}\prod_{\lambda}\left(1 - \frac{z}{\lambda}\right)e^{z/\lambda}.$$

Following a procedure already familiar to us, we construct from the product on the right a new entire function $\psi(z)$ *having only real zeros* (cf. §H.3 of Chapter III and the first half of the proof of the second Beurling–Malliavin theorem, §B.3, Chapter X).

Denote by Λ the set of zeros λ figuring in the above product with $\Re\lambda \neq 0$. For each $\lambda \in \Lambda$ we put

$$\frac{1}{\lambda'} = \Re\left(\frac{1}{\lambda}\right);$$

this gives us real numbers λ' with $|\lambda'| \geqslant |\lambda|$. (It is understood here that each λ' is to be taken with a multiplicity equal to the number of times that the corresponding $\lambda \in \Lambda$ figures as a zero of φ.) The *number N(r)* of points λ' having modulus $\leqslant r$ (counting multiplicities) is thus *at most* equal to the *total numbers of zeros with such modulus that φ has* (again counting multiplicities). The latter quantity has, however, *asymptotic behaviour for large r governed by Levinson's theorem* (§H.3, Chapter III), because $\varphi(x)$ must be *bounded* on $\mathbb{R}$, $W(x)$ being $\geqslant 1$. In this way we see that that quantity is $\sim 2A'r/\pi$ for $r \longrightarrow \infty$, where A' is some positive number $\leqslant$ the *type* of φ, and hence $\leqslant A$.

We therefore have

$$\frac{N(r)}{r} \leqslant \frac{2A'}{\pi} + o(1)$$

for large r.

This being so, the product

$$e^{z\,\Re\gamma} \prod_{\lambda \in \Lambda} \left(1 - \frac{z}{\lambda'}\right) e^{z/\lambda'}$$

(with each factor repeated according to the multiplicity of the corresponding λ') is *convergent* (see §A, Chapter III) and hence equal to some entire function $\psi(z)$. We know from §B of Chapter III, however, that the preceding relation involving $N(r)$ is insufficient to ensure ψ's being of exponential type. In order to show that, we resort to an indirect argument (cf. §H.3, Chapter III).

What the condition on $N(r)$ does give is the estimate $\log|\psi(z)| \leqslant O(|z|\log|z|)$, valid for large $|z|$ (§B, Chapter III); we thus have

$$\log|\psi(z)| \leqslant O(|z|^{1+\varepsilon})$$

(with arbitrary $\varepsilon > 0$) for z with large modulus. At the same time, $\psi(x)$ is bounded on the real axis. Indeed, for $\lambda \in \Lambda$,

$$\left|\left(1 - \frac{x}{\lambda}\right)e^{x/\lambda}\right| \geqslant \left|1 - \frac{x}{\lambda'}\right|e^{x/\lambda'}, \qquad x \in \mathbb{R},$$

whereas, for any *purely imaginary zero* λ of φ,

$$\left|\left(1 - \frac{x}{\lambda}\right)e^{x/\lambda}\right| = \sqrt{\left(1 + \frac{x^2}{|\lambda|^2}\right)} \geqslant 1, \qquad x \in \mathbb{R}.$$

Comparison of the above product equal to $\psi(z)$ with the Hadamard representation for φ thus shows at once that $|C\psi(x)| \leqslant |\varphi(x)|$ for $x \in \mathbb{R}$, yielding

$$|\psi(x)| \leqslant \text{const.}, \qquad x \in \mathbb{R},$$

since such a relation holds for $\varphi(x)$.

On the *imaginary axis*, the above estimate on $\log|\psi(z)|$ can be improved. We have:

$$\log|\psi(iy)| = \frac{1}{2}\sum_{\lambda \in \Lambda}\log\left(1 + \frac{y^2}{(\lambda')^2}\right) = \frac{1}{2}\int_0^\infty \log\left(1 + \frac{y^2}{r^2}\right)dN(r)$$

$$= |y|\int_0^\infty \frac{|y|}{t^2 + y^2}\frac{N(r)}{r}dr$$

(note that $N(r) = 0$ for $r > 0$ close to zero). Plugging the above inequality for $N(r)$ into the last integral, we see immediately that

$$\limsup_{y \to \pm\infty} \frac{\log|\psi(iy)|}{|y|} \leqslant A'.$$

Use this relation together with the two previous estimates on ψ to make a Phragmén–Lindelöf argument in each of the quadrants I, II, III and IV. One finds as in §H.3 of Chapter III that

$$|\psi(z)| \leqslant \text{const.}e^{A'|\Im z|}.$$

Thus, since $A' \leqslant A$, $\psi(z)$ is *of exponential type* $\leqslant A$ (as our original function φ was).

This argument has been given at length because it will be used again later on. Then we will simply *refer* to it, omitting the details.

Let us return to our weight $W(x)$. Since, as we have seen, $|\psi(x)| \leqslant \text{const.}|\varphi(x)|$ on $\mathbb{R}$, it is true that

$$W(x)|\psi(x)| \leqslant \text{const.}, \qquad x \in \mathbb{R}.$$

Knowing, then, of the existence of *any* entire function $\varphi(z) \not\equiv 0$ having exponential type $\leqslant A$ and satisfying this relation, *we can construct a new one, $\psi(z)$, with only real zeros, that also satisfies it.* Moreover, as the above work shows, we can get such a ψ with $\psi(0) = 1$.

We now rewrite the last relation using a Stieltjes integral. As in §B of Chapter X, it is convenient to introduce an increasing function $n(t)$, equal, for $t > 0$, to the *number of zeros λ' of ψ* (counting multiplicities) *in* $[0, t]$, and, for $t < 0$, to the *negative of the number of such λ' in* $[t, 0)$. This function $n(t)$ (N.B. *it should not be confounded with $N(r)$* !) is *integer-valued* and, since $\psi(0) = 1$, *identically zero in a neighborhood of the origin.* Application of the Levinson theorem from §H.2 of Chapter III to the entire function $\psi(z)$ shows that the limits of $n(t)/t$ for $t \longrightarrow \pm\infty$ exist, both being equal to a number $\leqslant A/\pi$. Thus,

$$\frac{n(t)}{t} \leqslant \frac{A}{\pi} + o(1) \qquad \text{for } t \longrightarrow \pm\infty.$$

Writing γ instead of $\Re\gamma$, the product representation for $\psi(z)$ can be put in the form

$$\log|\psi(z)| \;=\; \gamma\Re z \;+\; \int_{-\infty}^{\infty}\left(\log\left|1-\frac{z}{t}\right| + \frac{\Re z}{t}\right)dn(t).$$

The relation involving W and ψ can hence be expressed thus:

$$\gamma x \;+\; \int_{-\infty}^{\infty}\left(\log\left|1-\frac{x}{t}\right| + \frac{x}{t}\right)dn(t) \;+\; \log W(x) \;\leqslant\; \text{const.}, \qquad x\in\mathbb{R}.$$

The existence of our original multiplier φ for W, of exponential type $\leqslant A$, has in this way enabled us to get an increasing integer-valued function $n(t)$ having the above properties and fulfilling the last relation.

If, on the other hand, one *has* an integer-valued increasing function $n(t)$ meeting these conditions, it is easy to construct an entire function ψ of exponential type $\leqslant A$ making $W(x)|\psi(x)| \leqslant$ const. on $\mathbb{R}$. All one need do is put

$$\psi(z) \;=\; e^{\gamma z}\prod_{\lambda'}\left(1-\frac{z}{\lambda'}\right)e^{z/\lambda'}$$

with λ' running through the *discontinuities* of $n(t)$, each taken a number of times equal to the corresponding jump in $n(t)$. The boundedness of the product $W(x)\psi(x)$ then follows directly, and the Phragmén–Lindelöf argument used previously shows $\psi(z)$ to be of exponential type $\leqslant A$. *The existence of our multiplier φ is, in other words, equivalent to that of an increasing integer-valued function $n(t)$ satisfying the conditions just enumerated.*

Our regularity requirement is of course not *needed* for this equivalence, which holds for any weight bounded in a neighborhood of the origin. *What that requirement does is permit us*, when dealing with weights subject to it, *to drop* from the last statement *the condition that $n(t)$ be integer-valued.* The cost of this is that one ends with a multiplier φ of exponential type *several times larger than A* instead of one with type $\leqslant A$.

Some version of the lemma from §A.1 of Chapter X is needed for this reduction. If $W(x)$ were known to be *even* (with the increasing function involved odd!), the lemma could be used as it stands, and the proof of the next theorem made shorter (regarding this, the reader is again directed to the remark following the second of the next two theorems). The general situation requires a more elaborate form of that result. As in §B.2 of Chapter X, it is convenient to use $[p]$ to denote *the least integer $\geqslant p$ when* *p is negative, while maintaining the usual meaning of that symbol for $p \geqslant 0$. The following variant of the lemma is then sufficient for our purposes:*

Lemma. *Let $v(t)$ be increasing on $\mathbb{R}$, zero on $(-a, a)$, where $a > 0$, and $O(t)$ for $t \longrightarrow \pm \infty$. Then, for $\Im z \neq 0$, we have*

$$
c\Re z \;+\; \int_{-\infty}^{\infty} \left(\log \left| 1 - \frac{z}{t} \right| + \frac{\Re z}{t} \right) (\mathrm{d}[v(t)] - \mathrm{d}v(t))
$$

$$
\leqslant \;\; \log^{+} \left| \frac{\Re z}{\Im z} \right| \;+\; \log \left| 1 + \frac{|\Re z| + \mathrm{i}|\Im z|}{a} \right|,
$$

c being a certain real constant depending on v.

A proof of this estimate was already carried out for $a = 1$ and $\Im z = 1$ while establishing the *Little Multiplier Theorem* in §B.2 of Chapter X. The argument for the general case is not different from the one made there.*

We are now able to establish the promised reduction.

Theorem. *Let the weight $W(x) \geqslant 1$ meet our regularity requirement, with parameters L, C and α. Suppose there is an increasing function $\rho(t)$, zero on a neighborhood of the origin, with*

$$
\frac{\rho(t)}{t} \;\leqslant\; \frac{A}{\pi} + o(1) \qquad \text{for } t \longrightarrow \pm \infty
$$

and

$$
\gamma x \;+\; \int_{-\infty}^{\infty} \left(\log \left| 1 - \frac{x}{t} \right| + \frac{x}{t} \right) \mathrm{d}\rho(t) \;+\; \log W(x) \;\leqslant\; \text{const.}
$$

on the real axis, where γ is a real constant. Then there is a non-zero entire function $\psi(z)$ of exponential type $\leqslant 4A/\alpha$ with $W(x)\psi(x)$ bounded on $\mathbb{R}$.

Remark. The number 4 could be replaced by any other > 2 by refining one point in the following argument.

Proof of theorem. Put

$$
U(z) \;=\; \gamma \Re z \;+\; \int_{-\infty}^{\infty} \left(\log \left| 1 - \frac{z}{t} \right| + \frac{\Re z}{t} \right) \mathrm{d}\rho(t);
$$

our conditions on $\rho(t)$ make the right-hand integral have unambiguous

---

* By following the procedure indicated in the footnote on p. 186, one can, noting that $[v(t)] - v(t) \geqslant 0$ for $t < 0$, improve the upper bound furnished by the lemma to $\log |z/\Im z|$; this is *independent* of the size of the interval $(-a, a)$ on which $v(t)$ is known to vanish.

meaning for *all* complex z, taking, perhaps, the value $-\infty$ for some of these.* The *lack of evenness* of $W(x)$ and $U(z)$ will necessitate our attention to certain details.

$U(z)$ is *subharmonic* in the complex plane; it is, in other words, equal there to *the negative of a superharmonic function* having the properties taken up near the beginning of §A.1. According to the *first* of those we have in particular

$$\limsup_{z \to x_0} U(z) \;\leqslant\; U(x_0) \qquad \text{for } x_0 \in \mathbb{R}.$$

Our hypothesis, however, is that $U(x_0) + \log W(x_0) \leqslant K$, say, on $\mathbb{R}$, with $\log W(x_0) \geqslant 0$ there. Hence

$$\limsup_{z \to x_0} U(z) \;\leqslant\; K, \qquad x_0 \in \mathbb{R}.$$

Starting from this relation, one now *repeats* for $U(z)$ the Phragmén–Lindelöf argument made above for $\log|\psi(z)|$, using the properties of $\rho(t)$ in place of those of $N(r)$. In that way, it is found that

$$U(z) \;\leqslant\; K + A|\Im z|.$$

The function $U(z)$ is actually *harmonic*[†] for $\Im z > 0$, and we proceed to establish for it the Poisson representation

$$U(z) \;=\; A'\Im z \;+\; \frac{1}{\pi}\int_{-\infty}^{\infty} \frac{\Im z\, U(t)}{|z-t|^2}\,dt$$

in that half plane, with

$$A' \;=\; \limsup_{y \to \infty} \frac{U(iy)}{y} \;\leqslant\; A.$$

(This step could be avoided if $W(x)$ were known to be continuous; such continuity is, however, superfluous here.) Our *formula* for $U(z)$ shows $U(iy)$ to be $\geqslant 0$ for $y > 0$, so the quantity A' is certainly $\geqslant 0$. That it does not *exceed* A is guaranteed by the estimate on $U(z)$ just found. That estimate and the *fourth* theorem of §C, Chapter III, now show that in fact

$$U(z) \;\leqslant\; K + A'\Im z \qquad \text{for } \Im z > 0;$$

the function $U(z) - K - A'\Im z$ is thus *harmonic and* $\leqslant 0$ in the upper half plane.

---

* any such z must be *real* – $U(z)$ is *finite* for $\Im z \neq 0$

† and, in particular, *finite* (see preceding footnote) – the integral in the following Poisson representation is thus surely *convergent*.

By §F.1 of Chapter III we therefore have

$$U(z) \ - \ K \ - \ A'\Im z \ = \ -b\Im z \ - \ \frac{1}{\pi}\int_{-\infty}^{\infty} \frac{\Im z \, \mathrm{d}\sigma(t)}{|z-t|^2}$$

for $\Im z > 0$, with a constant $b \geqslant 0$ and a certain *positive measure* σ on $\mathbb{R}$. It is readily verified that b must equal zero. Our desired Poisson representation for $U(z)$ will now follow from an argument like the one in §G.1 of Chapter III *if we verify absolute continuity of σ.*

For this purpose, it is enough to show that when $y \longrightarrow 0$,

$$\int_{-M}^{M} |U(x+iy) - U(x)|\mathrm{d}x \ \longrightarrow \ 0$$

for each finite M. Given such an M, we can write

$$U(z) \ = \ \gamma\Re z \ + \ \left(\int_{-2M}^{2M} + \int_{|t|>2M}\right)\left(\log\left|1-\frac{z}{t}\right| + \frac{\Re z}{t}\right)\mathrm{d}\rho(t).$$

The *second* of the two integrals involved here clearly tends *uniformly* to

$$\int_{|t|>2M}\left(\log\left|1-\frac{x}{t}\right| + \frac{x}{t}\right)\mathrm{d}\rho(t)$$

as $z = x + iy$ tends to x, when $-M \leqslant x \leqslant M$. Hence, since $\rho(t)$ is zero on a neighborhood of the origin, the matter at hand boils down to checking that

$$\int_{-M}^{M}\left|\int_{-2M}^{2M}\left(\log|x+iy-t| \ - \ \log|x-t|\right)\mathrm{d}\rho(t)\right|\mathrm{d}x \ \longrightarrow \ 0$$

as $y \longrightarrow 0$. The inner integrand is already positive here, so the left-hand expression is just

$$\int_{-2M}^{2M}\int_{-M}^{M}\left(\log|x+iy-t| \ - \ \log|x-t|\right)\mathrm{d}x \, \mathrm{d}\rho(t).$$

In this last, however, the inner integral is easily seen – by direct calculation, if need be – to tend to zero uniformly for $-2M \leqslant t \leqslant 2M$ as $y \longrightarrow 0$. (Incidentally, $\int_{-M}^{M}\log|w-x|\mathrm{d}x$ is the negative of a logarithmic potential generated by a *bounded* linear density on a finite segment, and therefore continuous everywhere in w.) The preceding relation therefore holds, so σ is absolutely continuous, giving us the desired Poisson representation for $U(z)$.

Once that representation is available, we have, for $\Im z > 0$,

$$U(z) \;+\; \frac{1}{\pi}\int_{-\infty}^{\infty}\frac{\Im z\,\log W(t)}{|z-t|^{2}}\,dt$$

$$=\; A'\Im z \;+\; \frac{1}{\pi}\int_{-\infty}^{\infty}\frac{\Im z\,(U(t)+\log W(t))}{|z-t|^{2}}\,dt.$$

Since, however, $U(t) + \log W(t) \leqslant K$ on $\mathbb{R}$, the right side of this relation must be $\leqslant K + A'\Im z$, so we have

$$\gamma\Re z \;+\; \int_{-\infty}^{\infty}\left(\log\left|1-\frac{z}{t}\right| + \frac{\Re z}{t}\right)d\rho(t)$$

$$+\; \frac{1}{\pi}\int_{-\infty}^{\infty}\frac{\Im z\,\log W(t)}{|z-t|^{2}}\,dt \;\leqslant\; K \;+\; A'\Im z, \qquad \Im z > 0.$$

(Putting $z = i$, we see by the way that $\int_{-\infty}^{\infty}(\log W(t)/(1+t^{2}))\,dt \;<\; \infty$.)

By hypothesis, W meets our regularity requirement with parameters L, C, and α; this means that

$$\log W(t) \;\geqslant\; \alpha\log W(x) \;+\; \log C$$

for $t\in J_{x}$, an interval of length L containing x. Therefore, if

$$z \;=\; x+iL,$$

the *second* integral on the left in the preceding relation is $\geqslant (\alpha/4)\log W(x) + (1/4)\log C$. After multiplying the latter through by $4/\alpha$ we thus find, recalling that $A' \leqslant A$,

$$\frac{4\gamma}{\alpha}x \;+\; \int_{-\infty}^{\infty}\left(\log\left|1-\frac{x+iL}{t}\right| + \frac{x}{t}\right)d(4\rho(t)/\alpha)$$

$$+\; \log W(x) \;\leqslant\; K', \qquad x\in\mathbb{R},$$

where

$$K' \;=\; \frac{4K+4AL-\log C}{\alpha}.$$

It is at this point that we apply the last lemma, with

$$v(t) \;=\; \frac{4}{\alpha}\rho(t)$$

and $z = x+iL$. If $\rho(t)$, and hence $v(t)$, vanishes on the neighborhood $(-a, a)$ of the origin, we see on combining that lemma with the preceding relation that

$$\beta x \;+\; \int_{-\infty}^{\infty} \left(\log \left| 1 - \frac{x+iL}{t} \right| \;+\; \frac{x}{t} \right) d[v(t)] \;+\; \log W(x)$$

$$\leqslant \quad K' \;+\; \log^+ \left| \frac{x}{L} \right| \;+\; \log \left| 1 + \frac{|x|+iL}{a} \right|$$

on $\mathbb{R}$, with a certain constant β. From this we have, *a fortiori*,

$$\beta x \;+\; \int_{-\infty}^{\infty} \left(\log \left| 1 - \frac{x}{t} \right| \;+\; \frac{x}{t} \right) d[v(t)] \;+\; \log W(x)$$

$$\leqslant \quad K'' \;+\; 2 \log^+ |x| \qquad \text{for } x \in \mathbb{R},$$

K'' being a new constant. The first two terms on the left add up, however, to $\log |\varphi(x)|$, where

$$\varphi(z) \;=\; e^{\beta z} \prod_\lambda \left(1 - \frac{z}{\lambda} \right) e^{z/\lambda}$$

is the Hadamard product formed from the discontinuities λ of $[v(t)]$, each one taken with multiplicity equal to the height of the jump in that function corresponding to it. Since

$$\frac{v(t)}{t} \;=\; \frac{4\rho(t)}{\alpha t} \;\leqslant\; \frac{4A}{\alpha\pi} \;+\; o(1)$$

for $t \longrightarrow \pm \infty$ (hypothesis!), that product is certainly convergent in the complex plane, and φ is an entire function. In terms of it, the previous relation can be rewritten as

$$W(x)|\varphi(x)| \;\leqslant\; \text{const.}\,(x^2 + 1), \qquad x \in \mathbb{R}.$$

It is now claimed that $\varphi(z)$ *must have infinitely many zeros* λ, *unless* $W(x)$ *is already bounded on* $\mathbb{R}$ (in which case our theorem is trivially true). Because those λ are the discontinuities of $[v(t)] = [4\rho(t)/\alpha]$, the presence of infinitely many of them is equivalent to the *unboundedness* of $\rho(t)$ (either above or below). It is thus enough to show that *if* $|\rho(t)|$ *is bounded*, $W(x)$ *is also bounded.*

We do this by proving that if $|\rho(t)|$ is bounded, *the function* $U(z)$ *used above must be equal to zero.* For real y, we have

$$U(iy) \;=\; \frac{1}{2} \int_{-\infty}^{\infty} \log \left| 1 + \frac{y^2}{t^2} \right| d\rho(t) \;=\; \int_{-\infty}^{\infty} \frac{y^2}{y^2 + t^2} \, \frac{\rho(t)}{t} \, dt.$$

Here, $\rho(t)$ vanishes for $|t| < a$, so, if $|\rho(t)|$ is also *bounded*, the ratio $\rho(t)/t$

appearing in the last integral tends to *zero* for $t \longrightarrow \pm \infty$, besides being bounded on $\mathbb{R}$. That, however, makes

$$\int_{-\infty}^{\infty} \frac{y}{y^2 + t^2} \frac{\rho(t)}{t} \, dt \longrightarrow 0 \qquad \text{for } y \longrightarrow \pm \infty,$$

as one readily sees on breaking up the integral into two appropriate pieces. We thus have

$$\frac{U(iy)}{|y|} \longrightarrow 0 \qquad \text{for } y \longrightarrow \pm \infty,$$

and *the quantity A' figuring in the above examination of $U(z)$ is equal to zero.* By the estimate obtained there, we must then have

$$U(z) \leqslant K$$

for $\mathfrak{I}z \geqslant 0$, and exactly the same reasoning (or the evident equality of $U(\bar{z})$ and $U(z)$) shows this to also hold for $\mathfrak{I}z \leqslant 0$. The subharmonic function $U(z)$ is, in other words, *bounded above in the complex plane if $|\rho(t)|$ is bounded.*

Such a subharmonic function is, however, necessarily constant. That is a general proposition, set below as problem 52. In the present circumstances, we can arrive at the same conclusion by a simple *ad hoc* argument. Since $\rho(t)/t \geqslant 0$, the previous formula for $U(iy)$ yields, for $y > 0$,

$$U(iy) \geqslant \int_{-y}^{y} \frac{y^2}{y^2 + t^2} \frac{\rho(t)}{t} \, dt \geqslant \frac{1}{2} \int_{-y}^{y} \frac{\rho(t)}{t} \, dt.$$

If ever $\rho(t)$ is different from zero, there must be some k and y_0, both > 0, with either $\rho(t) \geqslant k$ for $y \geqslant y_0$ or $\rho(t) \leqslant -k$ for $y \leqslant -y_0$, and in both cases the last right-hand integral will be

$$\geqslant \frac{k}{2} \log \frac{y}{y_0}$$

for $y \geqslant y_0$. This, however, would make $U(iy) \longrightarrow \infty$ for $y \longrightarrow \infty$, *contradicting the boundedness of $U(z)$*, so we must have $\rho(t) \equiv 0$. But then

$$\gamma x \;=\; U(x) \;\leqslant\; K - \log W(x), \qquad x \in \mathbb{R},$$

which contradicts our assumption that $W(x) \geqslant 1$ (either for $x \longrightarrow \infty$ or for $x \longrightarrow -\infty$) *unless $\gamma = 0$.* Finally, then, the boundedness of $\rho(t)$ forces $U(x)$ to *reduce to zero*, whence

$$\log W(x) \;=\; U(x) + \log W(x) \;\leqslant\; K, \qquad x \in \mathbb{R},$$

i.e., $W(x)$ *is bounded*, as we claimed.

Thus, except for the latter trivial situation, $|\rho(t)|$ is unbounded and the entire function $\varphi(z)$ has infinitely many zeros. Dividing it by the factors $1 - z/\lambda$ corresponding to *any two such zeros*, we obtain a *new entire function*, $\psi(z)$, such that*

$$W(x)|\psi(x)| \leqslant \text{const.}, \qquad x \in \mathbb{R}.$$

We now repeat the Phragmén–Lindelöf argument applied previously to another function $\psi(z)$ and then, in the course of the present proof, to $U(z)$. Since

$$\frac{[v(t)]}{t} \leqslant \frac{v(t)}{t} = \frac{4\rho(t)}{\alpha t} \leqslant \frac{4A}{\alpha\pi} + o(1)$$

for $t \longrightarrow \pm\infty$, we find in that way that

$$|\psi(z)| \leqslant \text{const.}\,e^{4A|\Im z|/\alpha};$$

ψ is thus of exponential type $\leqslant 4A/\alpha$. We have $\psi(0) = 1$, so $\psi(z) \not\equiv 0$. Referring to the previous relation involving W and ψ, we see that the theorem is proved.

Let us now settle on a definite meaning for the notion of admitting multipliers, hitherto understood somewhat loosely, and agree to henceforth employ that term *only when actual boundedness on $\mathbb{R}$ is involved*.

Definition. A weight $W(x) \geqslant 1$ will be said to admit multipliers if there are entire functions $\varphi(z) \not\equiv 0$ of arbitrarily small exponential type for which

$$W(x)|\varphi(x)| \leqslant \text{const.}, \qquad x \in \mathbb{R}.$$

Combining the last theorem with the conclusion of the discussion preceding it, we then have the

Corollary. *A weight $W(x) \geqslant 1$ fulfilling our regularity requirement admits multipliers iff, corresponding to any $A > 0$, there is an increasing function $\rho(t)$, zero on some neighborhood of the origin, with*

$$\frac{\rho(t)}{t} \leqslant \frac{A}{\pi} + o(1) \qquad \text{for } t \longrightarrow \pm\infty$$

---

* $W(x)$ must be *bounded* in the neighborhood of each of the two zeros of φ just removed. Otherwise W would be identically infinite on an interval of length L by the first lemma in this article, and then the Poisson integral of $U(t) \leqslant K - \log W(t)$ would diverge. That, however, cannot happen, as we have already remarked in a footnote near the beginning of this proof.

and at the same time

$$\gamma x \; + \; \int_{-\infty}^{\infty} \left(\log\left|1 - \frac{x}{t}\right| + \frac{x}{t} \right) d\rho(t) \; + \; \log W(x) \;\; \leqslant \;\; \text{const.}$$

on $\mathbb{R}$ *for some real constant* γ.

In the case where $W(x)$ is equal to $|F(x)|$ for some *entire function* $F(z)$ of *exponential type*, the results just given hold *without any additional special assumption about the regularity of* W.

Theorem. *Let* $F(z)$ *be entire and of exponential type, with* $|F(x)| \geqslant 1$ *on* $\mathbb{R}$. *Suppose there is an increasing function* $\rho(t)$, *zero on a neighborhood of the origin, such that*

$$\frac{\rho(t)}{t} \;\; \leqslant \;\; \frac{A}{\pi} + \mathrm{o}(1) \quad \text{for } t \longrightarrow \pm\infty$$

and

$$\gamma x \; + \; \int_{-\infty}^{\infty} \left(\log\left|1 - \frac{x}{t}\right| + \frac{x}{t} \right) d\rho(t) \; + \; \log|F(x)| \;\; \leqslant \;\; \text{const.}$$

on $\mathbb{R}$ *for some real constant* γ. *Then there is an entire function* $\psi(z) \not\equiv 0$ *of exponential type* $\leqslant A$ (*sic!*) *with*

$$|F(x)\psi(x)| \;\; \leqslant \;\; \text{const.}, \qquad x \in \mathbb{R}.$$

Proof. Writing $|F(x)| = W(x)$, one starts out and proceeds as in the demonstration of the proceeding theorem, up to the point where the relation

$$U(z) \; + \; \frac{1}{\pi} \int_{-\infty}^{\infty} \frac{\Im z \log W(t)}{|z - t|^2} dt \;\; \leqslant \;\; K + A\Im z$$

is obtained for $\Im z > 0$, with

$$U(z) \;\; = \;\; \gamma \Re z \; + \; \int_{-\infty}^{\infty} \left(\log\left|1 - \frac{z}{t}\right| + \frac{\Re z}{t} \right) d\rho(t).$$

From this one sees in particular* that

$$\int_{-\infty}^{\infty} \frac{\log|F(t)|}{1 + t^2} dt \;\; = \;\; \int_{-\infty}^{\infty} \frac{\log W(t)}{1 + t^2} dt \;\; < \;\; \infty,$$

which enables us to use some results from Chapter III.

---

* cf. footnotes near beginning of proof of the preceding theorem.

We can, in the first place, assume that *all the zeros of F(z) lie in the lower half plane*, according to the *second* theorem of §G.3 in Chapter III. Then, however, by §G.1 of that chapter,

$$\frac{1}{\pi}\int_{-\infty}^{\infty}\frac{\Im z \log W(t)}{|z-t|^2}\,dt \;=\; \frac{1}{\pi}\int_{-\infty}^{\infty}\frac{\Im z \log|F(t)|}{|z-t|^2}\,dt$$

$$=\; \log|F(z)| \;-\; B\Im z \qquad \text{for } \Im z > 0,$$

where

$$B \;=\; \limsup_{y\to\infty}\frac{\log|F(iy)|}{y}.$$

Our previous relation involving U and W thus becomes

$$U(z) + \log|F(z)| \;\leqslant\; K + (A+B)\Im z, \qquad \Im z > 0.$$

In this we put $z = x + i$, getting

$$\gamma x \;+\; \int_{-\infty}^{\infty}\left(\log\left|1 - \frac{x+i}{t}\right| + \frac{x}{t}\right)d\rho(t)$$

$$+\; \log|F(x+i)| \;\leqslant\; \text{const.}, \qquad x \in \mathbb{R}.$$

Apply now the lemma used in the proof of the last theorem, but this time with

$$v(t) \;=\; \rho(t).$$

In that way one sees that

$$\beta x \;+\; \int_{-\infty}^{\infty}\left(\log\left|1 - \frac{x+i}{t}\right| + \frac{x}{t}\right)d[\rho(t)]$$

$$+\; \log|F(x+i)| \;\leqslant\; 2\log^+|x| + O(1), \qquad x \in \mathbb{R},$$

with a new real constant β. There is as before a certain entire function φ with $\log|\varphi(x+i)|$ equal to the *sum of the first two terms on the left*, and we have

$$|F(x+i)\varphi(x+i)| \;\leqslant\; \text{const.}(x^2+1), \qquad x \in \mathbb{R}.$$

It now follows as previously that $\varphi(z)$ *has infinitely many zeros*, unless $|F(x)|$ is *itself* bounded, in which case there is nothing to prove. Dividing out from $\varphi(z)$ the linear factors corresponding to *two* of those zeros gives us an entire function $\psi(z) \not\equiv 0$ with

$$|F(x+i)\psi(x+i)| \;\leqslant\; \text{const.}, \qquad x \in \mathbb{R}.$$

Here, our initial assumption that $|F(x)| \geqslant 1$ on $\mathbb{R}$ and the Poisson representation for $\log |F(z)|$ in $\{\Im z > 0\}$ already used imply that

$$|F(x+i)| \geqslant \text{const.} > 0 \quad \text{for } x \in \mathbb{R},$$

so by the preceding relation we have in particular

$$|\psi(x+i)| \leqslant \text{const.}, \quad x \in \mathbb{R}.$$

By hypothesis, we also have

$$\frac{[\rho(t)]}{t} \leqslant \frac{\rho(t)}{t} \leqslant \frac{A}{\pi} + o(1)$$

for $t \longrightarrow \pm \infty$, permitting us to use once again the Phragmén–Lindelöf argument made three times already in this article. In that way we see that

$$|\psi(z+i)| \leqslant \text{const.} \, e^{A|\Im z|},$$

meaning that ψ is of exponential type $\leqslant A$. The product $F(z+i)\psi(z+i)$ is then also of exponential type. Since that product is by the above relation *bounded for real z*, we have by the third theorem of §C in Chapter III, that

$$|F(x)\psi(x)| \leqslant \text{const.} \quad \text{for } x \in \mathbb{R}.$$

Our function ψ thus has all the properties claimed by the theorem, and we are done.

Remark. Suppose that we know of an increasing function $\rho(t)$, zero on a neighborhood of the origin, satisfying the conditions assumed for the above results with some number $A > 0$ and a weight $W(x) \geqslant 1$. For the increasing function $\mu(t) = \rho(t) - \rho(-t)$, also zero on a neighborhood of the origin, we then have

$$\frac{\mu(t)}{t} \leqslant \frac{2A}{\pi} + o(1) \quad \text{for } t \longrightarrow \infty,$$

as well as

$$\int_0^\infty \log \left| 1 - \frac{x^2}{t^2} \right| d\mu(t) + \log \{ W(x)W(-x) \} \leqslant \text{const.}$$

for $x \in \mathbb{R}$. In this relation, both terms appearing on the left are *even*; that enables us to simplify the argument made in proving the first of the preceding two theorems when applying it in the present situation.

If the weight $W(x)$ meets our regularity requirement* with parameters

---

* see also Remark 5 near the end of §E.2.

L, C, and α, we do have

$$\frac{1}{\pi}\int_{-\infty}^{\infty}\frac{L\log\{W(t)W(-t)\}}{(x-t)^2+L^2}\,dt \;\geqslant\; \frac{\alpha}{4}\log\{W(x)W(-x)\}$$

$$+ \;\frac{\log C}{2}\qquad\text{for } x\in\mathbb{R};$$

this one sees by writing the logarithm figuring in the left-hand member as a sum and then dealing separately with the two integrals thus obtained. The behaviour of the even subharmonic function

$$V(z) \;=\; \int_0^\infty \log\left|1 \,-\, \frac{z^2}{t^2}\right|d\mu(t)$$

is easier to investigate than that of the function $U(z)$ used in the above proofs (cf. §B of Chapter III). When $V(x+iL)$ has made its appearance, one may apply directly the lemma from §A.1 of Chapter X instead of resorting to the latter's more complicated variant given above.

By proceeding in this manner, one obtains an *even* entire function $\Psi(z)$ with

$$W(x)W(-x)|\Psi(x)| \;\leqslant\; \text{const.}, \qquad x\in\mathbb{R},$$

and thus, since $W(-x) \geqslant 1$ (!),

$$W(x)|\Psi(x)| \;\leqslant\; \text{const.}, \qquad x\in\mathbb{R}.$$

The function $\Psi(z)$ is of exponential type, but here that type turns out to be bounded above by $8A/\alpha$ rather than by $4A/\alpha$ as we found for the function $\psi(z)$ obtained previously.

Insofar as W's *admitting of multipliers* is concerned, the *extra factor of two* is of no importance. The reader may therefore prefer this approach (involving a preliminary reduction to the even case) which bypasses some fussy details of the one followed above, but yields less precise estimates for the exponential types of the multipliers obtained. Anyway, according to the remark following the statement of the first of the above two theorems, the estimate $4A/\alpha$ on the type of $\psi(z)$ is not very precise.

Problem 52

Show that a function $V(z)$ *superharmonic* in the whole complex plane and bounded *below* there is constant. (Hint: Referring to the first theorem of §A.2, take the means $(\Phi_r V)(z)$ considered there. Assuming wlog that $V(z)\not\equiv\infty$, each of those means is also superharmonic and bounded below in $\mathbb{C}$, and it is enough to establish the result *for them*. The Φ, V are also $\mathscr{C}_\infty$, so we may as well assume to begin with that $V(z)$ is $\mathscr{C}_\infty$.

That reduction made, observe that if $V(z)$ is actually *harmonic* in $\mathbb{C}$, the desired result boils down to Liouville's theorem, so it suffices to *establish* this harmonicity. For that purpose, fix any z_0 and look at the means

$$V_r(z_0) \;=\; \frac{1}{2\pi}\int_0^{2\pi} V(z_0 + re^{i\vartheta})\,\mathrm{d}\vartheta.$$

Consult the proof of the *second* lemma in §A.2, and then show that

$$\frac{\partial V_r(z_0)}{\partial \log r}$$

is a *decreasing* function of r, so that $V_r(z_0)$ *either remains constant for all* $r > 0$ – and hence equal to $V(z_0)$ – *or else tends to* $-\infty$ *as* $r \longrightarrow \infty$. In the second case, V could not be bounded below in $\mathbb{C}$. Apply Gauss' theorem from §A.1.)

Scholium. The regularity requirement for weights given in the 1962 paper of Beurling and Malliavin is much less stringent than the one we have been using. A relaxed version of the former can be stated thus:

There are four constants $C > 0$, $\alpha > 0$, $\beta < 1$ *and* $\gamma < 1$ *such that, to each* $x \in \mathbb{R}$ *corresponds an interval* I_x *of length* $e^{-|x|^\gamma}$ *(sic!) containing* x *with*

$$W(t) \;\geqslant\; C\,e^{-|x|^\beta}(W(x))^\alpha \qquad \text{for } t \in I_x.$$

The point we wish to make here is that the exponentials in $|x|^\gamma$ and $|x|^\beta$ are *in a sense* red herrings; a close analogue of the first of the above two theorems, *with practically the same proof*, is valid for weights meeting the more general condition. The only new ingredient needed is the elementary Paley–Wiener multiplier theorem.

Problem 53

Suppose that $W(x) \geqslant 1$ fulfills the condition just formulated, and that there is an increasing function $\rho(t)$, zero on a neighborhood of the origin, with

$$\frac{\rho(t)}{t} \;\leqslant\; \frac{A}{\pi} + \mathrm{o}(1) \qquad \text{for } t \longrightarrow \pm\infty$$

and

$$cx \;+\; \int_{-\infty}^{\infty}\left(\log\left|1 - \frac{x}{t}\right| + \frac{x}{t}\right)\mathrm{d}\rho(t) \;+\; \log W(x) \;\leqslant\; \text{const.}$$

on $\mathbb{R}$, where c is a certain real constant. Show that for any $\eta > 0$ there is an entire function $\psi(z) \not\equiv 0$ of exponential type $< 4A/\alpha + \eta$ making

$$W(x)|\psi(x)| \leqslant \text{const.}, \qquad x \in \mathbb{R}.$$

(Hint: Follow exactly the proof of the result referred to until arriving at the relation

$$U(z) + \frac{1}{\pi}\int_{-\infty}^{\infty} \frac{\Im z \log W(t)}{|z-t|^2}\,dt \leqslant K + A'\Im z, \qquad \Im z > 0.$$

In this, substitute $z = x + i e^{-|x|^\gamma}$ (!) and invoke the condition, finding, for that value of z,

$$\frac{4c}{\alpha}x + \int_{-\infty}^{\infty}\left(\log\left|1 - \frac{z}{t}\right| + \frac{x}{t}\right)d(4\rho(t)/\alpha)$$

$$+ \log W(x) \leqslant K' + \frac{4}{\alpha}|x|^\beta$$

with a new constant K'. Using the lemma (with z as above!) and continuing as before, we get an entire function $\varphi(z) \not\equiv 0$ such that

$$\frac{\log|\varphi(iy)|}{|y|} \leqslant \frac{4A}{\alpha} + o(1) \qquad \text{for } y \longrightarrow \pm\infty,$$

$$\log|\varphi(z)| \leqslant O(|z|^{1+\varepsilon})$$

for large $|z|$ ($\varepsilon > 0$ being arbitrary), and finally

$$W(x)|\varphi(x)| \leqslant \text{const.}(x^2+1)\exp\left(|x|^\gamma + \frac{4}{\alpha}|x|^\beta\right)$$

on the real axis. To the right side of the last relation, apply the theorem from §A.1 of Chapter X (and §D of Chapter IV!).).

2. The smallest superharmonic majorant

According to the results from the latter part of the preceding article (beginning with the second theorem therein), a weight $W(x) \geqslant 1$ having any one of various regularity properties *admits multipliers* if and only if, corresponding to any $A > 0$, there exists an increasing function $\rho(t)$, zero on some neighborhood of the origin, such that

$$\frac{\rho(t)}{t} \leqslant \frac{A}{\pi} + o(1) \qquad \text{for } t \longrightarrow \pm\infty$$

and

$$\gamma x \;+\; \int_{-\infty}^{\infty} \left(\log\left|1 - \frac{x}{t}\right| + \frac{x}{t} \right) d\rho(t) \;+\; \log W(x) \;\leqslant\; K \quad \text{for } x \in \mathbb{R}$$

with some constant γ. Hence, in keeping with the line of thought embarked on at the beginning of article 1, we regard the (hypothetical) *second* ('essential') *condition* for admittance of multipliers by a weight W as being *very close* (if not identical) to whatever requirement it must satisfy in order to guarantee existence of such increasing functions ρ. That requirement, and attempts to arrive at precise knowledge of it, will therefore be our main object of interest during the remainder of this chapter.

Suppose that for a given weight $W(x) \geqslant 1$ we *have* such a function $\rho(t)$ corresponding to some $A > 0$. The relation

$$\frac{1}{\pi} \int_{-\infty}^{\infty} \frac{|\Im z| \log W(t)}{|z - t|^2} \, dt \;-\; A|\Im z|$$

$$\leqslant\; K \;-\; \gamma \Re z \;-\; \int_{-\infty}^{\infty} \left(\log\left|1 - \frac{z}{t}\right| + \frac{\Re z}{t} \right) d\rho(t)$$

(with the left side *interpreted* as $\log W(x)$ for $z = x \in \mathbb{R}$) then holds throughout the complex plane. For $\Im z > 0$, this has indeed already been verified while proving the second theorem of article 1 (near the beginning of the proof). That, however, is enough, since both sides are unchanged when z is replaced by $\bar{z}$.

Now the *right side* of the last relation is obviously a *superharmonic function* of z, *finite* for z off of the real axis. The *existence* of our function ρ thus leads (in almost trivial fashion) to that of a *superharmonic majorant* $\not\equiv \infty$ *for*

$$\frac{1}{\pi} \int_{-\infty}^{\infty} \frac{|\Im z| \log W(t)}{|z - t|^2} \, dt \;-\; A|\Im z|$$

(interpreted as $\log W(x)$ for $z = x \in \mathbb{R}$) *in the whole complex plane*. The key to the proof of the Beurling–Malliavin multiplier theorem given below in §C lies in the observation that *the converse of this statement is true*, at least for *continuous* weights $W(x)$. That fact (which, from a certain point of view, is nearly tautological) will be established in the next article. For this purpose and the later applications as well, we will need the *smallest superharmonic majorant* of a continuous function together with some of its properties, to whose examination we now proceed.

Let $F(z)$ be any function *real-valued and continuous* in the whole complex plane. (In our applications, we will use a function $F(z)$ equal to the preceding expression – interpreted as $\log W(x)$ for $z = x \in \mathbb{R}$ – where $W \geqslant 1$ is *continuous* and such that $\int_{-\infty}^{\infty} (\log W(x)/(1 + x^2)) dx < \infty$.) We next take the *family* $\mathscr{F}$ *of functions superharmonic and* $\geqslant F$ (everywhere); our convention being to consider the function *identically equal to* $+\infty$ *as superharmonic* (see §A.1), $\mathscr{F}$ is *certainly not empty*. Then put

$$Q(z) = \inf \{ U(z) \colon U \in \mathscr{F} \}$$

for each complex z, and finally take

$$(\mathfrak{M}F)(z) = \liminf_{\zeta \to z} Q(\zeta);$$

$\mathfrak{M}F$ is the function we will be dealing with. (The reason for use of the symbol $\mathfrak{M}$ will appear in problems 55 and 56 below. $\mathfrak{M}F$ is a kind of *maximal function* for F.)

In our present circumstances, $Q(z)$ is $\geqslant$ the *continuous function* $F(z)$, so we must also have

$$(\mathfrak{M}F)(z) \geqslant F(z).$$

This certainly makes $(\mathfrak{M}F)(z) > -\infty$ everywhere, so $(\mathfrak{M}F)(z)$ is *itself superharmonic* (everywhere) by the last theorem of §A.1, *and must hence belong to* $\mathscr{F}$ in view of the relation just written. The same theorem also tells us, however, that $(\mathfrak{M}F)(z) \leqslant U(z)$ for *every* $U \in \mathscr{F}$; $\mathfrak{M}F$ is thus *a member of* $\mathscr{F}$ *and at the same time* $\leqslant$ *every member of* $\mathscr{F}$. $\mathfrak{M}F$ is, in other words, *the smallest superharmonic majorant of F*.

It may well happen, of course, that $(\mathfrak{M}F)(z) \equiv \infty$. However, if $\mathfrak{M}F$ is finite at just one point, it is finite everywhere. That is the meaning of the

Lemma. *If, for any* z_0, $(\mathfrak{M}F)(z_0) = \infty$, *we have* $(\mathfrak{M}F)(z) \equiv \infty$.

Proof. To simplify the writing, let us wlog consider the case where $z_0 = 0$. By continuity of F at 0, there is certainly some finite M such that

$$F(z) \leqslant M \quad \text{for } |z| \leqslant 1, \quad \text{say.}$$

Given, however, that $(\mathfrak{M}F)(0) = \infty$, there is an r, $0 < r < 1$, for which

$$(\mathfrak{M}F)(z) \geqslant M + 1, \quad |z| \leqslant r,$$

because the superharmonic function $\mathfrak{M}F$ has property (i) at 0 (§A.1).

It is now claimed that

$$\int_{-\pi}^{\pi} (\mathfrak{M}F)(re^{i\vartheta}) \, d\vartheta = \infty.$$

Reasoning by contradiction, *assume that the integral on the left is finite.* Then, since $(\mathfrak{M}F)(re^{i\vartheta})$ is *bounded below* for $0 \leqslant \vartheta \leqslant 2\pi$ (here, simply because $\mathfrak{M}F \geqslant F$, but see also the beginning of §A.1), we must have

$$\frac{1}{2\pi}\int_{-\pi}^{\pi}\frac{r^2-\rho^2}{r^2+\rho^2-2r\rho\cos(\varphi-\tau)}(\mathfrak{M}F)(re^{i\tau})\,d\tau \;\; < \;\; \infty$$

for $0 \leqslant \rho < r$ and $0 \leqslant \varphi \leqslant 2\pi$. Take now the function $V(z)$ *equal,* for $|z| \geqslant r$ to $(\mathfrak{M}F)(z)$ and, for $z = \rho e^{i\varphi}$ with $0 \leqslant \rho < r$, to the Poisson integral just written. This function $V(z)$ is superharmonic (everywhere) by the *second theorem* of §A.1.

We have

$$V(\rho e^{i\varphi}) \;\geqslant\; M+1 \quad \text{for } 0 \leqslant \rho < r,$$

since $(\mathfrak{M}F)(re^{i\tau}) \geqslant M+1$.   At the same time,

$$F(\rho e^{i\varphi}) \;\leqslant\; M \quad \text{for } 0 \leqslant \rho < r$$

because $r < 1$, so $V(z) \geqslant F(z)$ for $|z| < r$. This, however, is also true for $|z| \geqslant r$, where $V(z) = (\mathfrak{M}F)(z)$. We thus have in $V(z)$ a *superharmonic majorant* of $F(z)$, so

$$V(z) \;\geqslant\; (\mathfrak{M}F)(z).$$

Thence,

$$(\mathfrak{M}F)(0) \;\leqslant\; V(0) \;=\; \frac{1}{2\pi}\int_{-\pi}^{\pi}(\mathfrak{M}F)(re^{i\tau})\,d\tau \;\; < \;\; \infty.$$

But it was given that $(\mathfrak{M}F)(0) = \infty$. This contradiction shows that the *integral* in the last relation must be *infinite,* as claimed.

Apply now the *first* lemma of §A.2 to the function $(\mathfrak{M}F)(z)$, superharmonic everywhere. We find that

$$(\mathfrak{M}F)(z) \;\equiv\; \infty$$

for all z. The proof is complete.

Corollary. *The function $(\mathfrak{M}F)(z)$ is either finite everywhere or infinite everywhere.*

Henceforth, to indicate that the *first alternative* of the corollary holds, we will simply say *that $\mathfrak{M}F$ is finite.*

Lemma. *If $\mathfrak{M}F$ is finite and $F(z)$ is harmonic in any open set $\mathcal{O}$, $(\mathfrak{M}F)(z)$ is also harmonic in $\mathcal{O}$.*

Proof. Let $z_0 \in \mathcal{O}$ and take $r > 0$ so small that the *closed* disk of radius r about z_0 lies in $\mathcal{O}$; it suffices to show that $(\mathfrak{M}F)(z)$ is harmonic for $|z - z_0| < r$.

Supposing wlog that $z_0 = 0$, we take the superharmonic function $V(z)$ used in the proof of the preceding lemma. From the second theorem of §A.1, we have

$$V(z) \leqslant (\mathfrak{M}F)(z).$$

Here, however, we are assuming that $(\mathfrak{M}F)(z) < \infty$, so the Poisson integral

$$\frac{1}{2\pi} \int_{-\pi}^{\pi} \frac{r^2 - |z|^2}{|z - re^{i\tau}|^2} (\mathfrak{M}F)(re^{i\tau}) d\tau,$$

equal, for $|z| < r$, *to* $V(z)$, must be *absolutely convergent* for such z, $(\mathfrak{M}F)(re^{i\tau})$ being *bounded below*, as we know. $V(z)$ is thus *harmonic* for $|z| < r$.

Let $|z| < r$. Then, since $\{|z| \leqslant r\} \subseteq \mathcal{O}$, where $F(z)$ is given to be harmonic,

$$F(z) = \frac{1}{2\pi} \int_{-\pi}^{\pi} \frac{r^2 - |z|^2}{|z - re^{i\tau}|^2} F(re^{i\tau}) d\tau,$$

and the integral on the right is $\leqslant$ the preceding one, $\mathfrak{M}F$ being a *majorant* of F. Thus,

$$F(z) \leqslant V(z) \quad \text{for } |z| < r.$$

This, however, also holds for $|z| \geqslant r$ where $V(z) = (\mathfrak{M}F)(z)$. We see as in the proof of the last lemma that $V(z)$ is a *superharmonic majorant* of $F(z)$. Hence

$$V(z) \geqslant (\mathfrak{M}F)(z).$$

But the reverse inequality was already noted above. Therefore,

$$V(z) = (\mathfrak{M}F)(z).$$

Since $V(z)$ is harmonic for $|z| < r$, we are done.

Let us now look at the set E on which

$$(\mathfrak{M}F)(z) = F(z)$$

for some given continuous function F. E may, of course, be *empty*; it is, *in any event, closed*. Suppose, indeed, that we have a sequence of points $z_k \in E$ and that $z_k \xrightarrow[k]{} z_0$. Then, since $\mathfrak{M}F$ enjoys property (i) (§A.1),

we have

$$(\mathfrak{M}F)(z_0) \;\leqslant\; \liminf_{k\to\infty}(\mathfrak{M}F)(z_k) \;=\; \liminf_{k\to\infty} F(z_k) \;=\; F(z_0),$$

F being continuous at z_0. Because $\mathfrak{M}F$ is a majorant of F, we also have $(\mathfrak{M}F)(z_0) \geqslant F(z_0)$, and thus finally $(\mathfrak{M}F)(z_0) = F(z_0)$, making $z_0 \in E$.

This means that the set of z for which $(\mathfrak{M}F)(z) > F(z)$ is open. Regarding it, we have the important

Lemma. $(\mathfrak{M}F)(z)$, *if finite, is harmonic in the open set where it is* $> F(z)$.

Note. I became aware of this result while walking in Berkeley and thinking about a conversation I had just had with L. Dubins on the material of the present article, especially on the notions developed in problems 55 and 56 below. Dubins thus gave me considerable help with this work.

Proof of lemma. Is much like those of the two previous ones. Let us show that if $(\mathfrak{M}F)(z_0) > F(z_0)$ with $\mathfrak{M}F$ finite, then $(\mathfrak{M}F)(z)$ is harmonic in some small disk about z_0.

We can, wlog, take $z_0 = 0$; suppose, then, that

$$(\mathfrak{M}F)(0) \;>\; F(0) + 2\eta, \quad \text{say,}$$

where $\eta > 0$. Property (i) then gives us an $r > 0$ such that

$$(\mathfrak{M}F)(z) \;>\; F(0) + \eta$$

for $|z| \leqslant r$, and the continuity of F makes it possible for us to choose this r small enough so that we also have

$$F(z) \;<\; F(0) + \eta \quad \text{for } |z| \leqslant r.$$

Form now the superharmonic function $V(z)$ used in the proofs of the last two lemmas. As in the second of those, we certainly have

$$V(z) \;\leqslant\; (\mathfrak{M}F)(z),$$

according to our theorem from §A.1. In the present circumstances, for $|z| < r$,

$$V(z) \;=\; \frac{1}{2\pi}\int_{-\pi}^{\pi} \frac{r^2 - |z|^2}{|z - re^{i\tau}|^2}\,(\mathfrak{M}F)(re^{i\tau})\,d\tau$$

is $> F(0) + \eta$, whereas $F(z) < F(0) + \eta$ there; $V(z)$ is thus $\geqslant F(z)$ for $|z| < r$. When $|z| \geqslant r$, $V(z) = (\mathfrak{M}F)(z)$ is also $\geqslant F(z)$, so V is again a superharmonic majorant of F. Hence

$$V(z) \;\geqslant\; (\mathfrak{M}F)(z),$$

and we see finally that

$$V(z) = (\mathfrak{M}F)(z),$$

with the left side *harmonic* for $|z| < r$, just as in the proof of the preceding lemma. Done.

Lemma. *If* $\mathfrak{M}F$ *is finite, it is everywhere continuous.*

Proof. Depends on the Riesz representation for superharmonic functions.
Take the sets

$$E = \{z: (\mathfrak{M}F)(z) = F(z)\}$$

and

$$\mathcal{O} = \mathbb{C} \sim E;$$

as we have already observed, E is closed and $\mathcal{O}$ is open. By the preceding lemma, $(\mathfrak{M}F)(z)$ is *harmonic* in $\mathcal{O}$ and thus surely *continuous* therein. We therefore need only check continuity of $\mathfrak{M}F$ at the points of E.

Let, then, $z_0 \in E$ and consider any open disk Δ centered at z_0, say the one of radius $= 1$. In the open set $\Omega = \Delta \cap \mathcal{O}$ the function $(\mathfrak{M}F)(z)$ is *harmonic*, as just remarked and *on* $\Delta \sim \Omega = \Delta \cap E$, $(\mathfrak{M}F)(z) = F(z)$ *depends continuously on z.* The *restriction* of $\mathfrak{M}F$ to E is, in particular, *continuous* at the centre, z_0, of Δ.

The corollary to the *Evans-Vasilesco theorem* (at the end of §A.3) can now be invoked, thanks to the superharmonicity of $(\mathfrak{M}F)(z)$. After translating z_0 to the origin (and Δ to a disk about 0), we see by that result that $(\mathfrak{M}F)(z)$ is continuous at z_0. This does it.

Remark. These last two lemmas will enable us to use harmonic estimation to examine the function $(\mathfrak{M}F)(z)$ in §C.

It is a good idea at this point to exhibit two processes which generate $(\mathfrak{M}F)(z)$ when applied to a given continuous function F, although we will not make direct use of either in this book. These are described in problems 55 and 56. The first of those depends on

Problem 54

Let $U(z)$, defined and $> -\infty$ in a domain $\mathscr{D}$, satisfy

$$\liminf_{\zeta \to z} U(\zeta) \geqslant U(z)$$

for $z \in \mathscr{D}$. Show that $U(z)$ is then superharmonic in $\mathscr{D}$ iff, at each z therein, one has

$$\frac{1}{\pi r^2} \iint_{|\zeta - z| < r} U(\zeta)\,d\xi\,d\eta \;\leqslant\; U(z)$$

for all $r > 0$ sufficiently small. (As usual, $\zeta = \xi + i\eta$.)
(Hint: For the if part, the first theorem of §A.1 may be used.)

Problem 55

For Lebesgue measurable functions $F(z)$ defined on $\mathbb{C}$ and bounded below on each compact set, put

$$(MF)(z) \;=\; \sup_{r>0} \frac{1}{\pi r^2} \iint_{|\zeta - z| < r} F(\zeta)\,d\xi\,d\eta.$$

Then, starting with any F *continuous* on $\mathbb{C}$, form successively the functions $F^{(0)}(z) = F(z)$, $F^{(1)}(z) = (MF)(z)$, $F^{(2)}(z) = (MF^{(1)})(z)$, and so forth.

(a) Show that $F^{(0)}(z) \leqslant F^{(1)}(z) \leqslant F^{(2)}(z) \leqslant \cdots$.

(b) Show that $\lim_{n\to\infty} F^{(n)}(z) \leqslant (\mathfrak{M}F)(z)$.

(c) Show that $\lim_{n\to\infty} F^{(n)}(z)$ is superharmonic.
(Hint: For this, use problem 54.)

(d) Hence show that $\lim_{n\to\infty} F^{(n)}(z) = (\mathfrak{M}F)(z)$.

Remark. The function $\mathfrak{M}F$ was originally brought into the study of multiplier theorems through this construction.

The next problem involves *Jensen measures* (on $\mathbb{C}$). That term is used here to denote the *positive Radon measures μ of compact support* such that

$$\int_{\mathbb{C}} U(\zeta)\,d\mu(\zeta) \;\leqslant\; U(0)$$

for *each function U superharmonic on* $\mathbb{C}$. (Any such function $U(z)$ is certainly *Borel measurable*, for, by the *first* theorem of §A.2, it is the pointwise limit of an *increasing sequence* of $\mathscr{C}_\infty$ superharmonic functions.) Some simple Jensen measures are the ν_r given by

$$d\nu_r(\zeta) \;=\; \begin{cases} \dfrac{1}{\pi r^2}\,d\xi\,d\eta, & |\zeta| < r, \\[2mm] 0, & |\zeta| \geqslant r \end{cases}$$

(refer to problem 54 !).

For reasons which will soon become apparent, *we denote the collection of Jensen measures by* $\mathfrak{M}$. If U is any function superharmonic on $\mathbb{C}$, so are its translates, so, whenever $\mu \in \mathfrak{M}$ and $z \in \mathbb{C}$,

$$\int_{\mathbb{C}} U(z + \zeta)\,d\mu(\zeta) \;\leqslant\; U(z).$$

Problem 56

The purpose here is to show that if F is continuous on $\mathbb{C}$,

$$(\mathfrak{M}F)(z) \;=\; \sup_{\mu \in \mathfrak{M}} \int_{\mathbb{C}} F(z + \zeta)\,d\mu(\zeta).$$

(a) Show that $\int_{\mathbb{C}} F(z + \zeta)\,d\mu(\zeta) \leqslant (\mathfrak{M}F)(z)$ for each $\mu \in \mathfrak{M}$.

Denote now the set of Jensen measures *absolutely continuous with respect to two-dimensional Lebesgue measure by* $\mathfrak{L}$. As examples of some measures in $\mathfrak{L}$, we have, for instance, the ν_r described above. $\mathfrak{L}$ is of course a subset of $\mathfrak{M}$.

(b) Show that $\mathfrak{L}$ has a countable subset $\{\mu_k\}$, *dense therein with respect to L_1 convergence with bounded support*. This means that given any $\mu \in \mathfrak{L}$, with say $d\mu(\zeta) = \varphi(\zeta)\,d\xi\,d\eta$, where $\varphi(\zeta) = 0$ a.e. for $|\zeta| \geqslant$ some integer N, we can find a *subsequence* $\{\mu_{k_j}\}$ of the μ_k such that, if we write $d\mu_{k_j}(\zeta) = \varphi_{k_j}(\zeta)\,d\xi\,d\eta$, we also have $\varphi_{k_j}(\zeta) = 0$ a.e. for $|\zeta| \geqslant N$, and moreover

$$\iint_{|\zeta| \leqslant N} |\varphi(\zeta) - \varphi_{k_j}(\zeta)|\,d\xi\,d\eta \;\longrightarrow\; 0 \quad \text{as } j \longrightarrow \infty.$$

(Hint: For the open subsets of each of the spaces $L_1(|z| \leqslant N)$, $N = 1, 2, 3, \ldots$, there is a countable base, some of whose members contain densities belonging to measures from $\mathfrak{L}$. Select.)

(c) Taking the measures μ_k from (b), put

$$V_N(z) \;=\; \max_{1 \leqslant k \leqslant N} \int_{\mathbb{C}} F(z + \zeta)\,d\mu_k(\zeta)$$

for our given continuous function F. Fix any $z \in \mathbb{C}$ and $R > 0$. Show that there is a $\nu \in \mathfrak{L}$ (depending, in general, on z, R and N) such that

$$\frac{1}{2\pi} \int_0^{2\pi} V_N(z + Re^{i\vartheta})\,d\vartheta \;=\; \int_{\mathbb{C}} F(z + \zeta)\,d\nu(\zeta).$$

(Hint: First show how to get a Borel function $k(\vartheta)$ taking the values $1, 2; 3, \ldots, N$ such that

$$V_N(z + Re^{i\vartheta}) \;=\; \int_{\mathbb{C}} F(z + Re^{i\vartheta} + \zeta)\,d\mu_{k(\vartheta)}(\zeta).$$

Then define v by the formula

$$\int_C G(\zeta)\,dv(\zeta) \;=\; \frac{1}{2\pi}\int_0^{2\pi}\int_C G(\zeta + Re^{i\vartheta})\,d\mu_{k(\vartheta)}(\zeta)\,d\vartheta$$

and verify that it belongs to $\mathfrak{L}$.)

(d) Hence show that

$$V(z) \;=\; \sup_{\mu\in\mathfrak{L}}\int_C F(z + \zeta)\,d\mu(\zeta) \quad (sic!)$$

is superharmonic.

(Hint: Since F is continuous, we also have $V(z) \;=\; \sup_k\int_C F(z + \zeta)\,d\mu_k(\zeta)$
with the μ_k from (b). That is, $V(z) \;=\; \lim_{N\to\infty} V_N(z)$, where the V_N are the
functions from (c). But by (c),

$$\frac{1}{2\pi}\int_0^{2\pi} V_N(z + Re^{i\vartheta})\,d\vartheta \;\leqslant\; V(z)$$

for each N. Use monotone convergence.)

(e) Show that

$$\sup_{\mu\in\mathfrak{M}}\int_C F(z + \zeta)\,d\mu(\zeta) \;=\; (\mathfrak{M}F)(z).$$

(Hint: The left side is surely $\geqslant$ the function $V(z)$ from (d). Observe
that $V(z) \geqslant F(z)$; for this the measures v, specified above may be used.
This makes V a *superharmonic majorant* of F ! Refer to (a) and to the
definition of $\mathfrak{M}F$.)

Remark. The last problem exhibits $\mathfrak{M}F$ as a *maximal function* formed
from F by using the *Jensen measures*.

Each $\mu\in\mathfrak{M}$ acts as a *reproducing measure for functions harmonic on* $\mathbb{C}$.
We have, in other words,

$$\int_C H(z + \zeta)\,d\mu(\zeta) \;=\; H(z), \qquad z\in\mathbb{C},$$

for every function H harmonic on $\mathbb{C}$ and every Jensen measure μ. It is
important to realize that *not every positive measure μ of compact support
having this reproducing property is a Jensen measure*. The following example
was shown to me by T. Lyons:

Take

$$d\mu(\zeta) \;=\; \varphi(\zeta)\,d\xi\,d\eta \;+\; \tfrac{1}{4}d\delta_1(\zeta),$$

where δ_1 is the unit mass concentrated at the point 1 and

$$\varphi(\zeta) \quad = \quad \begin{cases} \dfrac{1}{4\pi}, & |\zeta| \leqslant 2 \text{ and } |\zeta - 1| \geqslant 1, \\[2ex] 0 \text{ otherwise.} \end{cases}$$

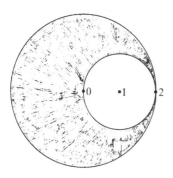

Figure 241

Then, since

$$\frac{1}{4}H(1) \quad = \quad \frac{1}{4\pi}\iint_{|\zeta-1|<1} H(\zeta)\,\mathrm{d}\xi\,\mathrm{d}\eta$$

for functions H harmonic on $\mathbb{C}$, we have

$$\int_{\mathbb{C}} H(\zeta)\,\mathrm{d}\mu(\zeta) \quad = \quad \frac{1}{4\pi}\iint_{|\zeta|\leqslant 2} H(\zeta)\,\mathrm{d}\xi\,\mathrm{d}\eta \quad = \quad H(0),$$

and similarly, by translation,

$$\int_{\mathbb{C}} H(z+\zeta)\,\mathrm{d}\mu(\zeta) \quad = \quad H(z)$$

for such functions H.

However, $U(z) = \log(1/|z-1|)$ is superharmonic in $\mathbb{C}$, and, with the present μ,

$$\int_{\mathbb{C}} U(\zeta)\,\mathrm{d}\mu(\zeta) \quad = \quad \infty \quad \text{although} \quad U(0) = 0.$$

This measure μ is therefore *not* in $\mathfrak{M}$.

The reader interested in a general treatment of the matters taken up in this article should consult a recent book by Gamelin (with *Jensen measures* in its title).

3.   **How $\mathfrak{M}F$ gives us a multiplier if it is finite**

Starting now with a continuous* weight $W(x) \geq 1$ for which $\int_{-\infty}^{\infty}(\log W(t)/(1+t^2))\mathrm{d}t < \infty$, we choose and fix an $A > 0$ and form the function

$$F(z) = \frac{1}{\pi}\int_{-\infty}^{\infty}\frac{|\Im z|\log W(t)}{|z-t|^2}\mathrm{d}t - A|\Im z|,$$

the expression on the right being interpreted as $\log W(x)$ when $z = x \in \mathbb{R}$. This function F is then continuous and the material of the preceding article applies to it; the smallest superharmonic majorant, $\mathfrak{M}F$, of F is thus at our disposal.

Our object in the present article is to establish a *converse* to the observation made near the beginning of the last one. This amounts to showing that if $\mathfrak{M}F$ is finite, one actually *has* an increasing function ρ, zero on a neighborhood of the origin, such that

$$\frac{\rho(t)}{t} \leq \frac{A}{\pi} + o(1) \qquad \text{for } t \longrightarrow \pm\infty$$

and that

$$\log W(x) + \gamma x + \int_{-\infty}^{\infty}\left(\log\left|1 - \frac{x}{t}\right| + \frac{x}{t}\right)\mathrm{d}\rho(t) \leq \text{const.} \quad \text{for } x \in \mathbb{R}$$

with a certain constant γ. We will do that by deriving a *formula*,

$$(\mathfrak{M}F)(z) = (\mathfrak{M}F)(0) - \gamma\Re z - \int_{-\infty}^{\infty}\left(\log\left|1 - \frac{z}{t}\right| + \frac{\Re z}{t}\right)\mathrm{d}\rho(t),$$

involving an increasing function ρ with (subject to an unimportant auxiliary condition on W) the first of the properties in question, and then by simply *using* the fact that $(\mathfrak{M}F)(x)$ is a *majorant* of $F(x) = \log W(x)$. That necessitates our making a preliminary examination of $\mathfrak{M}F$ for the present function F.

Lemma. *If, for $F(z)$ given by the above formula, $\mathfrak{M}F$ is finite, we have*

$$\int_{-\infty}^{\infty}\frac{(\mathfrak{M}F)(t)}{1+t^2}\mathrm{d}t < \infty,$$

---

* The regularity requirement for weights discussed in article 1 does not, *in itself*, imply their continuity. Nevertheless, in treating weights meeting the requirement, further restriction to the continuous ones (or even to those of class $\mathscr{C}_\infty$) does not constitute a serious limitation. See the first theorem of article 1.

and then

$$(\mathfrak{M}F)(z) \;=\; \frac{1}{\pi}\int_{-\infty}^{\infty}\frac{|\Im z|(\mathfrak{M}F)(t)}{|z-t|^2}\,dt \;-\; A|\Im z|$$

for $z \notin \mathbb{R}$.

Proof. Since $F(z) = F(\bar{z})$, we have $\min(U(z),\,U(\bar{z})) \geqslant F(z)$ for any superharmonic majorant U of F. By the next-to-the-last theorem of §A.1, the min just written is also superharmonic in z; it is, on the other hand, $\leqslant U(z)$, and does not change when z is replaced by $\bar{z}$. Therefore, for $\mathfrak{M}F$, the *smallest* superharmonic majorant of F, we have

$$(\mathfrak{M}F)(\bar{z}) \;=\; (\mathfrak{M}F)(z),$$

and for this reason it is necessary only to investigate $\mathfrak{M}F$ in the upper half plane.

The function $F(z)$ under consideration is *harmonic* for $\Im z > 0$, and thus, by the second lemma of the preceding article, $(\mathfrak{M}F)(z)$ *is too*, as long as it is *finite*. Because $(\mathfrak{M}F)(z) \geqslant F(z)$,

$$(\mathfrak{M}F)(z) \;+\; A\Im z \;\geqslant\; \frac{1}{\pi}\int_{-\infty}^{\infty}\frac{\Im z \,\log W(t)}{|z-t|^2}\,dt,$$

a quantity $\geqslant 0$, for $\Im z > 0$ ($W(t)$ being $\geqslant 1$). The function on the left is hence *harmonic and positive in $\Im z > 0$.*

According to Chapter III, §F.1, we therefore have

$$(\mathfrak{M}F)(z) \;+\; A\Im z \;=\; \alpha\Im z \;+\; \frac{1}{\pi}\int_{-\infty}^{\infty}\frac{\Im z \,d\mu(t)}{|z-t|^2}$$

in $\{\Im z > 0\}$, where $\alpha \geqslant 0$ and μ is some *positive measure* on $\mathbb{R}$, with $\int_{-\infty}^{\infty}(1+t^2)^{-1}d\mu(t) < \infty$. But $(\mathfrak{M}F)(z)$ is everywhere continuous by the fourth lemma of the last article; it is, in particular, continuous *up to the real axis.* Thus, $d\mu(t) = (\mathfrak{M}F)(t)dt$, and

$$(\mathfrak{M}F)(z) \;=\; (\alpha - A)\Im z \;+\; \frac{1}{\pi}\int_{-\infty}^{\infty}\frac{\Im z \,(\mathfrak{M}F)(t)}{|z-t|^2}\,dt$$

for $\Im z > 0$. Using the symmetry of $(\mathfrak{M}F)(z)$ with respect to the x-axis just noted, we see that

$$(\mathfrak{M}F)(z) \;=\; (\alpha - A)|\Im z| \;+\; \frac{1}{\pi}\int_{-\infty}^{\infty}\frac{|\Im z|(\mathfrak{M}F)(t)}{|z-t|^2}\,dt$$

(with the usual interpretation of the right side for $z \in \mathbb{R}$).

Here, $\alpha \geqslant 0$; it is claimed that α is in fact *zero*. Thanks to the sign of α,

$-\alpha|\Im z|$ is *superharmonic* (!), and the same is true of the difference

$$(\mathfrak{M}F)(z) \; - \; \alpha|\Im z|.$$

However, $(\mathfrak{M}F)(t) \geqslant F(t) = \log W(t)$, $\mathfrak{M}F$ being a majorant of F, so this difference must, by the preceding formula, be

$$\geqslant \; -A|\Im z| \; + \; \frac{1}{\pi}\int_{-\infty}^{\infty} \frac{|\Im z|\log W(t)}{|z-t|^2}\,dt \; = \; F(z).$$

$(\mathfrak{M}F)(z) \; - \; \alpha|\Im z|$ is thus a *superharmonic majorant* of $F(z)$, and therefore $\geqslant (\mathfrak{M}F)(z)$, the least such majorant. This makes $\alpha \leqslant 0$. Since $\alpha \geqslant 0$ as we know, we see that $\alpha = 0$, as claimed.

With $\alpha = 0$, the above formula for $(\mathfrak{M}F)(z)$ reduces to the desired representation. We are done.

Theorem. *Suppose that for a given continuous weight* $W(x) \geqslant 1$ *the function* $\mathfrak{M}F$ *corresponding to*

$$F(z) \;=\; \frac{1}{\pi}\int_{-\infty}^{\infty} \frac{|\Im z|\log W(t)}{|z-t|^2}\,dt \;-\; A|\Im z|$$

(where $A > 0$ *) is finite. If* $(\mathfrak{M}F)(z)$ *is also harmonic in a neighborhood of the origin, we have*

$$(\mathfrak{M}F)(z) \;=\; (\mathfrak{M}F)(0) \;-\; \gamma\Re z \;-\; \int_{-\infty}^{\infty}\left(\log\left|1 - \frac{z}{t}\right| + \frac{\Re z}{t}\right)d\rho(t),$$

with a constant γ *and a certain increasing function* $\rho(t)$, *zero on a neighborhood of the origin, such that*

$$\frac{\rho(t)}{t} \longrightarrow \frac{A}{\pi} \quad \text{for } t \longrightarrow \pm\infty.$$

Remark. The subsidiary requirement that $(\mathfrak{M}F)(z)$ be harmonic in a neighborhood of the origin serves merely to ensure $\rho(t)$'s vanishing in such a neighborhood; it can be lifted, but then the corresponding representation for $\mathfrak{M}F$ looks more complicated (see problem 57 below). Later on in this article, we will see that the harmonicity requirement does not really limit applicability of the boxed formula.

Proof of theorem. Is based on the Riesz representation from §A.2; to the superharmonic function $(\mathfrak{M}F)(z)$ we apply that representation as it is formulated in the remark preceding the last theorem of §A.2 (see the boxed formula there). For each $R > 0$ this gives us a positive measure μ_R on $\{|\zeta| \leqslant R\}$ and a function $H_R(z)$ harmonic *in the interior* of that disk, such that

$$(\mathfrak{M}F)(z) = \int_{|\zeta| \leqslant R} \log \frac{1}{|z-\zeta|} \, d\mu_R(\zeta) + H_R(z) \qquad \text{for } |z| < R.$$

By problem 48(c), the measures μ_R and $\mu_{R'}$ *agree* in $\{|\zeta| < R\}$ whenever $R' > R$; this means that we actually have a *single positive* (and in general *infinite*) *Borel measure* μ on $\mathbb{C}$ whose restriction to each open disk $\{|\zeta| < R\}$ is the corresponding μ_R (cf. problem 49). This enables us to rewrite the last formula as

$$(\mathfrak{M}F)(z) = \int_{|\zeta| \leqslant R} \log \frac{1}{|z-\zeta|} \, d\mu(\zeta) + H_R(z), \qquad |z| < R,$$

with, for each R, a certain function $H_R(z)$ (N.B. perhaps *not the same as the previous $H_R(z)$* !) *harmonic* in $\{|z| < R\}$.

We see by the preceding lemma that $(\mathfrak{M}F)(z)$ is itself harmonic both in $\{\Im z > 0\}$ and in $\{\Im z < 0\}$, so, according to the last theorem of §A.2, μ cannot have any mass in either of those half planes. By the same token, μ has no mass in a certain neighborhood of the origin, $\mathfrak{M}F$ being, by hypothesis, harmonic in such a neighborhood. There is thus an increasing function $\rho(t)$, zero on a neighborhood of the origin, such that

$$\mu(E) = \int_{E \cap \mathbb{R}} d\rho(t)$$

for Borel sets $E \subseteq \mathbb{C}$, and we have

$$(\mathfrak{M}F)(z) = \int_{-R}^{R} \log \frac{1}{|z-t|} \, d\rho(t) + H_R(z) \qquad \text{for } |z| < R,$$

with H_R harmonic there. Our desired representation will be obtained by making $R \longrightarrow \infty$ in this relation. For that purpose, we need to know the asymptotic behaviour of $\rho(t)$ as $t \longrightarrow \pm \infty$.

It is claimed that the ratio

$$\frac{\rho(r) - \rho(-r)}{r}$$

(which is certainly positive) *remains bounded* when $r \longrightarrow \infty$. Fixing any R, let us consider values of $r < R$. Using the preceding formula and reasoning as in the proof of the last theorem in §A.2, we easily find that

$$\frac{1}{2\pi} \int_{-\pi}^{\pi} (\mathfrak{M}F)(re^{i\vartheta}) d\vartheta = \int_{-R}^{R} \min\left(\log \frac{1}{|t|}, \log \frac{1}{r} \right) d\rho(t) + H_R(0),$$

and thence, subtracting $(\mathfrak{M}F)(0)$ from both sides, that

$$-\int_{-R}^{R}\log^{+}\frac{r}{|t|}\,\mathrm{d}\rho(t) \;=\; \frac{1}{2\pi}\int_{-\pi}^{\pi}(\mathfrak{M}F)(re^{i\vartheta})\mathrm{d}\vartheta \;-\; (\mathfrak{M}F)(0)$$

for $0 < r < R$. Here, $\rho(t)$ vanishes on a neighborhood of the origin, so we can integrate the left side by parts to get

$$\int_{0}^{r}\frac{\rho(t)\,-\,\rho(-t)}{t}\,\mathrm{d}t \;=\; (\mathfrak{M}F)(0) \;-\; \frac{1}{2\pi}\int_{-\pi}^{\pi}(\mathfrak{M}F)(re^{i\vartheta})\mathrm{d}\vartheta,$$

which is, of course, nothing but a version of Jensen's formula. In it, R no longer appears, so it is valid for all $r > 0$.

By the lemma, however,

$$-(\mathfrak{M}F)(z) \;=\; A|\Im z| \;-\; \frac{1}{\pi}\int_{-\infty}^{\infty}\frac{|\Im z|(\mathfrak{M}F)(t)}{|z-t|^{2}}\,\mathrm{d}t,$$

a quantity $\leqslant A|\Im z|$, since $(\mathfrak{M}F)(t) \geqslant \log W(t) \geqslant 0$. Using this in the previous relation, we get

$$\int_{0}^{r}\frac{\rho(t)\,-\,\rho(-t)}{t}\,\mathrm{d}t \;\leqslant\; (\mathfrak{M}F)(0) \;+\; \frac{2A}{\pi}r;$$

whence

$$\rho(r) \;-\; \rho(-r) \;\leqslant\; (\mathfrak{M}F)(0) \;+\; \frac{2A}{\pi}er$$

by the argument of problem 1(a) (!), $\rho(t)$ being increasing. Since $\rho(t)$ also vanishes in a neighborhood of 0, we see that

$$\frac{\rho(t)}{t} \;\leqslant\; \text{const.} \quad \text{on } \mathbb{R}.$$

Once this is known, it follows by reasoning like that of §A, Chapter III, that the integral

$$\int_{-\infty}^{\infty}\left(\log\left|1-\frac{z}{t}\right| \;+\; \frac{\Re z}{t}\right)\mathrm{d}\rho(t)$$

is convergent (*a priori*, to $-\infty$, possibly, when $z \in \mathbb{R}$) for all values of z — one needs here to again use the vanishing of $\rho(t)$ for t near 0. That integral, however, obviously differs from

$$\int_{-R}^{R}\log|z-t|\,\mathrm{d}\rho(t)$$

by a function *harmonic* for $|z| < R$. Referring to the previous representation of $(\mathfrak{M}F)(z)$ in that disk, we see that

$$G(z) \;=\; (\mathfrak{M}F)(z) \;+\; \int_{-\infty}^{\infty} \left(\log\left|1 - \frac{z}{t}\right| + \frac{\Re z}{t} \right) d\rho(t)$$

must be *harmonic* for $|z| < R$, and hence finally for *all* z, since the parameter R no longer occurs on the right. Our local Riesz representations for $(\mathfrak{M}F)(z)$ in the disks $\{|z| < R\}$ thus have a *global* version,

$$(\mathfrak{M}F)(z) \;=\; -\int_{-\infty}^{\infty} \left(\log\left|1 - \frac{z}{t}\right| + \frac{\Re z}{t} \right) d\rho(t) \;+\; G(z),$$

valid for all z, with G harmonic everywhere.

We proceed to investigate $G(z)$'s behaviour for large $|z|$. The lemma gives, first of all,

$$(\mathfrak{M}F)(z) \;\leqslant\; \frac{1}{\pi} \int_{-\infty}^{\infty} \frac{|\Im z|(\mathfrak{M}F)(t)}{|z - t|^2}\, dt,$$

$(\mathfrak{M}F)(t)$ being $\geqslant 0$. Therefore

$$[G(z)]^+ \;\leqslant\; \frac{1}{\pi} \int_{-\infty}^{\infty} \frac{|\Im z|(\mathfrak{M}F)(t)}{|z - t|^2}\, dt \;+\; \left(\int_{-\infty}^{\infty} \left(\log\left|1 - \frac{z}{t}\right| + \frac{\Re z}{t} \right) d\rho(t) \right)^{\!+}.$$

According to the discussion at the beginning of §B, Chapter III, our bound on the growth of $\rho(t)$ makes the *second* term on the right $\leqslant\ O(|z|\log|z|)$ for large values of $|z|$; we thus have

$$\int_{-\pi}^{\pi} [G(re^{i\vartheta})]^+\, d\vartheta \;\leqslant\; \text{const.}\, r\log r \;+\; \frac{1}{\pi} \int_{-\pi}^{\pi} \int_{-\infty}^{\infty} \frac{r|\sin\vartheta|(\mathfrak{M}F)(t)}{r^2 + t^2 - 2rt\cos\vartheta}\, dt\, d\vartheta$$

when r is large, and, desiring to estimate the integral on the *left*, we must study the one figuring on the *right*. Changing the order of integration converts the latter to

$$\frac{2}{\pi} \int_{-\infty}^{\infty} \frac{1}{t} \log\left| \frac{r + t}{r - t} \right| (\mathfrak{M}F)(t)\, dt,$$

which we handle by resorting to a trick.

Take the *average* of the expression in question for $R \leqslant r \leqslant 2R$, say, where $R > 0$ is arbitrary. That works out to

$$\frac{2}{\pi R} \int_{-\infty}^{\infty} \int_{R}^{2R} \frac{1}{t} \log\left| \frac{r + t}{r - t} \right| (\mathfrak{M}F)(t)\, dr\, dt \;=\; \frac{2}{\pi R} \int_{-\infty}^{\infty} \Psi\!\left(\frac{R}{|t|} \right) (\mathfrak{M}F)(t)\, dt,$$

where

$$\Psi(u) \;=\; \int_u^{2u} \log\left|\frac{s+1}{s-1}\right| \, ds.$$

The last integral can be directly evaluated, but here it is better to use power series and see how it acts when $u \longrightarrow 0$ and when $u \longrightarrow \infty$.

For $0 < u < 1$, expand the integrand in powers of s to get

$$\Psi(u) \;=\; 3u^2 \,+\, O(u^4), \qquad 0 < u < 1.$$

For $u > 1$, we expand the integrand in powers of $1/s$ and find that

$$\Psi(u) \;=\; 2\log 2 \,+\, O\!\left(\frac{1}{u^2}\right), \qquad u > 1.$$

$\Psi(R/|t|)/R$ thus behaves like $1/R$ for *small* values of $|t|/R$ and like R/t^2 for *large* ones, so, all in all,

$$\frac{2}{\pi R}\,\Psi\!\left(\frac{R}{|t|}\right) \;\leqslant\; \mathrm{const.}\,\frac{R}{R^2 + t^2} \qquad \text{for } t \in \mathbb{R}.$$

Substituting this into the previous relation, we see that

$$\frac{2}{\pi R}\int_R^{2R}\int_{-\infty}^{\infty} \frac{1}{t}\log\left|\frac{r+t}{r-t}\right| (\mathfrak{M}F)(t)\,dt\,dr \;\leqslant\; \mathrm{const.}\int_{-\infty}^{\infty} \frac{R}{R^2+t^2}\,(\mathfrak{M}F)(t)\,dt.$$

This, however, *implies the existence of an r', $R \leqslant r' \leqslant 2R$,* for which

$$\frac{2}{\pi}\int_{-\infty}^{\infty} \frac{1}{t}\log\left|\frac{r'+t}{r'-t}\right| (\mathfrak{M}F)(t)\,dt \;\leqslant\; \mathrm{const.}\int_{-\infty}^{\infty} \frac{R}{R^2+t^2}\,(\mathfrak{M}F)(t)\,dt.$$

Here, the right side is $\leqslant \mathrm{const.}\,R \leqslant \mathrm{const.}\,r'$ (and is even $o(R)$) for large R, since $\int_{-\infty}^{\infty} ((\mathfrak{M}F)(t)/(1 + t^2))\,dt < \infty$. Taking r equal to such an r' in our original relation involving G thus yields

$$\int_{-\pi}^{\pi} [G(r'e^{i\vartheta})]^+ \, d\vartheta \;\leqslant\; \mathrm{const.}\,(r'\log r' \,+\, r')$$

when R, and hence r', is large.

Letting R take successively the values 2^n with $n = 1, 2, 3, \ldots$, we obtain in this way a certain sequence of numbers r_n tending to ∞ for which

$$\int_{-\pi}^{\pi} [G(r_n e^{i\vartheta})]^+ \, d\vartheta \;\leqslant\; O(r_n \log r_n).$$

Since $G(z)$ is harmonic, we have on the other hand

$$\int_{-\pi}^{\pi} \left([G(r_n e^{i\vartheta})]^+ \,-\, [G(r_n e^{i\vartheta})]^-\right) d\vartheta \;=\; \int_{-\pi}^{\pi} G(r_n e^{i\vartheta})\,d\vartheta \;=\; 2\pi G(0),$$

so, subtracting this relation from *twice* the preceding, we get

$$\int_{-\pi}^{\pi} |G(r_n e^{i\vartheta})| \, d\vartheta \;\; \leqslant \;\; O(r_n \log r_n).$$

Now it follows that $G(z)$ must be of the form $A_0 + A_1 \Re z$. We have, indeed, $G(\bar{z}) = G(z)$, since $\mathfrak{M}F$ and the integral involving $d\rho$ have that property; the function $G(z)$, harmonic everywhere, is therefore given by a series development

$$G(re^{i\vartheta}) \;\; = \;\; \sum_{k=0}^{\infty} A_k r^k \cos k\vartheta.$$

For $k > 1$, we have

$$A_k \;\; = \;\; \frac{1}{\pi r^k} \int_{-\pi}^{\pi} G(re^{i\vartheta}) \cos k\vartheta \, d\vartheta.$$

Putting $r = r_n$ and making $n \longrightarrow \infty$, we see, using the estimate just found, that $A_k = 0$. The series thus boils down to *its first two terms*.

Going back to our global version of the Riesz representation for $(\mathfrak{M}F)(z)$ and using the description of G just found, we see that

$$(\mathfrak{M}F)(z) \;\; = \;\; A_0 \; + \; A_1 \Re z \; - \; \int_{-\infty}^{\infty} \left(\log \left| 1 - \frac{z}{t} \right| \; + \; \frac{\Re z}{t} \right) d\rho(t).$$

Because $\rho(t)$ vanishes for t near 0, it is obvious that $A_0 = (\mathfrak{M}F)(0)$. Denoting A_1 by $-\gamma$, we now have the formula we set out to establish.

In order to complete this proof, we must still refine the estimate

$$\frac{\rho(t)}{t} \;\; \leqslant \;\; \text{const.}$$

obtained and used above to the asymptotic relation

$$\frac{\rho(t)}{t} \;\; = \;\; \frac{A}{\pi} \; + \; o(1), \qquad t \longrightarrow \pm\infty.$$

For this, some version of Levinson's theorem (the one from Chapter III) must be used.

Write

$$V(z) \;\; = \;\; \gamma \Re z \; + \; \int_{-\infty}^{\infty} \left(\log \left| 1 - \frac{z}{t} \right| \; + \; \frac{\Re z}{t} \right) d\rho(t);$$

then, by the previous lemma and the representation formula just proved,

we have

$$V(z) - (\mathfrak{M}F)(0) = -(\mathfrak{M}F)(z) = A|\mathfrak{I}z| - \frac{1}{\pi}\int_{-\infty}^{\infty}\frac{|\mathfrak{I}z|(\mathfrak{M}F)(t)}{|z-t|^2}\,dt.$$

From this, we readily see that

$$\frac{V(iy)}{|y|} \longrightarrow A \quad\text{as}\quad y \longrightarrow \pm\infty,$$

whilst

$$V(z) \leqslant (\mathfrak{M}F)(0) + A|\mathfrak{I}z|$$

for all z.

   Take, as in the proofs of the last two theorems of article 1, an entire function φ such that

$$\log|\varphi(z)| = \beta\mathfrak{R}z + \int_{-\infty}^{\infty}\left(\log\left|1 - \frac{z}{t}\right| + \frac{\mathfrak{R}z}{t}\right)d[\rho(t)]$$

where β is constant; according to a lemma from that article, we have, for suitable choice of β, the inequality

$$\log|\varphi(z)| \leqslant V(z) + \log^+\left|\frac{\mathfrak{R}z}{\mathfrak{I}z}\right| + \log^+|z| + O(1).$$

Applying this first with $z = x \pm i$ and using the preceding estimate for V, we see, taking account of the fact that $|\varphi(z)|$ *diminishes* when $|\mathfrak{I}z|$ does, that

$$|\varphi(z)| \leqslant \text{const.}(|z|^2 + 1), \qquad |\mathfrak{I}z| \leqslant 1.$$

We next find from the same relations that

$$|\varphi(z)| \leqslant \text{const.}(|z|^2 + 1)e^{A|\mathfrak{I}z|}$$

when $|\mathfrak{I}z| > 1$; in view of the preceding inequality such an estimate (with perhaps a larger constant) must then hold *everywhere*. $\varphi(z)$ is thus of *exponential type*.

   A computation like one near the end of the next-to-the-last theorem in article 1 now yields, for $y \in \mathbb{R}$,

$$\log|\varphi(iy)| - V(iy) = \int_{-\infty}^{\infty}\frac{y^2}{y^2 + t^2}\frac{[\rho(t)] - \rho(t)}{t}\,dt.$$

Since $[\rho(t)] - \rho(t)$ is *bounded* (above and below!) and zero on a neighborhood of the origin, the integral on the right is $o(|y|)$ for

$y \longrightarrow \pm \infty$, and hence

$$\frac{\log |\varphi(iy)|}{|y|} \longrightarrow A \quad \text{as } y \longrightarrow \pm \infty,$$

in view of the above similar relation for $V(iy)$.

By the preceding estimates on $\varphi(z)$, we obviously have

$$\int_{-\infty}^{\infty} \frac{\log^{+} |\varphi(x)|}{1 + x^2} dx \quad < \quad \infty,$$

and the Levinson theorem from §H.2 of Chapter III can be applied to φ. Referring to the last of the above relations, we see in that way that

$$\frac{[\rho(t)]}{t} \longrightarrow \frac{A}{\pi} \quad \text{as } t \longrightarrow \pm \infty.$$

Therefore,

$$\frac{\rho(t)}{t} \longrightarrow \frac{A}{\pi} \quad \text{for } t \longrightarrow \pm \infty.$$

Our theorem is proved.

Problem 57

If $(\mathfrak{M}F)(z)$ is finite, but not necessarily harmonic in a neighborhood of 0, find a representation for it analogous to the one furnished by the result just obtained.

As stated previously, the last theorem has quite general utility in spite of its harmonicity requirement. Any situation involving a finite function $\mathfrak{M}F$ can be reduced to one for which the corresponding $\mathfrak{M}F$ *is* harmonic near 0. The easiest way of doing that is to use the following

Lemma. *Let $W(t)$, continuous and $\geqslant 1$ on $\mathbb{R}$, be $\equiv 1$ for $-h < t < h$, where $h > 0$, and suppose that for $|x| < h$, we have*

$$\frac{1}{\pi} \int_{-\infty}^{\infty} \frac{\log W(t)}{(x - t)^2} dt \quad > \quad A,$$

with the integral on the left convergent. Then the function

$$F(z) \quad = \quad \frac{1}{\pi} \int_{-\infty}^{\infty} \frac{|\Im z| \log W(t)}{|z - t|^2} dt \quad - \quad A|\Im z|$$

satisfies

$$\frac{1}{2\pi}\int_{-\pi}^{\pi} F(re^{i\vartheta})\,d\vartheta \;>\; F(0) \;=\; 0$$

for $0 < r < h$.

Proof. We have $F(z) = F(\bar z)$, so

$$\frac{1}{2\pi}\int_{-\pi}^{\pi} F(re^{i\vartheta})\,d\vartheta \;=\; \left|\,\frac{1}{\pi}\int_{0}^{\pi} F(re^{i\vartheta})\,d\vartheta.\right.$$

It will be convenient to denote the right-hand integral by $J(r)$ and to work with the function

$$G(z) \;=\; \frac{1}{\pi}\int_{-\infty}^{\infty} \frac{\Im z \log W(t)}{|z-t|^2}\,dt \;-\; A\,\Im z \quad \text{(sic!)}$$

instead of $F(z)$; we of course also have

$$J(r) \;=\; \frac{1}{\pi}\int_{0}^{\pi} G(re^{i\vartheta})\,d\vartheta.$$

In the present circumstances the function $G(z)$ is finite, and hence *harmonic*, in both the upper and the lower half planes. Moreover, since $\log W(t) \equiv 0$ for $|t| < h$, $G(z)$ (taken as *zero* on the real interval $(-h, h)$) is actually harmonic* in $\mathbb{C} \sim (-\infty, -h] \sim [h, \infty)$ and hence $\mathscr{C}_\infty$ in that region. There is thus no obstacle to differentiating under the integral sign so as to get

$$\frac{dJ(r)}{dr} \;=\; \frac{1}{\pi r}\int_{0}^{\pi} \frac{\partial G(re^{i\vartheta})}{\partial r}\,r\,d\vartheta, \qquad 0 < r < h.$$

Let $\mathscr{D}_r$ be the semi-circle of radius r lying in the upper half plane, having for diameter the real segment $[-r, r]$:

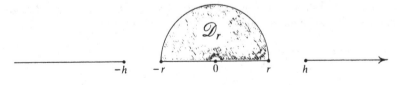

Figure 242

---

* by Schwarz' reflection principle, since $G(\bar z) = -G(z)$

When $r < h$, the function $G(z)$ is harmonic in a region including the *closure* of $\mathcal{D}_r$, so we can use Green's theorem to get

$$\int_{\partial\mathcal{D}_r} \frac{\partial G(\zeta)}{\partial n_\zeta} |d\zeta| \;\; = \;\; \int\!\!\int_{\mathcal{D}_r} (\nabla^2 G)(\zeta)\, d\xi\, d\eta \;\; = \;\; 0,$$

where $\partial/\partial n_\zeta$ denotes differentiation along the outward normal to $\partial\mathcal{D}_r$ at ζ. The left-hand expression is just

$$\int_0^\pi \frac{\partial G(re^{i\vartheta})}{\partial r}\, r\, d\vartheta \;\; - \;\; \int_{-r}^r G_y(x)\, dx,$$

so the previous relation yields

$$J'(r) \;\; = \;\; \frac{1}{\pi r}\int_{-r}^r G_y(x)\, dx.$$

Here, $G(z) = F(z)$ for $\Im z \geqslant 0$ with $G(x) = F(x) = \log W(x) = 0$ for $-h < x < h$, so, for such x,

$$G_y(x) \;\; = \;\; \lim_{\Delta y \to 0+} \frac{F(x + i\Delta y)}{\Delta y},$$

which, by our formula for F, is equal to

$$\frac{1}{\pi}\int_{-\infty}^\infty \frac{\log W(t)}{(x-t)^2}\, dt \;\; - \;\; A.$$

If, then, this expression is > 0 for $|x| < h$, we must, by the preceding formula, have

$$J'(r) \; > \; 0 \quad \text{for } 0 < r < h.$$

Obviously, $J(r) \longrightarrow F(0) = 0$ for $r \longrightarrow 0$. Therefore,

$$F(0) \;\; < \;\; J(r) \;\; = \;\; \frac{1}{2\pi}\int_{-\pi}^\pi F(re^{i\vartheta})\, d\vartheta$$

when $0 < r < h$, given that the hypothesis holds. We are done.

Corollary. *Given $W(x)$ continuous and $\geqslant 1$ with $\int_{-\infty}^\infty (\log W(t)/(1+t^2))\, dt < \infty$, and the number $A > 0$, form, for $h > 0$, the new weight*

$$W_h(x) \;\; = \;\; \begin{cases} 1, & |x| \leqslant h, \\ e^{2\pi A h} W(x), & |x| \geqslant 2h, \\ \text{linear for } -2h \leqslant x \leqslant -h \text{ and for } h \leqslant x \leqslant 2h. \end{cases}$$

Put then

$$F_h(z) = \frac{1}{\pi} \int_{-\infty}^{\infty} \frac{|\Im z| \log W_h(t)}{|z-t|^2} dt - A|\Im z|.$$

If $(\mathfrak{M}F_h)(z)$ *is finite, it is harmonic in a neighborhood of the origin.*

Proof. When $-h < x < h$,

$$\frac{1}{\pi} \int_{-\infty}^{\infty} \frac{\log W_h(t)}{(x-t)^2} dt \geqslant \frac{1}{\pi} \int_{2h}^{\infty} \left(\frac{2\pi Ah}{(t-x)^2} + \frac{2\pi Ah}{(t+x)^2} \right) dt$$

$$= \frac{8Ah^2}{4h^2 - x^2} \geqslant 2A > A.$$

The lemma, applied to W_h and F_h, thus yields

$$F_h(0) < \frac{1}{2\pi} \int_{-\pi}^{\pi} F_h(re^{i\vartheta}) d\vartheta$$

for $0 < r < h$. Since, however, $\mathfrak{M}F_h$ is a *superharmonic majorant* of F_h, the right-hand integral is

$$\leqslant \frac{1}{2\pi} \int_{-\pi}^{\pi} (\mathfrak{M}F_h)(re^{i\vartheta}) d\vartheta \leqslant (\mathfrak{M}F_h)(0),$$

i.e.,

$$F_h(0) < (\mathfrak{M}F_h)(0).$$

The corollary now follows by the third lemma of article 2.

The preceding results give us our desired converse to the statement from the last article.

Theorem. *Let* $W(x) \geqslant 1$ *be continuous, with*

$$\int_{-\infty}^{\infty} (\log W(t)/(1+t^2)) dt < \infty,$$

and put

$$F(z) = \frac{1}{\pi} \int_{-\infty}^{\infty} \frac{|\Im z| \log W(t)}{|z-t|^2} dt - A|\Im z|,$$

where $A > 0$, *interpreting the right side in the usual way when* $z \in \mathbb{R}$. *If the smallest superharmonic majorant,* $\mathfrak{M}F$, *of* F *is finite, there is an increasing function* ρ, *zero on a neighborhood of the origin, for which*

$$\log W(x) + \gamma x + \int_{-\infty}^{\infty}\left(\log\left|1 - \frac{x}{t}\right| + \frac{x}{t}\right)d\rho(t) \leqslant \quad \text{const.,} \quad x \in \mathbb{R}$$

(with a certain constant γ *), while*

$$\frac{\rho(t)}{t} \longrightarrow \frac{A}{\pi} \quad \text{as } t \longrightarrow \pm\infty.$$

Proof. With $h > 0$, form the functions W_h and F_h figuring in the preceding corollary. Since $\log W(t) \geqslant 0$, we have

$$\log W_h(t) \leqslant \log W(t) + 2\pi A h,$$

whence

$$F_h(z) \leqslant F(z) + 2\pi A h.$$

Thus, since $(\mathfrak{M}F)(z) \geqslant F(z)$,

$$F_h(z) \leqslant (\mathfrak{M}F)(z) + 2\pi A h.$$

In the last relation, the right-hand member is *superharmonic*, and, of course, *finite* if $\mathfrak{M}F$ is. Then, however, $\mathfrak{M}F_h$, the *least superharmonic majorant* of F_h, must *also* be finite.

This, according to the corollary, implies that $(\mathfrak{M}F_h)(z)$ is harmonic in a neighborhood of the origin. Once that is known, the previous theorem gives us an increasing function ρ having the required properties, such that

$$(\mathfrak{M}F_h)(z) = (\mathfrak{M}F_h)(0) - \gamma\Re z - \int_{-\infty}^{\infty}\left(\log\left|1 - \frac{z}{t}\right| + \frac{\Re z}{t}\right)d\rho(t),$$

γ being a certain constant. Thus, since $(\mathfrak{M}F_h)(x) \geqslant F_h(x) = \log W_h(x)$,

$$\log W_h(x) + \gamma x + \int_{-\infty}^{\infty}\left(\log\left|1 - \frac{x}{t}\right| + \frac{x}{t}\right)d\rho(t)$$

$$\leqslant (\mathfrak{M}F_h)(0) \quad \text{for } x \in \mathbb{R}.$$

Let now m_h denote the maximum of $W(x)$ for $-2h \leqslant x \leqslant 2h$. Then certainly

$$\log W(x) \leqslant \log m_h + \log W_h(x),$$

$W(x)$, and hence m_h, being $\geqslant 1$. This, substituted into the previous, yields finally

$$\log W(x) \; + \; \gamma x \; + \; \int_{-\infty}^{\infty} \left(\log\left|1 - \frac{x}{t}\right| \; + \; \frac{x}{t} \right) d\rho(t)$$

$$\leqslant \; (\mathfrak{M}F_h)(0) \; + \; \log m_h \qquad \text{for } x \in \mathbb{R}.$$

We are done.

The proof just given furnishes a more precise result which is sometimes useful.

Corollary. *If* $W(x)$, *satisfying the hypothesis of the theorem, is, in addition, 1 at the origin, and the function* $\mathfrak{M}F$ *corresponding to some given* $A > 0$ *is finite, we have, for any* $\eta > 0$, *an increasing function* $\rho(t)$ *with the properties affirmed by the theorem, such that*

$$\log W(x) \; + \; \gamma x \; + \; \int_{-\infty}^{\infty} \left(\log\left|1 - \frac{x}{t}\right| \; + \; \frac{x}{t} \right) d\rho(t) \leqslant (\mathfrak{M}F)(0) + \eta, \; x \in \mathbb{R}.$$

To verify this, we first observe that the continuity of $W(x)$ makes $m_h \longrightarrow 1$ and hence $\log m_h \longrightarrow 0$ when $h \longrightarrow 0$. On the other hand,

$$(\mathfrak{M}F_h)(0) \; \leqslant \; (\mathfrak{M}F)(0) \; + \; 2\pi Ah,$$

since $(\mathfrak{M}F)(z) \; + \; 2\pi Ah$ is a superharmonic majorant of $F_h(z)$, as remarked at the beginning of the proof. The desired relation involving ρ will therefore follow from the *last* one in the proof if we take $h > 0$ small enough so as to have

$$\log m_h \; + \; 2\pi Ah \; < \; \eta.$$

These results and the obvious converse noted in article 2 are used in conjunction with the material from article 1. Referring, for instance, to the corollary of the next-to-the-last theorem in article 1, we have the

Theorem. *Let* $W(x)$, *continuous and* $\geqslant 1$ *on the real axis, fulfill the regularity requirement formulated in article 1. In order that* W *admit multipliers, it is necessary and sufficient that*

$$\int_{-\infty}^{\infty} \frac{\log W(t)}{1+t^2} dt \; < \; \infty$$

and that then, for each $A > 0$, *the smallest superharmonic majorant of*

$$\frac{1}{\pi} \int_{-\infty}^{\infty} \frac{|\Im z| \log W(t)}{|z-t|^2} dt \; - \; A|\Im z|$$

be finite.

Looking at the *last* theorem of article 1 we see in the same way that such a result holds for any weight $W(x) \geqslant 1$ of the form $|F(x)|$, where F is *entire* and of *exponential type, without any additional assumption on the regularity of* W. This fact will be used in the next §.

The regularity requirement on W figuring in the above theorem may, of course, by replaced by the milder one discussed in the scholium to article 1.*

Let us hark back for a moment to the discussion at the beginning of article 1. Can one regard the condition that $(\mathfrak{M}F)(0)$ be *finite* for *each* of the functions

$$F(z) \;=\; \frac{1}{\pi} \int_{-\infty}^{\infty} \frac{|\Im z| \log W(t)}{|z-t|^2}\, dt \;-\; A|\Im z|, \qquad A > 0,$$

as one of *regularity* to be satisfied by the weight W? *In a sense*, one *can* – see especially problem 55. Is *this*, then, the presumed *second* ('essential') kind of regularity a weight must have in order to admit multipliers?

C.   Theorems of Beurling and Malliavin

We are going to apply the results from the end of the last § so as to obtain multiplier theorems for certain kinds of continuous weights W. Those are always assumed to be $\geqslant 1$ on the real axis, and only for the *unbounded* ones can there be any question about the existence of multipliers.

One can in fact work exclusively with weights $W(x)$ *tending to* ∞ *for* $x \longrightarrow \pm\infty$ without in any way lessening the generality of the results obtained. Suppose, indeed, that we are given an unbounded weight $W(x) \geqslant 1$; then

$$\Omega(x) \;=\; (1+x^2)W(x)$$

does tend to ∞ when $x \longrightarrow \pm\infty$, and it is claimed that there is a non-zero entire function of exponential type $\leqslant A$ *whose product with Ω is bounded on $\mathbb{R}$ if and only if* there is such an entire function whose *product with W is bounded there*.

It is clearly only the *if* part of this statement that requires checking. Consider, then, that we have an entire function $\varphi(z) \not\equiv 0$ of exponential type $\leqslant A$ making $\varphi(x)W(x)$ bounded on $\mathbb{R}$. Since $W(x)$ is unbounded, $|\varphi(x)|$ cannot be constant, so the Hadamard product for φ (Chapter III

---

* See also Remark 5 near the end of §E.2.

§A) must involve *linear factors* – there must in fact be *infinitely many* of those, for otherwise $|\varphi(x)|$ would *grow like a polynomial* in x when $|x| \longrightarrow \infty$. The function $\varphi(z)$ thus has infinitely many zeros, and, taking any two of them, say α and β, we can form a new entire function,

$$\psi(z) \;\;=\;\; \frac{\varphi(z)}{(z-\alpha)(z-\beta)},$$

also of exponential type $\leqslant A$, with $\psi(x)\Omega(x)$ bounded on the real axis.

The existence of multipliers for $W(x)$ is thus fully equivalent to existence thereof for $\Omega(x)$, a weight tending to ∞ for $x \longrightarrow \pm\infty$; that is fortunate, because weights having the latter property are easier to deal with. When working with a *given* weight W, it will sometimes be convenient to form from it the new one

$$\left(1 \;+\; \frac{x^2}{M^2}\right) W(x)$$

(using a *large* value of M) or

$$(1 + x^2)^\eta W(x)$$

(taking for η a *small* value > 0), instead of dealing with the weight $\Omega(x)$ just looked at. Any of these weights $\tilde{W}$ fulfills the condition

$$\int_{-\infty}^{\infty} \frac{\log \tilde{W}(x)}{1 + x^2}\,dx \;\;<\;\; \infty$$

as long as

$$\int_{-\infty}^{\infty} \frac{\log W(x)}{1 + x^2}\,dx \;\;<\;\; \infty\,;$$

unless the latter holds W cannot, as we know, admit any multipliers.

In order to establish the existence of multipliers for a weight $\geqslant 1$ satisfying the last condition, we first form from it a new one according to one of the above recipes* if that is necessary to ensure our having a weight tending to ∞ with $|x|$. Then, choosing a number $A > 0$ and using the *new* weight W, we take the function

$$F(z) \;\;=\;\; \frac{1}{\pi}\int_{-\infty}^{\infty} \frac{|\Im z|}{|z-t|^2}\log W(t)\,dt \;-\; A|\Im z|$$

studied in §B.3. According to results obtained there, the question of our weight's *admittance of multipliers* reduces in large part to a simple decision

---

* the new weight obviously meets the local regularity requirement of §B.1 iff the original one does

about the *finiteness* of $\mathfrak{M}F$, the smallest superharmonic majorant of F, for various initial choices of the number A. We know by the first lemma of §B.2 that the latter property is *equivalent to the finiteness of $\mathfrak{M}F$ at any one point*, say that of $(\mathfrak{M}F)(0)$. To evaluate this quantity we will use harmonic estimation, guided by the knowledge that $\mathfrak{M}F$, *if* finite, must be *harmonic* in both the upper and lower half planes (first lemma of §B.3), and *also harmonic* across any real interval on which it is $> F$ (by the third lemma of §B.2).

1. **Use of the domains from §C of Chapter VIII**

Starting, then, with a continuous weight $W(x) \geqslant 1$ tending to ∞ for $x \longrightarrow \pm \infty$, we take (using some given $A > 0$) the function $F(z)$ whose formula has just been written, and look at its *smallest superharmonic majorant $\mathfrak{M}F$,* our aim being to see whether or not $(\mathfrak{M}F)(0) < \infty$. The idea is to get at $\mathfrak{M}F$ by using *other* superharmonic majorants whose qualitative behaviour is known.

For each $N > 1$, let

$$W_N(x) \;=\; \min\left(W(x),\, N\right)$$

and then form the function

$$F_N(z) \;=\; \frac{1}{\pi} \int_{-\infty}^{\infty} \frac{|\Im z|}{|z - t|^2} \log W_N(t)\, \mathrm{d}t \;-\; A|\Im z|$$

corresponding to it in the way that $F(z)$ corresponds to W. Clearly,

$$F_N(z) \;\underset{N}{\uparrow}\; F(z);$$

it is claimed that also

$$(\mathfrak{M}F_N)(z) \;\underset{N}{\uparrow}\; (\mathfrak{M}F)(z).$$

For each N, we have, indeed,

$$(\mathfrak{M}F_{N+1})(z) \;\geqslant\; F_{N+1}(z) \;\geqslant\; F_N(z),$$

so $\mathfrak{M}F_{N+1}$ is a superharmonic *majorant of F_N*, and hence

$$(\mathfrak{M}F_{N+1})(z) \;\geqslant\; (\mathfrak{M}F_N)(z),$$

the *least* superharmonic majorant of $F_N(z)$. By the same token,

$$(\mathfrak{M}F)(z) \;\geqslant\; (\mathfrak{M}F_N)(z)$$

for each N, so we have

$$\lim_{N \to \infty} (\mathfrak{M} F_N)(z) \;\leqslant\; (\mathfrak{M} F)(z).$$

The sequence $\{\mathfrak{M} F_N\}$ is, as just shown, *increasing*, so $\lim_{N \to \infty} \mathfrak{M} F_N$ is *superharmonic* by the next-to-the-last theorem of §A.1. That limit must, however, be $\geqslant \lim_{N \to \infty} F_N = F$, so it is also a *majorant* for F. As a superharmonic majorant of F, $\lim_{N \to \infty} \mathfrak{M} F_N$ is therefore $\geqslant \mathfrak{M} F$. So, since the contrary relation holds, we in fact have *equality*, as asserted.

We thus have, in particular,

$$(\mathfrak{M} F)(0) \;=\; \lim_{N \to \infty} (\mathfrak{M} F_N)(0),$$

whence, in order to verify that $(\mathfrak{M} F)(0) < \infty$, *it suffices to obtain an upper bound independent of N on the values* $(\mathfrak{M} F_N)(0)$.

Each of the functions $\mathfrak{M} F_N$ is certainly *finite*. Indeed,

$$0 \;\leqslant\; \log W_N(t) \;\leqslant\; \log N,$$

so

$$F_N(z) \;\leqslant\; \log N \,-\, A|\Im z|.$$

But the right-hand expression in this last relation is *superharmonic*! Hence,

$$(\mathfrak{M} F_N)(z) \;\leqslant\; \log N \,-\, A|\Im z|.$$

Since, on the other hand,

$$F_N(z) \;\geqslant\; -A|\Im z|,$$

we also have

$$(\mathfrak{M} F_N)(z) \;\geqslant\; -A|\Im z|.$$

Thanks to our assumption that $W(x) \longrightarrow \infty$ for $x \longrightarrow \pm \infty$, there is a certain number L, depending on N, such that

$$W_N(x) \;=\; N \quad \text{for } |x| \geqslant L.$$

Therefore $F_N(x) = \log N$ for $|x| \geqslant L$, making $(\mathfrak{M} F_N)(x) \geqslant \log N$ for such x. By one of the previous relations, we have, however, $(\mathfrak{M} F_N)(x) \leqslant \log N$ on $\mathbb{R}$. Thus,

$$(\mathfrak{M} F_N)(x) \;=\; \log N \;=\; F_N(x) \quad \text{for } |x| \geqslant L.$$

We see that *on the real axis*, $(\mathfrak{M} F_N)(x)$ (a *continuous* function by the fourth lemma of §B.2) can *strictly exceed* the (continuous) function $F_N(x)$ *only on an open subset $\mathcal{O}$ of* $(-L, L)$. The *first* lemma of §B.3 and the *third*

1 Use of domains from §C of Chapter VIII

one of §B.2 then ensure that $(\mathfrak{M}F_N)(z)$ is *harmonic* in the region

$$\{\Im z > 0\} \ \cup \ \{\Im z < 0\} \ \cup \ \mathcal{O};$$

it is, moreover, *continuous up to the boundary* of that region (indeed, continuous everywhere), again by the fourth lemma of §B.2. The boundary certainly includes the two infinite segments $(-\infty, \ -L]$ and $[L, \ \infty)$ of the real axis, on which $(\mathfrak{M}F_N)(x) \ = \ F_N(x)$.

The open subset $\mathcal{O}$ of $\mathbb{R}$ *might*, however, be so complicated as to raise doubts about our being able to solve the Dirichlet problem in the region (bounded by $\mathbb{R} \sim \mathcal{O}$) just described, and it is thus not clear that one can do harmonic estimation there. We get around this difficulty by means of a simple device.

The difference

$$(\mathfrak{M}F_N)(x) \ - \ F_N(x)$$

is continuous on $\mathbb{R}$ and *identically zero* on $\mathbb{R} \sim \mathcal{O}$. The latter set includes $(-\infty, \ -L] \ \cup \ [L, \ \infty)$, so our difference is actually *uniformly continuous* on $\mathbb{R}$, and, given any $\varepsilon > 0$, we can find a $\delta > 0$ such that

$$(\mathfrak{M}F_N)(x) \ - \ F_N(x) \ \leqslant \ \varepsilon \qquad \text{if} \quad \text{dist}(x, \ \mathbb{R} \sim \mathcal{O}) \ \leqslant \ \delta.$$

The points x fulfilling the condition on the right make up a certain *closed* set

$$E_\delta \ = \ (\mathbb{R} \sim \mathcal{O}) + [-\delta, \ \delta]$$

possibly equal to $\mathbb{R}$ which, in any event, includes $(-\infty, \ -L+\delta] \ \cup \ [L-\delta, \ \infty)$ and *may*, in addition, contain some *disjoint closed intervals of length* $\geqslant \ 2\delta$ intersecting with $(-L+\delta, \ L-\delta)$. There can, of course, *be only finitely many of the latter* so E_δ, if not identical with $\mathbb{R}$, is simply a *finite union of disjoint closed intervals thereon including two of the form* $(-\infty, \ M]$, $[M', \ \infty)$. In the latter case, the *complement* $\mathbb{C} \sim E_\delta$ *is one of the domains* $\mathcal{D}$ considered in §C *of Chapter* VIII, and on $\partial\mathcal{D} \ = \ E_\delta$ we have

$$(\mathfrak{M}F_N)(x) \ \leqslant \ F_N(x) \ + \ \varepsilon$$

by construction.

Our object here is to estimate $(\mathfrak{M}F_N)(0)$. That quantity is of course $\geqslant \ F_N(0)$, and, as long as it is *equal* to $F_N(0)$, there is no problem, because $F_N(0) \ = \ \log W_N(0)$ is $\leqslant \ \log W(0)$ (is, in fact, equal to $\log W(0)$ for sufficiently large values of N). We thus need *only* look at the situation where

$$(\mathfrak{M}F_N)(0) \ > \ F_N(0).$$

Then, however, 0 *belongs* to our original open subset $\mathcal{O}$ of $\mathbb{R}$, and, given $\varepsilon > 0$, it is possible to take the $\delta > 0$ corresponding to it *small enough,* in our construction, so as to *still have*

$$0 \notin (\mathbb{R} \sim \mathcal{O}) + [-\delta, \delta] = E_\delta.$$

Doing so, we see that E_δ really is properly included in $\mathbb{R}$, making

$$\mathcal{D} = \mathbb{C} \sim E_\delta$$

a domain of the kind studied in Chapter VIII, §C, with $0 \in \mathcal{D}$. (We write $\mathcal{D}$ instead of the more logical $\mathcal{D}_\delta$ in order to *avoid* having to use subscripts of subscripts later on.)

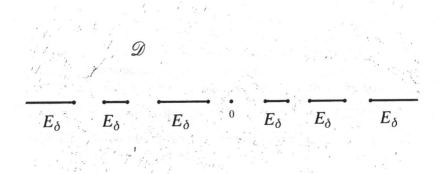

Figure 243

In the present circumstances, $\mathfrak{M}F_N$ is *harmonic* in $\mathcal{D}$ and *continuous* up to $\partial\mathcal{D}$, with

$$(\mathfrak{M}F_N)(t) \leqslant F_N(t) + \varepsilon = \log W_N(t) + \varepsilon \quad \text{for } t \in \partial\mathcal{D},$$

and also, in view of the previous estimates,

$$(\mathfrak{M}F_N)(z) = O(1) - A|\mathfrak{I}z|$$

(everywhere). These facts make it possible for us to follow the procedure indicated at the very beginning of §C, Chapter VIII, and in that way carry out the harmonic estimation of $(\mathfrak{M}F_N)(z)$ in $\mathcal{D}$.

Let us, as in Chapter VIII, §C, denote the Phragmén–Lindelöf function for $\mathcal{D}$ by $Y_\mathcal{D}(z)$ and harmonic measure for that domain by $\omega_\mathcal{D}(\ , z)$.

Then, by the last relations we have

$$(\mathfrak{M}F_N)(0) \ \leqslant \ \int_{\partial \mathscr{D}} (\mathfrak{M}F_N)(t)\, d\omega_{\mathscr{D}}(t,\ 0) \ - \ A\, Y_{\mathscr{D}}(0)$$

$$\leqslant \ \int_{\partial \mathscr{D}} (\log W_N(t) \ + \ \varepsilon)\, d\omega_{\mathscr{D}}(t,\ 0) \ - \ A\, Y_{\mathscr{D}}(0).$$

(The first two members in this chain of inequalities are in fact equal, $\mathfrak{M}F_N$ being harmonic in $\mathscr{D}$.) Because $\log W_N(t) \leqslant \log W(t)$ and $\omega_{\mathscr{D}}(\partial \mathscr{D},\ 0) = 1$, the estimate just written implies that

$$(\mathfrak{M}F_N)(0) \ \leqslant \ \varepsilon \ + \ \int_{\partial \mathscr{D}} \log W(t)\, d\omega_{\mathscr{D}}(t,\ 0) \ - \ A\, Y_{\mathscr{D}}(0);$$

this, then, *must hold, whenever* $(\mathfrak{M}F_N)(0)$ *is not simply equal to* $\log W_N(0)$ *and hence* $\leqslant \log W(0)$.

In the boxed formula (derived, we remind the reader, *under the assumption* that $W(t) \longrightarrow \infty$ for $t \longrightarrow \pm \infty$), the *integrand* appearing in the right-hand integral *no longer depends on N*. But the *right side as a whole certainly involves N* (and ε as well!) *through the domain* $\mathscr{D}$, whose very construction depended on our knowing that $W(x) \geqslant N$ for $|x|$ sufficiently large! *In principle*, it does not generally seem possible to actually *know* $\mathscr{D}$ precisely, because such knowledge would depend on information about the function $(\mathfrak{M}F_N)(z)$ which we are in fact *trying to estimate* (really, *to find*) by using $\mathscr{D}$.

The formula is useful nevertheless, on account of the results found in §§C.4 and C.5 of Chapter VIII. As we saw there, when dealing with *certain kinds* of weights W, one can, by using quantities involving *only* W, express the *entire dependence of*

$$\int_{\partial \mathscr{D}} \log W(t)\, d\omega_{\mathscr{D}}(t,\ 0)$$

on the domain $\mathscr{D}$ *in terms of* $Y_{\mathscr{D}}(0)$. That is the basis for the following applications.

2.   **Weight is the modulus of an entire function of exponential type**

We come to one of the main results of this chapter – indeed, of the present book. The proof, based on the matters discussed above and in §C of Chapter VIII, uses also a refinement of the Riesz–Fejér factorization theorem which has never been explicitly formulated up to now, although it is essentially contained in the material of Chapters III and VI. Here it is:

Lemma. *Let P(z), entire and of exponential type 2B, satisfy the condition*

$$\int_{-\infty}^{\infty} \frac{\log^+|P(x)|}{1+x^2}\,dx \;<\; \infty,$$

and suppose that P(x) $\geqslant$ 0 on $\mathbb{R}$. Then there is an entire function g(z) of exponential type B having all its zeros in $\Im z \leqslant 0$ and such that

$$g(z)\overline{g(\bar z)} \;=\; P(z).$$

Proof. Except for the specification of the exponential type of g, this is just a restatement of the Riesz–Fejér result (the *third* theorem of §G.3, Chapter III). There is thus an entire function $g_0(z)$ having the stipulated properties, but we do not know its type.

In particular, $g_0(z)\overline{g_0(\bar z)} = P(z)$. As long as λ is *real*, we then have

$$g_\lambda(z)\overline{g_\lambda(\bar z)} \;=\; P(z)$$

for the function

$$g_\lambda(z) \;=\; e^{i\lambda z}g_0(z).$$

However, $\log|g_\lambda(iy)| = -\lambda y + \log|g_0(iy)|$, so we can evidently adjust the real parameter λ so as to make

$$\limsup_{y\to\infty} \frac{\log|g_\lambda(iy)|}{y} \quad\text{and}\quad \limsup_{y\to-\infty} \frac{\log|g_\lambda(iy)|}{|y|}$$

equal. Do this, and denote the common value of the two limsups by A, taking, then, $g(z)$ as $g_\lambda(z)$ for that particular choice of λ.

Since P is of exponential type $2B$, we have

$$\limsup_{y\to\infty} \frac{\log|P(iy)|}{y} \;\leqslant\; 2B.$$

At the same time, $|P(iy)| = |g(iy)||g(-iy)|$ for real y, so

$$\frac{\log|g(iy)|}{y} + \frac{\log|g(-iy)|}{y} = \frac{\log|P(iy)|}{y}.$$

On the real axis, $|g(x)|^2 = P(x)$, whence

$$\int_{-\infty}^{\infty} \frac{\log^+|g(x)|}{1+x^2}\,dx \;<\; \infty$$

by hypothesis. Also, $g(z)$, like $g_0(z)$, has all its zeros in $\Im z \leqslant 0$; the remark at the end of §G.1, Chapter III, thus applies to it, and we actually have

$$\frac{\log|g(iy)|}{y} \longrightarrow A$$

as $y \longrightarrow \infty$. Now make $y \longrightarrow \infty$ through a sequence of values along which $\log|g(-iy)|/|y|$ *also* tends to A; referring to the previous relation we see that

$$A + A \leqslant \limsup_{y \to \infty} \frac{\log|P(iy)|}{y} \leqslant 2B,$$

so $A \leqslant B$.

By Chapter III, §E, we have

$$\log|g(z)| \leqslant A|\Im z| + \frac{1}{\pi}\int_{-\infty}^{\infty}\frac{|\Im z|\log^+|g(t)|}{|z-t|^2}\,\mathrm{d}t,$$

and from this it follows easily as in the *fourth* theorem of §E.2, Chapter VI (see also §B.2 of that chapter), that g is *of exponential type* A. Since $g(z)\overline{g(\bar{z})} = P(z)$, P must be of exponential type $\leqslant 2A$, i.e., $B \leqslant A$. We have, however, just shown that $A \leqslant B$. Thus, $A = B$, and we are done.

Now we can give the

Theorem on the Multiplier (Beurling and Malliavin, 1961). *If $f(z)$, entire and of exponential type, is such that*

$$\int_{-\infty}^{\infty}\frac{\log^+|f(x)|}{1+x^2}\,\mathrm{d}x < \infty,$$

there are entire functions $\varphi(z) \not\equiv 0$ of arbitrarily small exponential type with

$$(1 + |f(x)|)\varphi(x)$$

bounded on the real axis.

Proof. Given $A > 0$, we wish to find a non-zero entire φ of exponential type $\leqslant A$ having the desired property. Our plan is to invoke the second theorem from §B.3, referring to the last theorem in §B.1. This involves our showing that $(\mathfrak{M}F)(0) < \infty$ where

$$F(z) = \frac{1}{\pi}\int_{-\infty}^{\infty}\frac{|\Im z|}{|z-t|^2}\log W(t)\,\mathrm{d}t - A|\Im z|,$$

with $W(t) \geqslant 1$ an appropriate weight formed from f. For that purpose, the boxed formula at the end of the preceding article will be applied.

We proceed as at the beginning of §C.5, Chapter VIII, forming from f a new entire function $g_M(z)$ with $|g_M(x)| = |g_M(-x)| \geqslant 1$ on $\mathbb{R}$ and $g_M(0) = 1$. Our present construction differs slightly from the one made there.

Taking a large number M (whose value will depend on the type A of the multiplier we are seeking), we form the entire function

$$P_M(z) = \left(1 + \frac{z^2}{M^2}\right)\left(1 + \frac{z^2}{M^2}(f(z)\overline{f(\bar z)} + f(-z)\overline{f(-\bar z)})\right),$$

the purpose of the first factor on the right being to ensure that

$$P_M(x) \longrightarrow \infty \quad \text{for } x \longrightarrow \pm\infty.$$

Given that $f(z)$ is of exponential type B, $P_M(z)$ will be entire, and of exponential type $\leqslant 2B$. It is clear that $P_M(z)$ is *even*, that

$$P_M(x) \geqslant 1 \quad \text{for } x \in \mathbb{R},$$

with

$$P_M(0) = 1,$$

and that

$$P_M(x) \geqslant |f(x)|^2/M^2 \quad \text{for } |x| \geqslant 1.$$

From the hypothesis it follows also that

$$\int_{-\infty}^{\infty} \frac{\log P_M(x)}{1+x^2}\,dx < \infty;$$

we can, indeed, choose M so as to make

$$\int_{-\infty}^{\infty} \frac{\log P_M(x)}{x^2}\,dx \quad (sic\,!)$$

as small > 0 as we like. Here, the behaviour of the integrand near the origin is alright, because (for large M)

$$|\log P_M(x)| \leqslant \text{const.}\, x^2/M^2 \quad \text{for } |x| < 1$$

with a constant independent of M.

For any particular M, the lemma now gives us an entire function $g_M(z)$, of exponential type $\leqslant B$ (*half* that of P_M), having (here) all its zeros in the lower half plane, with

$$g_M(z)\overline{g_M(\bar z)} = P_M(z).$$

Thence, in particular,

$$|g_M(x)| \;=\; \sqrt{P_M(x)} \;\geqslant\; |f(x)|/M$$

for real x of modulus $\geqslant 1$, so, since $P_M(x) \geqslant 1$ on $\mathbb{R}$,

$$1 + |f(x)| \;\leqslant\; C_M|g_M(x)|, \qquad x \in \mathbb{R},$$

with a constant C_M depending on M. Our result will thus be established if, for a suitable value of M, we can find an entire $\varphi(z) \not\equiv 0$ of exponential type A with $\varphi(x)g_M(x)$ bounded on $\mathbb{R}$. To do this, we follow the procedure explained in the last article.

Fixing a value of M (in a way to be described shortly) we take the weight $W(x) = |g_M(x)|$ and then use it in the formula written at the beginning of this proof so as to obtain a function F. According to the last theorem of §B.1 and the second one of §B.3, a function φ having the desired properties exists provided that $(\mathfrak{M}F)(0) < \infty$. It is now claimed that for proper choice of M *we in fact have*

$$(\mathfrak{M}F)(0) \;=\; 0.$$

To see this we verify (in the notation of the last article) that $(\mathfrak{M}F_N)(0) = 0$ for every $N \geqslant 1$. In the present circumstances, $W(0) = 1$, so for $N \geqslant 1$,

$$F_N(0) \;=\; F(0) \;=\; \log W(0) \;=\; 0,$$

and it is enough to show that assuming

$$(\mathfrak{M}F_N)(0) \;>\; \varepsilon$$

for some $\varepsilon > 0$ leads to a contradiction.

In case the last relation holds, it is certainly true that

$$(\mathfrak{M}F_N)(0) \;>\; \log W(0),$$

so the boxed formula from the end of the preceding article is valid, $W(x) = \sqrt{P_M(x)}$ having been ensured by our construction to tend to ∞ for $x \longrightarrow \pm\infty$. Thus,

$$(\mathfrak{M}F_N)(0) \;\leqslant\; \varepsilon \;+\; \int_{\partial \mathscr{D}} \log W(t)\,d\omega_{\mathscr{D}}(t,\,0) \;-\; AY_{\mathscr{D}}(0),$$

where $\mathscr{D}$ is a certain (unknown) domain of the kind studied in §C of Chapter VIII.

Now *the second theorem of* §C.5 in Chapter VIII *can be used to estimate the quantity*

$$\int_{\partial \mathscr{D}} \log W(t)\,d\omega_{\mathscr{D}}(t,\,0) \;=\; \int_{\partial \mathscr{D}} \log |g_M(t)|\,d\omega_{\mathscr{D}}(t,\,0).$$

Our function g_M is of exponential type $\leqslant B$ and has otherwise the properties of the function G figuring in that theorem. Therefore,

$$\int_{\partial\mathscr{D}} \log|g_M(t)|\,d\omega_{\mathscr{D}}(t,\,0) \ \leqslant \ Y_{\mathscr{D}}(0)\{J \ + \ \sqrt{(2eJ(J+\pi B/4))}\},$$

where

$$J \ = \ \int_0^\infty \frac{\log|g_M(x)|}{x^2}\,dx \ = \ \frac{1}{4}\int_{-\infty}^\infty \frac{\log P_M(x)}{x^2}\,dx.$$

As observed above, the right-hand integral will, for large enough M, *be as small as we like.* We can hence choose (and *fix*) *a value of M for which J is small enough to render*

$$J \ + \ \sqrt{(2eJ(J+\pi B/4))} \ < \ A.$$

This having been done, the previous relations yield

$$(\mathfrak{M}F_N)(0) \ \leqslant \ \varepsilon$$

(whatever N may be), contradicting our assumption that $(\mathfrak{M}F_N)(0) > \varepsilon$. Thus, $(\mathfrak{M}F_N)(0) = 0$, so, since this holds for every N, we have $(\mathfrak{M}F)(0) = 0$ as claimed, and the theorem is proved.

We are done.

Scholium. The multiplier $\varphi(z)$ of exponential type $\leqslant A$ obtained by going from the conclusion of the above argument to the *second* theorem in §B.3 and thence to the *last* one in §B.1 has real zeros only. This is immediately apparent on glancing at the description of the function φ appearing towards the end of the latter result's proof – *that φ, by the way, is not the same as the multiplier we are talking about here*, which, in the theorem referred to, was called ψ.

In their 1967 *Acta* paper, Beurling and Malliavin made the important observation that the zeros of the multiplier φ can also, in the present circumstances, be taken to be *uniformly separated*, in other words, *that any two of those zeros are distant by at least a certain amount $h > 0$.* This can be readily seen by putting together some of the above results and then using a simple measure-theoretic lemma.

Let us look again at the *least superharmonic majorant* $(\mathfrak{M}F)(z)$ of the function

$$F(z) \ = \ \frac{1}{\pi}\int_{-\infty}^\infty \frac{|\Im z|\log|g_M(t)|}{|z-t|^2}\,dt \ - \ A|\Im z|$$

formed from the weight $W(x) = |g_M(x)|$ used in the preceding proof. Here, $|g_M(t)| = |g_M(-t)|$, so $F(z) = F(-z)$ and therefore $(\mathfrak{M}F)(z) = (\mathfrak{M}F)(-z)$ (cf. beginning of proof of first lemma, §B.3). We know that $\mathfrak{M}F$ is finite, but here, since $(\mathfrak{M}F)(0) = F(0) = \log|g_M(0)| = 0$, we cannot affirm that $(\mathfrak{M}F)(z)$ is harmonic in a neighborhood of 0 and thus are not able to directly apply the first theorem from §B.3. An analogous result is nevertheless available by problem 57. In the present circumstances, with $(\mathfrak{M}F)(z)$ *even*, that result takes the form

$$(\mathfrak{M}F)(z) = C - \int_0^1 \log|z^2 - t^2|\,d\rho(t) - \int_1^\infty \log\left|1 - \frac{z^2}{t^2}\right|d\rho(t),$$

where ρ is a certain *positive* measure on $[0, \infty)$ with

$$\frac{\rho([0,t])}{t} \longrightarrow \frac{A}{\pi} \quad \text{for } t \longrightarrow \infty.$$

Because $(\mathfrak{M}F)(0) < \infty$, we actually have

$$\int_0^1 \log(t^2)\,d\rho(t) > -\infty,$$

so, after changing the value of the constant C, we can just as well write

$$(\mathfrak{M}F)(z) = C - \int_0^\infty \log\left|1 - \frac{z^2}{t^2}\right|d\rho(t).$$

By the first lemma of §C.5, Chapter VIII (where the function corresponding to our present $g_M(z)$ was denoted by $G(z)$), we have

$$\log|g_M(z)| = \int_0^\infty \log\left|1 - \frac{z^2}{t^2}\right|d\nu(t) \quad \text{for } \Im z \geqslant 0,$$

$\nu(t)$ being a certain absolutely continuous (and *smooth*) increasing function. Taking the function $g_M(z)$ to be of exponential type exactly *equal* to B (so as not to bring in more letters!), we also have

$$\log|g_M(z)| = B\Im z + \frac{1}{\pi}\int_{-\infty}^\infty \frac{\Im z \log|g_M(t)|}{|z-t|^2}\,dt$$

for $\Im z > 0$ by §G.1 of Chapter III.* Referring to the above formula for $F(z) = F(\bar{z})$, we see from the last two relations that

$$F(z) = \int_0^\infty \log\left|1 - \frac{z^2}{t^2}\right|d\nu(t) - (A+B)|\Im z|.$$

---

* see also end of proof of lemma at beginning of this article

$(\mathfrak{M}F)(z)$ is, however, a *majorant of F(z)*. Hence,

$$F(z) - (\mathfrak{M}F)(z) = \int_0^\infty \log\left|1 - \frac{z^2}{t^2}\right|(d\nu(t) + d\rho(t))$$

$$- (A + B)|\Im z| - C \quad \text{is} \quad \leqslant 0.$$

Our statement about the zeros of $\varphi(z)$ will follow from this inequality.

The real line is the union of two disjoint subsets, an *open* one, Ω, on which

$$F(x) - (\mathfrak{M}F)(x) < 0,$$

and the *closed* set $E = \mathbb{R} \sim \Omega$, on which

$$F(x) - (\mathfrak{M}F)(x) = 0.$$

According to the *third* lemma of §B.2, $(\mathfrak{M}F)(z)$ is *harmonic* in a neighborhood of each $x_0 \in \Omega$, so the measure involved in the Riesz representation of $(\mathfrak{M}F)(z)$ can have no mass in such a neighborhood (last theorem, §A.2). This means that

$$d\rho(t) = 0 \quad \text{in } \Omega \cap [0, \infty).$$

It is now claimed that

$$d\nu(t) + d\rho(t) \leqslant \frac{A+B}{\pi} dt \quad \text{on } E \cap [0, \infty).$$

Once this is established, the separation of the zeros of our multiplier $\varphi(z)$ is immediate. That function is gotten by dividing out any two zeros from the even entire function $\varphi_1(z)$ given by

$$\log|\varphi_1(z)| = \int_0^\infty \log\left|1 - \frac{z^2}{t^2}\right| d[\rho(t)]$$

(as in the proof of the last theorem, §B.1). Because $d\nu(t) \geqslant 0$, the preceding two relations will certainly make

$$d\rho(t) \leqslant \frac{A+B}{\pi} dt \quad \text{for } t \geqslant 0,$$

and thus *any two zeros of $\varphi_1(z)$ will be distant by at least*

$$\frac{\pi}{A+B}$$

units, in conformity with Beurling and Malliavin's observation.

Verification of the claim remains, and it is there that we resort to the

Lemma. *Let μ be a finite positive measure on $\mathbb{R}$ without point masses. Then the derivative $\mu'(t)$ exists* (finite or infinite) *for all t save those belonging to a Borel set E_0 with $\mu(E_0) + |E_0| = 0$. If E is any compact subset of $\mathbb{R}$,*

$$|E| = \int_E \frac{1}{\mu'(t) + 1}(d\mu(t) + dt).$$

Proof. The initial statement is like that of Lebesgue's differentiation theorem which, however, only asserts the existence of a (finite) derivative $\mu'(t)$ almost everywhere (with respect to Lebesgue measure). The present result can nonetheless be deduced from the latter one by making a change of variable. Lest the reader feel that he or she is being hoodwinked by the juggling of notation, let us proceed somewhat carefully.

Put, as usual, $\mu(t) = \int_0^t d\mu(\tau)$, making the standard interpretation of the integral for $t < 0$. By hypothesis, $\mu(t)$ is bounded, increasing and without jumps, so

$$S(t) = \mu(t) + t$$

is a *continuous, strictly increasing* map of $\mathbb{R}$ onto itself. S therefore has a *continuous* (and also strictly increasing) *inverse* which we denote by T:

$$T(\mu(t) + t) = t.$$

If $\varphi(s)$ is continuous and of compact support we have the elementary substitution formula

$$(*) \qquad \int_{-\infty}^{\infty} \varphi(s)\,ds = \int_{-\infty}^{\infty} \varphi(S(t))(d\mu(t) + dt)$$

which is easily checked by looking at Riemann sums. The dominated convergence theorem shows that $(*)$ is valid as well for any function φ everywhere equal to the *pointwise limit* of a *bounded* sequence of continuous ones with *fixed* compact support. That is the case, in particular, for $\varphi = \chi_F$, the characteristic function of a compact set F, and we thus have

$$|F| = \int_{-\infty}^{\infty} \chi_F(s)\,ds = \int_{-\infty}^{\infty} \chi_F(S(t))(d\mu(t) + dt) = \mu(T(F)) + |T(F)|.$$

The quantity $|T(F)|$ is, of course, nothing other than the Lebesgue–Stieltjes measure $\int_F dT(s)$ generated by the increasing function $T(s)$ in the usual way. The last relation shows that

$$|T(F)| \leqslant |F|$$

for compact sets F; the measure on the left is thus *absolutely continuous* with respect to Lebesgue measure, and indeed

$$0 \;\leqslant\; \frac{T(s+h) - T(s)}{h} \;\leqslant\; 1 \quad \text{for } h \neq 0.$$

By the theorem of Lebesgue already referred to, we know that the derivative

$$T'(s) \;=\; \lim_{h \to 0} \frac{T(s+h) - T(s)}{h}$$

exists for all s outside some Borel set F_0 with $|F_0| = 0$. The image $E_0 = T(F_0)$ is also Borel (T being one-one and continuous both ways) and, for any *compact subset* C of E_0, the previous identity yields

$$\mu(C) \;+\; |C| \;=\; |S(C)| \;\leqslant\; |F_0| \;=\; 0,$$

since $C = T(S(C))$ and $S(C) \subseteq S(E_0) = F_0$. Therefore

$$\mu(E_0) \;+\; |E_0| \;=\; 0.$$

Suppose that $t \notin E_0$. Then $s = \mu(t) + t$ (for which $T(s) = t$) cannot lie in F_0, T being one-one, and thus $T'(s)$ exists. For $\delta \neq 0$ and any such t (and corresponding s), write

$$h(s, \delta) \;=\; \mu(t + \delta) \;-\; \mu(t) \;+\; \delta.$$

We have $s + h(s, \delta) = \mu(t + \delta) + t + \delta$, so by definition of T, $T(s + h(s, \delta)) = t + \delta$, and

$$(\dagger) \qquad \frac{T(s + h(s, \delta)) \;-\; T(s)}{h(s, \delta)} \;=\; \frac{\delta}{\mu(t + \delta) - \mu(t) + \delta}.$$

The function $\mu(t)$ is in any event continuous, so at each s (and corresponding t),

$$h(s, \delta) \;\longrightarrow\; 0 \quad \text{as } \delta \longrightarrow 0.$$

Therefore, when $t \notin E_0$, $\lim_{\delta \to 0}\big((\mu(t + \delta) - \mu(t))/\delta + 1\big)^{-1}$ must, by $(\dagger)$, *exist* and equal $T'(s)$. This shows that $\mu'(t)$ exists for such t (being infinite in case $T'(s) = 0$).

Take now any continuous function $\psi(s)$ of compact support. Because of the *absolute continuity* of the measure $\int_F dT(s)$ already noted, we have

$$\int_{-\infty}^{\infty} \psi(s)\, dT(s) \;=\; \int_{-\infty}^{\infty} \psi(s)\, T'(s)\, ds.$$

Here,

$$T'(s) = \lim_{\delta \to 0} \frac{T(s + h(s, \delta)) - T(s)}{h(s, \delta)} \quad \text{a.e.}$$

where $h(s, \delta)$ is the quantity introduced above. The difference quotients on the right lie, however, between 0 and 1. Hence $\int_{-\infty}^{\infty} \psi(s) T'(s) \, ds$ equals the limit, for $\delta \to 0$, of

$$\int_{-\infty}^{\infty} \psi(s) \frac{T(s + h(s, \delta)) - T(s)}{h(s, \delta)} \, ds$$

by dominated convergence. In this last expression the *integrand* is continuous and of compact support when $\delta \neq 0$. We may therefore use $(*)$ to make the substitution $s = \mu(t) + t$ therein; with the help of $(\dagger)$, that gives us

$$\int_{-\infty}^{\infty} \psi(\mu(t) + t) \frac{\delta}{\mu(t + \delta) - \mu(t) + \delta} \, (d\mu(t) + dt).$$

The quantity $\delta/(\mu(t + \delta) - \mu(t) + \delta)$ lies between 0 and 1 and, as we have just seen, tends to $1/(\mu'(t) + 1)$ for every t outside E_0 when $\delta \to 0$, where $\mu(E_0) + |E_0| = 0$. Another application of the dominated convergence theorem thus shows the integral just written to tend to $\int_{-\infty}^{\infty} \psi(\mu(t) + t)(\mu'(t) + 1)^{-1}(d\mu(t) + dt)$ as $\delta \to 0$. In this way, we see that

$$\int_{-\infty}^{\infty} \psi(s) \, dT(s) = \int_{-\infty}^{\infty} \frac{\psi(\mu(t) + t)}{\mu'(t) + 1} \, (d\mu(t) + dt)$$

when ψ is continuous and of compact support.

Extension of this formula to functions ψ of the form χ_F with F compact now proceeds as at the beginning of the proof. Given, then, any compact E, we put $F = S(E)$, making $T(F) = E$ and $\chi_F(\mu(t) + t) = \chi_F(S(t)) = \chi_E(t)$; using $\psi(s) = \chi_F(s)$ we thus find that

$$|E| = |T(F)| = \int_{-\infty}^{\infty} \chi_F(s) \, dT(s) = \int_{-\infty}^{\infty} \chi_E(t) \frac{d\mu(t) + dt}{\mu'(t) + 1}.$$

The lemma is established.

We proceed to the claim.

Problem 58

(a) Show that in our present situation, neither $v(t)$ nor $\rho(t)$ can have any point masses. (Hint: Concerning $\rho(t)$, recall that $(\mathfrak{M}F)(x)$ is continuous!)

(b) We take $v(0) = \rho(0) = 0$ and then extend the increasing *functions* $v(t)$ and $\rho(t)$ from $[0, \infty)$ to $\mathbb{R}$ by making them *odd*. Show that for $x \in E$ (the set on which $F(x) - (\mathfrak{M}F)(x) = 0$), we have

$$\frac{1}{2}\int_{-\infty}^{\infty} \log\left(1 + \frac{y^2}{(x-t)^2}\right)(dv(t) + d\rho(t)) - (A+B)y \leqslant 0 \quad \text{for } y > 0.$$

(c) Writing $\mu(t) = v(t) + \rho(t)$, show that for fixed $y > 0$,

$$\frac{1}{2}\int_{-\infty}^{\infty} \log\left(1 + \frac{y^2}{(x-t)^2}\right)d\mu(t) = \int_0^{\infty} \frac{y^2}{y^2 + \tau^2}\frac{\mu(x+\tau) - \mu(x-\tau)}{\tau}\,d\tau.$$

(Hint: Since $\mu(\{x\}) = 0$, the left hand integral is the limit, for $\delta \longrightarrow 0$, of

$$\frac{1}{2}\int_{|t-x|\geqslant\delta} \log\left(1 + \frac{y^2}{(t-x)^2}\right)d\mu(t).$$

Here we may integrate by parts to get

$$\int_{\delta}^{\infty} \frac{y^2}{y^2 + \tau^2}\frac{\mu(x+\tau) - \mu(x-\tau) - (\mu(x+\delta) - \mu(x-\delta))}{\tau}\,d\tau.$$

Now make $\delta \longrightarrow 0$ and use monotone convergence.)

(d) Hence show that *for each $x \in E$ where $\mu'(x)$ exists*, we have

$$\pi\mu'(x) \leqslant A + B.$$

(Hint: Refer to (b).)

(e) Show then that if F is any compact subset of E,

$$\pi\mu(F) \leqslant (A+B)|F|,$$

whence

$$v(F) + \rho(F) \leqslant \frac{A+B}{\pi}|F|.$$

(Hint: Apply the lemma.)

The reader who prefers a more modern treatment yielding the result of part (e) may, in place of (d), establish that

$$\pi(\underline{D}\mu)(x) \leqslant A + B \quad \text{for } x \in E,$$

where

$$(\underline{D}\mu)(x) = \liminf_{\Delta x \to 0} \frac{\mu(x + \Delta x) - \mu(x)}{\Delta x}.$$

Then, instead of using the lemma to get part (e), a suitable version of Vitali's covering theorem can be applied.

Remark. The original proof of the theorem of Beurling and Malliavin is different from the one given in this article, and the reader interested in working seriously on the subject of the present chapter should study it.

The first exposition of that proof is contained in a famous (and very rare) Stanford University preprint written by Malliavin in 1961, and the final version is in his joint *Acta* paper with Beurling, published in 1962. Other presentations can be found in Kahane's Seminaire Bourbaki lecture for 1961–62, and in de Branges' book. But the clearest explanation of the proof's *idea* is in a much later paper of Malliavin appearing in the 1979 *Arkiv*. Although some details are omitted in that paper, it is probably the best place to start reading.

3.   **A quantitative version of the preceding result.**

Theorem. *Let* $\Phi(z)$ *be entire and even, of exponential type B, with* $\Phi(x) \geqslant 0$ *on the real axis and*

$$\int_{-\infty}^{\infty} \frac{\log^+ \Phi(x)}{1 + x^2} \, dx \quad < \quad \infty.$$

For $M > 0$, *denote by* J_M *the quantity*

$$\int_0^{\infty} \frac{1}{x^2} \log\left(1 \ + \ x^2 \frac{\Phi(x)}{M} \right) dx.$$

Suppose that for some given $A > 0$, M *is large enough to make*

$$J_M \ + \ \frac{1}{\sqrt{\pi}} \sqrt{(J_M(J_M \ + \ \pi B))} \quad < \quad A.$$

Then there is an even entire function $\varphi(z)$ *of exponential type A with*

$$\varphi(0) \ = \ 1$$

and

$$|\varphi(x)| \, \Phi(x) \quad \leqslant \quad 2e^2(A + B)^2 M \quad \text{for } x \in \mathbb{R}.$$

Remark. There are actually functions φ having all the stipulated properties and satisfying a relation like the last one with the coefficient 2 replaced by any number > 1 – hence indeed by 1, as follows by a normal family argument. By that kind of argument one also sees that there are such φ

corresponding to a value of *M for which*

$$J_M \ + \ \frac{1}{\sqrt{\pi}}\sqrt{(J_M(J_M \ + \ \pi B))} \ = \ A.$$

Such improvements are not very significant.

Proof of theorem. We argue as in the last article, working this time with
the auxiliary entire function

$$P(z) \ = \ \left(1 \ + \ \frac{z^2}{R^2}\right)\left(1 \ + \ z^2\frac{\Phi(z)}{M}\right)$$

which involves a large constant R as well as the parameter M. An extra
factor has again been introduced on the right in order to make sure that

$$P(x) \ \longrightarrow \ \infty \quad \text{for } x \longrightarrow \pm \infty.$$

Like $\Phi(z)$, the function $P(z)$ is even, entire, and of exponential type B;
it is, moreover, $\geqslant 1$ on the real axis and we can use it as a *weight* thereon.
As long as M fulfills the condition in the hypothesis,

$$J \ = \ \int_0^\infty \frac{1}{x^2} \log P(x)\,dx$$

will satisfy the relation

$$J \ + \ \frac{1}{\sqrt{\pi}}\sqrt{(J(J \ + \ \pi B))} \ \leqslant \ A$$

for large enough R; we *choose and fix* such a value of that quantity.

Using the weight $W(x) \ = \ P(x)$ and the number A, we then form the
function $F(z)$ and the sequence of $F_N(z)$ corresponding to it as in the
previous two articles, and set out to show that $(\mathfrak{M}F)(0) \ = \ 0$.

This is done as before, by verifying that $(\mathfrak{M}F_N)(0) \ \leqslant \ 0$ for each N.
Assuming, on the contrary, that some $(\mathfrak{M}F_N)(0)$ is *strictly larger* than
some $\varepsilon > 0$, we have

$$(\mathfrak{M}F_N)(0) \ \leqslant \ \varepsilon \ + \ \int_{\partial\mathscr{D}} \log P(t)\,d\omega_{\mathscr{D}}(t, 0) \ - \ AY_{\mathscr{D}}(0)$$

with a domain $\mathscr{D}$ of the sort considered in §C of Chapter VIII,
since here

$$F_N(t) \ \leqslant \ F(t) \ = \ \log P(t), \quad t\in R,$$

where $\log P(t) \ \longrightarrow \ \infty$ for $t \longrightarrow \pm \infty$. We proceed to estimate the integral
on the right.

The entire function $P(z)$ is real and $\geqslant 1$ on R, so, by the lemma from the last article, we can get an entire function $g(z)$ of exponential type $B/2$ (*half* that of P), having all its zeros in the lower half plane, and such that

$$g(z)\overline{g(\bar{z})} \;=\; P(z).$$

For the entire function

$$G(z) \;=\; (g(z))^2$$

of exponential type B with all its zeros in $\Im z < 0$ we then have $|G(x)| = P(x)$ on $\mathbb{R}$, so that

$$\int_{\partial \mathscr{D}} \log P(t)\,d\omega_{\mathscr{D}}(t,\,0) \;=\; \int_{\partial \mathscr{D}} \log |G(t)|\,d\omega_{\mathscr{D}}(t,\,0).$$

Here, $G(z)$ satisfies the hypothesis of the second theorem in §C.5 of Chapter VIII, and that result can be used to get an upper bound for the last integral. We can, however, do somewhat better by first improving the theorem, using, at the very end of its proof, the estimate furnished by problem 28(c) in place of the one applied there. The effect of this is to replace the term $\sqrt{(2\mathrm{e}\,J(J + \pi B/4))}$ figuring in the theorem's conclusion by

$$\frac{1}{\sqrt{\pi}}\sqrt{(J(J \,+\, \pi B))}$$

with

$$J \;=\; \int_0^\infty \frac{1}{x^2}\,\log |G(x)|\,dx \;=\; \int_0^\infty \frac{1}{x^2}\,\log P(x)\,dx,$$

and in that way one finds that

$$\int_{\partial \mathscr{D}} \log |G(t)|\,d\omega_{\mathscr{D}}(t,\,0) \;\leqslant\; Y_{\mathscr{D}}(0)\left\{ J \,+\, \frac{1}{\sqrt{\pi}}\sqrt{(J(J + \pi B))} \right\}.$$

Substituted into the above relation, this yields

$$(\mathfrak{M}F_N)(0) \;\leqslant\; \varepsilon,$$

a contradiction, thanks to our initial assumption about M and our choice of R. It follows that $(\mathfrak{M}F_N)(0) \leqslant 0$ for every N and thus that $(\mathfrak{M}F)(0) = 0$.

Knowing that, we can, since $P(0) = 1$, apply the *corollary* to the second theorem of §B.3. That gives us, corresponding to any $\eta > 0$, an increasing function $\rho(t)$, *zero* on a neighborhood of the origin, such that

$$\frac{\rho(t)}{t} \longrightarrow \frac{A}{\pi} \quad \text{for } t \longrightarrow \pm\infty$$

and that

$$\log P(x) + \gamma x + \int_{-\infty}^{\infty} \left(\log\left|1 - \frac{x}{t}\right| + \frac{x}{t} \right) d\rho(t) \leqslant \eta$$

on $\mathbb{R}$, γ being a certain real constant. In the present circumstances $P(x) = P(-x)$, so, taking the increasing function

$$v(t) = \tfrac{1}{2}(\rho(t) - \rho(-t))$$

(also zero on a neighborhood of the origin), we have simply

$$\log P(x) + \int_0^{\infty} \log\left|1 - \frac{x^2}{t^2}\right| dv(t) \leqslant \eta \quad \text{for } x \in \mathbb{R}.$$

Our given function Φ is $\geqslant 0$ on the real axis. Therefore $P(x) \geqslant x^2 \Phi(x)/M$ there, and our last relation certainly implies that

$$\log\left(x^2 \frac{\Phi(x)}{M} \right) + \int_0^{\infty} \log\left|1 - \frac{x^2}{t^2}\right| dv(t) \leqslant \eta$$

on $\mathbb{R}$. Denote for the moment

$$\log\left| z^2 \frac{\Phi(z)}{M} \right| + \int_0^{\infty} \log\left|1 - \frac{z^2}{t^2}\right| dv(t)$$

by $U(z)$; this function is *subharmonic*, and, since Φ is of exponential type B while

$$\frac{v(t)}{t} \longrightarrow \frac{A}{\pi} \quad \text{for } t \longrightarrow \infty,$$

we have

$$U(z) \leqslant (B + A)|z| + o(|z|)$$

for large $|z|$. Because $U(x) \leqslant \eta$ on $\mathbb{R}$, we see by the third Phragmén–Lindelöf theorem of §C, Chapter III, that

$$U(z) \leqslant \eta + (A + B)\Im z \quad \text{for } \Im z \geqslant 0.$$

To the integral $\int_0^{\infty} \log|1 - (z^2/t^2)| dv(t)$ we now apply the lemma of

§A.1, Chapter X, according to which

$$\int_0^\infty \log\left|1 - \frac{z^2}{t^2}\right| (\mathrm{d}[\nu(t)] - \mathrm{d}\nu(t)) \;\leqslant\; \log\left\{\frac{\max(|x|,|y|)}{2|y|} + \frac{|y|}{2\max(|x|,|y|)}\right\}$$

(where, as usual, $z = x + iy$). Used together with the preceding inequality for $U(z)$, this yields

$$\log\left|z^2 \frac{\Phi(z)}{M}\right| \;+\; \int_0^\infty \log\left|1 - \frac{z^2}{t^2}\right| \mathrm{d}[\nu(t)]$$

$$\leqslant\; \eta \;+\; (A+B)y \;+\; \log\left\{\frac{\max(|x|,y)}{2y} + \frac{y}{2\max(|x|,y)}\right\}$$

for $\Im z = y > 0$.

There is clearly an entire function $\varphi(z)$ with

$$\log|\varphi(z)| \;=\; \int_0^\infty \log\left|1 - \frac{z^2}{t^2}\right| \mathrm{d}[\nu(t)];$$

φ is *even* and $\varphi(0) = 1$. Moreover, in view of the asymptotic behaviour of $\nu(t)$ for large t, $\varphi(z)$ *is of exponential type* A. In terms of φ, the preceding relation becomes

$$|\Phi(z)\varphi(z)| \;\leqslant\; M\mathrm{e}^{(A+B)y+\eta} \frac{\{\max(|x|,y)/y \;+\; y/\max(|x|,y)\}}{2(x^2 + y^2)}, \qquad y > 0.$$

The fraction on the right is just

$$\frac{1}{2y^2} \cdot \frac{\max(\xi,1) \;+\; 1/\max(\xi,1)}{\xi^2 + 1}$$

with $\xi = |x|/y$, and hence is $\leqslant 1/y^2$. Thus, putting $z = x + ih$ with $h > 0$, we see that

$$|\Phi(x+ih)\varphi(x+ih)| \;\leqslant\; \frac{M}{h^2} \mathrm{e}^{(A+B)h+\eta} \qquad \text{for } x \in \mathbb{R}.$$

Applying once more the third Phragmén–Lindelöf theorem of Chapter III, §C, this time to $\Phi(z)\varphi(z)$ (of exponential type $A + B$) in the half plane $\{\Im z \leqslant h\}$, we get finally

$$|\Phi(x)\varphi(x)| \;\leqslant\; \frac{M}{h^2} \mathrm{e}^{2(A+B)h+\eta}, \qquad x \in \mathbb{R},$$

and, putting $h = 1/(A+B)$, we have

$$\Phi(x)|\varphi(x)| \;\leqslant\; \mathrm{e}^\eta \mathrm{e}^2 (A+B)^2 M \qquad \text{on } \mathbb{R}.$$

The quantity $\eta > 0$ was arbitrary, so the desired result is established. We are done.

Scholium. Let us try to understand the rôle played by the parameter M in the result just proved. As long as

$$\int_{-\infty}^{\infty} \frac{\log^+ \Phi(x)}{1 + x^2} dx \;\; < \;\; \infty,$$

it is surely true that with

$$J_M \;\; = \;\; \int_0^{\infty} \frac{1}{x^2} \log\left(1 \; + \; x^2 \frac{\Phi(x)}{M} \right) dx,$$

the expression

$$J_M \;\; + \;\; \frac{1}{\sqrt{\pi}} \sqrt{(J_M(J_M \; + \; \pi B))}$$

eventually becomes less than any given $A > 0$ when M increases without limit; we *cannot*, however, tell *how large M must be taken for that to happen if only the value of the former integral and the type B of Φ are known.* Our result thus does not enable us to determine, *using that information alone*, how small $\sup_{x \in \mathbb{R}} \Phi(x) |\varphi(x)|$ can be rendered by taking a suitable even entire function φ of exponential type A with $\varphi(0) = 1$.

This is even the case for *polynomials* Φ (special kinds of functions of exponential type zero!).

Problem 59

Show that for the polynomials

$$\Phi_N(z) \;\; = \;\; \left(\frac{z^2}{2N^2} \; - \; 1 \right)^{2N}$$

one has

$$\int_0^{\infty} \frac{1}{x^2} \log^+ \Phi_N(x)\,dx \;\; \leqslant \;\; \text{const.},$$

but that for $J > 0$ small enough, there is no value of M which will make

$$\int_0^{\infty} \frac{1}{x^2} \log\left(1 \; + \; x^2 \frac{\Phi_N(x)}{M} \right) dx \;\; \leqslant \;\; J$$

for all N simultaneously.

$$\left(\text{Hint: Look at the values of } \int_0^\infty \frac{1}{x^2} \log^+ \left(\frac{\Phi_N(x)}{7^N} \right) dx. \right)$$

The parameter M, made to depend on A by requiring that

$$J_M + \frac{1}{\sqrt{\pi}} \sqrt{(J_M (J_M + \pi B))}$$

be $< A$ (or simply *equal to A*; see the remark following our theorem's statement) does nevertheless seem to be the *main factor governing how small*

$$\sup_{x \in \mathbb{R}} \Phi(x) |\varphi(x)|$$

can be for even entire functions φ of exponential type A with $\varphi(0) = 1$. The evidence for this is especially convincing when *entire functions* Φ of exponential type zero are concerned. Then the discrepancy between the above result and any best possible one essentially involves nothing more than *a constant factor affecting the type A of the multiplier φ* in question.

Problem 60

Suppose that $W(x) \geqslant 1$ is even and that there is an even entire function φ of exponential type A with $\varphi(0) = 1$ and

$$W(x)|\varphi(x)| \leqslant K \quad \text{for } x \in \mathbb{R}.$$

(a) Show that then

$$\int_{x_0}^\infty \frac{1}{x\sqrt{(x^2 - x_0^2)}} \log \left(\frac{W(x)}{K} \right) dx \leqslant \frac{\pi}{2} A$$

for any $x_0 > 0$. (Hint: Use harmonic estimation in

$$\mathscr{D} = \mathbb{C} \sim (-\infty, -x_0] \sim [x_0, \infty)$$

to get an upper bound for $\log|\varphi(0)|$. Note that $\omega_{\mathscr{D}}(\ , z)$ and $Y_{\mathscr{D}}(z)$ are explicitly available for this domain; for the latter, see, for instance, §A.2 of Chapter VIII).

Suppose now that φ, A and K are as in (a) *and that there is in addition an $x_0 > 0$ such that $W(x) \leqslant K$ for $|x| \leqslant x_0$ while $W(x) \geqslant K$ for $|x| \geqslant x_0$. Note that $W(x)$ need not be an increasing function of $|x|$ for this to hold for certain values of K:*

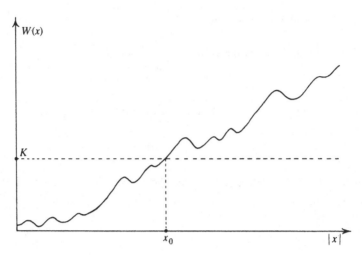

Figure 244

(b) Show that then, given any $\eta > 0$, there is a constant C depending only on η such that

$$\int_0^\infty \frac{1}{x^2} \log\left(1 + \frac{A^2 x^2}{C} \frac{W(x)}{K}\right) dx \; < \; \left(\frac{\pi}{2} + \eta\right) A.$$

(Hint: Observe that

$$\log\left(1 + \frac{A^2 x^2}{C} \frac{W(x)}{K}\right) \; \leqslant \; \log\left(1 + \frac{A^2 x^2}{C}\right) \; + \; \log^+\left(\frac{W(x)}{K}\right).$$

Refer to part (a).)

By this problem we see in particular that if $W(x)$ *is the restriction to* $\mathbb{R}$ *of an entire function of exponential type zero having, for some given* K, the behaviour described therein, then, for

$$M = CK/A^2,$$

the integral J_M corresponding to W *satisfies the condition of our theorem pertaining to multipliers of exponential type*

$$A' = \frac{\sqrt{\pi+1}}{\sqrt{\pi}}\left(\frac{\pi}{2} + \eta\right) A$$

(rather than to those of exponential type A). For suitable choice of the constant C, the right side is

$$< \quad 2.5 A$$

Thus, subject to the above proviso regarding $W(x)$ and K, *the theorem will furnish* an even entire function ψ with $\psi(0) = 1$, *of exponential type* 2.5A, for which

$$W(x)|\psi(x)| \;\leqslant\; 12.5e^2\, CK \quad \text{on } \mathbb{R}$$

whenever the existence of such an entire φ, *of exponential type* A, with

$$W(x)|\varphi(x)| \;\leqslant\; K \quad \text{on } \mathbb{R}$$

is otherwise known.

There may be certain even entire functions Φ, $\geqslant 1$ on $\mathbb{R}$ and of exponential type zero, such that, for some *arbitrarily large values* of x_0, $\Phi(x) \leqslant \Phi(x_0)$ for $|x| \leqslant x_0$ and $\Phi(x) \geqslant \Phi(x_0)$ for $|x| \geqslant x_0$ with, in addition, *the graph of* $\log\Phi(x)$ vs $|x|$ *having a sizeable hump immediately to the right of each abscissa* x_0:

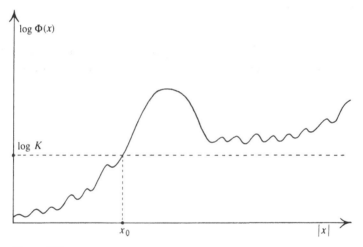

Figure 245

If we *have* such a function Φ and use the weight $W(x) = \Phi(x)$, the condition

$$\int_{x_0}^{\infty} \frac{1}{x\sqrt{(x^2-x_0^2)}} \log\!\left(\frac{\Phi(x)}{K}\right) dx \;\leqslant\; \frac{\pi}{2}A$$

obtained in part (a) of the problem would, for $K = \Phi(x_0)$ with any of the x_0 just described, give us

$$\int_0^{\infty} \frac{1}{x^2} \log^+\!\left(\frac{\Phi(x)}{K}\right) dx \;\leqslant\; cA$$

where c is a number *definitely smaller* than $\pi/2$, and thus make it possible to *bring down the bound found in part* (b) *from* $(\pi/2 + \eta)A$ *to* $(c+\eta)A$. It is conceivable that one could construct such an entire function Φ with humps large enough to make

$$c \;\leqslant\; \frac{\sqrt{\pi}}{\sqrt{\pi+1}}$$

for a sequence of values of K tending to ∞ and values of A corresponding to them (through the *first* of the above two integral inequalities) tending to zero. Denoting the first sequence by $\{K_n\}$ and the second by $\{A_n\}$, we see that *for the function* Φ (if there is one!), the *upper bound provided by the theorem* would, for $A = A_n$, *be proportional to* K_n and hence *exceed the actual value in question by at most a constant factor* (for such A). Although I do not think the value of c can be diminished that much, the construction is perhaps worth *trying*. I have no time for that now; this book must go to press.

Our result seems farther from the truth when functions Φ of exponential type $B > 0$ are in question. For those, the condition on J_M figuring in the statement is essentially of the form

$$J_M \;\leqslant\; \text{const. } A^2$$

when A is small.

It is not terribly difficult to build even functions Φ of exponential type > 0 whose graphs (for real x) contain infinitely many *very long and practically flat plateaux*, e.g.,

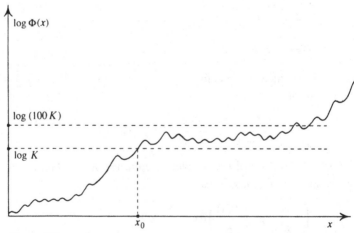

Figure 246

For this kind of function Φ one has arbitrarily large values of K (and of x_0 corresponding to them) such that

$$\int_0^\infty \frac{1}{x^2} \log^+ \left(\frac{\Phi(x)}{100\,K} \right) dx$$

(say) *is exceedingly small* in comparison to

$$\int_{x_0}^\infty \frac{1}{x\sqrt{(x^2 - x_0^2)}} \log \left(\frac{\Phi(x)}{K} \right) dx;$$

putting the *latter* integral *equal* to $(\pi/2)A$, we can thus (for values of A corresponding to these particular ones of K) have the *former* integral $\leqslant$ const. A^2, and even much smaller. What brings

$$\int_0^\infty \frac{1}{x^2} \log \left(1 + A^2 x^2 \frac{\Phi(x)}{100\,CK} \right) dx$$

back up to a constant multiple of A in this situation is *not* the presence of $\Phi(x)/K$ in the integrand, but rather that of x^2 ! In order to reduce this last integral to a multiple of A^2, the A^2 figuring in the integrand must be replaced by A^4, making M *a constant multiple* of

$$K/A^4$$

if J_M is to satisfy the condition in the theorem. We thus find a *discrepancy involving the factor* $1/A^4$ between the *upper bound*

$$2e^2(A + B)^2 M \;\; \cong \;\; \text{const } K/A^4$$

furnished by our result (for *small $A > 0$*) and the *correct value, at least equal to K* when

$$A \;\; = \;\; \frac{2}{\pi} \int_{x_0}^\infty \frac{1}{x\sqrt{(x^2 - x_0^2)}} \log \left(\frac{\Phi(x)}{K} \right) dx.$$

It frequently turns out in actual examples that the K related to A in this way (and such that $\Phi(x) \leqslant K$ for $|x| < x_0$ while $\Phi(x) \geqslant K$ for $|x| \geqslant x_0$) goes to infinity quite rapidly as $A \longrightarrow 0$; one commonly finds that $K \sim \exp(\text{const.}/A)$. Compared with such behaviour, a few factors of $1/A$ more or less are practically of no account. Considering especially the approximate nature of the bound on $\int_{\partial \mathscr{D}} \log|G(t)|\,d\omega_{\mathscr{D}}(t,\,0)$ that we have been using, it hardly seems possible to attain greater precision by the present method.

4.   **Still more about the energy. Description of the**
        Hilbert space $\mathfrak{H}$ used in Chapter VIII, §C.5

Beginning with §B.5 of Chapter VIII, we have been denoting

$$\int_0^\infty \int_0^\infty \log\left|\frac{x+t}{x-t}\right| d\rho(t)\,d\rho(x)$$

by $E(d\rho(t),\, d\rho(t))$ when dealing with real signed measures ρ on $[0,\, \infty)$ without point mass at the origin making the double integral absolutely convergent. In work with such measures ρ it is also convenient to write $U_\rho(x)$ for the Green potential

$$\int_0^\infty \log\left|\frac{x+t}{x-t}\right| d\rho(t)\,,$$

at least in cases where the integral is well defined for $x > 0$. In the latter circumstance $U_\rho(x)$ *cannot*, as remarked at the end of §C.3, Chapter VIII, *be identically zero on* $(0,\, \infty)$ (or, for that matter, vanish a.e. with respect to $|d\rho|$ there) *unless the measure ρ vanishes*. It thus makes sense to regard

$$\sqrt{(E(d\rho(t),\, d\rho(t)))}$$

as a *norm*, $\|U_\rho\|_E$, for the functions $U_\rho(x)$ arising in such fashion. This norm comes from a *bilinear form* $\langle\ \ ,\ \ \rangle_E$ on those functions U_ρ, defined by putting

$$\langle U_\rho,\ U_\sigma \rangle_E \ = \ \int_0^\infty \int_0^\infty \log\left|\frac{x+t}{x-t}\right| d\rho(t)\,d\sigma(x)$$

for any two of them, U_ρ and U_σ; the form's *positive definiteness* is a direct consequence of the results in §B.5, Chapter VIII. Since $\|U_\rho\|_E \ = \ \sqrt{(\langle U_\rho,\ U_\rho \rangle_E)}$, we obtain a certain *real Hilbert space* $\mathfrak{H}$ by forming the (abstract) *completion of the collection of functions U_ρ in the norm* $\|\ \ \|_E$.

The space $\mathfrak{H}$ was already used in the proof of the second theorem of §C.5, Chapter VIII. There, merely the *existence* of $\mathfrak{H}$ was needed, and we did not require any *concrete description* of its *elements*. One can indeed *make do* with just that existence and still proceed quite far. *Specific knowledge of* $\mathfrak{H}$ is, however, really necessary if one is to fully understand (and appreciate) the remaining work of this chapter. The present article is provided for that purpose.

It is actually better to use a wider collection of Green potentials U_ρ in forming the space $\mathfrak{H}$. One starts by showing that if ρ is a signed measure on $[0, \infty)$ without point mass at 0 making

$$\int_0^\infty \int_0^\infty \log\left|\frac{x+t}{x-t}\right| |d\rho(t)| |d\rho(x)| \;<\; \infty,$$

the integral,

$$\int_0^\infty \log\left|\frac{x+t}{x-t}\right| d\rho(t)$$

is absolutely convergent at least *almost everywhere* (but perhaps not everywhere!) for $0 < x < \infty$. In these more general circumstances we will continue to denote that integral by $U_\rho(x)$; we will also have occasion to use the *extension* of that function to the *complex plane* given by the formula

$$U_\rho(z) \;=\; \int_0^\infty \log\left|\frac{z+t}{z-t}\right| d\rho(t).$$

It turns out to be true for these functions U_ρ that $E(d\rho(t), d\rho(t))$ is *determined* when $U_\rho(x)$ is *specified a.e.* on $\mathbb{R}$ (indeed, on $(0, \infty)$); we will in fact obtain a *formula* for the former quantity involving just the function $U_\rho(x)$. This will justify our writing

$$\|U_\rho\|_E \;=\; \sqrt{(E(d\rho(t), d\rho(t)))};$$

the space $\mathfrak{H}$ *will then be taken as the completion of the present class of functions U_ρ in the norm* $\| \;\|_E$. It will follow from our work* that the Hilbert space $\mathfrak{H}$ thus defined coincides with the one initially referred to in this article which, *a priori*, could be a proper subspace of it. That fact is pointed out now; we shall not, however, insist on it during our discussion for *as such* it will not be used.

We shall see in a moment that our space $\mathfrak{H}$ consists of *actual Lebesgue measurable odd functions defined* a.e. *on* $\mathbb{R}$; those will need to be *characterized*.

Let's get down to work.

Lemma. *If ρ is a real signed measure on $[0, \infty)$ without point mass at 0, such that*

$$\int_0^\infty \int_0^\infty \log\left|\frac{x+t}{x-t}\right| |d\rho(t)| |d\rho(x)| \;<\; \infty,$$

---

* see the last theorem in this article and the remark following it

the integral

$$\int_0^\infty \log\left|\frac{x+t}{x-t}\right| d\rho(t)$$

is absolutely convergent for almost all real x, and equal a.e. on ℝ *to an odd Lebesgue measurable function which is locally* L_1.

Proof. For x and $t > 0$, $\log|(x+t)/(x-t)| > 0$, and the left-hand expression is simply changed to its *negative* if, in it, x is replaced by $-x$. The whole lemma thus follows if we verify that

$$\int_0^a \int_0^\infty \log\left|\frac{x+t}{x-t}\right| |d\rho(t)| dx \;\; < \;\; \infty$$

for each finite a. We will use Schwarz' inequality for this purpose.

Fixing $a > 0$, we take the *restriction λ of ordinary Lebesgue measure to* $[0, a]$, and easily verify by direct calculation that

$$\int_0^\infty \int_0^\infty \log\left|\frac{x+t}{x-t}\right| d\lambda(t) d\lambda(x) \;\; < \;\; \infty.$$

According, then, to the remark at the end of §B.5, Chapter VIII, the previous expression, nothing other than

$$E(|d\rho(t)|, \; d\lambda(t))$$

in the notation of that §, is

$$\leqslant \;\; \sqrt{(E(|d\rho(t)|, \; |d\rho(t)|)\cdot E(d\lambda(t), \; d\lambda(t)))},$$

a finite quantity (by the hypothesis). Done.

By almost the same reasoning we can show that *the Hilbert space* $\mathfrak{H}$ must consist of Lebesgue measurable and locally integrable functions on ℝ. In the logical development of the present material, that statement should come somewhat later. Let us, however, strike while the iron is hot:

Theorem. *Suppose that the signed measures ρ_n on $[0, \infty)$, each without point mass at the origin, are such that*

$$\int_0^\infty \int_0^\infty \log\left|\frac{x+t}{x-t}\right| |d\rho_n(t)||d\rho_n(x)| \;\; < \;\; \infty$$

and that furthermore,

$$E\big(d\rho_n(t) - d\rho_m(t), \; d\rho_n(t) - d\rho_m(t)\big) \;\; \xrightarrow[n,m]{} \;\; 0.$$

Then, for each compact subset K of $\mathbb{R}$, the functions

$$U_n(x) = \int_0^\infty \log\left|\frac{x+t}{x-t}\right| \mathrm{d}\rho_n(t)$$

(each defined a.e. by the lemma) form a Cauchy sequence in $L_1(K)$.

Proof. It is again sufficient to check this for sets $K = [0, a]$, where $a > 0$. Fixing any such a and focussing our attention on *some particular pair* (n, m), we take the function

$$\varphi(x) = \begin{cases} \operatorname{sgn}(U_n(x) - U_m(x)), & 0 \leqslant x \leqslant a, \\ 0 & \text{otherwise.} \end{cases}$$

Then,

$$\int_0^a |U_n(x) - U_m(x)| \,\mathrm{d}x = \int_0^\infty (U_n(x) - U_m(x))\,\varphi(x)\,\mathrm{d}x.$$

We have, however,

$$\int_0^\infty \int_0^\infty \log\left|\frac{x+t}{x-t}\right| |\varphi(t)|\,\mathrm{d}t\,|\varphi(x)|\,\mathrm{d}x \leqslant \int_0^a \int_0^a \log\left|\frac{x+t}{x-t}\right| \mathrm{d}t\,\mathrm{d}x$$

with the right side finite, as already noted. Thence, by the remark at the end of §B.5, Chapter VIII,

$$\int_0^\infty (U_n(x) - U_m(x))\,\varphi(x)\,\mathrm{d}x = E(\mathrm{d}\rho_n(t) - \mathrm{d}\rho_m(t),\ \varphi(t)\,\mathrm{d}t)$$

$$\leqslant \sqrt{\big(E(\mathrm{d}\rho_n(t) - \mathrm{d}\rho_m(t),\ \mathrm{d}\rho_n(t) - \mathrm{d}\rho_m(t)) \cdot E(\varphi(t)\mathrm{d}t,\ \varphi(t)\mathrm{d}t)\big)}.$$

Since $\log|(x+t)/(x-t)| > 0$ for x and $t > 0$,

$$E(\varphi(t)\,\mathrm{d}t,\ \varphi(t)\,\mathrm{d}t) \leqslant \int_0^\infty \int_0^\infty \log\left|\frac{x+t}{x-t}\right| |\varphi(t)||\varphi(x)|\,\mathrm{d}t\,\mathrm{d}x$$

which, as we have just seen, is bounded above by a finite quantity – call it C_a – *depending on a but completely independent of n and m* ! The preceding relation thus boils down to

$$\int_0^a |U_n(x) - U_m(x)|\,\mathrm{d}x \leqslant \sqrt{\big(C_a E(\mathrm{d}\rho_n(t) - \mathrm{d}\rho_m(t),\ \mathrm{d}\rho_n(t) - \mathrm{d}\rho_m(t))\big)},$$

and the theorem is proved.

Corollary. *Under the hypothesis of the theorem, a subsequence of the $U_n(x)$ converges a.e. to a locally integrable odd function $U(x)$ defined a.e. on $\mathbb{R}$. For* any bounded measurable function φ of compact support in $[0, \infty)$, we have

$$E(d\rho_n(t), \ \varphi(t)\,dt) \ \xrightarrow[n]{} \ \int_0^\infty U(x)\,\varphi(x)\,dx.$$

Proof. The first part of the statement follows by elementary measure theory from the theorem. A standard application of Fatou's lemma then shows that

$$\int_0^a |U(x) - U_n(x)|\,dx \ \xrightarrow[n]{} \ 0$$

for each finite a. Since the *left-hand* member of the limit relation to be proved is just

$$\int_0^\infty U_n(x)\,\varphi(x)\,dx,$$

we are done.

Remark. Later on, an important generalization of the corollary will be given.

If we *only knew* that the measures $\varphi(t)\,dt$ formed from *bounded* φ of *compact support* in $[0, \infty)$ were $\sqrt{(E(\ ,\ ))}$ *dense in the collection of signed measures* $d\rho(t)$ satisfying the hypothesis of the above lemma, *it would follow* from the results just proved that *any element of that collection's abstract completion in said norm is determined by the measurable function* $U(x)$ *associated to the element in the way described by the corollary.* The density in question is indeed not too hard to verify; we will not, however, proceed in this manner. Instead, the statement just made will be established as a consequence of a formula to be derived below which, for other reasons, is needed in our work.

Given a measure ρ satisfying the hypothesis of our lemma, we will write

$$U_\rho(x) \ = \ \int_0^\infty \log\left|\frac{x+t}{x-t}\right| d\rho(t).$$

The function $U_\rho(x)$ is thus *odd*, and defined at least a.e. on $\mathbb{R}$. Concerning extension of the function U_ρ to the *complex plane*, we observe that the integral

$$\int_0^\infty \log\left|\frac{z+t}{z-t}\right| d\rho(t)$$

converges *absolutely and uniformly* for z ranging over any compact subset of $\{\Im z > 0\}$ or of $\{\Im z < 0\}$.

It is sufficient to consider compact subsets K of the half-open quadrant

$$\{z:\ \Re z \geqslant 0\ (sic!) \text{ and } \Im z > 0\}.$$

We have, by the lemma,

$$\int_0^\infty \log\left|\frac{x_0+t}{x_0-t}\right| |d\rho(t)| \quad < \quad \infty$$

for almost all $x_0 > 0$; *fixing any one of them* gives us a number C_K corresponding to the compact subset K such that

$$\log\left|\frac{z+t}{z-t}\right| \quad \leqslant \quad C_K \log\left|\frac{x_0+t}{x_0-t}\right| \qquad \text{for } t > 0 \text{ and } z \in K.$$

The affirmed uniform convergence is now manifest.

The integral

$$\int_0^\infty \log\left|\frac{z+t}{z-t}\right| d\rho(t)$$

is thus very well defined when z lies *off the real axis*; we denote that expression by $U_\rho(z)$. The uniform convergence just established makes $U_\rho(z)$ harmonic in both *the upper* and *the lower half planes*. It is, moreover, *odd*, and *vanishes on the imaginary axis*. At real points x where the integral used to define $U_\rho(x)$ is absolutely convergent, we have

$$U_\rho(x) \quad = \quad \lim_{y \to 0} U_\rho(x + iy),$$

so *on* $\mathbb{R}$, the function U_ρ can be regarded as the *boundary data* (existing a.e.) for the *harmonic function* $U_\rho(z)$ defined in *either* of the *half planes* bounded by $\mathbb{R}$.

We turn to the proof of the formula mentioned above which, for measures ρ meeting the conditions of the lemma, enables us to express $E(d\rho(t),\ d\rho(t))$ in terms of ρ's Green potential $U_\rho(x)$. We have the good fortune to *already know what that formula should be*, for, if the *behaviour* of

$$\rho(t) \quad = \quad \int_0^t d\rho(\tau)$$

is *nice enough*, problem 23(a), from the beginning of §B.8, Chapter VIII, tells us that

$$E(\mathrm{d}\rho(t),\ \mathrm{d}\rho(t))\ =\ \frac{1}{4\pi^2}\int_{-\infty}^{\infty}\int_{-\infty}^{\infty}\left(\frac{U_\rho(x)-U_\rho(y)}{x-y}\right)^2\mathrm{d}x\,\mathrm{d}y.$$

What we have, then, to do is to *broaden the scope* of this relation, due to Jesse Douglas, so as to get it to apply to the whole class of measures ρ under consideration in this article. Instead to trying to extend the result directly, or to generalize the argument of problem 23(a)   (based on the first lemma of §B.5 in Chapter VIII which was proved there under quite restrictive conditions), we will undertake a new derivation using different ideas.

The machinery employed for this purpose consists of the L_2 theory of Hilbert transforms, sketched in the scholium at the end of §C.1, Chapter VIII. The reader may have already noticed a connection between Hilbert transforms and logarithmic (and Green) potentials, appearing, for instance, in the first lemma of §C.3, Chapter VIII, and in problem 29(b) (Chapter IX, §B.1).

As usual, we write

$$\rho(x)\ =\ \int_0^x \mathrm{d}\rho(t)\qquad\text{for }x\geqslant 0$$

when working with real signed measures ρ on $[0,\ \infty)$. It will also be convenient to extend the definition of such functions ρ to *all of $\mathbb{R}$ by making them even (sic!)* there.

Lemma. *Let*

$$\int_0^{\infty}\int_0^{\infty}\log\left|\frac{x+t}{x-t}\right||\mathrm{d}\rho(t)||\mathrm{d}\rho(x)|\ <\ \infty$$

for the real signed measure ρ on $[0,\ \infty)$ without point mass at the origin. Then $\rho(x)$ is $\mathrm{O}(\sqrt{x})$ for $x\longrightarrow\infty$.

Remark. This is a weak result.

Proof of lemma. Since

$$|\rho(x)|\ \leqslant\ \int_0^x|\mathrm{d}\rho(t)|\qquad\text{for }x>0,$$

it is just as well to assume to begin with that $\mathrm{d}\rho(t)$ is *positive*, and thus $\rho(x)$ *increasing* on $[0,\ \infty)$.

Then, by the second lemma of §B.5, Chapter VIII,

$$\int_0^\infty \int_0^\infty \left(\frac{\rho(x) - \rho(y)}{x - y}\right)^2 \frac{x^2 + y^2}{(x+y)^2}\, dx\, dy \;=\; E(d\rho(t),\, d\rho(t)) \;<\; \infty,$$

so

$$\int_0^\infty \left(\frac{\rho(x) - \rho(y)}{x - y}\right)^2 dx \;<\; \infty$$

for almost all $y > 0$. Fix such a y; we get

$$\int_{2y}^\infty \left(\frac{\rho(x)}{x}\right)^2 dx \;\leqslant\; 2\int_{2y}^\infty \left(\frac{\rho(x) - \rho(y)}{x - y}\right)^2 dx \;+\; 2\frac{(\rho(y))^2}{y} \;<\; \infty,$$

and thence, for $x > 2y$,

$$(\rho(x))^2 \int_x^\infty \frac{dt}{t^2} \;\leqslant\; \int_{2y}^\infty \left(\frac{\rho(t)}{t}\right)^2 dt \;<\; \infty,$$

ρ being increasing. Thus,

$$(\rho(x))^2 \;\leqslant\; \text{const.}\, x \qquad \text{for } x > 2y.$$

Done.

Lemma. *Let ρ satisfy the hypothesis of the preceding lemma. Then*

$$U_\rho(x) \;=\; -\int_0^\infty \left(\frac{1}{x-t} + \frac{1}{x+t}\right)\rho(t)\, dt \qquad \text{a.e., } x \in \mathbb{R}.$$

Proof. $U_\rho(x)$ is odd, so it is enough to establish the formula for almost all $x > 0$. Taking any such x for which the integral defining $U_\rho(x)$ converges absolutely, we have

$$U_\rho(x) \;=\; \lim_{\varepsilon \to 0}\left(\int_0^{x-\varepsilon} + \int_{x+\varepsilon}^\infty\right)\log\left|\frac{x+t}{x-t}\right|\, d\rho(t).$$

Fixing for the moment a small $\varepsilon > 0$, we treat the two integrals on the right by partial integration, very much as in the proof of the lemma in §C.3, Chapter VIII (but going in the opposite direction). The integrated term

$$\rho(t)\log\left|\frac{x+t}{x-t}\right|$$

which thus arises vanishes at $t = 0$ and also when $t \longrightarrow \infty$, the latter thanks to the preceding lemma. Subtraction of its values for $t = x \pm \varepsilon$ also gives a small result when $\varepsilon > 0$ is small, *as long as $\rho'(x)$ exists and is finite*, and therefore for almost all x. We thus end with the desired formula on making $\varepsilon \longrightarrow 0$,   Q.E.D.

Referring to our convention that $\rho(-t) = \rho(t)$, we immediately obtain the

Corollary

$$U_\rho(x) = -\int_{-\infty}^{\infty} \left(\frac{1}{x-t} + \frac{t}{t^2+1} \right) \rho(t)\, dt \qquad \text{a.e.,} \ x \in \mathbb{R}.$$

Remark. The dummy term $t/(t^2+1)$ is introduced in the integrand in order to guarantee absolute convergence of the integral near $\pm\infty$, and does so, $\rho(t)$ being $O(\sqrt{|t|})$ there. We see that, aside from a missing factor of $-1/\pi$, $U_\rho(x)$ is just the harmonic conjugate (Hilbert transform) of $\rho(x)$, which should be very familiar to anyone who has read up to here in the present book.

Lemma. *Let the signed measure ρ satisfy the hypothesis of the first of the preceding two lemmas. Then, for almost every real y, the function of x equal to $(\rho(x)-\rho(y))/(x-y)$ belongs to $L_2(-\infty,\infty)$, and*

$$\frac{U_\rho(x) - U_\rho(y)}{x-y} = -\int_{-\infty}^{\infty} \frac{1}{x-t} \frac{\rho(t)-\rho(y)}{t-y}\, dt \qquad \text{a.e.,} \ x \in \mathbb{R}.$$

Proof. As at the beginning of the proof of the first of the above two lemmas,

$$\int_0^\infty \int_0^\infty \left(\frac{\rho(x)-\rho(y)}{x-y} \right)^2 \frac{x^2+y^2}{(x+y)^2}\, dx\, dy \ < \ \infty,$$

so, for almost every $y > 0$, $(\rho(x)-\rho(y))/(x-y)$ belongs to $L_2(0,\infty)$ as a function of x. But since ρ is even,

$$\left| \frac{\rho(x) - \rho(y)}{x - y} \right| \leqslant \left| \frac{\rho(|x|) - \rho(|y|)}{|x| - |y|} \right|;$$

we thus see by the statement just made that as a function of x, $(\rho(x)-\rho(y))/(x-y)$ belongs in fact to $L_2(-\infty,\infty)$ for almost all $y \in \mathbb{R}$.

For any real number C, we have (trick!):

$$\int_{-\infty}^{\infty} \left(\frac{1}{x-t} + \frac{t}{t^2+1} \right) C\, dt \ = \ 0.$$

Adding this relation to the formula given by the last corollary, we thus get

$$U_\rho(x) \;=\; -\int_{-\infty}^{\infty} \left(\frac{1}{x-t} + \frac{t}{t^2+1} \right) (\rho(t) - C) \, dt$$

for almost all $x \in \mathbb{R}$, where the exceptional set *does not depend* on the number C. From this relation we subtract the similar one obtained on replacing x by any other value y for which it holds. That yields

$$U_\rho(x) - U_\rho(y) \;=\; -\int_{-\infty}^{\infty} \left(\frac{1}{x-t} - \frac{1}{y-t} \right) (\rho(t) - C) \, dt \quad \text{a.e., } x, y \in \mathbb{R}.$$

In the Cauchy principal value standing on the right, the integrand involves *two* singularities, at $t = x$ and at $t = y$. Consider, however, what happens when y takes one of the values for which $\rho'(y)$ *exists and is finite. Then, we can put $C = \rho(y)$ in the preceding relation* (!), and, after dividing by $x - y$, it becomes

$$\frac{U_\rho(x) - U_\rho(y)}{x - y} \;=\; -\int_{-\infty}^{\infty} \frac{1}{x-t} \frac{\rho(t) - \rho(y)}{t - y} \, dt,$$

in which the function $(\rho(t) - \rho(y))/(t - y)$ figuring on the right *remains bounded* for $t \longrightarrow y$. What we have on the right is thus just the *ordinary* Cauchy principal value involving an integrand with *one* singularity (at $t = x$), used in the study of Hilbert transforms.

We are, however, assured of the existence and finiteness of $\rho'(y)$ at almost every y. The last relation thus holds a.e. in both x and y, and we are done.

Theorem. *Let the real signed measure ρ on $[0, \infty)$, without point mass at the origin, be such that*

$$\int_0^{\infty} \log \left| \frac{x+t}{x-t} \right| |d\rho(t)| |d\rho(x)| \;<\; \infty,$$

and put

$$U_\rho(x) \;=\; \int_0^{\infty} \log \left| \frac{x+t}{x-t} \right| d\rho(t),$$

thus specifying the value of U_ρ almost everywhere on $\mathbb{R}$. Then we have Jesse Douglas' formula:

$$E(d\rho(t), d\rho(t)) \;=\; \frac{1}{4\pi^2} \int_{-\infty}^{\infty} \int_{-\infty}^{\infty} \left(\frac{U_\rho(x) - U_\rho(y)}{x - y} \right)^2 dx \, dy.$$

Proof. The last lemma exhibits the function of x equal to $(U_\rho(x) - U_\rho(y))/(x - y)$ as $-\pi$ times the *Hilbert transform* of the one equal to $(\rho(x) - \rho(y))/(x - y)$ for almost every $y \in \mathbb{R}$, and also tells us that the latter function of x is in $L_2(-\infty, \infty)$ for almost every such y. We may therefore apply to these functions *the L_2 theory of Hilbert transforms* taken up in the scholium at the end of §C.1, Chapter VIII. By that theory,

$$\int_{-\infty}^{\infty} \left(\frac{U_\rho(x) - U_\rho(y)}{\pi(x - y)} \right)^2 dx = \int_{-\infty}^{\infty} \left(\frac{\rho(x) - \rho(y)}{x - y} \right)^2 dx$$

for the values of y in question, i.e., almost everywhere in y.

Integrating now with respect to y, this gives

$$\frac{1}{\pi^2} \int_{-\infty}^{\infty} \int_{-\infty}^{\infty} \left(\frac{U_\rho(x) - U_\rho(y)}{x - y} \right)^2 dx\,dy = \int_{-\infty}^{\infty} \int_{-\infty}^{\infty} \left(\frac{\rho(x) - \rho(y)}{x - y} \right)^2 dx\,dy.$$

Because ρ is even, the right side is just

$$2 \int_0^{\infty} \int_0^{\infty} \left\{ \left(\frac{\rho(x) - \rho(y)}{x - y} \right)^2 + \left(\frac{\rho(x) - \rho(y)}{x + y} \right)^2 \right\} dx\,dy$$

$$= 4 \int_0^{\infty} \int_0^{\infty} \left(\frac{\rho(x) - \rho(y)}{x - y} \right)^2 \frac{x^2 + y^2}{(x + y)^2} dx\,dy.$$

Dividing by 4 and referring to the second lemma of §B.5, Chapter VIII, we immediately obtain the desired result.

Corollary. *For any measure ρ satisfying the hypothesis of the theorem,*

$$E(d\rho(t),\ d\rho(t))$$

is determined when the Green potential $U_\rho(x)$ is specified almost everywhere on $\mathbb{R}$, and $\rho = 0$ if $U_\rho(x) = 0$ a.e. in $(0, \infty)$. (Here $U_\rho(x)$ is determined by its values on $(0, \infty)$ because it is odd.)

This corollary finally gives us the right to denote $\sqrt{(E(d\rho(t), d\rho(t)))}$ by $\|U_\rho\|_E$ for *any* measure ρ fulfilling the conditions of the theorem; indeed, we simply have

$$\|U_\rho\|_E = \frac{1}{2\pi} \sqrt{\left(\int_{-\infty}^{\infty} \int_{-\infty}^{\infty} \left(\frac{U_\rho(x) - U_\rho(y)}{x - y} \right)^2 dx\,dy \right)}.$$

It will be convenient for us to use this formula for *arbitrary* real-valued Lebesgue measurable functions (*odd or not !*) defined on $\mathbb{R}$. Then, of course, it becomes a matter of

Notation. Given v, real-valued and Lebesgue measurable on $\mathbb{R}$, we write

$$\|v\|_E \;=\; \frac{1}{2\pi}\sqrt{\left(\int_{-\infty}^{\infty}\int_{-\infty}^{\infty}\left(\frac{v(x)-v(y)}{x-y}\right)^2 dx\,dy\right)}.$$

This clearly defines $\|\ \ \|_E$ as a *norm* on the collection of such functions v (modulo the constants); if $\|v\|_E \;=\; 0$ we must have

$$v(x) \;=\; \text{const.}\quad \text{a.e.,}\quad x \in \mathbb{R}.$$

Near the beginning of this article, we said how the Hilbert space $\mathfrak{H}$ was to be formed: $\mathfrak{H}$ was specified as *the abstract completion in norm* $\|\ \ \|_E$ *of the collection of Green potentials U_ρ coming from the measures ρ satisfying the conditions of the last theorem.* An *element* of $\mathfrak{H}$ is, in other words, defined by a *Cauchy sequence*, $\{U_{\rho_n}\}$, of such potentials. According, however, to the corollary of the *first* theorem in this article, such a Cauchy sequence has in it a *subsequence*, which we may as well *also*, for the moment, denote by $\{U_{\rho_n}\}$, with $U_{\rho_n}(x)$ *pointwise convergent* at almost every $x \in \mathbb{R}$. Writing

$$\lim_{n\to\infty} U_{\rho_n}(x) \;=\; U(x)$$

wherever the limit exists, we see that $U(x)$ is *defined* a.e. and *Lebesgue measurable*; it is also *odd*, because the individual Green potentials $U_{\rho_n}(x)$ are odd.

Fixing any index m, we have, making the usual application of Fatou's lemma,

$$\|U - U_{\rho_m}\|_E^2 \;=\; \frac{1}{4\pi^2}\int_{-\infty}^{\infty}\int_{-\infty}^{\infty}\left(\frac{U(x)-U_{\rho_m}(x)-U(y)+U_{\rho_m}(y)}{x-y}\right)^2 dx\,dy$$

$$\leqslant\; \liminf_{j\to\infty}\frac{1}{4\pi^2}\int_{-\infty}^{\infty}\int_{-\infty}^{\infty}\left(\frac{U_{\rho_j}(x)-U_{\rho_m}(x)-U_{\rho_j}(y)+U_{\rho_m}(y)}{x-y}\right)^2 dx\,dy$$

$$=\; \liminf_{j\to\infty}\|U_{\rho_j} - U_{\rho_m}\|_E^2.$$

Since we started with a Cauchy sequence, the last quantity is *small* if m is large. This in fact holds for all the U_{ρ_m} from our *original* sequence, for the last chain of inequalities is valid for any of those potentials as long as the U_{ρ_j} appearing therein run *through the subsequence* just described. *Corresponding to the Cauchy sequence* $\{U_{\rho_n}\}$, *we have thus found an odd measurable function U with*

$$\| U \, - \, U_{\rho_n} \|_E \xrightarrow[n]{} 0.$$

In this fashion we can associate *an odd measurable function U, approximable in the norm* $\| \ \|_E$ *by Green potentials* U_ρ *like the ones appearing in the last theorem, to each element of the space* $\mathfrak{H}$. It is, on the other hand, manifest that *each such function U does indeed correspond to some element of* $\mathfrak{H}$ – the Green potentials U_ρ approximating U in norm $\| \ \|_E$ *furnish us* with a Cauchy sequence of such potentials (in that norm)! There is thus a correspondence between *the collection of such functions U* and *the space* $\mathfrak{H}$.

It is necessary now to show that *this correspondence is one-one.* But that is easy. Suppose, in the first place that *two different odd functions*, say U and V, are associated to *the same* element of the space $\mathfrak{H}$ in the manner described. Then we have *two* Cauchy sequences of Green potentials, say $\{ U_{\rho_n} \}$ and $\{ U_{\sigma_n} \}$, with

$$\| U_{\rho_n} \, - \, U_{\sigma_n} \|_E \xrightarrow[n]{} 0,$$

and such that

$$\| U \, - \, U_{\rho_n} \|_E \xrightarrow[n]{} 0$$

while

$$\| V \, - \, U_{\sigma_n} \|_E \xrightarrow[n]{} 0.$$

It follows that

$$\| U \, - \, V \|_E \; = \; 0,$$

but then, as noted above,

$$U(x) \, - \, V(x) \; = \; \text{const.} \quad \text{a.e.,} \quad x \in \mathbb{R}.$$

Here, $U - V$ is *odd*, so the *constant* must be *zero*, and

$$U(x) \; = \; V(x) \quad \text{a.e.,} \quad x \in \mathbb{R}.$$

Given, on the other hand, *two* Cauchy sequences, $\{ U_{\rho_n} \}$ and $\{ U_{\sigma_n} \}$, of potentials associated to *the same odd function U*, we have

$$\| U \, - \, U_{\rho_n} \|_E \xrightarrow[n]{} 0$$

and

$$\| U \, - \, U_{\sigma_n} \|_E \xrightarrow[n]{} 0,$$

whence

$$\| U_{\rho_n} - U_{\sigma_n} \|_E \xrightarrow[n]{} 0.$$

Then, however, $\{U_{\rho_n}\}$ and $\{U_{\sigma_n}\}$ define *the same element* of the abstract completion $\mathfrak{H}$.

Our Hilbert space $\mathfrak{H}$ is thus in one-to-one correspondence with the collection of odd real measurable functions U approximable, in norm $\| \ \|_E$, by the potentials U_ρ under consideration here. There is hence nothing to keep us from *identifying the space $\mathfrak{H}$ with that collection of functions U, and we henceforth do so.*

We are now well enough equipped to give a strengthened version, promised earlier, of the corollary to the first theorem in this article.

Lemma. *Let the odd measurable function U be identified with an element of the space $\mathfrak{H}$ in the manner just described, and suppose that ρ is an absolutely continuous signed measure on $[0, \infty)$ with*

$$\int_0^\infty \int_0^\infty \log \left| \frac{x+t}{x-t} \right| \, d\rho(t) \, d\rho(x)$$

and

$$\int_0^\infty U(x) \, d\rho(x)$$

both absolutely convergent. Then

$$\int_0^\infty U(x) \, d\rho(x) = \langle U, U_\rho \rangle_E,$$

and especially

$$\left| \int_0^\infty U(x) \, d\rho(x) \right| \leqslant \| U \|_E \sqrt{(E(d\rho(t), \, d\rho(t)))}.$$

Remark. The second relation is very useful in certain applications.

Proof of lemma. We proceed to establish the first relation, using a somewhat repetitious crank-turning argument.

Starting with an absolutely continuous ρ fulfilling the conditions in the hypothesis, let us put, for $N \geqslant 1$,

$$d\rho_N(t) \;=\; \begin{cases} \rho'(t)\,dt & \text{if } |\rho'(t)| \leqslant N, \\ (N\,\mathrm{sgn}\,\rho'(t))\,dt & \text{otherwise}; \end{cases}$$

it is claimed that

$$\| U_\rho \;-\; U_{\rho_N} \|_E \;\longrightarrow\; 0$$

for $N \longrightarrow \infty$.

By breaking $\rho'(t)$ up into positive and negative parts, we can reduce the general situation to one in which

$$\rho'(t) \;\geqslant\; 0,$$

so we may as well assume this property. Then, for each $x > 0$, the potentials

$$U_{\rho_N}(x) \;=\; \int_0^\infty \log\left|\frac{x+t}{x-t}\right| \min(\rho'(t),\,N)\,dt$$

increase and tend to $U_\rho(x)$ *as* $N \longrightarrow \infty$. Hence, by monotone convergence,

$$\int_0^\infty U_{\rho_N}(x)\,d\rho(x) \;\xrightarrow[N]{}\; \int_0^\infty U_\rho(x)\,d\rho(x).$$

From this, we see that

$$\begin{aligned} \| U_\rho \;-\; U_{\rho_N} \|_E^2 \;&=\; E(d\rho(t) - d\rho_N(t),\; d\rho(t) - d\rho_N(t)) \\ &=\; \int_0^\infty (U_\rho(x) - U_{\rho_N}(x))(\rho'(x) - \min(\rho'(x),N))\,dx \\ &\leqslant\; \int_0^\infty (U_\rho(x) - U_{\rho_N}(x))\,d\rho(x) \quad (!) \end{aligned}$$

must *tend to zero* as $N \longrightarrow \infty$, verifying our assertion.

From what we have just shown, it follows that

$$\langle U,\, U_{\rho_N} \rangle_E \;\xrightarrow[N]{}\; \langle U,\, U_\rho \rangle_E.$$

But we clearly have

$$\int_0^\infty U(x)\,d\rho_N(x) \;\xrightarrow[N]{}\; \int_0^\infty U(x)\,d\rho(x)$$

by the given absolute convergence of the integral on the right. The desired first relation *will therefore follow* if we can prove that

$$\int_0^\infty U(x)\,d\rho_N(x) \;=\; \langle U,\, U_{\rho_N} \rangle_E$$

for each N; that, however, simply amounts to *verifying the relation* in question for *measures* ρ satisfying the hypothesis and *having, in addition, bounded densities* $\rho'(x)$. We have thus brought down by one notch the generality of what is to be proven.

Suppose, then, that ρ satisfies the hypothesis and that $\rho'(x)$ *is also bounded.* For each a, $\quad 0 < a < 1$, put

$$
\rho_a'(t) \quad = \quad \begin{cases} \rho'(t), & a \leqslant t \leqslant \dfrac{1}{a}, \\[2ex] 0 & \text{otherwise}, \end{cases}
$$

and then define a measure ρ_a (*not to be confounded with the ρ_N just used!*) by taking $d\rho_a(t) \;=\; \rho_a'(t)\,dt$. An argument very similar to the one made above now shows that

$$
\| U_\rho \;-\; U_{\rho_a} \|_E \;\longrightarrow\; 0 \quad \text{for } a \longrightarrow 0
$$

(it suffices as before to consider the case where $\rho'(t) \geqslant 0$) and hence that

$$
\langle U, \; U_{\rho_a} \rangle_E \;\longrightarrow\; \langle U, \; U_\rho \rangle_E \quad \text{as } a \longrightarrow 0.
$$

At the same time,

$$
\int_0^\infty U(x)\,d\rho_a(x) \;\longrightarrow\; \int_0^\infty U(x)\,d\rho(x),
$$

so it is enough to check that

$$
\int_0^\infty U(x)\,d\rho_a(x) \;=\; \langle U, \; U_{\rho_a} \rangle_E
$$

for each a.

Here, however, $d\rho_a(x) \;=\; \rho_a'(x)\,dx$ with $\rho_a'(x)$ *bounded and of compact support in* $(0, \infty)$, so the *corollary of the first theorem* in this article is applicable. Since U is identified with an element of $\mathfrak{H}$ there is, by the discussion preceding this lemma, a sequence of Green potentials U_{σ_n} of the kind used to form that space such that

$$
\| U \;-\; U_{\sigma_n} \|_E \;\xrightarrow[n]{}\; 0,
$$

and also

$$
U_{\sigma_n}(x) \;\xrightarrow[n]{}\; U(x) \quad \text{a.e., } x \in \mathbb{R}.
$$

Then we must have

$$E(\mathrm{d}\sigma_n(t), \ \mathrm{d}\rho_a(t)) \ = \ \langle U_{\sigma_n}, \ U_{\rho_a}\rangle_E \ \xrightarrow[n]{} \ \langle U, \ U_{\rho_a}\rangle_E,$$

and, by the corollary referred to,

$$E(\mathrm{d}\sigma_n(t), \ \mathrm{d}\rho_a(t)) \ \xrightarrow[n]{} \ \int_0^\infty U(x)\,\mathrm{d}\rho_a(x).$$

Thus,

$$\langle U, \ U_{\rho_a}\rangle_E \ = \ \int_0^\infty U(x)\,\mathrm{d}\rho_a(x),$$

what we needed to finish showing the first relation in the conclusion of the lemma.

From it, however, the *second* relation follows immediately by Schwarz' inequality and the preceding theorem, since

$$\|U_\rho\|_E \ = \ \sqrt{(E(\mathrm{d}\rho(t), \ \mathrm{d}\rho(t)))}.$$

Our lemma is proved.

What we have done so far still does not amount to a real *description* of the space $\mathfrak{H}$, for we do not yet know *which* odd measurable functions $U(x)$ with $\|U\|_E < \infty$ can be approximated in the norm $\| \ \|_E$ by potentials U_ρ of the kind appearing in the last theorem. The fact is that *all those functions U can be thus approximated.*

Theorem. $\mathfrak{H}$ *consists precisely of the real odd measurable functions* $U(x)$ *for which*

$$\|U\|_E^2 \ = \ \frac{1}{4\pi^2}\int_{-\infty}^\infty \int_{-\infty}^\infty \left(\frac{U(x)-U(y)}{x-y}\right)^2 \mathrm{d}x\,\mathrm{d}y$$

is finite.

Proof. That $\mathfrak{H}$ consists of such functions U was shown in the course of the previous discussion; what we have to do here is prove the *converse*, to the effect that *any* odd function U with $\|U\|_E < \infty$ *is in* $\mathfrak{H}$. This involves an approximation argument.

Starting, then, with an odd function U such that $\|U\|_E < \infty$, we must *obtain* signed measures ρ on $[0, \infty)$ meeting the conditions of the last theorem, *for which*

$$\|U \ - \ U_\rho\|_E$$

is as small as we please. That will be done in essentially *three steps.*

The first step makes use of the notion of *contraction*, brought into potential theory by Beurling and Deny. For $M > 0$, put

$$U_M(x) = \begin{cases} U(x) & \text{if } |U(x)| < M, \\ M \operatorname{sgn} U(x) & \text{if } |U(x)| \geqslant M. \end{cases}$$

Then

$$|U_M(x) - U_M(y)| \leqslant |U(x) - U(y)|$$

(this is the contraction property), so

$$\left(\frac{U(x) - U_M(x) - U(y) + U_M(y)}{x - y} \right)^2 \leqslant 4 \left(\frac{U(x) - U(y)}{x - y} \right)^2.$$

But $U_M(x) \longrightarrow U(x)$ as $M \longrightarrow \infty$, so the *left side* of this relation *tends to zero* a.e. in x and y (with respect to Lebesgue measure for $\mathbb{R}^2$) as $M \longrightarrow \infty$. (The *odd* function $U(x)$ *cannot be infinite* on a set of *positive measure*, for in that event U would be infinite on such a subset of $(-\infty, 0]$ or of $[0, \infty)$, and this would clearly make $\|U\|_E = \infty$.) Because $\|U\|_E < \infty$, the double integral of the *right side* of the relation over $\mathbb{R}^2$ is *finite*. Hence

$$\|U - U_M\|_E^2 \longrightarrow 0 \quad \text{for } M \longrightarrow \infty$$

by dominated convergence. *Any function U satisfying the hypothesis is thus $\| \quad \|_E$ - approximable by bounded ones.*

It therefore suffices to show how to do the desired approximation of *bounded* functions U fulfilling the conditions of this theorem. Taking such a one, for which

$$|U(x)| \leqslant M, \quad \text{say,}$$

on the real axis, we look at the products

$$U_H(x) = \frac{H^2}{H^2 + x^2} U(x)$$

where H is *large*. (*These should not be confounded with the contractions U_M used in the previous step.*) Denoting by $v_H(x)$ the (even!) function

$$\frac{H^2}{H^2 + x^2},$$

we have

$$U_H(x) - U_H(y) = v_H(x)(U(x) - U(y)) + U(y)(v_H(x) - v_H(y)),$$

so that

$$\left|\frac{U_H(x)-U_H(y)}{x-y}\right| \leqslant \left|\frac{U(x)-U(y)}{x-y}\right| + M\left|\frac{v_H(x)-v_H(y)}{x-y}\right|.$$

From this last, it is easily seen that

$$\|U_H\|_E^2 \leqslant 2\|U\|_E^2 + 2M^2\|v_H\|_E^2,$$

and we must compute the norms $\|v_H\|_E$.

To do that, we note that $v_H(x)$ has a harmonic extension to the upper half plane given by

$$v_H(z) \;=\; \frac{H(H+y)}{(H+y)^2+x^2} \;=\; -\Im\left(\frac{H}{z+iH}\right),$$

and that this function is sufficiently well behaved for the identities employed in the solution of problem 23(a) (§B.8, Chapter VIII) to apply to it. By the help of those, one finds that

$$\|v_H\|_E^2 \;=\; \frac{1}{2\pi}\int_0^\infty\int_{-\infty}^\infty \left\{\left(\frac{\partial v_H(z)}{\partial x}\right)^2 + \left(\frac{\partial v_H(z)}{\partial y}\right)^2\right\} dx\,dy.$$

Using the Cauchy–Riemann equations, we convert the integral on the right to

$$\frac{H^2}{2\pi}\int_0^\infty\int_{-\infty}^\infty \left|\frac{d}{dz}\left(\frac{1}{z+iH}\right)\right|^2 dx\,dy \;=\; \frac{H^2}{2\pi}\int_0^\infty\int_{-\infty}^\infty \frac{1}{|z+iH|^4}dx\,dy,$$

which becomes

$$\frac{1}{2\pi}\int_0^\infty\int_{-\infty}^\infty \frac{d\xi\,d\eta}{(\xi^2+(\eta+1)^2)^2}$$

after putting $z = H(\xi + i\eta)$. The further substitution $\xi = (\eta+1)\tau$ converts the last double integral to

$$\frac{1}{2\pi}\int_0^\infty \frac{d\eta}{(\eta+1)^3}\int_{-\infty}^\infty \frac{d\tau}{(\tau^2+1)^4} \;=\; \frac{1}{8},$$

so we have

$$\|v_H\|_E \;=\; \frac{1}{2\sqrt{2}}$$

independently of the number H.

Plugging this into the previous inequality, we get

$$\|U_H\|_E^2 \leqslant 2\|U\|_E^2 + \tfrac{1}{4}M^2,$$

so the *norms on the left remain bounded as $H \longrightarrow \infty$*. From this it follows that *the functions U_H corresponding to some suitable sequence of values of H going out to infinity tend weakly* (in the Hilbert space of *all* real *odd* functions with finite $\| \ \|_E$ norm!) *to some odd function W with* $\|W\|_E < \infty$. That in turn implies that *some sequence of* (finite) *convex linear combinations u_n of those U_H* (formed by using *ever larger values* of H from the sequence just mentioned) *actually tends in norm $\| \ \|_E$ to W.* It is clear, however, that

$$U_H(x) \longrightarrow U(x)$$

(wherever the value on the right is defined) as $H \longrightarrow \infty$. Hence

$$u_n(x) \xrightarrow[n]{} U(x) \quad \text{a.e.,} \quad x \in \mathbb{R},$$

so, since

$$\| W - u_n \|_E \xrightarrow[n]{} 0,$$

we see by the usual application of Fatou's lemma that

$$\| W - U \|_E = 0,$$

making (as in the previous discussion about the construction of $\mathfrak{H}$)

$$W(x) = U(x) \quad \text{a.e.,} \quad x \in \mathbb{R},$$

since W and U are *both odd*. Therefore, we in fact have

$$\| U - u_n \|_E \xrightarrow[n]{} 0,$$

and *we can approximate the function U in norm $\| \ \|_E$ by finite linear combinations u_n of the functions*

$$\frac{H^2}{H^2 + x^2} U(x).$$

Since the function $U(x)$ is *bounded*, we see that each function u_n, *besides being odd* (like U), *satisfies a condition of the form*

$$|u_n(x)| \leqslant \frac{K_n}{x^2 + 1}, \quad x \in \mathbb{R}$$

(where, of course, the constant K_n may be enormous, but we don't care about that).

For this reason it is enough if, using the potentials U_ρ, one can approximate in norm $\| \ \ \|_E$ any odd function u with $\|u\|_E < \infty$ and, in addition,

$$|u(x)| \leqslant \frac{K}{x^2 + 1} \quad \text{on } \mathbb{R}.$$

Showing how to do that is the last step in our proof.

Take any such u. For $\Im z > 0$, put

$$u(z) = \frac{1}{\pi} \int_{-\infty}^{\infty} \frac{\Im z}{|z - t|^2} u(t) \, dt.$$

The *size* of $|u(t)|$ on $\mathbb{R}$ is here well enough controlled so that the Fourier integral argument used in working problem 23(a) may be applied to $u(z)$. In that way, one readily verifies that

$$\|u\|_E^2 = \frac{1}{2\pi} \int_0^\infty \int_{-\infty}^\infty \{(u_x(z))^2 + (u_y(z))^2\} \, dx \, dy;$$

this, in particular, makes the Dirichlet integral appearing on the right *finite*.

For $h > 0$ (which in a moment will be made to tend to zero) we now put

$$u_h(z) = u(z + ih), \quad \Im z \geqslant 0.$$

An evident adaptation of the argument just referred to then shows that

$$\|u - u_h\|_E^2 = \frac{1}{2\pi} \iint_{\Im z > 0} \left\{ \left(\frac{\partial}{\partial x}(u(z) - u_h(z)) \right)^2 \right.$$
$$\left. + \left(\frac{\partial}{\partial y}(u(z) - u_h(z)) \right)^2 \right\} \, dx \, dy.$$

The right-hand integral is just

$$\frac{1}{2\pi} \iint_{\Im z > 0} \{(u_x(z) - u_x(z + ih))^2 + (u_y(z) - u_y(z + ih))^2\} \, dx \, dy,$$

and we see that *it must tend to zero when* $h \longrightarrow 0$ (by continuity of translation in $L_2(\mathbb{R}^2)$!), since

$$\iint_{\Im z > 0} (u_x(z))^2 \, dx \, dy \quad \text{and} \quad \iint_{\Im z > 0} (u_y(z))^2 \, dx \, dy$$

are *both finite*, according to the observation just made. Thus,

$$\|u - u_h\|_E \longrightarrow 0 \quad \text{as } h \longrightarrow 0$$

It is now claimed that *each function $u_h(x)$ is equal* (on $\mathbb{R}$) *to a potential* $U_\rho(x)$ of the required kind. Since $u(t)$ is odd, we have

$$u(x + iy) = u(-x + iy) \qquad \text{for } y > 0,$$

so $u_h(x)$ is odd. Our condition on $u(x)$ implies a similar one,

$$|u_h(x)| \leqslant \frac{\text{const}}{x^2 + 1}, \qquad x \in \mathbb{R},$$

on u_h, so that function is (and by far!) in $L_2(-\infty, \infty)$, and we can apply to it the L_2 theory of Hilbert transforms from the scholium at the end of §C.1, Chapter VIII. In the present circumstances, $u_h(x) = u(x + ih)$ is $\mathscr{C}_\infty$ in x, so the Hilbert transform

$$\tilde{u}_h(x) = \frac{1}{\pi} \int_{-\infty}^{\infty} \frac{u_h(t)}{x - t} \, dt = \frac{1}{\pi} \int_0^{\infty} \frac{u_h(x - \tau) - u_h(x + \tau)}{\tau} \, d\tau$$

is defined and continuous at each real x, the last integral on the right being absolutely convergent. From the Hilbert transform theory referred to (even a watered-down version of it will do here!) we thence get, by the inversion formula,

$$u_h(x) = -\frac{1}{\pi} \int_{-\infty}^{\infty} \frac{\tilde{u}_h(t)}{x - t} \, dt \qquad \text{a.e., } x \in \mathbb{R}.$$

This relation, like the one preceding it, holds in fact at *each* $x \in \mathbb{R}$, for $\tilde{u}_h(x)$ is nothing but the value of a *harmonic conjugate to* $u(z)$ at $z = x + ih$ and is hence (like $u_h(x)$) $\mathscr{C}_\infty$ in x. Wishing to integrate the right-hand member by parts, we look at the behaviour of $\tilde{u}_h'(t)$.

By the Cauchy–Riemann equations,

$$\tilde{u}_h'(x) = \tilde{u}_x(x + ih) = -u_y(x + ih).$$

After differentiating the (Poisson) formula for $u(z)$ and then plugging in the given estimate on $|u(t)|$, we get (for small $h > 0$)

$$|u_y(x + ih)| \leqslant \frac{\text{const.}}{h(x^2 + 1)},$$

so

$$|\tilde{u}_h'(t)| \leqslant \frac{\text{const.}}{t^2 + 1} \qquad \text{for } t \in \mathbb{R}.$$

As we have noted, $u_h(x)$ is *odd*. Its Hilbert transform $\tilde{u}_h(t)$ is therefore *even*, and the preceding formula for u_h can be written

$$u_h(x) = -\frac{1}{\pi}\int_0^\infty \left(\frac{1}{x-t} + \frac{1}{x+t}\right)\tilde{u}_h(t)\,dt.$$

Here, we integrate by parts as in proving the lemma of §C.3, Chapter VIII and the third lemma of the present article. By the last inequality we actually have

$$\int_0^\infty |\tilde{u}_h'(t)|\,dt < \infty,$$

so, $\tilde{u}_h(t)$ being $\mathscr{C}_\infty$, the partial integration readily yields the formula

$$u_h(x) = \frac{1}{\pi}\int_0^\infty \log\left|\frac{x+t}{x-t}\right|\tilde{u}_h'(t)\,dt,$$

valid for all real x.

This already exhibits $u_h(x)$ as a Green potential $U_\rho(x)$ with

$$d\rho(t) = \frac{1}{\pi}\,\tilde{u}_h'(t)\,dt,$$

and in order to complete this last step of the proof, it is only necessary to check that

$$\int_0^\infty\int_0^\infty \log\left|\frac{x+t}{x-t}\right||\tilde{u}_h'(t)||\tilde{u}_h'(x)|\,dt\,dx < \infty.$$

That, however, follows in straightforward fashion from the above estimate on $|\tilde{u}_h'(t)|$. Breaking up (for $x > 0$)

$$\int_0^\infty \log\left|\frac{x+t}{x-t}\right|\frac{dt}{1+t^2}$$

as $\int_0^{2x} + \int_{2x}^\infty$, we have

$$\int_0^{2x} = \int_0^2 \log\left|\frac{1+\tau}{1-\tau}\right|\frac{x\,d\tau}{1+x^2\tau^2} \leqslant \frac{1}{2}\int_0^2 \log\left|\frac{1+\tau}{1-\tau}\right|\frac{d\tau}{\tau},$$

a finite constant, whilst

$$\int_{2x}^\infty \leqslant \int_{2x}^\infty O\left(\frac{x}{t}\right)\frac{dt}{1+t^2} = O\left(x\log\frac{1+x^2}{x^2}\right) = O\left(\frac{1}{1+x}\right).$$

The double integral in question is thus

$$\leqslant \text{const.}\int_0^\infty\left(1 + \frac{1}{1+x}\right)\frac{dx}{1+x^2} < \infty,$$

showing that the measure ρ given by the above formula has the required property.

The three steps of our approximation have thus been carried out, and the theorem completely proved.

Remark. The Green potentials U_ρ furnished by this proof and approximating, in norm $\|\ \ \|_E$, a given odd function $U(x)$ with $\|U\|_E < \infty$, are formed from signed measures ρ on $[0, \infty)$ having, in addition to the properties enumerated at the beginning of this article, the following special one:

each ρ is absolutely continuous, with $\mathscr{C}_\infty$ density satisfying a relation of the form

$$|\rho'(t)| \ \leqslant \ \frac{\text{const.}}{t^2 + 1}, \qquad t \geqslant 0.$$

The corresponding potentials $U_\rho(x)$ are also $\mathscr{C}_\infty$, and especially,

$$|U_\rho(x)| \ \leqslant \ \frac{\text{const.}}{x^2 + 1}, \qquad x \in \mathbb{R}.$$

This property will be used to advantage in the next article.

Scholium. Does the space ⑤ actually *consist entirely* of Green potentials

$$U_\mu(x) \ = \ \int_0^\infty \log\left|\frac{x + t}{x - t}\right| d\mu(t)$$

formed from certain signed Borel measures μ on $(0, \infty)$, perhaps more general than the measures ρ considered in this article?

One easily proves that if the *positive* measures $d\rho_n$ (without point mass at 0) form a Cauchy sequence in the norm $\sqrt{(E(\ ,\ ))}$, *then at least a subsequence of them* (and in fact the original sequence) *does converge w** *on every compact subinterval* $[a, 1/a]$ *of* $(0, \infty)$, thus yielding a *positive Borel measure* μ on $(0, \infty)$ (perhaps with $\mu((0, 1)) = \infty$ as well as $\mu([1, \infty)) = \infty$). It is thence not hard to show that

$$U_\mu(x) \ = \ \lim_{n \to \infty} U_{\rho_n}(x) \qquad \text{a.e., } x \in \mathbb{R},$$

and U_μ may hence be identified with the limit of the U_{ρ_n} in the space ⑤.

Verification of the w^* convergence statement goes as follows: by the first lemma of this article, the integrals

$$\int_0^1 U_{\rho_n}(x)\,dx$$

are surely bounded. However,

$$\int_0^1 U_{\rho_n}(x)\,dx \;=\; \int_0^\infty \int_0^1 \log\left|\frac{x+t}{x-t}\right|\,dx\,d\rho_n(t)$$

with

$$\int_0^1 \log\left|\frac{x+t}{x-t}\right|\,dx \;\geqslant\; 0 \qquad \text{for } t \geqslant 0$$

and clearly *bounded below by a number* > 0 *on any segment of the form* $\{a \leqslant t \leqslant 1/a\}$ with $a > 0$. Therefore, since $d\rho_n(t) \geqslant 0$ for each n, the quantities

$$\rho_n([a,\ 1/a])$$

must stay bounded as $n \longrightarrow \infty$ for each $a > 0$. The existence of a subsequence of the ρ_n having the stipulated property now follows by the usual application of Helly's selection principle and the Cantor diagonal process.

As soon, however, as the measures $d\rho_n$ *are allowed to be of variable sign*, the argument just made, and its conclusion as well, *cease to be valid.* There are thus plently of functions U in $\mathfrak{H}$ which are *not* of the form U_μ (unless one accepts to bring in certain *Schwartz distributions* μ). This fact is even familiar from physics: *lots* of functions $U(z)$, harmonic in the upper half plane and continuous up to the real axis, with $U(x)$ odd thereon, *cannot be obtained as logarithmic potentials of charge distributions on* $\mathbb{R}$, *even though they have finite Dirichlet integrals*

$$\iint_{\Im z > 0} \{(U_x(z))^2 + (U_y(z))^2\}\,dx\,dy.$$

Instead, physicists are obliged to resort to what they call a *double-layer distribution* on $\mathbb{R}$ (formed from 'dipoles'); mathematically, this simply amounts to using the Poisson representation

$$U(z) \;=\; \frac{1}{\pi}\int_{-\infty}^\infty \frac{\Im z}{|z - t|^2}\,U(t)\,dt$$

in place of the formula

$$U(z) = -\frac{1}{\pi}\int_0^\infty \log\left|\frac{z+t}{z-t}\right| U_y(t+i0)\,dt,$$

which is not available unless $\partial U(z)/\partial y$ is sufficiently well behaved for $\Im z \longrightarrow 0$. ($U(x)$ may be *continuous* and the above Dirichlet integral *finite*, and yet the boundary value $U_y(x+i0)$ *exist almost nowhere* on $\mathbb{R}$. This is most easily seen by first mapping the upper half plane conformally onto the unit disk and then working with lacunary Fourier series.)

Problem 61

Let $V(x)$ be *even and* > 0, with $\|V\|_E < \infty$. Given $U \in \mathfrak{H}$, define a function $U_V(x)$ by putting

$$U_V(x) = \begin{cases} U(x) & \text{if } |U(x)| < V(x), \\ V(x)\,\mathrm{sgn}\,U(x) & \text{if } |U(x)| \geqslant V(x); \end{cases}$$

the formation of U_V is illustrated in the following figure:

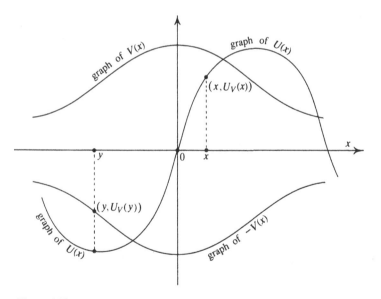

Figure 247

Show that $U_V \in \mathfrak{H}$.
(Hint: Show that

$$|U_V(x) - U_V(y)| \leqslant \max(|U(x)-U(y)|,\ |V(x)-V(y)|).$$

To check this, it is enough to look at six cases:

(i) $|U(x)| < V(x)$ and $|U(y)| < V(y)$

(ii) $U(x) \geqslant V(x)$ and $U(y) \geqslant V(y)$

(iii) $U(x) \geqslant V(x)$ and $U(y) \leqslant -V(y)$

(iv) $0 \leqslant U(x) < V(x)$ and $U(y) \leqslant -V(y)$

(v) $0 \leqslant U(x) < V(x)$, $V(y) \leqslant U(x)$ and $U(y) \geqslant V(y)$

(vi) $0 \leqslant U(x) < V(x)$, $V(y) > U(x)$ and $U(y) \geqslant V(y)$.   )

Problem 62

(Beurling and Malliavin) Let $\omega(x)$ be *even*, $\geqslant 0$, and *uniformly* Lip 1, with .

$$\int_0^\infty \frac{\omega(x)}{x^2}\,dx \quad < \quad \infty.$$

Show that then $\omega(x)/x$ belongs to $\mathfrak{H}$.

(Hint: The function $\omega(x)$ is certainly *continuous*, so, if the integral condition on it is to hold, we must have $\omega(0) = 0$. Thence, by the Lip 1 property, $\omega(x) \leqslant C|x|$, i.e.,

$$\left| \frac{\omega(x)}{x} \right| \leqslant C, \qquad x \in \mathbb{R}.$$

In the circumstances of this problem,

$$2\pi^2 \left\| \frac{\omega(x)}{x} \right\|_E^2 = \int_0^\infty \int_0^\infty \left(\frac{\dfrac{\omega(x)}{x} - \dfrac{\omega(y)}{y}}{x-y} \right)^2 dx\,dy$$

$$+ \int_0^\infty \int_0^\infty \left(\frac{\dfrac{\omega(x)}{x} + \dfrac{\omega(y)}{y}}{x+y} \right)^2 dx\,dy.$$

Using the inequality $(A+B)^2 \leqslant 2(A^2+B^2)$, the second double integral is immediately seen by symmetry to be

$$\leqslant 4\int_0^\infty \int_0^\infty \frac{1}{(x+y)^2}\left(\frac{\omega(x)}{x} \right)^2 dy\,dx \leqslant 4C\int_0^\infty \frac{\omega(x)}{x^2}\,dx.$$

The first double integral, by symmetry, is

$$2 \int_0^\infty \int_y^\infty \left(\frac{\dfrac{\omega(x)}{x} - \dfrac{\omega(y)}{y}}{x - y} \right)^2 dx\,dy \;\leqslant\; 4 \int_0^\infty \int_y^\infty \frac{1}{x^2} \left(\frac{\omega(x) - \omega(y)}{x - y} \right)^2 dx\,dy$$

$$+ \; 4 \int_0^\infty \int_y^\infty \left(\frac{\dfrac{1}{x} - \dfrac{1}{y}}{x - y} \right)^2 (\omega(y))^2 \, dx\,dy.$$

The *second* of the expressions on the right boils down to

$$4 \int_0^\infty \int_y^\infty \left(\frac{\omega(y)}{y} \right)^2 \frac{dx}{x^2}\,dy$$

which is handled by reasoning already used; we are thus left with the *first* expression on the right.

That one we break up further as

$$4 \int_0^\infty \int_y^{y + \omega(y)} \;+\; 4 \int_0^\infty \int_{y + \omega(y)}^\infty \;;$$

again, the first of these terms is readily estimated, and the second is

$$\leqslant\; 8 \int_0^\infty \int_{y + \omega(y)}^\infty \left\{ \frac{(\omega(y))^2}{y^2} \frac{1}{(x - y)^2} + \frac{(\omega(x))^2}{x^2} \frac{1}{(x - y)^2} \right\} dx\,dy.$$

Integration of the first term in { } still does not give any problem, and we only need to deal with

$$8 \int_0^\infty \int_{y + \omega(y)}^\infty \frac{(\omega(x))^2}{x^2} \frac{1}{(x - y)^2}\,dx\,dy.$$

By reversing the order of integration, show that this is

$$\leqslant\; 8 \int_0^\infty \frac{(\omega(x))^2}{x^2} \cdot \frac{1}{\omega(Y(x))}\,dx,$$

where $Y(x)$ denotes the *largest value* of y for which $y + \omega(y) \leqslant x$. Then use the Lip 1 property of ω to get

$$|\omega(x) - \omega(Y(x))| \;\leqslant\; C\omega(Y(x)),$$

whence $1/\omega(Y(x)) \leqslant (C + 1)/\omega(x)$. This relation is then substituted into the last integral.)

5.   **Even weights W with $\| \log W(x)/x \|_E < \infty$**

Theorem (Beurling and Malliavin). *Let $W(x) \geqslant 1$ be continuous and even, with*

$$\int_0^\infty \frac{\log W(x)}{x^2}\, dx < \infty,$$

and suppose that the odd function $\log W(x)/x$ belongs to the Hilbert space $\mathfrak{H}$ discussed in the preceding article. Then, given any $A > 0$, there is an increasing function $v(t)$, zero on a neighborhood of the origin, such that

$$\frac{v(t)}{t} \longrightarrow \frac{A}{\pi} \quad \text{as } t \longrightarrow \infty$$

and

$$\log W(x) + \int_0^\infty \log\left|1 - \frac{x^2}{t^2}\right| dv(t) \leqslant \text{const.} \quad \text{on } \mathbb{R}.$$

Proof. The argument, again based on the procedure explained in article 1, is much like those made in articles 2 and 3. For that reason, certain of its details may be omitted.

In order to have a weight going to ∞ for $x \longrightarrow \pm\infty$, we first take

$$W_\eta(x) = (1 + x^2)^\eta\, W(x)$$

with a small number $\eta > 0$. Then, given a value of $A > 0$, we form the function*

$$F(z) = \frac{1}{\pi} \int_{-\infty}^\infty \frac{|\Im z|}{|z - t|^2} \log W_\eta(t)\, dt - A|\Im z|.$$

As before, the proof of our theorem boils down to showing that

$$(\mathfrak{M}F)(0) < \infty.$$

This, in turn, reduces to the *determination of an upper bound, independent of $\mathscr{D}$*, on

$$\int_{\partial\mathscr{D}} \log W_\eta(t)\, d\omega_\mathscr{D}(t,\, 0) - A Y_\mathscr{D}(0),$$

where $\mathscr{D}$ is any domain of the kind studied in §C of Chapter VIII. We

---

* as usual, we put $F(x)$ equal to $\log W_\eta(x)$ on the real axis

see that what is needed is a comparison of

$$\int_{\partial \mathscr{D}} \log W_\eta(t) \, d\omega_{\mathscr{D}}(t, \, 0)$$

with the quantity $Y_{\mathscr{D}}(0)$.

We have, in the first place, to take care of the factor $(1 + x^2)^\eta$ used in forming $W_\eta(x)$. That is easy. Writing, for $t \geqslant 0$,

$$\Omega_{\mathscr{D}}(t) \;=\; \omega_{\mathscr{D}}(\partial \mathscr{D} \cap ((-\infty, \, -t] \cup [t, \infty)), \, 0)$$

as in §C of Chapter VIII, we have

$$\int_{\partial \mathscr{D}} \eta \log(1 + t^2) \, d\omega_{\mathscr{D}}(t, \, 0) \;=\; -\eta \int_0^\infty \log(1 + t^2) \, d\Omega_{\mathscr{D}}(t)$$

$$=\; \eta \int_0^\infty \frac{2t}{1 + t^2} \Omega_{\mathscr{D}}(t) \, dt.$$

By the fundamental result of §C.2, Chapter VIII,

$$\Omega_{\mathscr{D}}(t) \;\leqslant\; \frac{Y_{\mathscr{D}}(0)}{t},$$

so the last expression is

$$\leqslant\; 2\eta \, Y_{\mathscr{D}}(0) \int_0^\infty \frac{dt}{1 + t^2} \;=\; \pi \eta \, Y_{\mathscr{D}}(0).$$

Thence,

$$\int_{\partial \mathscr{D}} \log W_\eta(t) \, d\omega_{\mathscr{D}}(t, \, 0)$$

$$=\; \int_{\partial \mathscr{D}} \eta \log(1 + t^2) \, d\omega_{\mathscr{D}}(t, \, 0) \;+\; \int_{\partial \mathscr{D}} \log W(t) \, d\omega_{\mathscr{D}}(t, \, 0)$$

$$\leqslant\; \pi \eta \, Y_{\mathscr{D}}(0) \;+\; \int_{\partial \mathscr{D}} \log W(t) \, d\omega_{\mathscr{D}}(t, \, 0),$$

and our main work is the estimation of the integral in the last member.

For that purpose, we may as well make full use of the third theorem in the preceding article, having done the work to get it. The reader wishing to avoid use of that theorem will find a similar alternative procedure sketched in problems 63 and 64 below. According to our hypothesis, $\log W(x)/x \in \mathfrak{H}$ so, by the theorem referred to, there is, for any $\eta > 0$, a potential $U_\rho(x)$

of the sort considered in the last article, with

$$\left\| \frac{\log W(x)}{x} - U_\rho(x) \right\|_E < \eta$$

and also (by the remark to that theorem)

$$|U_\rho(x)| \leqslant \frac{K_\eta}{1 + x^2} \quad \text{for } x \in \mathbb{R}.$$

Let us now proceed as in proving the theorem of §C.4, Chapter VIII, trying, however, to make use of the difference $(\log W(x)/x) - U_\rho(x)$.

We have

$$\int_{\partial \mathcal{D}} \log W(t) \, d\omega_{\mathcal{D}}(t, \, 0) = \int_{\partial \mathcal{D}} t U_\rho(t) \, d\omega_{\mathcal{D}}(t, \, 0)$$

$$+ \int_{\partial \mathcal{D}} (\log W(t) - t U_\rho(t)) \, d\omega_{\mathcal{D}}(t, \, 0).$$

Because $|t U_\rho(t)| \leqslant K_\eta/2$ by the last inequality, and $\omega_{\mathcal{D}}(\;\; , \; 0)$ is a positive measure of total mass 1, the *first* integral on the right is

$$\leqslant \;\; K_\eta/2.$$

In terms of $\Omega_{\mathcal{D}}(t)$, the *second* right-hand integral is

$$-\int_0^\infty (\log W(t) - t U_\rho(t)) \, d\Omega_{\mathcal{D}}(t);$$

to this we now apply the trick used in the proof just mentioned, rewriting the last expression as

$$\int_0^\infty \left(\frac{\log W(t)}{t} - U_\rho(t) \right) \Omega_{\mathcal{D}}(t) \, dt - \int_0^\infty \left(\frac{\log W(t)}{t} - U_\rho(t) \right) d(t \Omega_{\mathcal{D}}(t)).$$

The *first* of these terms can be disposed of immediately. Taking a large number L, we break it up as

$$\int_0^L \frac{\log W(t)}{t} \Omega_{\mathcal{D}}(t) \, dt - \int_0^\infty U_\rho(t) \, \Omega_{\mathcal{D}}(t) \, dt + \int_L^\infty \frac{\log W(t)}{t} \Omega_{\mathcal{D}}(t) \, dt.$$

We use the inequality $\Omega_{\mathcal{D}}(t) \leqslant 1$ in the first two integrals, plugging the above estimate on $U_\rho(t)$ into the second one. In the third integral, the relation $\Omega_{\mathcal{D}}(t) \leqslant Y_{\mathcal{D}}(0)/t$ is once again employed. In that way, the sum of these integrals is seen to be

$$\leqslant \;\; L \int_0^\infty \frac{\log W(t)}{t^2} \, dt + \frac{\pi}{2} K_\eta + Y_{\mathcal{D}}(0) \int_L^\infty \frac{\log W(t)}{t^2} \, dt.$$

We come to

$$\int_0^\infty \left(\frac{\log W(t)}{t} - U_\rho(t) \right) \mathrm{d}(t\Omega_{\mathscr{D}}(t)),$$

the *second* of the above terms. According to §C.3 of Chapter VIII, the double integral

$$\int_0^\infty \int_0^\infty \log \left| \frac{x+t}{x-t} \right| \, \mathrm{d}(t\Omega_{\mathscr{D}}(t)) \, \mathrm{d}(x\Omega_{\mathscr{D}}(x))$$

is absolutely convergent, and its value,

$$E(\, \mathrm{d}(t\Omega_{\mathscr{D}}(t)), \ \mathrm{d}(t\Omega_{\mathscr{D}}(t))),$$

is

$$\leqslant \quad \pi (Y_{\mathscr{D}}(0))^2.$$

The measure $\mathrm{d}(t\Omega_{\mathscr{D}}(t))$ is *absolutely continuous* (albeit with *unbounded* density!), and acts like $\mathrm{const.}(\mathrm{d}t/t^3)$ for large t. Near the origin, $\mathrm{d}(t\Omega_{\mathscr{D}}(t)) = \mathrm{d}t$, for $\Omega_{\mathscr{D}}(t) \equiv 1$ in a neighborhood of that point. These properties together with our given conditions on $W(t)$ and the above estimate for $U_\rho(t)$ ensure absolute convergence of

$$\int_0^\infty ((\log W(t)/t) - U_\rho(t)) \, \mathrm{d}(t\Omega_{\mathscr{D}}(t)),$$

which may hence be estimated by the *fifth* lemma of the last article. In that way this integral is found to be in absolute value

$$\leqslant \quad \left\| \frac{\log W(t)}{t} - U_\rho(t) \right\|_E \sqrt{(E(\mathrm{d}(t\Omega_{\mathscr{D}}(t)), \ \mathrm{d}(t\Omega_{\mathscr{D}}(t))))}.$$

Referring to the previous relation, we see that for our choice of $U_\rho(t)$, the quantity just found is

$$\leqslant \quad \sqrt{\pi} \ \eta \, Y_{\mathscr{D}}(0),$$

and we have our upper bound for the second term in question.

Combining this with the estimate already obtained for the *first* term, we get

$$-\int_0^\infty (\log W(t) - t U_\rho(t)) \, \mathrm{d}\Omega_{\mathscr{D}}(t)$$

$$\leqslant \quad L\int_0^\infty \frac{\log W(t)}{t^2} \, \mathrm{d}t \ + \ \frac{\pi}{2} K_\eta \ + \ Y_{\mathscr{D}}(0) \int_L^\infty \frac{\log W(t)}{t^2} \, \mathrm{d}t \ + \ \sqrt{\pi} \ \eta \, Y_{\mathscr{D}}(0),$$

whence, by an earlier computation,

$$\int_{\partial \mathscr{D}} \log W(t)\, d\omega_{\mathscr{D}}(t, 0)$$

$$\leqslant \quad L\int_0^\infty \frac{\log W(t)}{t^2}\, dt \ + \ \frac{\pi + 1}{2} K_\eta \ + \ \left(\sqrt{\pi}\, \eta \ + \ \int_L^\infty \frac{\log W(t)}{t^2}\, dt \right) Y_{\mathscr{D}}(0),$$

and thus finally

$$\int_{\partial \mathscr{D}} \log W_\eta(t)\, d\omega_{\mathscr{D}}(t, 0)$$

$$\leqslant \quad \left((\pi + \sqrt{\pi})\eta \ + \ \int_L^\infty \frac{\log W(t)}{t^2}\, dt \right) Y_{\mathscr{D}}(0)$$

$$+ \ L\int_0^\infty \frac{\log W(t)}{t^2}\, dt \ + \ \frac{\pi + 1}{2} K_\eta.$$

Wishing now to have the initial term on the right outweighed by $-AY_{\mathscr{D}}(0)$ we first, for our given value of $A > 0$, pick

$$\eta \quad \leqslant \quad \frac{A}{2(\pi + \sqrt{\pi})} \quad \text{(say)},$$

and then choose (and fix!) L large enough so as to have

$$\int_L^\infty \frac{\log W(t)}{t^2}\, dt \quad \leqslant \quad \frac{A}{2}.$$

For these particular values of η and L, it will follow that

$$\int_{\partial \mathscr{D}} \log W_\eta(t)\, d\omega_{\mathscr{D}}(t, 0) \ - \ AY_{\mathscr{D}}(0) \quad \leqslant \quad L\int_0^\infty \frac{\log W(t)}{t^2}\, dt \ + \ \frac{\pi + 1}{2} K_\eta$$

for any of the domains $\mathscr{D}$, and hence that

$$(\mathfrak{M}F)(0) \quad \leqslant \quad L\int_0^\infty \frac{\log W(t)}{t^2}\, dt \quad + \quad \frac{\pi + 1}{2} K_\eta,$$

by the method used in articles 2 and 3. This, however, proves the theorem. We are done.

Referring now to the corollary of the next-to-the last theorem in §B.1, we immediately obtain the

Corollary. *Let the continuous weight* $W(x) \geqslant 1$ *satisfy the hypothesis of the theorem, and also fulfill the regularity requirement formulated in* §B.1. *Then* W *admits multipliers.*

This result and the one obtained in problem 62 (last article) give us once again a proposition due to Beurling and Malliavin, already deduced from their Theorem on the Multiplier (of article 2) in §C.1, Chapter X. That proposition may be stated in the following form:

Theorem. *Let* $W(x) \geqslant 1$ *be even, with* $\log W(x)$ *uniformly* Lip 1 *on* $\mathbb{R}$, *and*

$$\int_0^\infty \frac{\log W(x)}{x^2} dx < \infty.$$

Then W *admits multipliers.*

It suffices to observe that the regularity requirement of §B.1 is certainly met by weights W with $\log W(x)$ uniformly Lip 1.

Originally, this theorem was essentially derived in such fashion from the preceding one by Beurling and Malliavin.

Problem 63

(a) Let ρ be a positive measure on $[0, \infty)$ without point mass at the origin, such that $E(d\rho(t), d\rho(t)) < \infty$. Show that there is a sequence of positive measures σ_n of *compact support* in $(0, \infty)$ with $d\rho(t) - d\sigma_n(t) \geqslant 0$, $U_{\sigma_n}(x)$ *bounded* on $\mathbb{R}$ for each n, and $\| U_\rho - U_{\sigma_n} \|_E \xrightarrow[n]{} 0$. (Hint: First argue as in the proof of the *fifth* lemma of the last article to verify that if ρ_n denotes the restriction of ρ to $[1/n, n]$, then $\| U_\rho - U_{\rho_n} \|_E \xrightarrow[n]{} 0$. Then, for each n, take σ_n as the restriction of ρ_n to the closed subset of $[1/n, n]$ on which $U_{\rho_n}(x) \leqslant$ some sufficiently large number M_n.)

(b) Let σ be a positive measure of *compact support* $\subseteq (0, \infty)$ with $\| U_\sigma \|_E < \infty$ and $U_\sigma(x)$ *bounded on* $\mathbb{R}$. Show that, corresponding to any $\varepsilon > 0$, there is a *signed* measure τ on $[0, \infty)$, without point mass at the origin, such that $U_\tau(x)$ is also bounded on $\mathbb{R}$, that $\| U_\sigma - U_\tau \|_E < \varepsilon$, and that $U_\tau(x) = 0$ for all sufficiently large x. (Hint: We have $U_\sigma(x) \longrightarrow 0$ for $x \longrightarrow \infty$. Taking a very large $R > 0$, far beyond the support K of σ, consider the domain $\mathscr{D}_R = \{\Re z > 0\} \sim [R, \infty)$, and the harmonic measure $\omega_R(\ , z)$ for $\mathscr{D}_R$. Define an absolutely continuous measure σ_R on $[R, \infty)$ by putting, for $t > R$,

$$\frac{d\sigma_R(t)}{dt} = \int_K \frac{d\omega_R(t, \xi)}{dt} d\sigma(\xi).$$

Show that $U_{\sigma_R}(x) = U_{\sigma}(x)$ for $x > R$, that U_{σ_R} is bounded on $\mathbb{R}$, and that $\|U_{\sigma_R}\|_E < \varepsilon$ if R is taken large enough. Then put $\tau = \sigma - \sigma_R$. *Note*: Potential theorists say that σ_R has been obtained from σ by *balayage* (sweeping) onto the set $[R, \infty)$.)

(c) Hence show that if ρ is any *signed* measure on $[0, \infty)$ without point mass at 0 making $E(|d\rho(t)|, |d\rho(t)|) < \infty$, there is another such signed measure μ on $[0, \infty)$ with $\|U_\rho - U_\mu\|_E < \varepsilon$, $U_\mu(x)$ *bounded* on $\mathbb{R}$, and $U_\mu(x) = 0$ for all $x > R$, a number depending on ε. (Here, parts (a) and (b) are applied in turn to the *positive* part of ρ and to its *negative* part.)

Problem 64

Prove the first theorem of this article using the result of problem 63. (Hint: Given that $\log W(x)/x \in \mathfrak{H}$, take first a signed measure ρ on $[0, \infty)$ like the one in problem 63(c) such that

$$\|(\log W(x)/x) - U_\rho(x)\|_E < \eta/2,$$

and then a μ, furnished by that problem, with $\|U_\rho - U_\mu\|_E < \eta/2$. Argue as in the proof given above, working with the difference

$$\frac{\log W(x)}{x} - U_\mu(x),$$

and taking the number L figuring there to be *larger* than the R obtained in problem 63 (c).)

D.   Search for the presumed essential condition

At the beginning of §B.1, it was proposed to limit a good part of the considerations of this chapter to weights $W(x) \geq 1$ satisfying a mild local regularity requirement:

There are three constants C, α and $L > 0$ (depending on W) such that, for each real x, one has an interval J_x of length L containing x with

$$W(t) \geq C(W(x))^\alpha \quad \text{for } t \in J_x.$$

That restriction was accepted because, while leaving us with room enough to accommodate many of the weights arising in different circumstances, it serves, we believe, to rule out accidental and, so to say, *trivial* irregularities in a weight's behaviour that could spoil the existence of multipliers which might otherwise be forthcoming. Admittance of multipliers by a weight W was thought to be *really* governed *by some other* condition on its behaviour – an 'essential' one, probably not of strictly local character –

acting in conjunction with the growth requirement

$$\int_{-\infty}^{\infty} \frac{\log W(x)}{1+x^2}\,dx \quad < \quad \infty.$$

In adopting this belief, we of course made a tacit asumption that another condition regarding the weight (besides convergence of its logarithmic integral) *is in fact involved*. Up to now, however, we have not seen any reasons why that should be the case. *It is still quite conceivable that the integral condition and the local regularity requirement are, by themselves, sufficient to guarantee admittance of multipliers.*

Such a conclusion would be most satisfying, and indeed make a fitting end to this book. If its truth seemed likely, we would have to abandon our present viewpoint and think instead of looking for a proof. We have arrived at the place where one must decide which path to take.

It is for that purpose that the example given in the first article was constructed. This shows that an additional condition on our weights – what we are thinking of as the 'essential' one – *is really needed*. Our aim during the succeeding articles of this § will then be to find out *what that condition is* or at least arrive at some partial knowledge of it.

In working towards that goal, we will be led to the construction of a *second* example, actually quite similar to the one of the first article, but yielding a weight that *admits* multipliers although the weight furnished by the latter *does not*. Comparison of the two examples will enable us to form an idea of what the 'essential' condition on weights must look like, and, eventually, lead us to the *necessary and sufficient conditions for admittance of multipliers* (on weights meeting the local regularity requirement) formulated in the theorem of §E.

Before proceeding to the first example, it is worthwhile to see what the *absence* of an additional condition on our weights *would have* entailed. The local regularity requirement quoted at the beginning of this discussion is certainly satisfied by weights $W(x) \geqslant 1$ with

$$|\log^+ \log W(x) \; - \; \log^+ \log W(x')| \quad \leqslant \quad \text{const.}\,|x - x'|$$

on $\mathbb{R}$. Absence of an additional condition would therefore make

$$\int_{-\infty}^{\infty} \frac{\log W(x)}{1+x^2}\,dx \quad < \quad \infty$$

necessary and sufficient for the admittance of multipliers by such W. This would in turn have an obvious but quite interesting corollary: if, for a weight $W(x)$ with uniformly Lip 1 iterated logarithm, there is *even one* entire function $\Phi(z) \not\equiv 0$ of *some* (finite) *exponential type* making $W(x)\Phi(x)$ *bounded* on $\mathbb{R}$, there must be such functions $\varphi(z) \not\equiv 0$ of *arbitrarily small exponential type* having the *same property*. The example given in the first article will show that *even this corollary is false*.

The absence of an additional condition on *just* the weights with uniform Lip 1 iterated logarithms *would*, by the way, *imply* that absence for *all* weights meeting our (less stringent) local regularity requirement. Indeed, if $W(x) \geqslant 1$ fulfills the latter (with constants C, α and L), and

$$\int_{-\infty}^{\infty} \frac{\log W(x)}{1 + x^2}\, dx \;\; < \;\; \infty,$$

we *know* from the proof of the first theorem in §B.1 that W *admits multipliers* (for which the last relation is at least *necessary*), *if and only if* the weight

$$W_1(x) \;\; = \;\; \exp\left\{ \frac{4}{\pi\alpha} \int_{-\infty}^{\infty} \frac{L \log W(t)}{(x - t)^2 + L^2}\, dt \right\}$$

also does. We see, however, that $|\,\mathrm{d} \log W_1(x)/dx\,| \;\leqslant\; (1/L) \log W_1(x)$, i.e.,

$$\left| \frac{\mathrm{d} \log \log W_1(x)}{dx} \right| \;\leqslant\; \frac{1}{L},$$

so W_1 *does* have a uniformly Lip 1 iterated logarithm.

Let us go on to the first example.

1.   **Example. Uniform Lip 1 condition on $\log\log W(x)$ not sufficient**

Take the points

$$x_p \;\; = \;\; e^{p^{1/3}}, \quad p \;\; = \;\; 8, 9, 10, \ldots,$$

and put

$$\Delta_p \;\; = \;\; \begin{cases} x_8, & p = 8, \\ x_p - x_{p-1}, & p > 8. \end{cases}$$

Let then

$$F(z) \;\; = \;\; \prod_{p=8}^{\infty} \left(1 - \frac{z^2}{x_p^2} \right)^{[\Delta_p]},$$

where $[\Delta_p]$ denotes the largest integer $\leqslant \Delta_p$; it is not hard to see – is, indeed, a particular consequence of the following work – that the product is convergent, making $F(z)$ an entire function with a zero of order $[\Delta_p]$ at each of the points $\pm x_p$, $p \geqslant 8$, and no other zeros.

According to custom, we write $n(t)$ for the *number of zeros* of $F(z)$ in $[0, t]$ (counting multiplicities) when $t \geqslant 0$. Thus,

$$n(t) = 0 \quad \text{for } 0 \leqslant t < x_8$$

and

$$n(t) = [\Delta_8] + [\Delta_9] + \cdots + [\Delta_p] \quad \text{for } x_p \leqslant t < x_{p+1}.$$

The *right side* of the last relation lies between

$$\Delta_8 + \Delta_9 + \cdots + \Delta_p - p = x_p - p$$

and

$$\Delta_8 + \Delta_9 + \cdots + \Delta_p = x_p,$$

so, since

$$p = (\log x_p)^3,$$

we have

$$t - \Delta_{p+1} - (\log x_p)^3 \leqslant n(t) \leqslant t \quad \text{for } x_p \leqslant t < x_{p+1},$$

with the *second* inequality actually valid for *all* $t \geqslant 0$. Here,

$$\Delta_{p+1} = e^{(p+1)^{1/3}} - e^{p^{1/3}} = (\tfrac{1}{3}p^{-2/3} + O(p^{-5/3}))x_p$$

is

$$\sim \frac{x_p}{3(\log x_p)^2},$$

making

$$t\left(1 - \frac{1}{3(\log t)^2} - O\left(\frac{1}{\log t}\right)^5\right) \leqslant n(t) \leqslant t$$

for $x_p \leqslant t < x_{p+1}$. We thus certainly have

$$t - \frac{t}{(\log t)^2} \leqslant n(t) \leqslant t$$

for large values of t, with the *upper bound* in fact holding for all $t \geqslant 0$, as just noted.

We can write

$$\log|F(z)| = \int_0^\infty \log\left|1 - \frac{z^2}{t^2}\right| dn(t),$$

and the reader should now refer to problem 29 (§B.1, Chapter IX). Reasoning as in part (a) of that problem, one readily concludes that

$$\frac{\log|F(iy)|}{y} \longrightarrow \pi \quad \text{for } y \longrightarrow \infty,$$

since

$$\frac{n(t)}{t} \longrightarrow 1 \quad \text{as } t \longrightarrow \infty$$

by the previous relation. Clearly,

$$|F(z)| \leqslant F(i|z|),$$

so our function $F(z)$ *is of exponential type* π.

To estimate $|F(x)|$ for real x, we refer to part (c) of the same problem, according to which

$$\log|F(x)| \leqslant 2n(x)\log\frac{1}{\lambda} + 2\int_0^\lambda \frac{\dfrac{n(xt)}{t} - tn\left(\dfrac{x}{t}\right)}{1 - t^2} dt$$

for $x > 0$, where for λ we may take any number between 0 and 1. Assuming x *large*, we put

$$\lambda = 1 - \frac{1}{(\log x)^2}$$

and plug the above relation for $n(t)$ into the integral (using, of course, the upper bound with $n(xt)/t$ and the lower one with $tn(x/t)$). We thus find that

$$\log|F(x)| \leqslant 2n(x)\log\frac{1}{\lambda} + 2x\int_0^\lambda \frac{dt}{(\log(x/t))^2(1 - t^2)}$$

$$\leqslant \text{const.}\frac{x}{(\log x)^2} + x\frac{\log 2 + 2\log\log x}{(\log x)^2} \leqslant C\frac{x\log\log x}{(\log x)^2}$$

for large values of x.

The quantity on the right is *increasing* when $x > 0$ is large enough, and satisfies

$$\int_e^\infty \frac{1}{x^2} C \, \frac{x \log \log x}{(\log x)^2} \, dx \;\; = \;\; C \int_1^\infty \frac{\log u}{u^2} \, du \;\; = \;\; C \;\; < \;\; \infty.$$

Therefore, since $\log |F(x)|$ is *even* and bounded above by that quantity when x is large, we can conclude by the elementary Paley–Wiener multiplier theorem of Chapter X, §A.1 (obtained by a different method far back in §D of Chapter IV!) that there is, corresponding to any $\eta > 0$, a non-zero entire function $\psi(z)$ of exponential type $\leqslant \eta$ with $F(x)\psi(x)$ *bounded* on the real axis. The function $\psi(z)$ obtained in Chapter X is in fact of the form $\varphi(z + \mathrm{i})$, where

$$\varphi(z) \;\; = \;\; \prod_k \left(1 \; - \; \frac{z^2}{\lambda_k^2} \right)$$

is *even* and has only the *real* zeros $\pm \lambda_k$; it is thus clear that for $x \in \mathbb{R}$,

$$|\psi(x)| \;\; = \;\; |\varphi(x + \mathrm{i})| \;\; \geqslant \;\; |\varphi(x)|.$$

(Observe that for $\zeta \; = \; \xi + \eta \mathrm{i}$,

$$|1 - \zeta^2| \;\; = \;\; |1 + \zeta||1 - \zeta| \;\; \geqslant \;\; |1 + \xi||1 - \xi| \; !)$$

Hence,

$$|F(x)\varphi(x)| \;\; \leqslant \;\; \text{const.} \quad \text{on } \mathbb{R}$$

with an *even* function $\varphi(z)$ of exponential type $\leqslant \eta$ *having only real zeros*. Fixing a constant $c > 0$ for which

$$c|F(x)\varphi(x)| \;\; \leqslant \;\; 1, \quad x \in \mathbb{R},$$

we put

$$\Psi(z) \;\; = \;\; cF(z)\varphi(z),$$

getting a certain *even* entire function Ψ, with *only real zeros*, having exponential type equal to a number B lying between π (the type of F) and $\pi + \eta$. For this function the Poisson representation

$$\log|\Psi(z)| \;\; = \;\; B\Im z \; + \; \frac{1}{\pi} \int_{-\infty}^\infty \frac{\Im z \log|\Psi(t)|}{|z - t|^2} \, dt$$

from §G.1 of Chapter III is valid for $\Im z > 0$, the integral on the right being absolutely convergent. In particular,

$$\log\left| \frac{e^B}{\Psi(x + \mathrm{i})} \right| \;\; = \;\; \frac{1}{\pi} \int_{-\infty}^\infty \frac{1}{(x - t)^2 + 1} \log \frac{1}{|\Psi(t)|} \, dt \qquad \text{for } x \in \mathbb{R},$$

where the integral is obviously $\geqslant 0$.

We now take

$$W(x) \;=\; \frac{e^B}{|\Psi(x+i)|}, \qquad x \in \mathbb{R}.$$

Then $|W(x)| \geqslant 1$ and differentiation of the preceding formula immediately yields

$$\left| \frac{\mathrm{d} \log W(x)}{\mathrm{d}x} \right| \;\leqslant\; \log W(x),$$

making

$$|\log\log W(x) \;-\; \log\log W(x')| \;\leqslant\; |x-x'| \qquad \text{for } x,\, x' \in \mathbb{R}.$$

From the same formula we also see that

$$\int_{-\infty}^{\infty} \frac{\log W(x)}{1+x^2}\,\mathrm{d}x \;<\; \infty.$$

It is claimed, however, that the weight W has no mutipliers of exponential type $< \pi$. Suppose, indeed, that there *is* a non-zero entire function $f(z)$ of exponential type $A' < \pi$ with $f(x)W(x)$ bounded on $\mathbb{R}$. According to the discussion following the *first* theorem of §B.1 we can, from f, obtain another non-zero entire function g of exponential type $A \leqslant A'$ (hence $A < \pi$ $-$ A is in fact *equal* to A'), *having only real zeros*, with

$$|g(x)| \;\leqslant\; |f(x)| \quad \text{on } \mathbb{R},$$

so that $g(x)W(x)$ is also bounded for $x \in \mathbb{R}$. This function $g(z)$ (denoted by $C\psi(z)$ in the passage referred to) is a constant multiple of a product like

$$e^{bz} \prod_{\lambda'} \left(1 \;-\; \frac{z}{\lambda'} \right) e^{z/\lambda'}$$

formed from *real* numbers b and λ', and hence has the important property that

$$|g(\Re z)| \;\leqslant\; |g(z)|,$$

which we shall presently have occasion to use.

There is no loss of generality in our assuming that

$$|g(x)W(x)| \;\leqslant\; e^{-B} \qquad \text{for } x \in \mathbb{R}.$$

Referring to our definition of W, we see that this is the same as the relation

$$|g(x)| \;\leqslant\; e^{-2B}|\Psi(x+i)|, \qquad x \in \mathbb{R}.$$

To the function g, having all its zeros on the real axis (and surely *bounded* there!) we may apply the Poisson representation from §G.1 of Chapter III to get

$$\log|g(z)| \;=\; A\Im z \;+\; \frac{1}{\pi}\int_{-\infty}^{\infty}\frac{\Im z\,\log|g(t)|}{|z-t|^2}\,dt, \qquad \Im z > 0.$$

In like manner,

$$\log|\Psi(z+\mathrm{i})| \;=\; B\Im z \;+\; \frac{1}{\pi}\int_{-\infty}^{\infty}\frac{\Im z\,\log|\Psi(t+\mathrm{i})|}{|z-t|^2}\,dt, \qquad \Im z > 0,$$

so that, for $\Im z > 0$,

$$\log\left|\frac{g(z)}{\Psi(z+\mathrm{i})}\right| \;=\; (A-B)\Im z \;+\; \frac{1}{\pi}\int_{-\infty}^{\infty}\frac{\Im z}{|z-t|^2}\log\left|\frac{g(t)}{\Psi(t+\mathrm{i})}\right|\,dt.$$

Since $A < \pi \leqslant B$, the right side in the last relation is $\leqslant -2B$ by the preceding inequality, so we have in particular

$$|g(x+\mathrm{i})| \;\leqslant\; \mathrm{e}^{-2B}|\Psi(x+2\mathrm{i})|, \qquad x\in\mathbb{R}.$$

Let us note moreover that $\mathrm{e}^{-2B}|\Psi(x+2\mathrm{i})| \leqslant 1$ on the real axis by the third Phragmén–Lindelöf theorem of Chapter III, §C, $\Psi(z)$ being of exponential type B and of modulus $\leqslant 1$ for real z. Another application of the same Phragmén–Lindelöf theorem thence shows that

$$\mathrm{e}^{-2B}|\Psi(z+2\mathrm{i})| \;\leqslant\; \mathrm{e}^{B|\Im z|}$$

This estimate will also be of use to us.

Our idea now is to show that $|g(x)|$ must get so small near the zeros $\pm x_p$ of our original function $F(z)$ as to make

$$\int_{-\infty}^{\infty}\frac{\log^-|g(x)|}{1+x^2}\,dx \;=\; \infty$$

and thus imply that $g(z) \equiv 0$ (*a contradiction!*) by §G.2 of Chapter III. We start by looking at $|g(x_p+\mathrm{i})|$, which a previous relation shows to be $\leqslant \mathrm{e}^{-2B}|\Psi(x_p+2\mathrm{i})|$. The latter quantity we estimate by Jensen's formula.

For the moment, let us denote by $N(r, z_0)$ the *number of zeros of* $\Psi(z)$ inside any closed disk of the form $\{|z-z_0| \leqslant r\}$. Then we have, for any $R > 0$,

$$\log|\mathrm{e}^{-2B}\Psi(x_p+2\mathrm{i})| \;=\; \frac{1}{2\pi}\int_{-\pi}^{\pi}\log|\mathrm{e}^{-2B}\Psi(x_p+2\mathrm{i}+R\mathrm{e}^{\mathrm{i}\vartheta})|\,d\vartheta$$

$$-\int_0^R\frac{N(r,\,x_p+2\mathrm{i})}{r}\,dr.$$

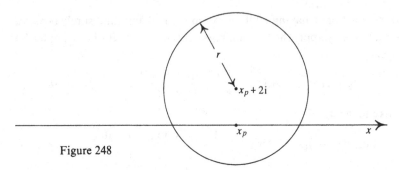

Figure 248

Substituting the *last* of the above relations involving Ψ into the first integral on the right and noting that $\Psi(z)$ has *at least a* $[\Delta_p]$ *–fold zero at* x_p, we see that for $R > 2$,

$$\log|e^{-2B}\Psi(x_p + 2i)| \;\leqslant\; \frac{2}{\pi}BR \;-\; [\Delta_p]\log\frac{R}{2}.$$

Write now $v(r, z_0)$ for the *number of zeros of* $g(z)$ in the closed disk $\{|z - z_0| \leqslant r\}$. Application of Jensen's formula to g then yields

$$\log|g(x_p + i)| \;=\; \frac{1}{2\pi}\int_{-\pi}^{\pi}\log|g(x_p + i + Re^{i\vartheta})|\,\mathrm{d}\vartheta \;-\; \int_0^R \frac{v(r,\,x_p + i)}{r}\,\mathrm{d}r.$$

Using the inequality $|g(x_p + i)| \leqslant e^{-2B}|\Psi(x_p + 2i)|$ and referring to the preceding relation we find, after transposing, that

$$\frac{1}{2\pi}\int_{-\pi}^{\pi}\log|g(x_p + i + Re^{i\vartheta})|\,\mathrm{d}\vartheta$$

$$\leqslant\; \frac{2}{\pi}BR \;+\; \int_0^R \frac{v(r,\,x_p + i)}{r}\,\mathrm{d}r \;-\; [\Delta_p]\log\frac{R}{2} \qquad \text{for } R > 2.$$

Since all the zeros of $g(z)$ are real, $v(r, x_p + i)$ is certainly zero for $r < 1$, whence

$$\int_0^R \frac{v(r,\,x_p + i)}{r}\,\mathrm{d}r \;\leqslant\; v(R, x_p + i)\log R, \qquad R > 1,$$

which, together with the last, gives

$$\frac{1}{2\pi}\int_{-\pi}^{\pi}\log|g(x_p + i + Re^{i\vartheta})|\,\mathrm{d}\vartheta$$

$$\leqslant\; \frac{2}{\pi}BR \;+\; (v(R, x_p + i) - [\Delta_p])\log R \;+\; [\Delta_p]\log 2, \qquad R > 2.$$

We want to use this to show that for a certain R_p, $\int_{-R_p}^{R_p} \log|g(x_p + t)|\,dt$ comes out *very negative*.

To do that, we simply (trick!) plug the inequality $|g(\Re z)| \leqslant |g(z)|$ noted above into the left side of the last relation. We are, in other words, *flattening* the circle involved in Jensen's formula to its horizontal diameter which is then moved down to the real axis:

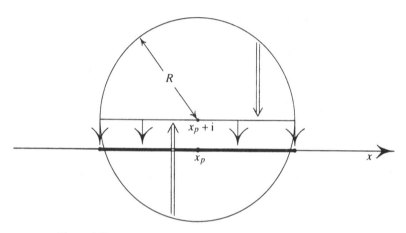

Figure 249

That causes $\log|g(x_p + \mathrm{i} + R\mathrm{e}^{\mathrm{i}\vartheta})|$ to be replaced by $\log|g(x_p + R\cos\vartheta)|$ in the integral appearing in the relation in question; the resulting integral then becomes

$$\frac{1}{\pi}\int_{-R}^{R} \frac{\log|g(x_p + s)|}{\sqrt{(R^2 - s^2)}}\,ds$$

on making the substitution $R\cos\vartheta = s$. What we have just written is hence

$$\leqslant \quad \frac{2}{\pi}BR \;+\; \big(v(R,\, x_p + \mathrm{i}) \;-\; [\Delta_p]\big)\log R \;+\; [\Delta_p]\log 2 \qquad \text{for } R > 2.$$

Our reasoning at this point is much like that in §D.1 of Chapter IX. Taking

$$R_p \;=\; \frac{\Delta_p}{2},$$

we multiply the preceding integral and the expression immediately following it by $R\,dR$ and integrate from $R_p/2$ to R_p. That yields

$$\frac{1}{\pi} \int_{R_p/2}^{R_p} \int_{-R}^{R} \frac{R \log |g(x_p + s)|}{\sqrt{(R^2 - s^2)}} \, ds \, dR$$

$$\leqslant \frac{7B}{12\pi} R_p^3 + \frac{3}{8} R_p^2 \nu(R_p, \, x_p + i) \log R_p$$

$$- \frac{3}{8} R_p^2 [\Delta_p] \log \frac{R_p}{2} + \frac{3 \log 2}{8} [\Delta_p] R_p^2 .$$

An integral like the one on the left (involving $\log |\hat{\mu}(c + t)|$ instead of $\log |g(x_p + s)|$) has already figured in the proof of the theorem from the passage just referred to. Here, we may argue as in that proof (reversing the order of integration), for $\log |g(x)| \leqslant 0$ on the real axis, as follows from the inequalities $|g(x)| W(x) \leqslant e^{-B}$ and $W(x) \geqslant 1$, valid thereon. In that way, one finds the left-hand integral to be

$$\geqslant \frac{\sqrt{3}}{2\pi} R_p \int_{-R_p}^{R_p} \log |g(x_p + s)| \, ds.$$

After dividing by $R_p x_p^2$ and clearing out some coefficients, we see that

$$\frac{1}{x_p^2} \int_{-R_p}^{R_p} \log |g(x_p + s)| \, ds$$

$$\leqslant \frac{7B}{6\sqrt{3}} \frac{R_p^2}{x_p^2} + \frac{\sqrt{3}\pi}{4} \{\nu(R_p, \, x_p + i) - [\Delta_p]\} \frac{R_p \log R_p}{x_p^2}$$

$$+ \frac{\sqrt{3}\pi \log 2}{2} [\Delta_p] \frac{R_p}{x_p^2}$$

(the actual values of the numerical coefficients on the right are not so important).

The quantities $R_p = \Delta_p/2 = (x_p - x_{p-1})/2$ are *increasing* when $p > 8$ because $x_p = \exp(p^{1/3})$, and the latter function has a *positive second derivative* for $p > 8$. (*Now* we see why we use the sequence of points x_p beginning with x_8 !) The intervals $[x_p - R_p, \, x_p + R_p]$ *therefore do not overlap* when $p > 8$, so our desired conclusion, namely, that

$$\int_{-\infty}^{\infty} \frac{\log^- |g(x)|}{1 + x^2} \, dx = \infty,$$

will surely follow if we can establish that

$$\sum_{p > 8} \frac{1}{x_p^2} \int_{-R_p}^{R_p} \log |g(x_p + s)| \, ds = -\infty$$

with the help of the preceding relation.

Here, we are guided by a simple idea. Everything turns on the *middle term* figuring on the right side of our relation, for the sums of the *first* and *third* terms are readily seen to be *convergent*. To see how the middle term behaves, we observe that by *Levinson's theorem* (!), the function $g(z)$, bounded on the real axis and of exponential type A, should, *on the average*, have about

$$\frac{2}{\pi} A R_p \;\; = \;\; \frac{A}{\pi} \Delta_p$$

zeros on the interval $(x_p - R_p, \; x_p + R_p]$, for all of g's zeros are real. The quantity $\nu(R_p, \; x_p + i)$ is clearly *not more* than that number of zeros, so the factor in $\{ \;\; \}$ from our middle term should, *on the average*, be

$$\leqslant \;\; - \frac{\pi - A}{\pi} \Delta_p$$

(approximately). Straightforward computation easily shows, however, that

$$\frac{\Delta_p R_p \log R_p}{x_p^2} \;\; \sim \;\; \frac{1}{18p}$$

for large values of p. It is thus quite plausible that the series

$$\sum_p \{ \nu(R_p, \; x_p + i) \; - \; [\Delta_p] \} \frac{R_p \log R_p}{x_p^2}$$

should diverge to $-\infty$. This inference is in fact *correct*, but for its justification we must resort to a technical device.

Picking a number $\gamma > 1$ *close* to 1 (the exact manner of choosing it will be described presently), we form the sequence

$$X_m \;\; = \;\; \gamma^m, \quad m \; = \; 1, 2, 3, \ldots .$$

We think of $\{X_m\}$ as a *coarse* sequence of points, amongst which those of $\{x_p\}$ – regarded as a *fine* sequence – are interspersed:

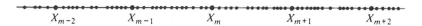

    X_{m-2} X_{m-1} X_m X_{m+1} X_{m+2}

    Figure 250

It is convenient to denote by $\nu(t)$ the *number of zeros* of $g(t)$ in $[0, \; t]$ for $t \geqslant 0$; then, as remarked above,

$$\nu(R_p, \; x_p + i) \;\; \leqslant \;\; \nu(x_p + R_p) \; - \; \nu(x_p - R_p).$$

For any large value of m we thus have, recalling that $R_p = \Delta_p/2$,

$$\sum_{X_m < x_p \leqslant X_{m+1}} v(R_p, x_p + \mathrm{i}) \frac{R_p \log R_p}{x_p^2}$$

$$\leqslant \sum_{X_m < x_p \leqslant X_{m+1}} \left\{ v\left(x_p + \frac{\Delta_p}{2}\right) - v\left(x_p - \frac{\Delta_p}{2}\right) \right\} \cdot \sup_{X_m < x_p \leqslant X_{m+1}} \frac{\Delta_p \log \Delta_p}{2x_p^2}.$$

Denote by h_m the value of $\Delta_p/2$ corresponding to the *smallest* $x_p > X_m$, and by h'_{m+1} the value of that quantity corresponding to the *largest* $x_p \leqslant X_{m+1}$. Since, for $p > 8$, the intervals $[x_p - \frac{1}{2}\Delta_p, \; x_p + \frac{1}{2}\Delta_p]$ don't overlap, we have

$$\sum_{X_m < x_p \leqslant X_{m+1}} \left\{ v\left(x_p + \frac{\Delta_p}{2}\right) - v\left(x_p - \frac{\Delta_p}{2}\right) \right\} \leqslant v(X_{m+1} + h'_{m+1})$$

$$- v(X_m - h_m).$$

According to Levinson's theorem (the simpler version from §H.2 of Chapter III is adequate here), we have

$$\frac{A}{\pi} t - \mathrm{o}(t) \leqslant v(t) \leqslant \frac{A}{\pi} t + \mathrm{o}(t)$$

for t tending to ∞. Since, in our construction, $\Delta_p = \mathrm{o}(x_p)$ for large p, it follows that $h_m = \mathrm{o}(X_m)$ and $h'_{m+1} = \mathrm{o}(X_{m+1})$ for $m \longrightarrow \infty$; the preceding relation thus implies that

$$v(X_{m+1} + h'_{m+1}) - v(X_m - h_m) \leqslant \frac{A}{\pi}(X_{m+1} - X_m) + \mathrm{o}(X_{m+1})$$

when $m \longrightarrow \infty$. Hence, since $X_{m+1} - X_m = (1 - 1/\gamma)X_{m+1}$, we have

$$\sum_{X_m < x_p \leqslant X_{m+1}} \left\{ v\left(x_p + \frac{\Delta_p}{2}\right) - v\left(x_p - \frac{\Delta_p}{2}\right) \right\} \leqslant \left(\frac{A}{\pi} + \varepsilon\right)(X_{m+1} - X_m)$$

for any given $\varepsilon > 0$, as long as m is sufficiently large.

We require an estimate of $(\Delta_p \log \Delta_p)/x_p^2$ for $X_m < x_p \leqslant X_{m+1}$. As p tends to ∞,

$$\Delta_p \log \Delta_p \;\sim\; \tfrac{1}{3} p^{-2/3} \mathrm{e}^{p^{1/3}} \log(\tfrac{1}{3} p^{-2/3} \mathrm{e}^{p^{1/3}}) \;\sim\; \frac{x_p}{3p^{1/3}},$$

so

$$\frac{\Delta_p \log \Delta_p}{2x_p^2} \;\sim\; \frac{1}{6x_p \log x_p},$$

a *decreasing* function of p. Therefore, since $X_m = \gamma^m$,

$$\sup_{X_m < x_p \leqslant X_{m+1}} \frac{\Delta_p \log \Delta_p}{2x_p^2} \leqslant \left(\frac{1}{6} + o(1)\right) \frac{1}{mX_m \log \gamma}$$

for large values of m.

Use this estimate together with the preceding one in the above relation. It is found that

$$\sum_{X_m < x_p \leqslant X_{m+1}} v(R_p, x_p + i) \frac{R_p \log R_p}{x_p^2} \leqslant \frac{1}{6}\left(\frac{A}{\pi} + 2\varepsilon\right)\frac{\gamma-1}{\log\gamma}\cdot\frac{1}{m}$$

for sufficiently large values of m, where $\varepsilon > 0$ is arbitrary.

We turn to the sum

$$\sum_{X_m < x_p \leqslant X_{m+1}} \frac{[\Delta_p] R_p \log R_p}{x_p^2},$$

for which a *lower bound* is needed. We have

$$\frac{[\Delta_p] R_p \log R_p}{x_p^2} \sim \frac{1}{2x_p^2}(\tfrac{1}{3}p^{-2/3}x_p)^2 \log(\tfrac{1}{6}p^{-2/3}x_p) \sim \frac{1}{18p^{4/3}}\log x_p$$

$$= \frac{1}{18p} \qquad \text{for } p \longrightarrow \infty.$$

Thence, calling p_m the *smallest* value of p for which $x_p > X_m$ and p'_{m+1} the *largest* such value with $x_p \leqslant X_{m+1}$, the preceding sum works out to

$$\left(\frac{1}{18} + o(1)\right)\log\frac{p'_{m+1}}{p_m}$$

when m is large. In that case, $x_{p_m} = \exp(p_m^{1/3})$ is nearly $X_m = \gamma^m$, and $\exp((p'_{m+1})^{1/3})$ nearly γ^{m+1}. So then, p'_{m+1}/p_m is practically equal to $((m+1)/m)^3 \sim 1 + 3/m$, and

$$\log\frac{p'_{m+1}}{p_m} \sim \frac{3}{m}.$$

Thus,

$$\sum_{X_m < x_p \leqslant X_{m+1}} \frac{[\Delta_p] R_p \log R_p}{x_p^2} \geqslant \left(\frac{1}{6} - o(1)\right)\cdot\frac{1}{m}$$

for large values of m.

Now combine this result with the one obtained previously. One gets

$$\sum_{X_m < x_p \leqslant X_{m+1}} \{v(R_p, x_p+i) - [\Delta_p]\} \frac{R_p \log R_p}{x_p^2}$$

$$\leqslant \frac{1}{6}\left\{\left(\frac{A}{\pi} + 2\varepsilon\right)\frac{\gamma-1}{\log\gamma} - 1 + \varepsilon\right\} \cdot \frac{1}{m}$$

(with $\varepsilon > 0$ arbitrary) for large m. We had, however, $A/\pi < 1$. It is thus possible to *choose $\gamma > 1$ so as to make*

$$\frac{A}{\pi}\frac{\gamma-1}{\log\gamma} < 1 - \delta,$$

say, where δ is a certain number > 0. *Fixing* such a γ, we can *then* take an $\varepsilon > 0$ small enough to ensure that

$$\frac{1}{6}\left\{\left(\frac{A}{\pi} + 2\varepsilon\right)\frac{\gamma-1}{\log\gamma} - 1 + \varepsilon\right\} < -\frac{\delta}{7}.$$

The left-hand sum in the previous relation is then

$$\leqslant -\frac{\delta}{7m}$$

for sufficiently large m, so

$$\frac{\sqrt{3}\pi}{4}\sum_p \{v(R_p, x_p+i) - [\Delta_p]\} \frac{R_p \log R_p}{x_p^2}$$

does diverge to $-\infty$.

Aside from the general term of this series, our upper bound on

$$\frac{1}{x_p^2}\int_{-R_p}^{R_p} \log|g(x_p+s)|\,ds$$

involved *two other terms*, each of which is $\sim \text{const.}\Delta_p^2/x_p^2$ when p is large. But

$$\frac{\Delta_p^2}{x_p^2} \sim \frac{1}{9}p^{-4/3} \qquad \text{for } p \longrightarrow \infty,$$

so *the sum of those remaining terms is certainly convergent*. The divergence just established therefore does imply that

$$\sum_{p > 8} \frac{1}{x_p^2}\int_{-R_p}^{R_p} \log|g(x_p+s)|\,ds = -\infty,$$

and hence that

$$\int_{-\infty}^{\infty} \frac{\log^- |g(x)|}{1 + x^2} \mathrm{d}x = \infty$$

as claimed, yielding finally our desired contradiction.

The weight

$$W(x) = \frac{e^B}{|\Psi(x + i)|} \geqslant 1$$

constructed above thus *admits no multipliers f of exponential type $< \pi$,
even though* it enjoys the regularity property

$$|\log\log W(x) - \log\log W(x')| \leqslant |x - x'|, \qquad x, x' \in \mathbb{R},$$

and satisfies the condition

$$\int_{-\infty}^{\infty} \frac{\log W(x)}{1 + x^2} \mathrm{d}x < \infty.$$

2.   **Discussion**

We see that our local regularity requirement and the convergence of the logarithmic integral do not, by themselves, ensure admittance of multipliers. Some other property of the weight is thus really involved.

For the weight W constructed in the example just given we actually had

$$|\log\log W(x) - \log\log W(x')| \leqslant |x - x'|$$

on $\mathbb{R}$. By this we are reminded that another regularity condition of similar appearance *has* previously been shown to *be* sufficient when combined with the requirement that $\int_{-\infty}^{\infty} (\log W(x)/(1 + x^2))\mathrm{d}x < \infty$. The theorem proved in §C.1 of Chapter X (and reestablished by a different method at the end of §C.5 in this chapter) states that a weight W *does* admit multipliers if its logarithmic integral converges and

$$|\log W(x) - \log W(x')| \leqslant \mathrm{const.}|x - x'|$$

on $\mathbb{R}$. A uniform Lipschitz condition on $\log W(x)$ thus *gives us* enough regularity, although such a requirement on $\log\log W(x)$ *does not*.

An *intermediate property* is in fact *already sufficient*. Consider a continuous weight W with $\log W(x) = O(x^2)$ near the origin (not a real restriction), and suppose, besides, that $W(x)$ is *even*. As remarked near the end of §B.1, that involves no loss of generality *either*, because $W(x)$ admits multipliers if and only if $W(x)W(-x)$ does. In these circumstances, convergence of the logarithmic integral is equivalent to the condition that

$$\int_0^\infty \frac{\log W(x)}{x^2}\,dx \quad < \quad \infty.$$

When this holds, we know, however, by the corollary at the end of §C.5 that $W(x)$, if it meets the local regularity requirement, *admits multipliers* as long as $\log W(x)/x$ belongs to the Hilbert space $\mathfrak{H}$ studied in §C.4, i.e., that

$$\| \log W(x)/x \|_E \quad < \quad \infty.$$

Problem 62 tells us on the other hand that an even weight $W(x)$ will *have* that property when the above integral is convergent and $\log W(x)$ uniformly Lip 1. These last conditions are thus *more stringent* than the *sufficient* ones furnished by the corollary of §C.5.

This fact leads us to believe, or at least to hope, that the intermediate property just spoken of could serve as basis for the formulation of *necessary and sufficient conditions for admittance of multipliers* by weights satisfying the local regularity requirement. But how one could set out to accomplish that is not immediately apparent, because pointwise behaviour *of the weight itself* seems at the same time *to be involved* and *not to be involved* in the matter.

Behaviour of the weight *itself* seems to *not* be directly involved (beyond the local regularity requirement), because, if $W(x)$ *admits multipliers, so does any weight* $W_1(x)$ with $1 \leq W_1(x) \leq W(x)$. Even when such a $W_1(x)$ meets the local regularity requirement, its behaviour may be very wild in comparison to whatever we may otherwise stipulate for $W(x)$. Were one, for instance to prescribe that $\| \log W(x)/x \|_E < \infty$, there would be weights W_1 *failing* to meet that criterion even though W *answered* to it – see the formula provided by the last theorem of §C.4.

Nevertheless, *some* condition on the behaviour of our weights *does* appear to be involved! Support for this point of view is obtained by putting the theorem on the multiplier (from §C.2) together with those of Szegő (from Chapter II!) and de Branges (Chapter VI, §F).

Consider *any* continuous function $W(x) \geq 1$ tending to ∞ for

$x \longrightarrow \pm \infty$, and such that

$$\int_{-\infty}^{\infty} \frac{\log W(x)}{1 + x^2} \, dx \quad < \quad \infty.$$

Then the weight $\Omega(x) = (1 + x^2)W(x)$ also has convergent logarithmic integral, so, by the version of Szegő's theorem set as problem 2 (in Chapter II), there is *no finite sum*

$$s(x) \quad = \quad \sum_{\lambda \geqslant 1} a_\lambda e^{i\lambda x}$$

which can make

$$\int_{-\infty}^{\infty} \frac{|1 - s(x)|}{\Omega(x)} \, dx$$

smaller than a certain $\delta > 0$. Hence, for any such $s(x)$, we must have

$$\sup_{x \in \mathbb{R}} \frac{|1 - s(x)|}{W(x)} \quad \geqslant \quad \frac{\delta}{\pi}.$$

Given any $L > 0$, this holds a fortiori for sums $s(x)$ of the form

$$s(x) \quad = \quad \sum_{1 \leqslant \lambda \leqslant 2L+1} a_\lambda e^{i\lambda x}.$$

Problem 65

(a) Show that in this circumstance there is, corresponding to any L, an entire function $\Phi(z)$ of exponential type L such that

$$\int_{-\infty}^{\infty} \frac{\log^+ |\Phi(x)|}{1 + x^2} \, dx \quad < \quad \infty$$

and

$$W(x_n) \leqslant |\Phi(x_n)|$$

at the points x_n of a *two-way* real sequence Λ with $x_n \neq x_m$ for $n \neq m$ and

$$\frac{n_\Lambda(t)}{t} \longrightarrow \frac{L}{\pi} \quad \text{for } t \longrightarrow \pm \infty.$$

Here, $n_\Lambda(t)$ denotes (as usual) the number of points of Λ in $[0, \ t]$ when $t \geqslant 0$ and *minus* the number of such points in $[t, \ 0)$ when $t < 0$. (Hint: See §F.3, Chapter VI.)

(b) Show that for the sequence $\{x_n\} = \Lambda$ obtained in (a) we also have

$$\tilde{D}_{\Lambda_+} = \tilde{D}_{\Lambda_-} = \frac{L}{\pi}$$

for $\Lambda_+ = \Lambda \cap [0, \infty)$ and $\Lambda_- = (-\Lambda) \cap (0, \infty)$, with $\tilde{D}$ the Beurling–Malliavin effective density defined in §D.2 of chapter IX. (Hint: See the very end of §E.2, Chapter IX.)

(c) Show that for any given $A > 0$ there is a non-zero entire function $f(z)$ of exponential type $\leqslant A$, bounded on $\mathbb{R}$, with

$$W(x_n)|f(x_n)| \leqslant \text{const.}$$

at the points x_n of the sequence from part (a).

The result obtained in part (c) of this problem holds on the *mere assumptions that* $W(x) \geqslant 1$ *is* continuous *and tends to* ∞ *for* $x \longrightarrow \pm\infty$, and that

$$\int_{-\infty}^{\infty} \frac{\log W(x)}{1 + x^2} dx < \infty.$$

The points x_n on which any of the products $W(x)f(x)$ is *bounded* behave, however, *rather closely* like the ones of the *arithmetic progression*

$$\frac{\pi}{L}n, \quad n = 0, \pm 1, \pm 2, \ldots$$

which, for large enough L, seem to 'fill out' the real axis. From this standpoint it appears to be plausible that *some regularity property* of the weight $W(x)$ would be both *necessary and sufficient* to ensure *boundedness of the products* $W(x)f(x)$ *on* $\mathbb{R}$.

These considerations illustrate our present difficulty, but also suggest a way out of it, which is to look for an additional condition *pertaining to a majorant of* $W(x)$ rather than *directly* to the latter. That such an approach is *reasonable* is shown by the first theorem of §B.1, according to which a weight $W(x) \geqslant 1$ meeting the local regularity requirement (with constants C, α and L) and satisfying $\int_{-\infty}^{\infty} (\log W(x)/(1 + x^2))dx < \infty$ *admits multipliers if and only if a certain* $\mathscr{C}_\infty$ *majorant of it also does so.* For that majorant one may take the weight

$$\Omega(x) = M \exp\left\{ \frac{4}{\pi\alpha} \int_{-\infty}^{\infty} \frac{L \log W(t)}{(x-t)^2 + L^2} dt \right\}$$

where M is a large constant, and then $|d(\log\log\Omega(x))/dx| \leqslant 1/L$ on $\mathbb{R}$.

This idea actually underlies much of what is done in §B.1. One may of course use the *even $\mathscr{C}_\infty$ majorant* $\Omega(x)\Omega(-x)$ *instead* – see the remark just preceding problem 52.

Let us try then to characterize a weight's admittance of multipliers by the *existence for it of some even majorant also admitting multipliers and having, in addition, some specific kind of regularity*. What we have in mind at present is essentially the regularity embodied in the *intermediate property* described earlier in this article. We think the criterion should be that $W(x)$ have an even $\mathscr{C}_\infty$ majorant $\Omega(x)$ with $\log^+\log\Omega(x)$ uniformly Lip 1, $\int_0^\infty (\log\Omega(x)/x^2)\,\mathrm{d}x < \infty$, and $\|\log\Omega(x)/x\|_E < \infty$.

A minor hitch encountered at this point is easily taken care of. The trouble is that neither of the last two of the conditions on Ω is compatible with Ω's being a *majorant* of W when $W(x) > 1$ on a neighborhood of the origin. That, however, should not present a real problem because admittance of multipliers by a finite weight W meeting the local regularity requirement *does not depend* on the behaviour of $W(x)$ near 0 – according to the first lemma of §B.1, $W(x)$, if *not bounded* on finite intervals, would have to be *identically infinite* on one of length > 0. We can thus allow majorants $\Omega(x)$ which are merely $\geqslant W(x)$ *for* $|x|$ *sufficiently large*, instead of for *all* real x. In that way we arrive at a statement having (we hope) some chances of being true:

A finite weight $W(x) \geqslant 1$ meeting the local regularity requirement admits multipliers if and only if there exists an even $\mathscr{C}_\infty$ function $\Omega(x) \geqslant 1$ with $\Omega(0) = 1$ (making $\log\Omega(x) = O(x^2)$ near 0),

$\log^+\log\Omega(x)$ *uniformly Lip 1 on* $\mathbb{R}$,

$\Omega(x) \geqslant W(x)$ *whenever* $|x|$ *is sufficiently large,*

$$\int_0^\infty \frac{\log\Omega(x)}{x^2}\,\mathrm{d}x < \infty,$$

and

$$\|\log\Omega(x)/x\|_E < \infty.$$

According to what we already know, the 'if' part of this proposition *is* valid, because a weight Ω with the stipulated properties *does* admit multipliers (it enjoys the *intermediate property*), and hence W must *also* do so. But the 'only if' part is still just a *conjecture*.

Support for believing 'only if' to be *correct* comes from a review of how the energy norm $\|\log W(x)/x\|_E$ entered into the argument of §C.5. There, as in §C.4 of Chapter VIII, that was through the use of Schwarz' inequality for the inner product $\langle \ , \ \rangle_E$. This encourages us to look for a proof of the 'only if' part based on the Schwarz inequality's being best possible.

There is, on the other hand, nothing to prevent anyone's *doubting* the truth of 'only if'. We have again to choose between two approaches – to look for a proof or try constructing a counterexample. The *second* approach proves fruitful here.

In article 4 we give an example showing that the *existence* of an Ω having the properties enumerated above is *not necessary* for the admittance of multipliers by a weight W. The 'essential' condition we are seeking turns out to be more elusive than at first thought.

The reader who is still following the present discussion is urged not to lose patience with this §'s chain of seesaw arguments and interspersed seemingly artificial examples. By going on in such fashion we will arrive at a clear vision of the object of our search. See the first paragraph of article 5.

Our example's construction depends on an auxiliary result relating the norm $\| \ \|_E$ of a certain kind of Green potential to the same norm of a majorant for it. This we attend to in the next article.

3.   **Comparison of energies**

The weight W to be presently constructed is similar to the one considered in article 1, being of the form

$$W(x) \ = \ \frac{\text{const.}}{\exp F(x + \mathrm{i})},$$

where $F(z)$, bounded above on the real axis, is given by the formula

$$F(z) \ = \ \int_0^\infty \log \left| 1 \ - \ \frac{z^2}{t^2} \right| \, \mathrm{d}\mu(t)$$

with $\mu(t)$ *increasing* and $O(t)$ (for both large and small values of t) on $[0, \infty)$. $W(x)$ is thus much like the *reciprocal* of the modulus of an entire function of exponential type.

From $\mu(t)$ one can, as in §C.5 of Chapter VIII, form another increasing function $v(t)$, this one defined* and infinitely differentiable on $\mathbb{R}$, $O(|t|)$

---

* by the formula $v'(t) \ = \ (1/\pi)\int_0^\infty \{((t + s)^2 + 1)^{-1} \ + \ ((t - s)^2 + 1)^{-1}\} \, \mathrm{d}\mu(t)$; see next article, about 3/4 of the way through.

there and *odd*, such that

$$F(x+i) - F(i) = \int_0^\infty \log\left|1 - \frac{x^2}{t^2}\right| dv(t)$$

for $x \in \mathbb{R}$. The right-hand integral can in turn be converted to

$$-x \int_0^\infty \log\left|\frac{x+t}{x-t}\right| d\left(\frac{v(t)}{t}\right),$$

and our weight $W(x)$ thereby expressed in the form

$$\text{const.} + x \int_{-\infty}^\infty \log\left|\frac{x+t}{x-t}\right| d\left(\frac{v(t)}{t}\right).$$

The reader should take care to distinguish between this representation and the one which has frequently been used in this book for certain entire functions $G(z)$ of exponential type. The latter also involves a function $v(t)$, increasing and $O(t)$ on $[0, \infty)$, but reads

$$\log|G(x)| = -x \int_0^\infty \log\left|\frac{x+t}{x-t}\right| d\left(\frac{v(t)}{t}\right)$$

with a *minus sign* in front of the integral. It will eventually become clear that this *difference in sign* is very important for the matter under discussion.

The weight W we will be working with in the next article is closely related to the *Green potentials* studied in §C.4, since

$$\frac{1}{x}\log\left(\frac{W(x)}{W(0)}\right) = \int_0^\infty \log\left|\frac{x+t}{x-t}\right| d\left(\frac{v(t)}{t}\right).$$

We will want to be able to affirm that this expression belongs to the Hilbert space $\mathfrak{H}$ considered in §C.4 provided that there is some even $\Omega(x) \geqslant 1$ with $\log\Omega(x)/x$ in $\mathfrak{H}$ and $\int_0^\infty (\log\Omega(x)/x^2)dx$ finite, such that $W(x) \leqslant \Omega(x)$ for all x of sufficiently large modulus.

This kind of comparison is well known for the simpler circumstance involving *pure potentials*. Those are the potentials

$$U_\rho(x) = \int_0^\infty \log\left|\frac{x+t}{x-t}\right| d\rho(t)$$

corresponding to *positive* measures ρ. *Cartan's lemma* says that if for two of them, U_ρ and U_σ, we have $U_\rho(x) \leqslant U_\sigma(x)$ for $x \geqslant 0$, then

$\|U_\rho\|_E \leqslant \|U_\sigma\|_E.$ Proof:

$$\|U_\rho\|_E^2 = \int_0^\infty U_\rho(x)\mathrm{d}\rho(x) \leqslant \int_0^\infty U_\sigma(x)\mathrm{d}\rho(x) = \int_0^\infty U_\rho(x)\mathrm{d}\sigma(x)$$

$$\leqslant \int_0^\infty U_\sigma(x)\mathrm{d}\sigma(x) = \|U_\sigma\|_E^2 \ !$$

The result obviously depends greatly on the positivity of ρ and σ.

For our weight W, $(1/x)\log(W(x)/W(0))$ is of the form $U_\rho(x)$, but the measure ρ is *not positive*. Instead,

$$\mathrm{d}\rho(t) = \mathrm{d}\left(\frac{v(t)}{t}\right) = \frac{\mathrm{d}v(t)}{t} - \frac{v(t)}{t}\frac{\mathrm{d}t}{t},$$

and all that the properties of our v give us is the relation

$$\mathrm{d}\rho(t) \geqslant -\,\mathrm{const.}\,\frac{\mathrm{d}t}{t}.$$

As we know,

$$\int_0^\infty \log\left|\frac{x+t}{x-t}\right|\frac{\mathrm{d}t}{t} = \frac{\pi^2}{2} \quad \text{for } x > 0;$$

the measure σ on $(0,\infty)$ with $\mathrm{d}\sigma(t) = \mathrm{d}t/t$ thus *just misses* having finite energy. *Finiteness* of $\|U_\rho\|_E$, *if realized*, must hence be due to *interference* between $\mathrm{d}v(t)/t$ and $v(t)\mathrm{d}t/t^2$. A version of Cartan's result is nevertheless still available in this situation.

In order to deal with the measure $\mathrm{d}t/t$ we will use the following two elementary lemmas.

Lemma. *For $A > 0$, we have:*

$$\int_0^A\int_A^\infty \log\left|\frac{x+t}{x-t}\right|\frac{\mathrm{d}t}{t}\frac{\mathrm{d}x}{x} = 2 + \frac{2}{3^3} + \frac{2}{5^3} + \cdots;$$

$$\frac{\mathrm{d}}{\mathrm{d}x}\int_A^\infty \log\left|\frac{x+t}{x-t}\right|\frac{\mathrm{d}t}{t} = \frac{1}{x}\log\left|\frac{x+A}{x-A}\right| \quad \text{for } x > 0,\ x \neq A;$$

$$\frac{\mathrm{d}}{\mathrm{d}x}\int_0^A \log\left|\frac{x+t}{x-t}\right|\frac{\mathrm{d}t}{t} = -\frac{1}{x}\log\left|\frac{x+A}{x-A}\right| \quad \text{for } x > 0,\ x \neq A.$$

Proof. To establish the first relation, make the changes of variable

$$\xi = \frac{x}{A}, \quad \tau = \frac{t}{A}$$

and expand the logarithm in powers of ξ/τ, then integrate term by term.

For the last two relations, we use a different change of variable, putting $s = t/x$. Then the left side of the *second* relation becomes

$$\frac{d}{dx}\int_{A/x}^{\infty}\log\left|\frac{1+s}{1-s}\right|\frac{ds}{s},$$

and this may be worked out for $x \neq A$ by the fundamental theorem of calculus. The third relation follows in like manner.

Lemma. Let $\rho(t) = \int_0^t d\rho(\tau)$ be bounded for $0 \leqslant t < \infty$. Then, for $A > 0$, the two expressions

$$\int_0^A \int_A^{\infty} \log\left|\frac{x+t}{x-t}\right|\frac{dt}{t}\,d\rho(x),$$

$$\int_0^A \int_A^{\infty} \log\left|\frac{x+t}{x-t}\right|d\rho(t)\frac{dx}{x},$$

are bounded in absolute value by quantities independent of A.

Proof. Considering the *second* expression, we have, for large $M > A$ and any $M' > M$,

$$\int_M^{M'}\log\left|\frac{x+t}{x-t}\right|d\rho(t)$$

$$= \rho(M')\log\left|\frac{1+(x/M')}{1-(x/M')}\right| - \rho(M)\log\left|\frac{1+(x/M)}{1-(x/M)}\right| + 2x\int_M^{M'}\frac{\rho(t)}{t^2-x^2}dt$$

whenever $0 < x < A$. Because $|\rho(t)|$ is bounded, the right side is equal to *x times a quantity uniformly small for* $0 < x < A$ when M and M' are both large. The second expression is therefore equal to the *limit*, for $M \to \infty$, of the double integrals

$$\int_0^A \int_A^M \log\left|\frac{x+t}{x-t}\right|d\rho(t)\frac{dx}{x}.$$

Any one of these is equal to

$$\int_A^M\left(\int_0^A\log\left|\frac{x+t}{x-t}\right|\frac{dt}{t}\right)d\rho(x);$$

here we use partial integration on the *outer* integral and refer to the third formula provided by the preceding lemma. In that way we get

$$\rho(M) \int_0^A \log\left|\frac{M+t}{M-t}\right| \frac{dt}{t} \quad - \quad \rho(A) \int_0^A \log\left|\frac{A+t}{A-t}\right| \frac{dt}{t}$$

$$+ \quad \int_A^M \log\left|\frac{x+A}{x-A}\right| \frac{\rho(x)}{x} dx.$$

Remembering that

$$\int_0^\infty \log\left|\frac{x+t}{x-t}\right| \frac{dt}{t} \quad = \quad \frac{\pi^2}{2} \quad \text{for } x > 0,$$

we see that the last expression is $\leqslant 3\pi^2 K/2$ in absolute value if $|\rho(t)| \leqslant K$ on $[0, \infty)$. Thence,

$$\left| \int_0^A \int_A^\infty \log\left|\frac{x+t}{x-t}\right| d\rho(t) \frac{dx}{x} \right| \quad \leqslant \quad \frac{3\pi^2}{2} K$$

independently of $A > 0$.

Treatment of the *first* expression figuring in the lemma's statement is similar (and easier). We are done.

Now we are ready to give our version of Cartan's lemma. So as not to obscure its main idea with fussy details, we avoid insisting on more generality than is needed for the next article. An alternative formulation is furnished by problem 68 below.

Theorem. *Let $\omega(x)$, even and tending to ∞ for $x \longrightarrow \pm \infty$, be given by a formula*

$$\omega(x) \quad = \quad -\int_0^\infty \log\left|1 - \frac{x^2}{t^2}\right| dv(t),$$

where $v(t)$, odd and increasing, is $\mathscr{C}_\infty$ on $\mathbb{R}$, with $v(t)/t$ bounded there. Suppose there is an even function $\Omega(x) \geqslant 1$, with

$$\int_0^\infty \frac{\log \Omega(x)}{x^2} dx \quad < \quad \infty$$

and $\log \Omega(x)/x$ in the Hilbert space $\mathfrak{H}$ of §C.4, such that

$$\omega(x) \leqslant \log \Omega(x)$$

for all x of sufficiently large absolute value. Then $\omega(x)/x$ also belongs to

$\mathfrak{H}$, *and*

$$\int_0^\infty \frac{\omega(x)}{x^2}\,dv(x) \;\; < \;\; \infty.$$

Proof. If there is an Ω meeting the stipulated conditions, there is an L such that

$$\omega(x) \;\leqslant\; \log\Omega(x) \quad \text{for } x \geqslant L.$$

Because $\omega(x) \longrightarrow \infty$ for $x \longrightarrow \infty$, we can take (and *fix*) L large enough to also make

$$\omega(x) \;\geqslant\; 0 \quad \text{for } x \geqslant L.$$

The given properties of $v(t)$ make $\omega(0) = 0$ and $\omega(x)$ infinitely differentiable* on $\mathbb{R}$. Therefore, since $\omega(x)$ is even, we have $\omega(x) = O(x^2)$ near 0, and, having chosen L, we can find an M such that

$$-x^2 M \;\leqslant\; \omega(x) \;\leqslant\; x^2 M \quad \text{for } 0 \leqslant x \leqslant L.$$

According to the first lemma of §B.4, Chapter VIII, the given formula for $\omega(x)$ can be rewritten

$$\omega(x) \;=\; x \int_0^\infty \log\left|\frac{x+t}{x-t}\right| d\left(\frac{v(t)}{t}\right), \quad x > 0.$$

We put

$$\rho(t) \;=\; \frac{v(t)}{t},$$

making $\rho(t) \geqslant 0$ and bounded by hypothesis, with

$$d\rho(t) \;\geqslant\; -\frac{v(t)}{t}\frac{dt}{t} \;\geqslant\; -C\,\frac{dt}{t}.$$

---

* To check infinite differentiability of $\omega(x)$ in $(-A, A)$, say, take any *even* $\mathscr{C}_\infty$ function $\varphi(t)$ equal to 1 for $|t| \leqslant A$ and to 0 for $|t| \geqslant 2A$. Then, since $v'(t)$ is also even, we have

$$\omega(x) \;=\; \int_A^\infty \log|1 - x^2/t^2|(1-\varphi(t))v'(t)\,dt \;+\; \int_{-2A}^{2A} \log|x-t|\,\varphi(t)v'(t)\,dt$$
$$-\; \int_{-2A}^{2A} \log|t|\,\varphi(t)v'(t)\,dt.$$

The *first* integral on the right is clearly $\mathscr{C}_\infty$ in x for $|x| < A$. When $|x| < A$, the *second* one can be rewritten as $\int_{-3A}^{3A} \varphi(x-s)v'(x-s)\log|s|\,ds$, and this, like φ and v', is $\mathscr{C}_\infty$ (in x), since $\log|s| \in L_1(-3A,3A)$.

In order to keep our notation simple, let us, without real loss of generality, assume that $C = 1$, i.e., that

$$0 \leqslant \rho(t) \leqslant 1$$

and

$$d\rho(t) \geqslant -\frac{dt}{t}.$$

We now consider the Green potentials

$$U_A(x) = \int_0^A \log\left|\frac{x+t}{x-t}\right| d\rho(t),$$

where $A > L$. Since $v(t)$ is $\mathscr{C}_\infty$ and $\rho(t) = v(t)/t$ bounded, it is readily verified with the help of l'Hôpital's rule that $\rho(t)$ (taken as $v'(0)$ for $t = 0$) is *differentiable right down to the origin*, and that

$$\rho'(t) = \frac{tv'(t) - v(t)}{t^2}$$

stays bounded as $t \longrightarrow 0$. The quantity $|\rho'(t)|$ is thus bounded on each of the finite segments $[0, A]$, and the double integrals

$$\int_0^A \int_0^A \log\left|\frac{x+t}{x-t}\right| |d\rho(t)| \, |d\rho(x)|$$

hence finite. Each of the potentials U_A therefore belongs to $\mathfrak{H}$. We proceed to obtain an upper bound on $\| U_A \|_E$ which, for $A > L$, is *independent* of A.

By the absolute convergence just noted we have, according to §C.4,

$$\| U_A \|_E^2 = \int_0^A U_A(x) \, d\rho(x).$$

In view of the above formula for $\omega(x)$, we can write

$$U_A(x) = \frac{\omega(x)}{x} - \int_A^\infty \log\left|\frac{x+t}{x-t}\right| d\rho(t).$$

When $A \longrightarrow \infty$, the integral on the right tends to zero *uniformly* for $0 \leqslant x \leqslant L$ (see beginning of the proof of the *second* of the above lemmas). Therefore, $|\rho'(x)|$ being bounded for $0 \leqslant x \leqslant L$,

$$\int_0^L U_A(x) \, d\rho(x) = \int_0^L \frac{\omega(x)}{x} \rho'(x) \, dx + o(1)$$

for $A \longrightarrow \infty$. Referring to one of the initial inequalities for $\omega(x)$, we see that

$$\int_0^L U_A(x)\,d\rho(x) \;\leqslant\; \int_0^L xM|\rho'(x)|\,dx \;+\; o(1)$$

for large A.

Our main work is with the integral $\int_L^A U_A(x)\,d\rho(x)$. Since $d\rho(t) \geqslant -dt/t$, it is convenient to put

$$d\rho(t) \;+\; \frac{dt}{t} \;=\; d\sigma(t),$$

getting a *positive* measure σ on $[0, \infty)$. The above relation connecting $U_A(x)$ and $\omega(x)/x$ then gives us

$$U_A(x) \;\leqslant\; \frac{\omega(x)}{x} \;+\; \int_A^\infty \log\left|\frac{x+t}{x-t}\right| \frac{dt}{t}, \qquad x > 0;$$

$$U_A(x) \;\geqslant\; \frac{\omega(x)}{x} \;-\; \int_A^\infty \log\left|\frac{x+t}{x-t}\right| d\sigma(t), \qquad x > 0.$$

Thence, by our initial relations for $\omega(x)$,

$$U_A(x) \;\leqslant\; \frac{\log \Omega(x)}{x} \;+\; \int_A^\infty \log\left|\frac{x+t}{x-t}\right| \frac{dt}{t}, \qquad x > L;$$

$$U_A(x) \;\geqslant\; -\int_A^\infty \log\left|\frac{x+t}{x-t}\right| d\sigma(t), \qquad x \geqslant L.$$

Since $\log \Omega(x) \geqslant 0$ by hypothesis and $\log|(x+t)/(x-t)| \geqslant 0$ for x and $t \geqslant 0$, the first of these inequalities yields

$$\int_L^A U_A(x)\,d\sigma(x) \;\leqslant\; \int_0^A \frac{\log \Omega(x)}{x}\,d\sigma(x) \;+\; \int_0^A \int_A^\infty \log\left|\frac{x+t}{x-t}\right| \frac{dt}{t}\,d\sigma(x)$$

$$\leqslant\; \int_0^\infty \frac{\log \Omega(x)}{x^2}\,dx \;+\; \int_0^A \frac{\log \Omega(x)}{x}\,d\rho(x) \;+\; \int_0^A \int_A^\infty \log\left|\frac{x+t}{x-t}\right| \frac{dt}{t}\,d\rho(x)$$

$$+\; \int_0^A \int_A^\infty \log\left|\frac{x+t}{x-t}\right| \frac{dt}{t}\frac{dx}{x}.$$

Similarly, from the second inequality,

$$-\int_L^A U_A(x)\,\frac{dx}{x} \;\leqslant\; \int_0^A \int_A^\infty \log\left|\frac{x+t}{x-t}\right| d\sigma(t)\,\frac{dx}{x}$$

$$=\; \int_0^A \int_A^\infty \log\left|\frac{x+t}{x-t}\right| d\rho(t)\,\frac{dx}{x} \;+\; \int_0^A \int_A^\infty \log\left|\frac{x+t}{x-t}\right| \frac{dt}{t}\frac{dx}{x}.$$

Putting these results together and then adding on the one obtained previously, we find that for large $A > L$,

$$\| U_A \|_E^2 = \int_0^A U_A(x)\,d\rho(x)$$

$$\leqslant o(1) + \int_0^L xM|\rho'(x)|\,dx + \int_0^\infty \frac{\log \Omega(x)}{x^2}\,dx + \int_0^A \frac{\log \Omega(x)}{x}\,d\rho(x)$$

$$+ \int_0^A \int_A^\infty \log\left|\frac{x+t}{x-t}\right| \frac{dt}{t}\,d\rho(x) + \int_0^A \int_A^\infty \log\left|\frac{x+t}{x-t}\right| d\rho(t)\,\frac{dx}{x}$$

$$+ 2\int_0^A \int_A^\infty \log\left|\frac{x+t}{x-t}\right| \frac{dt}{t}\frac{dx}{x}.$$

It is part of our hypothesis that

$$\int_0^\infty \frac{\log \Omega(x)}{x^2}\,dx < \infty.$$

Because $0 \leqslant \rho(t) \leqslant 1$, the *fourth* and *fifth* of the right-hand integrals are *bounded* (by quantities independent of A) according to the second of the above lemmas. By the first of those lemmas, the *sixth* integral is equal to a finite constant independent of A. We thus have a constant c independent of A such that

$$\| U_A \|_E^2 \leqslant c + \int_0^A \frac{\log \Omega(x)}{x}\,d\rho(x)$$

for large $A > L$.

Recalling, however, that

$$U_A(x) = \int_0^A \log\left|\frac{x+t}{x-t}\right| d\rho(t)$$

and that $\log \Omega(x)/x$ is in $\mathfrak{H}$ by hypothesis, we see from the fifth lemma of §C.4 that

$$\int_0^A \frac{\log \Omega(x)}{x}\,d\rho(x) = \left\langle \frac{\log \Omega(x)}{x}, U_A(x) \right\rangle_E \leqslant \left\| \frac{\log \Omega(x)}{x} \right\|_E \| U_A \|_E.$$

The preceding relation thus becomes

$$\| U_A \|_E^2 \leqslant c + \| \log \Omega(x)/x \|_E \| U_A \|_E.$$

Knowing, then, that $\| U_A \|_E < \infty$, we get by 11th grade algebra (!) that

$$\| U_A \|_E \leqslant \frac{1}{2}\left(\| \log \Omega(x)/x \|_E + \sqrt{(\| \log \Omega(x)/x \|_E^2 + 4c)}\right)$$

for large $A > L$, *with the bound on the right independent of A.*

Now it is easy to show that $\omega(x)/x$ belongs to $\mathfrak{H}$. Since $\omega(x)/x$ is *odd*, we need, according to the last theorem of §C.4, merely check that $\|\omega(x)/x\|_E < \infty$ where, for $\|\ \|_E$, the general definition adopted towards the middle of §C.4 is taken. As observed earlier,

$$U_A(x) \longrightarrow \frac{\omega(x)}{x} \qquad \text{u.c.c. in } [0, \infty)$$

for $A \longrightarrow \infty$. Thence, by the second theorem of §C.4 and Fatou's lemma,

$$\|\omega(x)/x\|_E^2 \leqslant \liminf_{A \to \infty} \|U_A\|_E^2.$$

(Cf. the discussion of how $\mathfrak{H}$ is formed, about half way into §C.4.) The result just found therefore implies that

$$\|\omega(x)/x\|_E \leqslant \frac{1}{2}\left(\|\log\Omega(x)/x\|_E + \sqrt{(\|\log\Omega(x)/x\|_E^2 + 4c)} \right),$$

making $\omega(x)/x \in \mathfrak{H}$. (Appeal to the *last* theorem of §C.4 can be avoided here. A sequence of the U_A with $A \longrightarrow \infty$ certainly *converges weakly* to *some* element, say U, of $\mathfrak{H}$. Some convex linear combinations of those U_A then *converge in norm* $\|\ \|_E$ to U, which then can be easily identified with $\omega(x)/x$, reasoning as in the discussion towards the middle of §C.4.)

Once it is known that $\omega(x)/x \in \mathfrak{H}$, the rest of the theorem is almost immediate. The relations for $\omega(x)$ given near the beginning of this proof make

$$\int_0^\infty \frac{|\omega(x)|}{x^2}\,dx \leqslant \int_0^L M\,dx + \int_L^\infty \frac{\log\Omega(x)}{x^2}\,dx$$

$$\leqslant ML + \int_0^\infty \frac{\log\Omega(x)}{x^2}\,dx < \infty,$$

so, since (here) $0 \leqslant v(x)/x \leqslant 1$, we have

$$\int_0^\infty \frac{|\omega(x)|}{x} \frac{v(x)}{x^2}\,dx < \infty.$$

It is thus enough to verify that

$$\int_0^\infty \frac{\omega(x)}{x}\,d\!\left(\frac{v(x)}{x}\right) < \infty$$

in order to show that

$$\int_0^\infty \frac{\omega(x)}{x^2}\,dv(x)$$

is finite. Since $|d(v(x)/x)/dx|$ is bounded for $0 \leqslant x \leqslant L$ with $|\omega(x)/x^2| \leqslant M$ there, while $\omega(x) \geqslant 0$ for $x > L$, the *first* of these integrals is *perfectly unambiguous* according to the observation just made, and equal to the limit, for $A \longrightarrow \infty$, of

$$\int_0^A \frac{\omega(x)}{x} \, d\left(\frac{v(x)}{x}\right).$$

Here we may again resort to the fifth lemma of §C.4, according to which any one of the last integrals, identical with $\int_0^A (\omega(x)/x) \, d\rho(x)$, is just the inner product

$$\left\langle \frac{\omega(x)}{x}, \ U_A(x) \right\rangle_E \ \leqslant \ \left\| \frac{\omega(x)}{x} \right\|_E \|U_A\|_E.$$

Plugging in the bounds found above, we see that for large $A > L$,

$$\int_0^A \frac{\omega(x)}{x} \, d\left(\frac{v(x)}{x}\right) \ \leqslant \ \frac{1}{4}\left(\left\| \frac{\log \Omega(x)}{x} \right\|_E + \sqrt{\left(\left\| \frac{\log \Omega(x)}{x} \right\|_E^2 + 4c\right)}\right)^2,$$

and, making $A \longrightarrow \infty$, we arrive at the desired conclusion.
The theorem is proved.

The variant of this result referred to earlier, to be given in problem 68, applies to functions $\omega(x)$ of the form

$$\omega(x) \ = \ x \int_0^\infty \log\left|\frac{x+t}{x-t}\right| d\rho(t),$$

where the *only* assumptions on the measure ρ are that it is absolutely continuous, with $\rho'(t)$ bounded on each finite interval, and that

$$d\rho(t) \ \geqslant \ -C\frac{dt}{t} \quad \text{for } t \geqslant 1.$$

That generalization is related to some material of independent interest taken up in problems 66 and 67.

Let us, as usual, write

$$U_\rho(x) \ = \ \int_0^\infty \log\left|\frac{x+t}{x-t}\right| d\rho(t).$$

Under our assumption on ρ, the integral on the right is certainly unambiguously defined because

$$\int_0^\infty \log\left|\frac{x+t}{x-t}\right| \min(1, \ 1/t) \, dt$$

is finite for $x > 0$ and, if K is large enough,

$$d\sigma(t) = d\rho(t) + K \min(1, 1/t)\, dt$$

is $\geqslant 0$ for $t \geqslant 0$.* The preceding integral is indeed $O(x\log(1/x))$ for small values of $x > 0$, so, by applying Fubini's theorem separately to

$$\int_0^A \int_0^\infty \log\left|\frac{x+t}{x-t}\right| d\sigma(t)\, \frac{dx}{x}$$

and to the similar expression with $d\sigma(t) - d\rho(t)$ standing in place of $d\sigma(t)$, we see that for each finite $A > 0$, $\int_0^A (U_A(x)/x)\, dx$ is well defined and equal to

$$\int_0^\infty \int_0^A \log\left|\frac{x+t}{x-t}\right| \frac{dx}{x}\, d\rho(t).$$

By writing $d\rho(t)$ one more time as the difference of the two positive measures $d\sigma(t)$ and $K\min(1, 1/t)\, dt$, one verifies that the last expression is in turn equal to

$$\lim_{M\to\infty} \int_0^M \int_0^A \log\left|\frac{x+t}{x-t}\right| \frac{dx}{x}\, d\rho(t).$$

Problem 66

In this problem, we suppose that the above assumptions on the measure ρ hold, and that in addition the integrals

$$\int_0^A \frac{U_\rho(x)}{x}\, dx$$

are *bounded* as $A \longrightarrow \infty$. The object is to then obtain a preliminary grip on the magnitude of $|\rho(t)|$.

(a) Show that for each M and A.

$$\int_0^M \int_0^A \log\left|\frac{x+t}{x-t}\right| \frac{dx}{x}\, d\rho(t) = \rho(M) \int_0^A \log\left|\frac{x+M}{x-M}\right| \frac{dx}{x}$$

$$+ \int_0^M \log\left|\frac{t+A}{t-A}\right| \frac{\rho(t)}{t}\, dt.$$

(Hint: cf. proof of second lemma, beginning of this article.)

---

* Only *this* property of ρ is used in problems 66 and 67; absolute continuity of that measure plays no rôle in them (save that $\rho(t)$ should be replaced by $\rho(t) - \rho(0+)$ throughout if ρ has point mass at the origin).

(b) Hence show that

$$\frac{\rho(M)}{M} \longrightarrow 0 \quad \text{as } M \longrightarrow \infty.$$

(Hint: $\rho(t) \geqslant -K(1 + \log^+ t)$ for $t > 0$, making

$$-\int_0^M \log\left|\frac{t+A}{t-A}\right| \frac{\rho(t)}{t} dt \;\leqslant\; \text{const.} \log A$$

with a constant independent of M, for $A > e$, say. Deduce that for fixed large A and $M \longrightarrow \infty$,

$$2\frac{\rho(M)}{M} A \;\leqslant\; O(1) + \text{const.} \log A. \quad)$$

(c) Then show that

$$\int_0^A \frac{U_\rho(x)}{x} dx \;=\; \int_0^\infty \log\left|\frac{t+A}{t-A}\right| \frac{\rho(t)}{t} dt.$$

(d) Show that for large $t > 1$ we not only have $\rho(t) \geqslant -\text{const.} \log t$ but also $\rho(t) \leqslant \text{const.} \log t$. (Hint: Wlog, $d\rho(t) \geqslant -dt/t$ for $t > 1$. Assuming that for some large A we have $\rho(A) \geqslant k \log A$ with a number $k > 0$, it follows that

$$\rho(t) \;\geqslant\; k \log A - \log\frac{t}{A} \quad \text{for } t > A.$$

At the same time, $\rho(t) \geqslant -O(1) - \log^+ t$ for $0 < t < A$. Use result of (c) with these relations to get a *lower bound* on $\int_0^A (U_\rho(x)/x) dx$ involving k and $\log A$, thus arriving at an *upper bound* for k.)

Problem 67

Continuing with the material of the preceding problem, we now assume that

$$\lim_{A \to \infty} \int_0^A \frac{U_\rho(x)}{x} dx$$

exists (and is finite). It is proposed to show by means of an elementary Tauberian argument that $\rho(t)$ then *also has a limit* (equal to $2/\pi^2$ times the preceding one) for $t \longrightarrow \infty$. Essentially this result was used by Beurling and Malliavin* in their original proof of the Theorem on the

---

* under the milder condition on ρ pointed out in the preceding footnote – they in fact assumed only that the measure ρ on $[0, \infty)$ satisfies $d\rho(t) \geqslant -\text{const.} dt/t$ there, but then the conclusion of problem 67 holds just as well because the existence

Multiplier.

(a) Show that for a and $b > 0$,

$$\int_0^\infty \log\left|\frac{x+a}{x-a}\right| \log\left|\frac{x+b}{x-b}\right| dx = \pi^2 \min(a, b).$$

(Hint: We have

$$\frac{1}{\pi}\log\left|\frac{x+a}{x-a}\right| = \frac{1}{\pi}\int_{-a}^a \frac{dt}{x-t}.$$

Apply the L_2 theory of Hilbert transforms sketched at the end of §C.1, Chapter VIII.)

(b) Hence derive the formula

$$\int_0^\infty \log\left|\frac{t+x}{t-x}\right|\left\{2\log\left|\frac{x+A}{x-A}\right| - \log\left|\frac{x+(1+\delta)A}{x-(1+\delta)A}\right| - \log\left|\frac{x+(1-\delta)A}{x-(1-\delta)A}\right|\right\}dx$$

$$= \pi^2(\delta A - |t-A|)^+,$$

valid for $t > 0$, $A > 0$ and $0 < \delta < 1$.

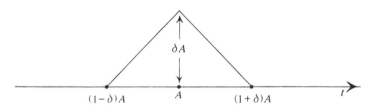

Figure 251

(c) Then, referring to part (c) of the previous problem, prove that

$$\frac{1}{A}\int_0^\infty \left(\int_0^x \frac{U_\rho(\xi)}{\xi}d\xi\right)\left\{2\log\left|\frac{x+A}{x-A}\right| - \log\left|\frac{x+(1+\delta)A}{x-(1+\delta)A}\right|\right.$$

$$\left. - \log\left|\frac{x+(1-\delta)A}{x-(1-\delta)A}\right|\right\}dx = \pi^2\delta\int_{(1-\delta)A}^{(1+\delta)A}\left(1 - \frac{|t-A|}{\delta A}\right)\frac{\rho(t)}{t}dt.$$

(Hint: Use the formula obtained in (b) together with Fubini's theorem. In justifying application of the latter, the bound on $|\rho(t)|$ found at the end of the preceding problem comes in handy. One should also observe that the expression in { } involved in the left-hand integrand belongs to $L_1(0, \infty)$ and is in fact $O(1/x^3)$ for large x.)

of $\lim_{A\to\infty}\int_0^A (U_\rho(x)/x)dx$ is not affected when ρ is replaced by its restriction to $[1, \infty)$

(d) Assume now that

$$\int_0^x \frac{U_\rho(\xi)}{\xi}\,d\xi \longrightarrow l$$

for $x \longrightarrow \infty$. Show that then the *right side* of the relation establish in (c) *tends to a limit*, equal to $(2\delta^2 + O(\delta^3))l$, as $A \longrightarrow \infty$. (Hint: The *left side* of the relation referred to can be rewritten as

$$\int_0^\infty \left(\int_0^{uA} \frac{U_\rho(\xi)}{\xi}\,d\xi \right) \varphi_\delta(u)\,du,$$

where $\varphi_\delta(u)$ is a certain L_1 function not involving A. To compute $\int_0^\infty \varphi_\delta(u)\,du$, look at $\int_0^M \varphi_\delta(u)\,du$. By making appropriate changes of variable, the last integral is thrown into the form

$$(1 + \delta) \int_{M/(1+\delta)}^M \log\left|\frac{v+1}{v-1}\right|\,dv \;-\; (1-\delta) \int_M^{M/(1-\delta)} \log\left|\frac{v+1}{v-1}\right|\,dv,$$

and this is readily evaluated for large M by expanding the integrands in powers of $1/v$.)

(e) Hence show that under the assumption in (d),

$$\rho(t) \longrightarrow \frac{2}{\pi^2}l \quad \text{for } t \longrightarrow \infty.$$

(Hint: Picking a small $\delta > 0$, assume that for some large A we have

$$\rho((1-\delta)A) > m,$$

a number $> 2l/\pi^2$. Recalling that $d\rho(t) \geq -\,dt/t$ for $t > 1$, we then get

$$\rho(t) > m - \log\left(\frac{1+\delta}{1-\delta}\right) \quad \text{for } (1-\delta)A \leq t \leq (1+\delta)A,$$

and this will contradict the result in (d) if A is large, and m much bigger than $2l/\pi^2$.

In case we have

$$\rho((1+\delta)A) < m',$$

a number $< 2l/\pi^2$ for some large A, we get

$$\rho(t) < m' + \log\left(\frac{1+\delta}{1-\delta}\right) \quad \text{for } (1-\delta)A \leq t \leq (1+\delta)A,$$

and then the same kind of argument can be made.)

Problem 68

Formulate and prove a theorem analogous to the one of this article for functions $\omega(x)$ of the form

$$\omega(x) = x \int_0^\infty \log\left|\frac{x+t}{x-t}\right| \, d\rho(t),$$

where ρ is a measure subject to the assumption* stated above. (Hint: Work directly in terms of $\rho(t)$ and $d\rho(t)$. *Boundedness* of $\rho(t)$ – needed to apply the second lemma of this article – is guaranteed by the last problem.)

4.   **Example. The finite energy condition not necessary**

The construction starts out as in article 1; again we take

$$x_p = \exp(p^{1/3}) \quad \text{for } p = 8, 9, 10, \ldots.$$

and put

$$\Delta_8 = x_8$$
$$\Delta_p = x_p - x_{p-1} \quad \text{for } p \geq 9.$$

We also use a sequence of strictly positive numbers $\lambda_p < 1$ tending monotonically to 1, but at so slow a rate that

$$\sum_8^\infty \frac{(1-\lambda_p)^2}{p} = \infty.$$

It will turn out to be convenient to specify the λ_p explicitly near the end of this article.

Based on these sequences, we form an increasing function $v(t)$, defined for $t \geq 0$ according to the rule

$$v(t) = \begin{cases} \lambda_8 t, & 0 \leq t < x_8, \\ x_{p-1} + \lambda_p(t - x_{p-1}) & \text{for } x_{p-1} \leq t < x_p \text{ with } p \geq 9. \end{cases}$$

This function has a jump of magnitude $(1-\lambda_p)\Delta_p$ at each of the points x_p; its behaviour is shown by the following figure:

* in its original form, including absolute continuity of ρ and boundedness of $\rho'(t)$ on finite intervals

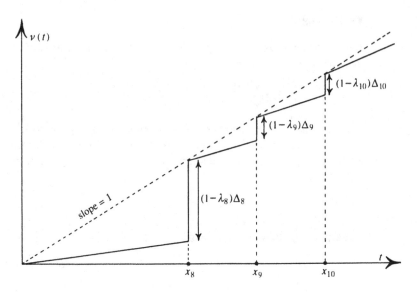

Figure 252

Obviously,

$$v(t) \leqslant t.$$

Also, since

$$\Delta_p \;\sim\; \tfrac{1}{3}p^{-2/3}x_p \;=\; \frac{x_p}{3(\log x_p)^2}$$

and

$$\frac{x_p}{x_{p-1}} \longrightarrow 1$$

as $p \longrightarrow \infty$, it is evident that

$$v(t) \;\geqslant\; t \;-\; \frac{t}{(\log t)^2} \qquad \text{for large } t.$$

Thence, putting

$$F_1(z) \;=\; \int_0^\infty \log\left|1 \;-\; \frac{z^2}{t^2}\right| \, dv(t),$$

we see by computations just like those at the beginning of article 1 that

$$F_1(z) \;\leqslant\; F_1(\mathrm{i}|z|) \;=\; \pi|z| \;+\; \mathrm{o}(|z|)$$

for $|z|$ large and that moreover, for real values of x with sufficiently large absolute value,

$$F_1(x) \;\leqslant\; C\,\frac{|x|\log\log|x|}{(\log|x|)^2}$$

where C is a certain constant.

The right side of the last relation is an increasing function of $|x|$ when that quantity is large. Choosing, then, a large number l in a manner to be described presently, we take

$$T(x) \;=\; \begin{cases} 0, & 0 \leqslant x < l, \\[2mm] C\,\dfrac{x\log\log x}{(\log x)^2}, & x \geqslant l, \end{cases}$$

thus getting an increasing function T such that

$$\int_0^\infty \frac{T(x)}{x^2}\,dx \;<\; \infty.$$

By the above two inequalities for F_1, we then have

$$F_1(x) \;\leqslant\; T(|x|) \;+\; \text{const.} \qquad \text{for } x \in \mathbb{R}.$$

We now follow the procedure used in Chapter X, §A.1, to prove the elementary multiplier theorem of Paley and Wiener. Using a constant B to be determined shortly, write

$$\mu(t) \;=\; Bt\int_t^\infty \frac{T(\tau)}{\tau^2}\,d\tau$$

and then let

$$F_2(z) \;=\; \int_0^\infty \log\left|1 \,-\, \frac{z^2}{t^2}\right|\,d\mu(t).$$

Since $T(t)$ is increasing, we have

$$\mu'(t) \;=\; B\int_t^\infty \frac{T(\tau)}{\tau^2}\,d\tau \,-\, B\frac{T(t)}{t} \;\geqslant\; 0$$

for $t > 0$, and at the same time,

$$\frac{\mu(t)}{t} \;\longrightarrow\; 0 \quad \text{as } t \to \infty.$$

The first lemma of Chapter VIII, §B.4, may thus be applied to the right side of our formula for F_2, yielding

$$F_2(x) \;=\; -x \int_0^\infty \log\left|\frac{x+t}{x-t}\right| d\!\left(\frac{\mu(t)}{t}\right)$$

$$=\; Bx \int_0^\infty \log\left|\frac{x+t}{x-t}\right| \frac{T(t)}{t^2}\, dt \qquad \text{for } x \geqslant 0.$$

Therefore, because $T(t)$ is increasing, we have

$$F_2(x) \;\geqslant\; BT(x) \int_1^\infty \log\left|\frac{1+\tau}{1-\tau}\right| \frac{d\tau}{\tau^2}, \qquad x \geqslant 0.$$

The integral on the right is just a certain strictly positive numerical quantity. We can thus pick B large enough (independently of the value of the large number l used in the specification of T) so as to ensure that

$$F_2(x) \;\geqslant\; 2T(x) \qquad \text{for } x \geqslant 0.$$

Fix such a value of B – it will be clear later on why we want the coefficient 2 on the right. Then, taking

$$F(z) \;=\; F_1(z) \,-\, F_2(z),$$

we will have

$$F(x) \;\leqslant\; -T(|x|) + \text{const.} \;\leqslant\; \text{const.}$$

for real values of x.

The function F is given by the formula

$$F(z) \;=\; \int_0^\infty \log\left|1 \,-\, \frac{z^2}{t^2}\right| d(v(t) - \mu(t)),$$

in which $v(t) \,-\, \mu(t)$ is *increasing*, provided that the parameter l entering into the definition of T is chosen properly. Because $v(t)$ and $\mu(t)$ are each increasing, with the second function absolutely continuous, this may be verified by looking at $v'(t) \,-\, \mu'(t)$. For $x_{p-1} < t < x_p$ with $p > 8$, we have

$$v'(t) \,-\, \mu'(t) \;=\; \lambda_p \,+\, B\frac{T(t)}{t} \,-\, B\int_t^\infty \frac{T(\tau)}{\tau^2}\, d\tau,$$

and an analogous relation holds in the interval $(0, \, x_8)$. *Choose, therefore, l large enough to make*

$$B\int_0^\infty \frac{T(\tau)}{\tau^2}\, d\tau \;=\; BC \int_l^\infty \frac{\log\log\tau}{(\log\tau)^2\,\tau}\, d\tau \;<\; \lambda_8.$$

Then, the sequence $\{\lambda_p\}$ being increasing, we will have $v'(t) - \mu'(t) > 0$ for $t > 0$ different from any of the points x_p, and $v(t) - \mu(t)$ will be increasing.

It is also clear that

$$\frac{v(t) - \mu(t)}{t} \longrightarrow 1 \quad \text{as } t \longrightarrow \infty.$$

Hence

$$F(z) \;\leqslant\; F(\mathrm{i}|z|) \;=\; \pi|z| + \mathrm{o}(|z|)$$

for large $|z|$. $F(x)$ is, on the other hand, *bounded above* for real x. From these two properties and the formula for $F(z)$ we can now deduce the representation of §G.1, Chapter III,

$$F(z) \;=\; \pi|\mathfrak{J}z| \;+\; \frac{1}{\pi}\int_{-\infty}^{\infty} \frac{|\mathfrak{J}z|\, F(t)}{|z - t|^2}\,\mathrm{d}t,$$

by an argument like one used in the proof of the second theorem of §B.1.

Let K be any upper bound for $F(x)$ on $\mathbb{R}$, and then, proceeding much as in article 1, put

$$W(x) \;=\; \frac{\mathrm{e}^{\pi + K}}{\exp F(x + \mathrm{i})}, \qquad x \in \mathbb{R}.$$

From the preceding relation, we get

$$\log W(x) \;=\; \frac{1}{\pi}\int_{-\infty}^{\infty} \frac{(K - F(t))}{(x - t)^2 + 1}\,\mathrm{d}t,$$

and from this we see that

$$W(x) \;\geqslant\; 1,$$

besides which

$$\left|\frac{\mathrm{d}\log W(x)}{\mathrm{d}x}\right| \;\leqslant\; \log W(x),$$

making $\log\log W(x)$ uniformly Lip 1 on $\mathbb{R}$. *The present weight W thus meets the local regularity requirement from §B.1*, quoted at the beginning of this §. Since $F(t)$ is even, so is $W(x)$, and the relation $F(t) \leqslant -T(|t|) + \text{const.}$, together with $T(t)$'s tending to ∞ for $t \longrightarrow \infty$, implies that

$$W(x) \;\longrightarrow\; \infty \qquad \text{for } x \longrightarrow \pm\infty.$$

(*That's why we chose B so as to have $F_2(x) \geqslant 2T(|x|)$ with a factor of 2.*)

It will now be shown that $W(x)$ *admits multipliers*, but that *there can be no even function* $\Omega(x) \geqslant 1$ *with*

$$\int_0^\infty \frac{\log \Omega(x)}{x^2}\,dx \;<\; \infty$$

and $\log \Omega(x)/x$ *in* $\mathfrak{H}$ *such that*

$$W(x) \;\leqslant\; \Omega(x)$$

for large values of $|x|$.

To show that W *admits multipliers*, we start from the relation

$$\log W(x) \;=\; \pi + K - F_1(x+i) + F_2(x+i)$$

and deal separately with the terms $F_1(x+i)$ and $F_2(x+i)$ standing on the right. One handles each of those by first moving down to the real axis and working with $F_1(x)$ and $F_2(x)$; afterwards, one goes back up to the line $z = x+i$.

The function F_1 is easier to take care of on account of $v(t)$'s special form. Knowing that

$$-F_1(x) \;=\; -\int_0^\infty \log\left|1 - \frac{x^2}{t^2}\right| dv(t),$$

we proceed, for given arbitrary $\eta > 0$, to build an increasing $\sigma_1(t)$ with $\sigma_1(t)/t \leqslant \eta/2$ having *jumps* that will *cancel out* most of v's, making, indeed, $\sigma_1(t) - v(t)$ a *constant multiple* of t for large values of that variable. The property that $\lambda_p \longrightarrow 1$ as $p \longrightarrow \infty$ enables us to do this.

Given the quantity $\eta > 0$, there is a number $p(\eta)$ such that

$$\lambda_p \;>\; 1 - \frac{\eta}{2} \quad \text{for } p > p(\eta).$$

We put

$$\sigma_1(t) \;=\; \begin{cases} 0, \; t < x_{p(\eta)}, \\ \dfrac{\eta}{2}x_{p-1} + \left\{\lambda_p - \left(1-\dfrac{\eta}{2}\right)\right\}(t-x_{p-1}) \quad \text{for} \\ \qquad\qquad x_{p-1} \leqslant t < x_p \text{ with } p > p(\eta). \end{cases}$$

This increasing function $\sigma_1(t)$ is related to $v(t)$ in the way shown by the following diagram:

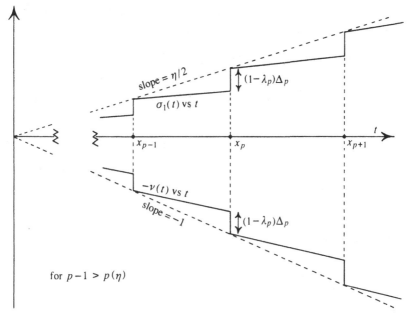

Figure 253

It is clear that

$$\sigma_1(t) \; - \; v(t) \;\; = \;\; -\left(1 \; - \; \frac{\eta}{2}\right)t \qquad \text{for } t \;\geqslant\; x_{p(\eta)}.$$

Take now

$$G_1(z) \;\; = \;\; \int_0^\infty \log\left|1 \; - \; \frac{z^2}{t^2}\right| d\sigma_1(t).$$

We have

$$\frac{\sigma_1(t)}{t} \;\; \longrightarrow \;\; \frac{\eta}{2} \quad \text{as } t \longrightarrow \infty,$$

so for large values of $|z|$,

$$G_1(z) \;\; \leqslant \;\; G_1(i|z|) \;\; = \;\; \frac{\pi\eta}{2}|z| \; + \; o(|z|).$$

The first lemma of §B.4, Chapter VIII, tells us that

$$G_1(x) \; - \; F_1(x) \;\; = \;\; \int_0^\infty \log\left|1 \; - \; \frac{x^2}{t^2}\right| d(\sigma_1(t) - v(t))$$

$$= \;\; x \int_0^\infty \log\left|\frac{x+t}{x-t}\right| d\left(\frac{v(t)-\sigma_1(t)}{t}\right), \qquad x \in \mathbb{R}.$$

As we have just seen, $(v(t) - \sigma_1(t))/t$ is *constant* for $t \geqslant x_{p(\eta)}$; the last expression on the right thus reduces to

$$x \int_0^{x_{p(\eta)}} \log\left|\frac{x+t}{x-t}\right| d\left(\frac{v(t) - \sigma_1(t)}{t}\right).$$

This, however, is clearly bounded (above *and* below!) for $|x| \geqslant 2x_{p(\eta)}$, say. Therefore

$$G_1(x) - F_1(x) \leqslant \text{const.}, \qquad |x| \geqslant 2x_{p(\eta)}.$$

This relation does not hold *everywhere* on $\mathbb{R}$; $G_1(x) - F_1(x)$ is indeed *infinite* at each of the points $\pm x_p$ with $8 \leqslant p < p(\eta)$. But at those places (corresponding to the points where $\sigma_1(t) - v(t)$ jumps *downwards*) the infinities of $G_1(z) - F_1(z)$ are *logarithmic*, and hence *harmless* as far as we are concerned. Besides becoming $-\infty$ (logarithmically again) at $\pm x_{p(\eta)}$, the function $G_1(x) - F_1(x)$ is otherwise well behaved on $\mathbb{R}$, and belongs to $L_1(-2x_{p(\eta)}, 2x_{p(\eta)})$. We can now reason once again as in the proof of the second theorem, §B.1, and deduce from the properties of $G_1(x) - F_1(x)$ just noted, and from those of $G_1(z)$ and $F_1(z)$ in the complex plane, pointed out previously, that

$$G_1(z) - F_1(z) = -\pi\left(1 - \frac{\eta}{2}\right)|\Im z| + \frac{1}{\pi}\int_{-\infty}^{\infty} \frac{|\Im z|(G_1(t) - F_1(t))}{|z - t|^2} dt.$$

Keeping in mind the behaviour of $G_1(t) - F_1(t)$ on the real axis, we see by this relation that

$$G_1(x+i) - F_1(x+i) \leqslant \text{const.}, \qquad x \in \mathbb{R}.$$

We turn to the function $F_2(x)$, equal, as we have seen, to

$$Bx \int_0^\infty \log\left|\frac{x+t}{x-t}\right| \frac{T(t)}{t^2} dt$$

for $x \in \mathbb{R}$ (both $F_2(x)$ and this expression being even). Here we proceed just as in the passage from the function $Cx(\log\log x)/(\log x)^2$ to $F_2(z)$. A change of variable shows that

$$F_2(x) = B \int_0^\infty \log\left|\frac{1+\tau}{1-\tau}\right| \frac{T(x\tau)}{\tau^2} d\tau \qquad \text{for } x \geqslant 0,$$

from which it is manifest that $F_2(x)$, like $T(x)$, is *increasing* on $[0, \infty)$. Again, by Fubini's theorem,

$$\int_0^\infty \frac{F_2(x)}{x^2} dx = B \int_0^\infty \int_0^\infty \log\left|\frac{x+t}{x-t}\right| \frac{dx}{x} \frac{T(t)}{t^2} dt =$$

$$= \frac{\pi^2 B}{2} \int_0^\infty \frac{T(t)}{t^2} dt \quad < \quad \infty.$$

Pick, then, a large number m (in a way to be described in a moment), and put

$$\Theta(t) \;=\; \begin{cases} 0, & 0 \leqslant t < m, \\ F_2(t), & t \geqslant m. \end{cases}$$

Bringing in once more the given quantity $\eta > 0$, we form the function

$$\sigma_2(t) \;=\; \frac{\eta}{2} t \;-\; Bt \int_t^\infty \frac{\Theta(\tau)}{\tau^2} d\tau$$

and observe that it is *increasing provided that m is chosen large enough* – verification of this statement is just like that of the corresponding one about $v(t) \;-\; \mu(t)$. *Fixing* once and for all such a value of m, we take

$$G_2(z) \;=\; \int_0^\infty \log\left|1 \;-\; \frac{z^2}{t^2}\right| d\sigma_2(t).$$

The function $\sigma_2(t)$, besides being increasing, has the property that

$$\frac{\sigma_2(t)}{t} \;\longrightarrow\; \frac{\eta}{2} \quad \text{as } t \longrightarrow \infty.$$

Thence,

$$G_2(z) \;\leqslant\; G_2(i|z|) \;=\; \frac{\pi\eta}{2}|z| \;+\; o(|z|)$$

for large values of $|z|$.

For $x > 0$, by the first lemma of §B.4, Chapter VIII,

$$G_2(x) \;=\; -x \int_0^\infty \log\left|\frac{x+t}{x-t}\right| d\left(\frac{\sigma_2(t)}{t}\right)$$

$$=\; -Bx \int_0^\infty \log\left|\frac{x+t}{x-t}\right| \frac{\Theta(t)}{t^2} dt.$$

Thanks to our choice of B, the last quantity is $\leqslant -2\Theta(x)$ (cf. the previous examination of $F_2(x)$'s relation to $T(x)$). Thus,

$$G_2(x) \;\leqslant\; -2F_2(x) \qquad \text{for } |x| \geqslant m,$$

and we certainly have

$$G_2(x) \;+\; F_2(x) \;\leqslant\; 0$$

for such real x, $F_2(x)$ being clearly positive.

Since $\mu(t)/t \longrightarrow 0$ for $t \longrightarrow \infty$, with $\mu(t)$ increasing, we must have

$$F_2(z) \;\leqslant\; F_2(i|z|) \;=\; o(|z|)$$

for large $|z|$. Using this estimate and the corresponding one on $G_2(z)$ given above we deduce from the behaviour of $G_2(x) + F_2(x)$ on $\mathbb{R}$ just found that

$$G_2(z) + F_2(z) \;=\; \frac{\pi\eta}{2}|\mathfrak{J}z| \;+\; \frac{1}{\pi}\int_{-\infty}^{\infty} \frac{|\mathfrak{J}z|(G_2(t)+F_2(t))}{|z-t|^2}\,dt\,;$$

this is done by the procedure followed twice already. It then follows from this relation and from the fact that $G_2(t) + F_2(t) \leqslant 0$ for $|t| \geqslant m$ that

$$G_2(x+i) + F_2(x+i) \;\leqslant\; \text{const.}, \qquad x \in \mathbb{R}.$$

Going back to $\log W(x)$ we find, recalling the above formula for it and using the two results now obtained, that

$$\log W(x) + G(x+i) \;\leqslant\; \text{const.}, \qquad x \in \mathbb{R},$$

where

$$G(z) \;=\; G_1(z) + G_2(z) \;=\; \int_0^{\infty} \log\left|1 - \frac{z^2}{t^2}\right| d(\sigma_1(t) + \sigma_2(t)).$$

Consider now the entire function $\varphi(z)$ given by the formula

$$\log|\varphi(z)| \;=\; \int_0^{\infty} \log\left|1 - \frac{z^2}{t^2}\right| d[\sigma_1(t) + \sigma_2(t)].$$

Because

$$\frac{[\sigma_1(t) + \sigma_2(t)]}{t} \;=\; \frac{\sigma_1(t)}{t} + \frac{\sigma_2(t)}{t} + o(1) \;\longrightarrow\; \eta$$

for $t \longrightarrow \infty$, we have

$$\log|\varphi(z)| \;\leqslant\; \log|\varphi(i|z|)| \;=\; \pi\eta|z| + o(|z|)$$

for z of large modulus, making $\varphi(z)$ *of exponential type* $\pi\eta$. The lemma from Chapter X, §A.1, now yields

$$\log|\varphi(x+i)| \;\leqslant\; G(x+i) + \log^+|x|, \qquad x \in \mathbb{R},$$

which, with the previous relation, gives

$$W(x)|\varphi(x+i)| \;\leqslant\; \text{const.}\sqrt{(x^2+1)}, \qquad x \in \mathbb{R}.$$

However, $\sigma_1(t) + \sigma_2(t) \longrightarrow \infty$ for $t \longrightarrow \infty$, so $\varphi(z)$ certainly *has* zeros.

Dividing out any one of them then yields a new entire function, $\psi(z)$, *also of exponential type* $\pi\eta$, with

$$W(x)|\psi(x+i)| \leqslant \text{const.}, \qquad x \in \mathbb{R}.$$

The number $\eta > 0$ was, however, *arbitrary.* Our weight $W(x)$ therefore *admits multipliers,* as claimed.

To see that there is *no* even function $\Omega(x) \geqslant 1$ with $\log \Omega(x)/x$ in $\mathfrak{H}$ and

$$\int_0^\infty \frac{\log \Omega(x)}{x^2} \, dx \quad < \quad \infty$$

such that

$$W(x) \leqslant \Omega(x)$$

for large $|x|$, we use the theorem from the last article. Because $W(0) \geqslant 1$, the relation just written would make

$$\frac{W(x)}{W(0)} \leqslant \Omega(x) \qquad \text{for } |x| \text{ large,}$$

so, since $W(x) \longrightarrow \infty$ for $x \longrightarrow \pm\infty$ as we have noted, the theorem referred to is applicable *provided that*

$$\log\left(\frac{W(x)}{W(0)}\right) \;=\; -\int_0^\infty \log\left|1 - \frac{x^2}{t^2}\right| d\rho(t),$$

where $\rho(t)$ is an *increasing, infinitely differentiable* odd function *defined on* $\mathbb{R}$, with $\rho(t)/t$ *bounded* for $t > 0$.

In our present circumstances,

$$\log\left(\frac{W(x)}{W(0)}\right) \;=\; F(i) - F(x+i)$$

where, as already pointed out,

$$F(z) \;=\; \int_0^\infty \log\left|1 - \frac{z^2}{t^2}\right| d(v(t) - \mu(t))$$

with $v(t) - \mu(t)$ increasing and $O(t)$ on $[0, \infty)$. Taking note of the identity

$$\left|1 - \frac{z+i}{t}\right| \;=\; \left|1 - \frac{z}{t-i}\right|\left|1 - \frac{i}{t}\right|, \qquad t \in \mathbb{R},$$

we see that

$$F(z+\mathrm{i}) \; - \; F(\mathrm{i}) \;\; = \;\; \int_0^\infty \log\left|\left(1 \, - \, \frac{z}{t-\mathrm{i}}\right)\!\left(1 \, + \, \frac{z}{t+\mathrm{i}}\right)\right| \mathrm{d}(v(t) - \mu(t)).$$

Now for any particular z, $\Im z \geqslant 0$, the function of w equal to $\log|1 + (z/w)|$ is *harmonic* for $\Im w > 0$. We can thence conclude, just as in proving the first lemma of §C.5, Chapter VIII, that the right-hand integral in the preceding relation is equal to

$$\int_0^\infty \log\left|1 \, - \, \frac{z^2}{t^2}\right| \mathrm{d}\rho(t) \qquad \text{for } \Im z \geqslant 0,$$

with an absolutely continuous increasing function $\rho(t)$ defined on $\mathbb{R}$, having there the derivative

$$\frac{\mathrm{d}\rho(t)}{\mathrm{d}t} \;\; = \;\; \frac{1}{\pi}\int_0^\infty \left(\frac{1}{(t-\tau)^2+1} \, + \, \frac{1}{(t+\tau)^2+1}\right)\mathrm{d}(v(\tau) - \mu(\tau)).$$

Infinite differentiability of $\rho(t)$ is manifest from the last formula. Taking

$$\rho(0) \;\; = \;\; 0$$

(which makes $\rho(t)$ *odd*), we can also verify boundedness of $\rho(t)/t$ in $(0, \infty)$ without much difficulty. One way is to simply refer to the *second* lemma of §C.5, Chapter VIII, using the formula

$$F(z+\mathrm{i}) \; - \; F(\mathrm{i}) \;\; = \;\; \int_0^\infty \log\left|1 \, - \, \frac{z^2}{t^2}\right| \mathrm{d}\rho(t), \qquad \Im z \geqslant 0,$$

just established together with the fact noted above that $F(z) \leqslant \pi|z| + \mathrm{o}(|z|)$ for large $|z|$. We see in this way that the *hypothesis of the theorem from the preceding article is fulfilled* for the function

$$\omega(x) \;\; = \;\; \log\!\left(\frac{W(x)}{W(0)}\right) \;\; = \;\; F(\mathrm{i}) \, - \, F(x+\mathrm{i}).$$

According to that theorem, *if an Ω having the properties described above did exist, we would have*

$$\int_0^\infty \frac{\omega(x)}{x^2} \, \mathrm{d}\rho(x) \;\; < \;\; \infty,$$

or, what comes to the same thing,

$$\int_1^\infty \frac{F(x+\mathrm{i})}{x^2} \, \mathrm{d}\rho(x) \;\; > \;\; -\infty,$$

$\rho(x)$ being increasing and $O(x)$. It thus *suffices to prove that*

$$\int_1^\infty \frac{F(x+\mathrm{i})}{x^2}\,\mathrm{d}\rho(x) \;=\; -\infty$$

in order to show that no such function Ω *can exist.*

For this purpose, we first obtain an *upper* bound on $F(x+\mathrm{i})$ for x near one of the points x_p, arguing somewhat as in article 1. Given $x > 0$ and $0 < r \leqslant x$, denote by $N(r,\ x+\mathrm{i})$ the quantity

$$\int_J \mathrm{d}(v(t) - \mu(t)),$$

where J is the *intersection* of the *disk* of radius r about $x+\mathrm{i}$ with the *real axis*:

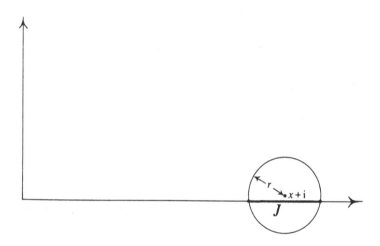

Figure 254

Keeping in mind the relation

$$F(z) \;=\; \int_0^\infty \log\left|1 \;-\; \frac{z^2}{t^2}\right| \mathrm{d}(v(t) - \mu(t)),$$

we see then, by an evident adaptation of Jensen's formula (cf. near the beginning of the proof of the *first* theorem, §B.3), that

$$F(x+\mathrm{i}) \;\leqslant\; \frac{1}{2\pi}\int_{-\pi}^{\pi} F(x+\mathrm{i}+R\mathrm{e}^{\mathrm{i}\vartheta})\,\mathrm{d}\vartheta \;-\; \int_0^R \frac{N(r,\ x+\mathrm{i})}{r}\,\mathrm{d}r$$

as long as $R \leqslant x$. (Some such restriction on R is *necessary* in order to ensure that the disk of radius R about $x + i$ *not intersect* with the *negative real axis*.) We use this formula for

$$x_p - 1 \leqslant x \leqslant x_p + 1$$

with p large, remembering that the increasing function $v(t) - \mu(t)$ *jumps upwards by* $(1 - \lambda_p)\Delta_p$ *units at* $t = x_p$. That makes

$$N(r, \ x + i) \geqslant (1 - \lambda_p)\Delta_p$$

for such x as soon as r exceeds $\sqrt{2}$. Since $\Delta_p/x_p \longrightarrow 0$ for $p \longrightarrow \infty$ we may, for large p, take $R = \sqrt{2}\,\Delta_p$ in the formula, which, in view of the relation just written, then yields

$$F(x + i) \ \leqslant \ \frac{1}{2\pi}\int_{-\pi}^{\pi} F(x + i + \sqrt{2}\,\Delta_p e^{i\vartheta})\,\mathrm{d}\vartheta \ - \ (1 - \lambda_p)\Delta_p\log\Delta_p$$

for $x_p - 1 \leqslant x \leqslant x_p + 1$.

From the relations $F(x) \leqslant$ const., $x \in \mathbb{R}$, and $F(z) \leqslant \pi|z| + o(|z|)$ it now follows by the third Phragmén–Lindelöf theorem of §C, Chapter III, that

$$F(x + i + \sqrt{2}\,\Delta_p e^{i\vartheta}) \ \leqslant \ \text{const.} + \sqrt{2}\,\pi\Delta_p|\sin\vartheta|.$$

Plugging this into the preceding inequality, we find that

$$F(x + i) \ \leqslant \ \text{const.} + 2\sqrt{2}\,\Delta_p - (1 - \lambda_p)\Delta_p\log\Delta_p$$

for $x_p - 1 \leqslant x \leqslant x_p + 1$, p being large.

We need also to see how much $\rho(t)$ increases on the interval $[x_p - 1, \ x_p + 1]$. Because $v(t) - \mu(t)$ has a jump of magnitude $(1 - \lambda_p)\Delta_p$ at x_p, we see by the above formula for $\rho'(t)$ that

$$\frac{\mathrm{d}\rho(t)}{\mathrm{d}t} \ \geqslant \ \frac{(1 - \lambda_p)\Delta_p}{\pi((t - x_p)^2 + 1)}.$$

Integrating, we get

$$\int_{x_p-1}^{x_p+1} \mathrm{d}\rho(x) \ \geqslant \ \frac{1}{2}(1 - \lambda_p)\Delta_p.$$

We can simplify our work at this point by specifying the sequence $\{\lambda_p\}$ *in precise fashion. Take, namely,*

$$\lambda_p \ = \ 1 - \frac{1}{\sqrt{(\log p)}}, \quad p \geqslant 8.$$

Then, for large p,

$$(1 - \lambda_p)\Delta_p \log \Delta_p \quad \sim \quad \frac{\exp(p^{1/3})}{3p^{1/3}(\log p)^{1/2}}$$

is *much bigger* than

$$2\sqrt{2}\,\Delta_p \quad \sim \quad \frac{2\sqrt{2}\,\exp(p^{1/3})}{3p^{2/3}},$$

and thus *the third term in our last estimate for $F(x + \mathrm{i})$ will greatly outweigh both of the first two*. When $p \geqslant$ a certain p_0, the estimate therefore reduces to

$$F(x + \mathrm{i}) \quad \leqslant \quad -\frac{1}{2}(1 - \lambda_p)\Delta_p \log \Delta_p, \qquad x_p - 1 \leqslant x \leqslant x_p + 1.$$

Use this relation together with the one just found involving ρ. That gives

$$\int_{x_p - 1}^{x_p + 1} F(x + \mathrm{i})\,\mathrm{d}\rho(x) \quad \leqslant \quad -\frac{1}{4}(1 - \lambda_p)^2 \Delta_p^2 \log \Delta_p, \qquad p \geqslant p_0.$$

Here, $F(x + \mathrm{i})$ is, as we know, bounded above for real x, and the increasing function $\rho(x)$ is $\mathrm{O}(x)$. The *divergence* of

$$\int_1^\infty \frac{F(x + \mathrm{i})}{x^2}\,\mathrm{d}\rho(x)$$

to $-\infty$ is hence *implied* by that of the *sum*

$$\sum_{p \geqslant p_0} \frac{1}{x_p^2} \int_{x_p - 1}^{x_p + 1} F(x + \mathrm{i})\,\mathrm{d}\rho(x).$$

By the preceding inequality,

$$\frac{1}{x_p^2} \int_{x_p - 1}^{x_p + 1} F(x + \mathrm{i})\,\mathrm{d}\rho(x) \quad \leqslant \quad -\frac{1}{4}(1 - \lambda_p)^2 \left(\frac{\Delta_p}{x_p}\right)^2 \log \Delta_p$$

for $p \geqslant p_0$, and the right side is

$$\sim \quad -\frac{1}{4}(1 - \lambda_p)^2 \cdot \frac{1}{9} p^{-4/3} \cdot p^{1/3} \quad = \quad -\frac{1}{36 p \log p}$$

for $p \longrightarrow \infty$. So, since $\sum_p (1/p \log p)$ is divergent, we do indeed have

$$\int_1^\infty \frac{F(x + \mathrm{i})}{x^2}\,\mathrm{d}\rho(x) \quad = \quad -\infty.$$

Therefore, *no* even function $\Omega(x) \geqslant 1$ having the properties stated above and with

$$W(x) \leqslant \Omega(x)$$

for large $|x|$ *can exist. Nevertheless, $W(x)$ admits multipliers.*

5.   **Further discussion and a conjecture**

At this point, the reader may have the impression that we have been merely raising up straw men in order to knock them down again, but that is not so. Considerable insight about the nature of the 'essential' condition we are seeking may be gained by studying the examples constructed above.

Suppose that we have an even weight $W(x) \geqslant 1$ meeting our local regularity requirement, with

$$\int_{-\infty}^{\infty} \frac{\log W(x)}{1 + x^2} \, dx \quad < \quad \infty.$$

Let us, for purposes of discussion, also assume $W(x)$ to be infinitely differentiable. In these circumstances, the odd function

$$u(x) \quad = \quad \frac{1}{x} \log\left(\frac{W(x)}{W(0)}\right)$$

has a $\mathscr{C}_\infty$ Hilbert transform*

$$\tilde{u}(x) \quad = \quad \frac{1}{\pi} \int_{-\infty}^{\infty} \frac{u(t)}{x - t} \, dt,$$

and it is frequently possible to justify the formula

$$u(x) \quad = \quad \frac{1}{\pi} \int_0^{\infty} \log\left|\frac{x + t}{x - t}\right| \tilde{u}'(t) \, dt$$

by an argument like the one made near the end of the proof of the *last* theorem in §C.4. Provided that $|\tilde{u}(t)|$ does not get very big for $t \to \infty$, further manipulation will yield

$$\log\left(\frac{W(x)}{W(0)}\right) \quad = \quad xu(x) \quad = \quad -\frac{1}{\pi} \int_0^{\infty} \log\left|1 - \frac{x^2}{t^2}\right| \, d(t\tilde{u}(t)).$$

---

* regarding the infinite differentiability of $\tilde{u}(x)$, cf. initial footnote to the *third* lemma of §E.1 below

In this article, let us not worry further about the restrictions on W needed in order to justify these transformations; what we *have* is a representation of the form

$$\log W(x) \;=\; \log W(0) \;-\; \int_0^\infty \log\left|1 \,-\, \frac{x^2}{t^2}\right| \mathrm{d}\lambda(t)$$

for a fairly general collection of weights W, involving *signed* (and very smooth) measures λ on $[0, \infty)$. How is it for the admittance of multipliers by such weights?

We can see already from the work of §C.2 that *negative measures λ are 'good'* insofar as this question is concerned. In the case of a weight with convergent logarithmic integral given by such a measure λ, one readily shows with help of the argument in §H.2, Chapter III, that the *increasing* function

$$-\lambda(t) \;=\; -\int_0^t \mathrm{d}\lambda(\tau)$$

must be $O(t)$ on $[0, \infty)$. The proof of the Theorem on the Multiplier in §C.2 may then be taken over, essentially without change, to conclude that $W(x)$ admits multipliers.*

From this point of view, *positive measures λ are 'bad'*; the example in article 1 shows that weights with convergent logarithmic integrals given by *positive λ's need not* admit multipliers.

How bad is bad? The example in article 4 *does*, after all, furnish a weight admitting multipliers and given by a positive measure λ. The first thing to be observed is that absolutely continuous λ's with $\lambda'(t)$ *bounded above* on $[0, \infty)$ are *just as good* as the *negative* ones. For, since

$$\int_0^\infty \log\left|1 \,-\, \frac{x^2}{t^2}\right| \mathrm{d}t \;=\; 0, \qquad x \in \mathbb{R},$$

we have, for any weight $W(x)$ given by such a λ with $\lambda'(t) \leqslant K$, say,

$$\log W(x) \;=\; \log W(0) \;-\; \int_0^\infty \log\left|1 \,-\, \frac{x^2}{t^2}\right| \mathrm{d}(\lambda(t) - Kt),$$

showing that W is also given by the *negative measure* ρ with $\mathrm{d}\rho(t) = \mathrm{d}\lambda(t) - K\,\mathrm{d}t$. Things can hence go wrong only for measures λ with $\lambda'(t)$ *very large* in certain places. It is therefore reasonable, when trying to find out *how* the *positive part* of a *signed measure λ* can bring about *failure of the weight given by it to admit multipliers*, to *slough off*

---

* See also the footnote on p. 556.

from λ *its portions having densities bounded above by ever larger constants,* and then *look each time at what is left.* That amounts to examining the behaviour of

$$\max(\lambda'(t), \ K) \ - \ K$$

on $[0, \infty)$ for larger and larger values of K.

The weights constructed in articles 1 and 4 (one *admitting* multipliers and the other *not*) are given by positive measures λ so similar in behaviour that something should be learned by treating those measures in the way described. It is better to first look at the measure giving the weight *of article 4.*

For that weight W we had

$$\log W(x) \ = \ \log W(0) \ - \ \int_0^\infty \log\left|1 \ - \ \frac{x^2}{t^2}\right| d\rho(t)$$

with the absolutely continuous (indeed, $\mathscr{C}_\infty$) positive measure ρ furnished by the formula

$$\frac{d\rho(t)}{dt} \ = \ \frac{1}{\pi} \int_0^\infty \left(\frac{1}{(t-\tau)^2 \ + \ 1} \ + \ \frac{1}{(t+\tau)^2 + 1}\right) d(v(\tau) - \mu(\tau)).$$

Here, $v(\tau)$ and $\mu(\tau)$, as well as the difference $v(\tau) \ - \ \mu(\tau)$ figuring in the integral, are increasing functions. The function $\mu(t)$, equal, in the notation of the last article, to

$$Bt \int_t^\infty \frac{T(\tau)}{\tau^2} \, d\tau,$$

is *absolutely continuous,* with *bounded derivative,* and the behaviour of $v(t)$ is shown by the figure at the beginning of article 4. The latter consists of an *absolutely continuous part,* again with *bounded derivative,* together with a *singular* part having *jumps of magnitude* $(1 - \lambda_p)\Delta_p$ *at the points* x_p, $p \geqslant 8$. The difference $v(t) \ - \ \mu(t)$ has therefore the *same description,* and, since $(1 - \lambda_p)\Delta_p$ and $x_p - x_{p-1}$ both tend to ∞ with p in our example, the function

$$\rho(t) \ = \ \int_0^t \rho'(\tau) \, d\tau,$$

really nothing but a regularized version of that difference, shows almost the same behaviour as the latter for large t, except for being somewhat smoother.

Thus, when K is big, a good representation of the graph of the residual function

$$\rho_K(t) \;=\; \int_0^t \bigl(\max(\rho'(\tau),\, K) \;-\; K\bigr)\,d\tau$$

will, for large values of t, be provided by one simply showing the *jumps* of $v(t)$ that go to make up the singular part of $v(t) \;-\; \mu(t)$.

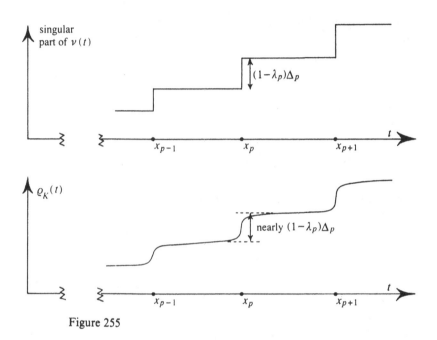

Figure 255

We turn to the weight W considered in *article 1*. In the notation of that article, it is given by the formula

$$W(x) \;=\; \frac{\text{const.}}{|F(x+i)\varphi(x+i)|},$$

where $F(z)$ and $\varphi(z)$ are certain even entire functions, of exponential type π and η respectively, having only real zeros. For the first of these, we had

$$\log|F(z)| \;=\; \int_0^\infty \log\left|1 \;-\; \frac{z^2}{t^2}\right|\,dn(t)$$

with a function $n(t)$, increasing by a *jump* of magnitude $[\Delta_p]$ at each of the points x_p, $p \geqslant 8$, and constant on the intervals separating those points (as well as on $[0, x_8)$). The function $\varphi(z)$, obtained from §A.1 of

Chapter X, has the representation

$$\log|\varphi(z)| = \int_0^\infty \log\left|1 - \frac{z^2}{t^2}\right| d[s(t)],$$

with

$$s(t) = \left(\frac{\eta}{\pi}t - \mu_1(t) - 1\right)^+$$

an increasing function formed from a certain $\mu_1(t)$ very much like the $\mu(t)$ appearing in the example of article 4. Thus, although $[s(t)]$ is composed exclusively of *jumps*, it is *based* on the function $(\eta/\pi)t - \mu_1(t)$ which increases *quite uniformly*, having *derivative* between 0 and η/π in value at each $t \geq 0$.

Referring to the first lemma of §C.5, Chapter VIII, we see that for the weight W of article 1,

$$\log W(x) = \log W(0) - \int_0^\infty \log\left|1 - \frac{x^2}{t^2}\right| d\sigma(t),$$

where $\sigma(t)$ is an absolutely continuous increasing function determined by the relation

$$\frac{d\sigma(t)}{dt} = \frac{1}{\pi}\int_0^\infty \left(\frac{1}{(t-\tau)^2 + 1} + \frac{1}{(t+\tau)^2 + 1}\right) d(n(\tau) + [s(\tau)]).$$

By feeding just the increasing function $[s(\tau)]$ into the integral on the right (which has the effect of smoothing out the former's jumps), one obtains an *increasing function* having a *bounded derivative* (given by the integral in question), thanks to the moderate behaviour of $(\eta/\pi)t - \mu_1(t)$ just noted. Therefore, when K is *big*, the residual function

$$\sigma_K(t) = \int_0^t (\max(\sigma'(\tau, K) - K) d\tau$$

acts, for large t, essentially like $n(t)$, which has the quite substantial jumps of height $[\Delta_p]$ at the points x_p. In this respect, the present situation is much like the one described above corresponding to the *weight from article 4*, involving the functions $\rho_K(t)$ and $v(t)$.

If now we compare the graph of $\rho_K(t)$, corresponding to the weight *admitting* multipliers, with the one for $\sigma_K(t)$, corresponding to the *weight that does not*, only *one difference* is apparent, and that is in the *relative heights of the steps*. Wishing to arrive at a quantitative notion of this difference, one soon thinks of performing the F. Riesz construction on

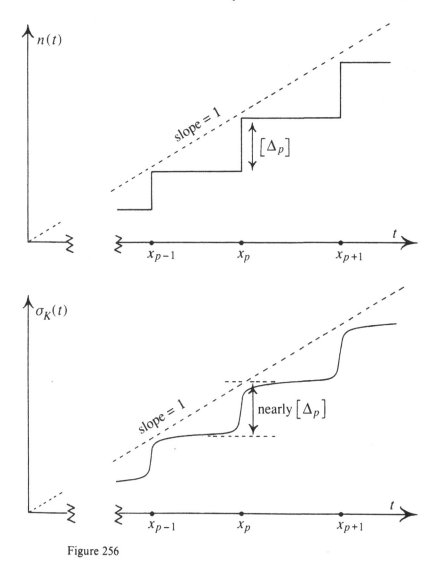

Figure 256

both graphs, letting light shine downwards on each of them from the right along a direction of small positive slope. On account of the great similarity just described between the graphs of $\sigma_K(t)$ and $n(t)$ for large t, and between those of $\rho_K(t)$ and the singular part of $v(t)$, it seems quite certain that we will (for large t) arrive at the same results by instead carrying out the Riesz construction for $n(t)$ and for the singular part of $v(t)$. This we do in order to save time, simply *assuming*, without bothering to verify the fact, that the results thus obtained really are the same as those that would be gotten (for large t), were the constructions to be made for $\sigma_K(t)$ and for $\rho_K(t)$. We are, after all, trying to *find* a theorem and not to *prove* one!

Taking, then, any small $\delta > 0$, we look at the set of *large t* with the property that

$$\frac{\text{sing. part of } v(t') - \text{sing. part of } v(t)}{t' - t} > \delta$$

for some $t' > t$ (depending, of course, on t). Here a crucial rôle is played by the fact that

$$\lambda_p \longrightarrow 1 \quad \text{as } p \longrightarrow \infty.$$

That makes $1 - \lambda_p < \delta$ for large enough p, and then the *jump* which $v(t)$ has at x_p, equal to $(1 - \lambda_p)\Delta_p$, will be $< \delta\Delta_p$, with $\Delta_p = x_p - x_{p-1}$, the distance from x_p to the *preceding* point of discontinuity for $v(t)$. Therefore the t fulfilling the last condition will, beyond a certain point, all lie in a collection of *disjoint intervals* (x'_p, x_p) with

$$x_p - x'_p = \frac{1 - \lambda_p}{\delta} \Delta_p < \Delta_p$$

and

$$\frac{\text{sing. part of } v(x_p) - \text{sing. part of } v(x'_p)}{x_p - x'_p} = \delta.$$

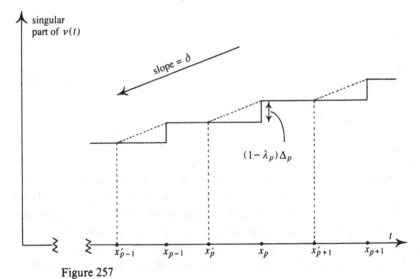

Figure 257

How *big* are the intervals (x'_p, x_p) ? In the present circumstances,

$$\Delta_p \quad \sim \quad \frac{1}{3} p^{-2/3} x_p \qquad \text{for } p \longrightarrow \infty,$$

so then

$$\frac{x_p - x'_p}{x'_p} \quad < \quad \frac{\Delta_p}{x_p - \Delta_p} \quad \sim \quad \frac{1}{3} p^{-2/3},$$

and we have

$$\sum_p \left(\frac{x_p - x'_p}{x'_p} \right)^2 \quad < \quad \infty ;$$

the intervals (x'_p, x_p) satisfy the Beurling condition that has played such an important rôle in this book!

What can we (with almost certain confidence) conclude from this about the residual functions $\rho_K(t)$? The function $\rho(t)$ is, after all, $\mathscr{C}_\infty$, so a large enough K will *swamp out the derivative $\rho'(t)$* for *all save the very large values of t*. The residual $\rho_K(t)$ will, in other words, *stay equal to zero* until t gets so large that the singular part of $\nu(t)$ shows the behaviour just described; *thereafter*, however, $\rho_K(t)$ and the latter function have *almost the same behaviour*, as we have seen. This means that *for given $\delta > 0$, we can, by making K sufficiently large, ensure that $\rho'_K(t) \leqslant \delta$ for all $t \geqslant 0$ save those belonging to a certain collection of disjoint intervals (a_n, b_n) (like the (x'_p, x_p)), with*

$$\frac{\rho_K(b_n) - \rho_K(a_n)}{b_n - a_n} = \delta$$

and

$$\sum_n \left(\frac{b_n - a_n}{a_n} \right)^2 \quad < \quad \infty.$$

Now what distinguishes the functions $\sigma_K(t)$ from the $\rho_K(t)$ is that *the analogous statement does not hold for the former when $\delta < 1$.* This is evident if we look at the graph of $n(t)$ which, for large enough t, is almost the same as that of any of the $\sigma_K(t)$. When $\delta < 1$, the Riesz construction, applied to $n(t)$, *will not even yield an infinite sequence of disjoint intervals like the (x'_p, x_p)*; instead, one simply obtains a single big interval of infinite length. That's because at each x_p, $n(t)$, instead of jumping by a *small multiple* of Δ_p, *jumps by* $[\Delta_p]$, which is, for all intents and purposes, *the same as* $\Delta_p = x_p - x_{p-1}$ when p is large.

The *size* of these jumps of $n(t)$ was, by the way, the *key property ensuring* that the weight constructed in *article 1 did not admit multipliers of exponential type* $< \pi$. Cutting the jumps down to $(1 - \lambda_p)\Delta_p$ for the construction in *article 4* was also what *made* the weight obtained there *admit multipliers*; it did so because $\lambda_p \longrightarrow 1$ as $p \longrightarrow \infty$. That, however, is just what guarantees the truth of the above statement about the $\rho_K(t)$! It thus seems likely that the distinction we have observed between behaviour of the $\rho_K(t)$ and that of the $\sigma_K(t)$ is the *source* of the corresponding two weights' difference in behaviour regarding the admittance of multipliers. *The 'essential' condition we have been seeking may well involve a requirement that the above statement hold for the ρ_K corresponding to a certain function ρ, associated with whatever weight one may have under consideration.*

Having been carried thus far by inductive reasoning, let us continue on grounds of pure speculation. We have been looking at $\mathscr{C}_\infty$ weights $W(x) \geqslant 1$ corresponding to monotone $\mathscr{C}_\infty$ functions $\lambda(t)$ according to the formula

$$\log W(x) \;\; = \;\; \log W(0) \;\; - \;\; \int_0^\infty \log\left|1 \; - \; \frac{x^2}{t^2}\right| \mathrm{d}\lambda(t).$$

Mostly, we have been considering *increasing* functions λ, and we have come around to the view that a weight W corresponding to one of these admits multipliers if (and, in *some* sense, *only if*) the above statement holds, with the functions

$$\lambda_K(t) \;\; = \;\; \int_0^t \left(\max\left(\lambda'(\tau), \; K\right) \; - \; K\right)\mathrm{d}\tau$$

standing in place of the $\rho_K(t)$. Insofar as *decreasing* functions $\lambda(t)$ were concerned, we simply observed near the beginning of this article that they were *good*, for a weight W given by any of *those admits* multipliers as long as

$$\int_{-\infty}^\infty \frac{\log W(x)}{1 + x^2}\mathrm{d}x \;\; < \;\; \infty.$$

Let us now *drop* any requirement that the function $\lambda(t)$ be monotone, but *keep* the criterion that *the above statement hold for the $\lambda_K(t)$*.

The *increase* of $\lambda(t)$ is thereby *limited*, but *not its decrease*! Observe that for any $\mathscr{C}_\infty$ function $\omega(x)$ of the form

$$\omega(x) \;\; = \;\; \omega(0) \;\; - \;\; \int_0^\infty \log\left|1 \; - \; \frac{x^2}{t^2}\right| \mathrm{d}\lambda(t)$$

with

$$\int_{-\infty}^{\infty} \frac{|\omega(x)|}{1+x^2} dx \quad < \quad \infty,$$

the *Hilbert transform* $\tilde{\omega}(x)$ is defined (everywhere) and infinitely differenti-able*, and *it differs from* $\pi\lambda(x)$ *by a constant multiple of* x. The statement involving the $\lambda_K(t)$ may thus be rephrased in terms of the Hilbert transform of $\log W(x)$, *eliminating* any direct reference to a *particular representation* for W.

Let us go one step further and guess at a criterion applicable to *any* weight $W(x)$ meeting the local regularity requirement of §B.1. Here, we *give up* trying to have the Hilbert transform of $\log W(x)$ fit the above statement. Instead, we let the latter apply to $\tilde{\omega}(x)$, where $\exp\omega(x)$ is *some even* $\mathscr{C}_\infty$ *majorant* of $W(x)$, as is in keeping with the guiding idea of this §. In that way, we arrive at the following

Conjecture. *A weight* $W(x) \geqslant 1$ *meeting the local regularity requirement admits multipliers iff it has an even* $\mathscr{C}_\infty$ *majorant* $\Omega(x)$ *with the following properties:*

(i) $\quad \displaystyle\int_{-\infty}^{\infty} \frac{\log\Omega(x)}{1+x^2} dx \quad < \quad \infty,$

(ii) *To any* $\delta > 0$ *corresponds a* K *such that the* ($\mathscr{C}_\infty$) *Hilbert transform* $\tilde{\omega}(x)$ *of* $\omega(x) = \log\Omega(x)$ *has derivative* $\leqslant K + \delta$ *at all positive* x *save those contained in a set of disjoint intervals* (a_n, b_n), *with*

$$\sum_n \left(\frac{b_n - a_n}{a_n}\right)^2 \quad < \quad \infty$$

and

$$\int_{a_n}^{b_n} \left(\max(\tilde{\omega}'(x), K) - K\right) dx \quad \leqslant \quad \delta(b_n - a_n)$$

for each n.

E.   A necessary and sufficient condition for weights meeting the local regularity requirement

The conjecture advanced at the end of the last § is true. One may look on its statement as an expression of the 'essential' condition for the admittance of multipliers that we had set out in that § to bring to light. A proof, which turns out to be not all that difficult, involves techniques

---

* cf. initial footnote to the *third* lemma of §E.1 below

like those employed in the determination of the completeness radius for a set of imaginary exponentials, carried out in Chapters IX and X. That proof is given below in article 2. Some auxiliary results are needed for it; we attend to those first.

1.   Five lemmas

Lemma. *Let $v(t)$ be increasing on $[0, \infty)$ with $v(t)/t$ bounded for $t > 0$, and put*

$$F(x) \;=\; \int_0^\infty \log\left|1 \,-\, \frac{x^2}{t^2}\right| dv(t) \qquad \text{for } x \in \mathbb{R}.$$

Suppose that $\rho(\xi)$, positive and infinitely differentiable, has compact support in $(0, \infty)$. Then the function

$$F_\rho(x) \;=\; \int_0^\infty F(\xi x)\, \frac{\rho(\xi)}{\xi}\, d\xi$$

is equal to

$$\int_0^\infty \log\left|1 \,-\, \frac{x^2}{t^2}\right| dv_\rho(t),$$

where

$$v_\rho(t) \;=\; \int_0^\infty v(\xi t)\, \frac{\rho(\xi)}{\xi}\, d\xi$$

is increasing and $\mathscr{C}_\infty$ in $(0, \infty)$, with

$$v_\rho'(t) \;\leqslant\; \text{const.} \quad \text{for } t > 0.$$

Proof. Is essentially an exercise about multiplicative convolution. Because the function $\log|1 \,-\, (x^2/t^2)|$ is neither bounded above nor below the justification of the transformations involved is a bit tricky.

Let us, as usual, write

$$F(z) \;=\; \int_0^\infty \log\left|1 \,-\, \frac{z^2}{t^2}\right| dv(t)$$

for complex z, noting that $F(x + iy)$, for fixed $x \in \mathbb{R}$, is an *increasing* function of y when $y \geqslant 0$ (because $v(t)$ increases), and that

$$F(z) \;\leqslant\; \text{const.}\,|z|$$

(because $v(t)$ is also $O(t)$ for $t \geqslant 0$). Supposing, then, that $\rho(\xi)$ has its support in $[a, b]$, $0 < a < b < \infty$, we have, for each fixed $x \in \mathbb{R}$,

$$F_\rho(x) = \int_a^b F(\xi x) \frac{\rho(\xi)}{\xi} \, d\xi = \lim_{y \to 0} \int_a^b F(\xi(x + iy)) \frac{\rho(\xi)}{\xi} \, d\xi$$

by monotone convergence*.

For $z = x + iy$ with $y \neq 0$ it is easy, thanks to the properties of v, to show by partial integration that

$$F(z) = 2\Re \int_0^\infty \frac{z^2}{z^2 - t^2} \frac{v(t)}{t} \, dt.$$

Hence

$$\int_a^b F(\xi z) \frac{\rho(\xi)}{\xi} \, d\xi = 2\Re \int_a^b \int_0^\infty \frac{\xi^2 z^2}{\xi^2 z^2 - t^2} \frac{v(t)}{t} \frac{\rho(\xi)}{\xi} \, dt \, d\xi.$$

On making the change of variable $t/\xi = \tau$, this becomes

$$2\Re \int_a^b \int_0^\infty \frac{z^2}{z^2 - \tau^2} \frac{v(\xi \tau)}{\tau} \frac{\rho(\xi)}{\xi} \, d\tau \, d\xi.$$

Here, it is legitimate to change the order of integration, for the double integral is absolutely convergent. The last expression is thus equal to

$$2\Re \int_0^\infty \frac{z^2}{z^2 - \tau^2} \frac{v_\rho(\tau)}{\tau} \, d\tau,$$

where

$$v_\rho(\tau) = \int_a^b \frac{v(\xi \tau)}{\xi} \rho(\xi) \, d\xi.$$

Since $\rho(\xi) \geqslant 0$ and $v(t)$ is increasing, $v_\rho(\tau)$ is also increasing. For $\tau > 0$, we can make the change of variable $\xi \tau = s$ in the preceding integral, getting

$$v_\rho(\tau) = \int_{a\tau}^{b\tau} \rho\left(\frac{s}{\tau}\right) \frac{v(s)}{s} \, ds.$$

Noting that $\rho(\xi)$ is $\mathscr{C}_\infty$ and vanishes (together with all its derivatives) for $\xi = a$ and $\xi = b$, we see that the expression on the right can be differentiated with respect to τ as many times as we want, making $v_\rho(\tau)$ $\mathscr{C}_\infty$ for $\tau > 0$. For the first derivative, we find

$$v'_\rho(\tau) = -\int_{a\tau}^{b\tau} \frac{s}{\tau^2} \rho'\left(\frac{s}{\tau}\right) \frac{v(s)}{s} \, ds, \qquad \tau > 0.$$

We have $v(s)/s \leqslant C$, say, for $s > 0$, so, denoting the maximum value of $|\rho'(\xi)|$ for $a \leqslant \xi \leqslant b$ by K, we get for the right side of the last relation a

---

* the integrand on the right being *bounded above* by the preceding inequality

value

$$\leqslant \ \frac{1}{\tau^2}\int_{a\tau}^{b\tau} KCs\, ds \ \leqslant \ \tfrac{1}{2}KCb^2.$$

The increasing function $v_\rho(\tau)$ thus has a *bounded derivative* in $(0, \infty)$.

We may at this point integrate

$$2\Re\int_0^\infty \frac{z^2}{z^2-\tau^2}\frac{v_\rho(\tau)}{\tau}\,d\tau$$

by parts in the direction opposite to the one taken previously, to get

$$\int_0^\infty \log\left|1\ -\ \frac{z^2}{\tau^2}\right| dv_\rho(\tau);$$

this, then, is equal to

$$\int_a^b F(\xi z)\frac{\rho(\xi)}{\xi}\,d\xi$$

by the above work. Making $y \longrightarrow 0$ now causes the first of these two integrals to tend to

$$\int_0^\infty \log\left|1\ -\ \frac{x^2}{\tau^2}\right| dv_\rho(\tau)$$

(that is especially easy to see here, where $0 \leqslant dv_\rho(\tau) \leqslant \text{const.}\, d\tau$). The integral just written is therefore equal to $F_\rho(x)$ according to what was observed initially. We are done.

Lemma. *Let $v(t)$ be odd and increasing, with $v(t)/t$ bounded on $(0, \infty)$, and put*

$$F(x) \ = \ \int_0^\infty \log\left|1\ -\ \frac{x^2}{t^2}\right| dv(t) \qquad \text{for } x\in\mathbb{R}.$$

Suppose that

$$\int_{-\infty}^\infty \frac{|F(x)|}{1+x^2}\,dx \ < \ \infty.$$

Then, for the Hilbert transform

$$\tilde{F}(x) \ = \ \frac{1}{\pi}\int_{-\infty}^\infty \left(\frac{1}{x-t}\ +\ \frac{t}{t^2+1}\right)F(t)\,dt$$

of F, we have, with a certain constant A,

$$\tilde{F}(x) \ = \ Ax\ -\ \pi v(x) \qquad \text{a.e.,} \quad x\in\mathbb{R}.$$

Proof. Write

$$F(z) = \int_0^\infty \log\left|1 - \frac{z^2}{t^2}\right| dv(t)$$

for complex z. The given conditions on v then make

$$F(z) \leqslant F(i|z|) \leqslant \text{const.}|z|.$$

Based on this relation and on the property of $|F(x)|$ assumed in the hypothesis we can, by an argument like one made during the proof of the *second* theorem in §B.1, establish for F the representation from §G.1 of Chapter III:

$$F(z) = A|\Im z| + \frac{1}{\pi}\int_{-\infty}^\infty \frac{|\Im z|\,F(t)}{|z-t|^2}\,dt.$$

A *harmonic conjugate* $G(z)$ for $F(z)$ in the upper half plane is therefore given by the formula

$$G(z) = -A\Re z + \frac{1}{\pi}\int_{-\infty}^\infty \left(\frac{\Re z - t}{|z-t|^2} + \frac{t}{t^2+1}\right)F(t)\,dt.$$

This, then, must, to within an additive constant, agree with the *obvious* harmonic conjugate for $\Im z > 0$ of the original logarithmic potential defining $F(z)$. In other words,

$$G(z) = C + \int_0^\infty \arg\left(1 - \frac{z^2}{t^2}\right) dv(t)$$

where, for $\Im z > 0$, we take the determination of the argument having $\arg 1 = 0$ in order to ensure convergence of the integral on the right.

For $\Im z > 0$ and $t > 0$, $1 - (z^2/t^2)$ lies in the following domain:

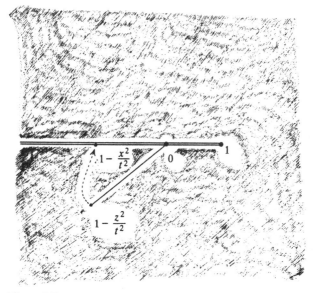

Figure 258

The branch of the argument we are using has in this domain a value between $-\pi$ and π. When z tends from the upper half plane to a given real $x > 0$, $\arg(1 - (z^2/t^2))$ tends to a *boundary value* equal to $-\pi$ for $0 < t < x$ (see the preceding figure) and to *zero* for $t > x$. As long, then, as $v(t)$ is *continuous* at such an x,

$$\int_0^\infty \arg\left(1 - \frac{z^2}{t^2}\right) dv(t)$$

will tend to $-\pi v(x)$ as $z \longrightarrow x$ from the upper half plane. At the same time, $G(z)$ will, at almost every such x, tend to

$$-Ax + \tilde{F}(x)$$

as we know (see the scholium in §H.1 of Chapter III and problem 25 at the end of §C.1, Chapter VIII). Thus,

$$-Ax + \tilde{F}(x) \;=\; C - \pi v(x) \qquad \text{a.e.,} \quad x > 0.$$

On the negative real axis we find in the same way that

$$-Ax + \tilde{F}(x) \;=\; C + \pi v(|x|) \qquad \text{a.e.;}$$

the right side is, however, equal to $C - \pi v(x)$ there, v being odd. Hence,

$$-Ax + \tilde{F}(x) \;=\; C - \pi v(x) \qquad \text{a.e.,} \quad x \in \mathbb{R}.$$

But $F(x)$ is even, making $\tilde{F}(x)$ odd, like $v(x)$. Therefore C must be zero, and

$$\tilde{F}(x) \;=\; Ax - \pi v(x) \qquad \text{a.e. on } \mathbb{R},$$

$$\text{Q.E.D.}$$

Lemma. *Let $F(x)$, even and $\mathscr{C}_2$, satisfy the condition*

$$\int_{-\infty}^\infty \frac{|F(x)|}{1 + x^2}\, dx \;<\; \infty,$$

and suppose that there is an increasing continuous odd function $\mu(x)$, $O(x)$ on $[0, \infty)$, such that $\tilde{F}(x) + \mu(x)$ is also increasing on $\mathbb{R}$, $\tilde{F}(x)$ being the Hilbert transform of F. Suppose, moreover, that $|\tilde{F}(x)/x|$ is bounded on $\mathbb{R}$. Then

$$F(x) \;=\; F(0) \;-\; \frac{1}{\pi} \int_0^\infty \log\left|1 - \frac{x^2}{t^2}\right| d\tilde{F}(t) \qquad \text{for } x \in \mathbb{R},$$

with the integral on the right (involving the signed measure $d\tilde{F}(t)$) absolutely convergent.

Remark. Boundedness of $\tilde{F}(x)/x$ on $\mathbb{R}$ actually *follows* from the rest of the hypothesis. That is the conclusion of the next lemma.

Proof of lemma. The given properties of $F(x)$ ensure* that $\tilde{F}(x)$ is at least $\mathscr{C}_1$, and that

$$f(z) = \frac{i}{\pi} \int_{-\infty}^{\infty} \left(\frac{1}{z-t} + \frac{t}{t^2+1} \right) F(t) \, dt,$$

analytic for $\mathfrak{I}z > 0$, is continuous up to the real axis, where it takes the boundary value

$$f(x) = F(x) + i\tilde{F}(x).$$

Since $F(x)$ is even, $\tilde{F}(x)$ is odd. Put

$$v(t) = \tilde{F}(t) + \mu(t),$$

then $v(t)$, like $\mu(t)$, is odd and continuous and, by hypothesis, increasing and $O(t)$ on $[0, \infty)$. For $\mathfrak{I}z \neq 0$ we can thus form the function

$$V(z) = \int_0^{\infty} \log \left| 1 - \frac{z^2}{t^2} \right| dv(t);$$

it is harmonic in both the upper and the lower half planes. The same is true for

$$U(z) = \int_0^{\infty} \log \left| 1 - \frac{z^2}{t^2} \right| d\mu(t).$$

These functions have, in $\mathfrak{I}z > 0$, the harmonic conjugates

$$\tilde{V}(z) = \int_0^{\infty} \arg\left(1 - \frac{z^2}{t^2} \right) dv(t),$$

$$\tilde{U}(z) = \int_0^{\infty} \arg\left(1 - \frac{z^2}{t^2} \right) d\mu(t)$$

---

* To show that $\tilde{F}(x)$ is $\mathscr{C}_1$ for $|x| < A$, say, take an even $\mathscr{C}_\infty$ function $\varphi(t)$ equal to 1 for $|t| \leqslant A$ and to 0 for $|t| \geqslant 2A$. Then, since $F(t)$ is also even, we have, for $|x| < A$,

$$\tilde{F}(x) = (2x/\pi) \int_A^{\infty} ((1 - \varphi(t))F(t)/(x^2 - t^2)) \, dt + (1/\pi) \int_{-2A}^{2A} (\varphi(t)F(t)/(x-t)) \, dt.$$

The *first* expression on the right is clearly $\mathscr{C}_\infty$ in x for $|x| < A$. To the *second*, we apply the partial integration technique used often in this book, and get $(1/\pi) \int_{-2A}^{2A} \log|x - t| \, d(\varphi(t)F(t))$. Reason now as in the footnote to the theorem of §D.3. Since $d(\varphi(t)F(t))/dt$ is $\mathscr{C}_1$, the integral is also $\mathscr{C}_1$ (in x) for $|x| < A$.

(where the argument is determined so as to make arg $1 = 0$). Here, where $v(t)$ and $\mu(t)$ are continuous, we can argue as in the proof of the last lemma to show that $\tilde{V}(z)$ and $\tilde{U}(z)$ are continuous up to the real axis, where they take the boundary values

$$\tilde{V}(x) = -\pi v(x), \qquad \tilde{U}(x) = -\pi \mu(x).$$

Thus,

$$\tilde{V}(x) - \tilde{U}(x) = -\pi \tilde{F}(x).$$

Our assumptions on $\mu(x)$ are not strong enough to yield as much information about the behaviour of $U(z)$ (or of $V(z)$) at the points of $\mathbb{R}$. Consider, however, the *difference*

$$G(z) = V(z) - U(z).$$

Since $\tilde{F}(t) = v(t) - \mu(t)$, we can write

$$G(z) = \int_0^\infty \log\left|1 - \frac{z^2}{t^2}\right| d\tilde{F}(t)$$

for $\Im z \neq 0$; the integral on the right is, however, absolutely convergent *even when z is real*. To check this, take any $R > |z|$ and break up that integral into two pieces, the first over $[0, 2R]$ and the second over $[2R, \infty)$. Regarding the *first* portion, note that $d\tilde{F}(t) = \tilde{F}'(t)dt$ with $\tilde{F}'(t)$ continuous and hence *bounded* on finite intervals ($\tilde{F}(t)$ being $\mathscr{C}_1$); for the *second*, just use $|d\tilde{F}(t)| \leqslant dv(t) + d\mu(t)$. In this way we also verify without difficulty that $G(z)$ is continuous up to (and *on*) $\mathbb{R}$, and takes there the boundary value

$$G(x) = \int_0^\infty \log\left|1 - \frac{x^2}{t^2}\right| d\tilde{F}(t).$$

By this observation and the one preceding it we see that the function

$$g(z) = G(z) + i(\tilde{V}(z) - \tilde{U}(z)),$$

analytic for $\Im z > 0$, is continuous up to the real axis where it has the boundary value

$$g(x) = G(x) - i\pi \tilde{F}(x).$$

Bringing in now the function $f(z)$ described earlier, we can conclude that $\pi f(z) + g(z)$, analytic in $\Im z > 0$, is continuous up to $\mathbb{R}$ and assumes there the boundary value

$$\pi f(x) + g(x) = \pi F(x) + G(x).$$

The right side is obviously *real*, so we may use Schwarz reflection to continue $\pi f(z) + g(z)$ analytically across $\mathbb{R}$ and thus obtain an *entire function*. The latter's *real part*, $H(z)$, is hence *everywhere harmonic*, with

$$H(x) = \pi F(x) + G(x) \quad \text{on } \mathbb{R}.$$

For $\Im z \neq 0$, we have $H(z) = H(\bar{z})$, so*

$$H(z) = \int_{-\infty}^{\infty} \frac{|\Im z| F(t)}{|z-t|^2} dt + V(z) - U(z).$$

It is now claimed that $H(z)$ is a *linear function* of $\Re z$ and $\Im z$; this we verify by estimating the integrals $\int_{-\pi}^{\pi} (H(re^{i\vartheta}))^+ d\vartheta$ for certain large values of r.

By the last relation, we have

$$(H(z))^+ \leqslant \int_{-\infty}^{\infty} \frac{|\Im z| |F(t)|}{|z-t|^2} dt + (V(z))^+ + (U(z))^-$$

for $\Im z \neq 0$. Consider first the *second* term on the right. Since $v(t)$ is increasing,

$$V(z) \leqslant \int_0^{\infty} \log\left(1 + \frac{|z|^2}{t^2}\right) dv(t).$$

Here, $v(t) \leqslant \text{const.} t$ on $[0, \infty)$ by hypothesis, from which we deduce by the usual integration by parts that $V(z) \leqslant \text{const.}|z|$, and thence that

$$\int_{-\pi}^{\pi} (V(re^{i\vartheta}))^+ d\vartheta \leqslant \text{const.} r.$$

To estimate the circular means of the *third* term on the right we use the formula

$$\int_{-\pi}^{\pi} (U(re^{i\vartheta}))^- d\vartheta = \int_{-\pi}^{\pi} (U(re^{i\vartheta}))^+ d\vartheta - \int_{-\pi}^{\pi} U(re^{i\vartheta}) d\vartheta$$

together with the inequality

$$U(z) \leqslant \text{const.}|z|,$$

analogous to the one for $V(z)$ just mentioned. For this procedure, a *lower* bound on $\int_{-\pi}^{\pi} U(re^{i\vartheta}) d\vartheta$ is needed, and that quantity is indeed $\geqslant 0$, as we now show[†]. When $0 < t \leqslant r$,

---

[*] the *integral* in the next formula is just $\pi \Re f(z)$ when $\Im z > 0$

[†] one may also just refer to the subharmonicity of $U(z)$

$|1 - (re^{i\vartheta}/t)^2|^2 = (1 - r^2/t^2)^2 + 4(r/t)^2 \sin^2 \vartheta \geqslant |1 - e^{2i\vartheta}|^2$, so, since $\mu(t)$ increases,

$$U(re^{i\vartheta}) = \int_0^\infty \log\left|1 - \frac{r^2 e^{2i\vartheta}}{t^2}\right| d\mu(t)$$

$$\geqslant \mu(r)\log|1 - e^{2i\vartheta}| + \int_r^\infty \log\left|1 - \frac{r^2 e^{2i\vartheta}}{t^2}\right| d\mu(t).$$

Integration of the two right-hand terms from $-\pi$ to π now presents no difficulty (Fubini's theorem being applicable to the second one), and both of the results are zero. Thus, $\int_{-\pi}^\pi U(re^{i\vartheta}) d\vartheta \geqslant 0$ which, substituted with $(U(re^{i\vartheta}))^+ \leqslant \text{const.} r$ into the previous relation, yields

$$\int_{-\pi}^\pi (U(re^{i\vartheta}))^- d\vartheta \leqslant \text{const.} r.$$

Examination of the *first* right-hand term in the above inequality for $(H(z))^+$ remains. To estimate the circular means of that term – call it $P(z)$ – one argues as in the proof of the first theorem from §B.3, leaning heavily on the convergence of $\int_{-\infty}^\infty (|F(t)|/(1 + t^2)) dt$ (without which, it is true, $P(z)$ would be infinite!). In that way, one finds that

$$\int_{-\pi}^\pi P(r_n e^{i\vartheta}) d\vartheta \leqslant \text{const.} r_n$$

for a certain sequence of r_n tending to ∞.

Combining our three estimates, we get

$$\int_{-\pi}^\pi (H(r_n e^{i\vartheta}))^+ d\vartheta \leqslant \text{const.} r_n,$$

and from this we can deduce as in the proof just referred to that

$$H(z) = H(0) + A\Re z + B\Im z,$$

thus verifying the above claim.

For $x \in \mathbb{R}$, the last relation reduces to $\pi F(x) + G(x) = H(0) + Ax$. Here, $F(x)$ and $G(x)$ are both even, so $A = 0$, and

$$\pi F(x) = H(0) - \int_0^\infty \log\left|1 - \frac{x^2}{t^2}\right| d\tilde{F}(t).$$

The integral on the right vanishes for $x = 0$, so $H(0) = \pi F(0)$, and finally

$$F(x) = F(0) - \frac{1}{\pi}\int_0^\infty \log\left|1 - \frac{x^2}{t^2}\right| d\tilde{F}(t)$$

for $x \in \mathbb{R}$, as required.

Lemma. *Let $F(x)$ be as in the preceding lemma, and suppose that for a certain continuous and increasing odd function $\mu(x)$, $O(x)$ on $[0, \infty)$, the sum $\tilde{F}(x) + \mu(x)$ is also increasing. Then $|\tilde{F}(x)/x|$ is bounded on $\mathbb{R}$.*

Proof. Is for the most part nothing but a crude version of the argument made in §H.2 of Chapter III.

It is really only the boundedness of $|\tilde{F}(x)/x|$ for *large* $x > 0$ that requires proof. That's because our assumptions on F make $\tilde{F}(x)$ odd and $\mathscr{C}_1$, and hence $\tilde{F}(x)/x$ even, and bounded near 0.

In order to see what happens for large values of x, we resort to Kolmogorov's theorem from Chapter III, §H.1, according to which

$$\int_{|\tilde{F}(x)| > \lambda} \frac{dx}{1 + x^2} \;\leqslant\; \frac{K}{\lambda} \quad \text{for } \lambda > 0,$$

K being a certain constant depending on F. In this relation, put $\lambda = 5K \cdot 2^n$ with $n \geqslant 1$; then, since

$$\int_{2^n}^{2^{n+1}} \frac{dx}{1 + x^2} \;>\; \frac{1}{5 \cdot 2^n},$$

there must, in each open interval $(2^n, 2^{n+1})$, be a point x_n with

$$|\tilde{F}(x_n)| \;\leqslant\; 5K \cdot 2^n.$$

By hypothesis, the functions $\mu(x)$ and $\tilde{F}(x) + \mu(x)$ are increasing, so for $x_n \leqslant x \leqslant x_{n+1}$ we have

$$-5K \cdot 2^n + \mu(2^n) \;\leqslant\; \tilde{F}(x_n) + \mu(x_n) \;\leqslant\; \tilde{F}(x) + \mu(x)$$
$$\leqslant\; \tilde{F}(x_{n+1}) + \mu(x_{n+1}) \;\leqslant\; 5K \cdot 2^{n+1} + \mu(2^{n+2}),$$

whence

$$-5K \cdot 2^n - \mu(x) \;\leqslant\; \tilde{F}(x) \;\leqslant\; 5K \cdot 2^{n+1} + \mu(2^{n+2}),$$

from which

$$-5K - \frac{\mu(x)}{x} \;\leqslant\; \frac{\tilde{F}(x)}{x} \;\leqslant\; 10K + 4\frac{\mu(2^{n+2})}{2^{n+2}}$$

in view of the relation $2^n < x_n < x_{n+1} < 2^{n+2}$.

It was also given that $\mu(t) \leqslant Ct$ on $[0, \infty)$. Thence,

$$\left| \frac{\tilde{F}(x)}{x} \right| \;\leqslant\; 10K + 4C$$

for $x_n \leqslant x \leqslant x_{n+1}$, and thus finally for all $x \geqslant x_1$.

Done.

We will need, finally, a simple extension of the Jensen formula for confocal ellipses derived in §C of Chapter IX.

Lemma. *Let $F(z)$ be subharmonic in and on a simply connected closed region $\bar{\Omega}$ containing the ellipse*

$$z = \frac{1}{2}\left(Re^{i\vartheta} + \frac{e^{-i\vartheta}}{R}\right), \quad 0 \leqslant \vartheta \leqslant 2\pi,$$

in its interior, where $R > 1$. Suppose that μ is the positive measure on $\bar{\Omega}$ figuring in the Riesz representation of the superharmonic function $-F(z)$ in Ω, the interior of $\bar{\Omega}$, in other words, that

$$F(z) = \int_{\bar{\Omega}} \log|z - \zeta| \, d\mu(\zeta) + h(z) \quad \text{for } z \in \Omega,$$

where $h(z)$ is harmonic in Ω (see problem 49, §A.2). If, then, $M(r)$ denotes, for $1 < r \leqslant R$, the total mass μ has inside or on the ellipse

$$z = \frac{1}{2}\left(re^{i\vartheta} + \frac{e^{-i\vartheta}}{r}\right),$$

we have

$$\int_1^R \frac{M(r)}{r} \, dr = \frac{1}{2\pi} \int_{-\pi}^{\pi} F\left(\frac{1}{2}\left(Re^{i\vartheta} + \frac{e^{-i\vartheta}}{R}\right)\right) d\vartheta - \frac{1}{\pi} \int_{-1}^{1} \frac{F(x)}{\sqrt{(1-x^2)}} \, dx.$$

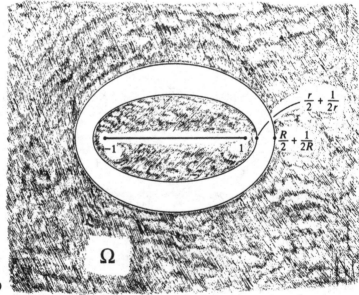

Figure 259

Proof. It is simplest to just *derive* this result from the theorem of Chapter IX, §C by double integration.

Fix, for the moment, any point $\zeta \in \bar{\Omega}$ with

$$\zeta = \frac{1}{2}\left(\rho e^{i\varphi} + \frac{e^{-i\varphi}}{\rho}\right)$$

where $\rho \geqslant 1$, and observe that (in case $\rho > 1$), we have

$$\rho = |\zeta + \sqrt{(\zeta^2 - 1)}|,$$

taking the proper determination of the square root for ζ outside the segment $[-1, 1]$.

Apply now the theorem referred to with the *analytic* function of *z equal to* $z - \zeta$ (!), getting

$$\int_{|\zeta + \sqrt{(\zeta^2 - 1)}|}^{R} \frac{dr}{r} = \int_{\rho}^{R} \frac{dr}{r} = \frac{1}{2\pi}\int_{-\pi}^{\pi} \log\left|\frac{1}{2}\left(Re^{i\vartheta} + \frac{e^{-i\vartheta}}{R}\right) - \zeta\right| d\vartheta$$

$$- \frac{1}{\pi}\int_{-1}^{1} \frac{\log|x - \zeta|}{\sqrt{(1 - x^2)}} dx,$$

the very first integral on the left being understood as zero for

$$|\zeta + \sqrt{(\zeta^2 - 1)}| \geqslant R.$$

Multiply the last relation by $d\mu(\zeta)$ and integrate over $\bar{\Omega}$. On the *left* we will get

$$\int_{\bar{\Omega}}\int_{|\zeta + \sqrt{(\zeta^2 - 1)}|}^{R} \frac{dr}{r} d\mu(\zeta) = \int_{1}^{R}\int_{1 \leqslant |\zeta + \sqrt{(\zeta^2 - 1)}| \leqslant r} d\mu(\zeta) \frac{dr}{r} = \int_{1}^{R} \frac{M(r)}{r} dr,$$

and on the *right*, after changing the order of integration in both integrals,

$$\frac{1}{2\pi}\int_{-\pi}^{\pi} \Phi\left(\frac{1}{2}\left(Re^{i\vartheta} + \frac{e^{-i\vartheta}}{R}\right)\right) d\vartheta - \frac{1}{\pi}\int_{-1}^{1} \frac{\Phi(x)}{\sqrt{(1 - x^2)}} dx,$$

where

$$\Phi(z) = \int_{\bar{\Omega}} \log|z - \zeta| d\mu(\zeta).$$

Our given subharmonic function $F(z)$ is equal to $\Phi(z) + h(z)$. Since $h(z)$ is harmonic in the *simply connected* region Ω, it has a harmonic conjugate $\tilde{h}(z)$ there, and the function

$$f(z) = \exp(h(z) + i\tilde{h}(z))$$

is *analytic and without zeros in* Ω. Apply the theorem of §C, Chapter IX, once more, this time to $f(z)$. Because $\log|f(z)| = h(z)$, we get

$$0 = \frac{1}{2\pi}\int_{-\pi}^{\pi} h\left(\frac{1}{2}\left(Re^{i\vartheta} + \frac{e^{-i\vartheta}}{R}\right)\right)d\vartheta - \frac{1}{\pi}\int_{-1}^{1}\frac{h(x)}{\sqrt{(1-x^2)}}dx.$$

Adding the right side of this relation to the previous similar expression involving Φ equal, as we have seen, to

$$\int_{1}^{R}\frac{M(r)}{r}dr$$

now gives us the desired result.

$$\text{Done.}$$

2.   Proof of the conjecture from §D.5

This book is coming to an end. Let us get on with the establishment of our conjecture which, after rephrasing, reads a bit more smoothly. One can, of course, write

$$\max(\tilde{\omega}'(x), K) - K$$

as

$$(\tilde{\omega}'(x) - K)^{+}.$$

Then the result we have in mind may be stated as follows:

Theorem. *Let* $W(x) \geqslant 1$ *be a weight meeting the local regularity requirement of* §B.1. *A necessary and sufficient condition for* W *to admit multipliers is that there exist an even* $\mathscr{C}_{\infty}$ *function* $\omega(x)$, *defined on* $\mathbb{R}$, *with*

$$\log W(x) \leqslant \omega(x)$$

there, such that

(i) $\displaystyle\int_{-\infty}^{\infty}\frac{\omega(x)}{1+x^2}dx < \infty;$

(ii) *the* $(\mathscr{C}_{\infty})$ *Hilbert transform* $\tilde{\omega}(x)$ *of* ω *has the following property:*

to any $\delta > 0$ *there corresponds a* K *with*

$$\tilde{\omega}'(x) \leqslant K$$

everywhere in $(0, \infty)$ *outside a set of disjoint intervals* (a_n, b_n), $a_n > 0$, *such that*

$$\sum_n \left(\frac{b_n - a_n}{a_n}\right)^2 \;<\; \infty,$$

for each of which

$$\int_{a_n}^{b_n} (\tilde{\omega}'(x) \;-\; K)^+ \, dx \;\leqslant\; \delta(b_n - a_n).$$

Remark. For any $\mathscr{C}_\infty$ function $\omega(x) \geqslant \log W(x)$ (and thus $\geqslant 0$) satisfying (i), we have

$$\tilde{\omega}'(x) \;=\; \frac{1}{\pi}\int_0^\infty \frac{2\omega(x) - \omega(x+t) - \omega(x-t)}{t^2}\,dt.$$

This is practically immediate, for then the functions

$$\omega(z) \;=\; \frac{1}{\pi}\int_{-\infty}^\infty \frac{\Im z\,\omega(t)}{|z-t|^2}\,dt,$$

$$\tilde{\omega}(z) \;=\; \frac{1}{\pi}\int_{-\infty}^\infty \left(\frac{\Re z - t}{|z-t|^2} + \frac{t}{t^2+1}\right)\omega(t)\,dt,$$

both harmonic in $\Im z > 0$, have there partial derivatives *continuous up to* $\mathbb{R}$ (in the present circumstances). Thence, by the Cauchy–Riemann equations,

$$\tilde{\omega}'(x) \;=\; \lim_{y\to 0+} \frac{\partial\tilde{\omega}(x+iy)}{\partial x} \;=\; -\lim_{y\to 0+} \frac{\partial\omega(x+iy)}{\partial y}$$

$$=\; \lim_{y\to 0+} \frac{\omega(x) - \omega(x+iy)}{y},$$

and the last limit is clearly equal to the integral in question.

Proof of theorem

1^0 **The necessity.** As we saw at the very beginning of §C, there is no loss of generality incurred in taking

$$W(x) \longrightarrow \infty \quad \text{for } x \longrightarrow \pm\infty\,;$$

this property we henceforth assume.

If $W(x)$ *admits multipliers* there is, corresponding to *any $a > 0$*, a non-zero entire function $f(z)$ of exponential type $\leqslant a$ with

$$W(x)|f(x)| \;\leqslant\; \text{const.} \quad \text{for } x \in \mathbb{R}.$$

If peradventure $f(0) = 0$, the quotient $f(x)/x$ satisfies the same kind of relation, for $W(x)$ must be bounded on finite intervals by the first lemma of §B.1. We can thus first divide $f(z)$ by whatever power of z is needed to get an entire function different from zero at the origin, and may therefore just as well *assume to begin with that*

$$f(0) = 1.$$

The *even* entire function $g(z) = f(z)f(-z)$ is then also 1 at the origin. It has exponential type $\leqslant 2a$, and we have

$$W(x)\,W(-x)\,|g(x)| \;\leqslant\; \text{const.,} \qquad x \in \mathbb{R}.$$

We know by the discussion following the first theorem in §B.1 that $g(z)$ may be taken to have *all its zeros real*, thus being of the form

$$g(z) \;=\; \prod_{1}^{\infty}\left(1 - \frac{z^2}{\lambda_k^2}\right),$$

where the λ_k are certain numbers > 0. The preceding relation then reads

$$\log W(x) \,+\, \log W(-x) \,+\, \sum_{1}^{\infty} \log\left|1 - \frac{x^2}{\lambda_k^2}\right| \;\leqslant\; \text{const.,} \qquad x \in \mathbb{R};$$

it is from the expressions

$$\log|g(x)| \;=\; \sum_{1}^{\infty} \log\left|1 - \frac{x^2}{\lambda_k^2}\right|$$

corresponding to entire functions g having smaller and smaller exponential type that we will construct a function $\omega(x)$ having the properties affirmed by our theorem.

Take, then, a sequence of entire functions $g_n(z)$, each of the form just indicated, such that

$$\log W(x) \,+\, \log W(-x) \,+\, \log|g_n(x)| \;\leqslant\; \Gamma_n, \qquad \text{say,}$$

for $x \in \mathbb{R}$, while the exponential type α_n of g_n tends monotonically to zero as $n \longrightarrow \infty$. By passing to a subsequence if necessary, we arrange matters so as to also have

$$\sum_{n} \alpha_n \;<\; \infty.$$

Denoting by $\mu_n(t)$ the *number of zeros of $g_n(z)$ on the segment* $[0, t]$

(which makes $\mu_n(t) = O(t)$ on $[0, \infty)$ for each n), we have

$$\log W(x) + \log W(-x) + \int_0^\infty \log\left|1 - \frac{x^2}{t^2}\right| d\mu_n(t) \leqslant \Gamma_n, \qquad x \in \mathbb{R}.$$

Here, $\log W(-x) \longrightarrow \infty$ for $x \longrightarrow \pm\infty$, so for each n there is a number $A_n > 0$ with

$$\log W(-x) \geqslant \Gamma_n \qquad \text{for } |x| \geqslant A_n.$$

It follows then from the last relation that

$$\log W(x) + \int_0^\infty \log\left|1 - \frac{x^2}{t^2}\right| d\mu_n(t) \leqslant 0 \qquad \text{for } |x| \geqslant A_n.$$

We can evidently choose the A_n successively so as to have

$$A_{n+1} > 2A_n;$$

this property will be assumed to hold from now on.

For each n, let

$$v_n(t) = \begin{cases} 0, & 0 \leqslant t < A_n/\sqrt{2}, \\ \mu_n(t), & t \geqslant A_n/\sqrt{2}; \end{cases}$$

like the $\mu_n(t)$, the $v_n(t)$ are each *increasing* and $O(t)$ on $[0, \infty)$. We then put

$$F_n(x) = -\int_0^\infty \log\left|1 - \frac{x^2}{t^2}\right| dv_n(t) \qquad (\textit{sic!}),$$

and claim in the first place that

$$F_n(x) \geqslant 0 \qquad \text{for } x \in \mathbb{R}.$$

This is true when $|x| \leqslant A_n$, for then $|1 - (x^2/t^2)| \leqslant 1$ for the values of t (all $\geqslant A_n/\sqrt{2}$) that are actually involved in the preceding integral. For $|x| > A_n$, use the evident formula

$$F_n(x) = \int_0^{A_n/\sqrt{2}} \left\{\log\left(\frac{x^2}{t^2} - 1\right) - \log\left(\frac{2x^2}{A_n^2} - 1\right)\right\} d\mu_n(t)$$

$$- \int_0^\infty \log\left|1 - \frac{x^2}{t^2}\right| d\mu_n(t).$$

The *first* integral on the right is here clearly $\geqslant 0$, and the *second* $\geqslant \log W(x) \geqslant 0$ by the above inequality. This establishes the claim, and shows besides that

$$F_n(x) \;\geqslant\; \log W(x) \qquad \text{for } |x| \geqslant A_n.$$

In order to get the function $\omega(x)$, we first smooth out each of the $F_n(x)$ by multiplicative convolution, relying on the fulfillment by $W(x)$ of the local regularity requirement*. According to the latter, there are three constants C, L and $k \geqslant 0$, the first two > 0, such that, for any $x \in \mathbb{R}$,

$$\log W(x) \;\leqslant\; C \log W(t) \;+\; k$$

for the t belonging to a certain interval of length L containing the point x.

Choose, for each n, a small number $\eta_n > 0$ less than both of the quantities

$$\frac{A_n}{2A_{n+1}} \qquad \text{and} \qquad \frac{L}{4A_{n+1}} \;;$$

it is convenient to also have the η_n tend monotonically towards 0 as $n \longrightarrow \infty$. Take then a sequence of *infinitely differentiable functions* $\rho_n(\xi) \geqslant 0$ with ρ_n supported on the interval $[1 - \eta_n, \ 1 + \eta_n]$, such that

$$\int_{1-\eta_n}^{1} \frac{\rho_n(\xi)}{\xi}\,\mathrm{d}\xi \;=\; \int_{1}^{1+\eta_n} \frac{\rho_n(\xi)}{\xi}\,\mathrm{d}\xi \;=\; 1.$$

When

$$0 \;\leqslant\; x \;\leqslant\; \frac{L}{2\eta_n},$$

the points ξx with $1 - \eta_n \leqslant \xi \leqslant 1$ are included in the segment $[x - L/2, \ x]$ and the ones with $1 \leqslant \xi \leqslant 1 + \eta_n$ in the segment $[x, \ x + L/2]$. One of those segments surely lies in the interval of length L containing x on which the preceding relation involving $\log W(t)$ does hold. By that relation and the specifications for ρ_n we thus have

$$\log W(x) \;\leqslant\; C \int_0^\infty \log W(\xi x)\,\frac{\rho_n(\xi)}{\xi}\,\mathrm{d}\xi \;+\; k \qquad \text{for } 0 \leqslant x \leqslant \frac{L}{2\eta_n}.$$

If, however, x is also $\geqslant 2A_n$, $\log W(\xi x)$ will, by the inequality found above, be $\leqslant F_n(\xi x)$ for $1 - \eta_n \leqslant \xi \leqslant 1 + \eta_n$, since then

---

* This is not really needed here. See next footnote.

$\xi x \geqslant 2A_n - 2A_{n+1}\eta_n \geqslant A_n$. The *right side* of the last relation is therefore

$$\leqslant \; C \int_0^\infty F_n(\xi x) \frac{\rho_n(\xi)}{\xi} d\xi \;\; + \;\; k$$

for such x. This expression is hence $\geqslant \log W(x)$ for $2A_n \leqslant x \leqslant 2A_{n+1}$ because $2A_{n+1} \leqslant L/2\eta_n$. Exactly the same reasoning* can be used for negative real x, and we have

$$\log W(x) \;\; \leqslant \;\; C \int_0^\infty F_n(\xi x) \frac{\rho_n(\xi)}{\xi} d\xi \;\; + \;\; k \quad \text{for } 2A_n \leqslant |x| \leqslant 2A_{n+1}.$$

Put now

$$G_n(x) \;\; = \;\; 2C \int_0^\infty F_n(\xi x) \frac{\rho_n(\xi)}{\xi} d\xi.$$

We have $G_n(0) = 0$, and

$$G_n(x) \geqslant 0, \qquad x \in \mathbb{R}.$$

Also, since $W(x) \longrightarrow \infty$ for $x \longrightarrow \pm \infty$, the preceding relation implies that

$$\log W(x) \leqslant G_n(x) \qquad \text{for } 2A_n \leqslant |x| \leqslant 2A_{n+1}$$

as long as n exceeds a certain number n_0. We can, of course, arrange to have this inequality hold for *all* n by simply *throwing away* the G_n and A_n with $n \leqslant n_0$ and re-indexing the *remaining* ones. *This we henceforth suppose done.*

By the first lemma of article 1,

$$G_n(x) \;\; = \;\; - \int_0^\infty \log \left| 1 - \frac{x^2}{t^2} \right| d\sigma_n(t)$$

with an increasing function $\sigma_n(t)$, *infinitely differentiable* in $(0, \infty)$, given by the formula

$$\sigma_n(t) \;\; = \;\; 2C \int_0^\infty v_n(\xi t) \frac{\rho_n(\xi)}{\xi} d\xi.$$

---

* As long as $\log W(x)$ is *uniformly continuous* on $[A_n, 3A_{n+1}]$, say, the argument just made will go through (with $C = 1$) for small enough η_n. The necessity proof now under way *therefore works whenever* $W(x) \geqslant 1$ *is continuous on* $\mathbb{R}$. For this, we do not even *need* the entire functions $g_n(z)$ but *only* a sequence of (not necessarily integer-valued) functions $\mu_n(t)$ increasing on $[0, \infty]$ and $O(t)$ there, with $\limsup_{t \to \infty} (\mu_n(t)/t) = \alpha_n/\pi$ going to zero as $n \longrightarrow \infty$ and $\log W(x) + \log W(-x) + \int_0^\infty \log|1 - (x^2/t^2)|d\mu_n(t)$ bounded above on $\mathbb{R}$ for each n.

In the present circumstances,

$$v_n(t) = 0 \quad \text{for } 0 \leqslant t < A_n/\sqrt{2},$$

so

$$\sigma_n(t) = 0 \quad \text{for } 0 \leqslant t \leqslant A_n/(\sqrt{2}\,(1+\eta_n)),$$

ρ_n being supported on $[1-\eta_n,\ 1+\eta_n]$. Therefore, if we extend $\sigma_n(t)$ to *the whole real axis* by making it *odd*, we get a function which is actually $\mathscr{C}_\infty$ on $\mathbb{R}$. It follows also by the lemma referred to that $\sigma'_n(t)$ is, for each n, *bounded in* $(0, \infty)$, indeed, *bounded on* $\mathbb{R}$ after $\sigma_n(t)$ is extended in the way just mentioned.

For our function $\omega(x)$ we will take the *sum of the* $G_n(x)$ and an additive constant. In order to verify that that function enjoys the properties it should, we shall need some bounds on the $G_n(x)$ and the $\sigma_n(t)$. To obtain those bounds, we must go back and look again at the entire functions $g_n(z)$ of exponential type α_n with which we started.

Because $W(x) \geqslant 1$, we certainly have

$$|g_n(x)| \leqslant e^{\Gamma_n} \quad \text{for } x \in \mathbb{R},$$

and Levinson's theorem from §H.2 of Chapter III can be applied to the g_n to yield

$$\frac{\mu_n(t)}{t} \longrightarrow \frac{\alpha_n}{\pi}, \quad t \longrightarrow \infty.$$

(One could in fact make do with a less elaborate result here.) Hence, since $v_n(t) = \mu_n(t)$ for $t \geqslant A_n/\sqrt{2}$,

$$\frac{v_n(t)}{t} \longrightarrow \frac{\alpha_n}{\pi} \quad \text{for } t \longrightarrow \infty$$

and thence, by definition of σ_n,

$$\limsup_{t \to \infty} \frac{\sigma_n(t)}{t} \leqslant \frac{4}{\pi} C(1+\eta_n)\alpha_n,$$

in view of ρ_n's vanishing outside of $[1-\eta_n,\ 1+\eta_n]$ and the condition that

$$\int_0^\infty \frac{\rho_n(\xi)}{\xi}\,d\xi = 2.$$

Let us now extend the definition of our function G_n to the whole complex plane by taking

$$G_n(z) = -\int_0^\infty \log\left|1 - \frac{z^2}{t^2}\right| d\sigma_n(t);$$

this function is harmonic both in $\{\Im z > 0\}$ and in $\{\Im z < 0\}$, and, thanks to the smoothness of $\sigma_n(t)$, *continuous* up to the real axis. The previous relation for $\sigma_n(t)/t$ makes

$$G_n(z) \geqslant -4C(1+\eta_n)\alpha_n|z| - o(|z|)$$

for large $|z|$, so, putting

$$\beta_n = \limsup_{y\to\infty}\left(\frac{-G_n(iy)}{y}\right),$$

we have

$$\beta_n \leqslant 4C(1+\eta_n)\alpha_n.$$

As we know, $G_n(x) \geqslant 0$ on the real axis. This, together with the other properties of $G_n(z)$, ensures that an analogue of the representation from §G.1 of Chapter III is valid for that function, viz.,

$$G_n(z) = -\beta_n\Im z + \frac{1}{\pi}\int_{-\infty}^\infty \frac{\Im z\, G_n(t)}{|z-t|^2}dt, \quad \Im z > 0.$$

(Cf. proof of the first lemma in §B.3. The more elaborate argument made during the proof of the second theorem in §B.1 is not needed here.)

According to the above observations about $\sigma_n(t)$, that function is certainly *zero* in a neighborhood of the origin. That, however, makes

$$G_n(z) = O(|z|^2) \quad \text{for } z \to 0,$$

whence

$$\frac{G_n(iy)}{y} \longrightarrow 0 \quad \text{as } y \to 0.$$

Referring to the preceding Poisson representation, we see from this that

$$\frac{1}{\pi}\int_{-\infty}^\infty \frac{G_n(t)}{t^2}dt = \beta_n$$

for the *positive* function $G_n(t)$.

It is now easy to get a *uniform* bound on $\sigma_n(t)/t$. Since $G_n(x) \geqslant 0$ on the real axis and $G_n(z) = G_n(\bar z)$, the above Poisson representation tells us in particular that $G_n(z) \geqslant -\beta_n|\Im z|$, in other words, that

$$\int_0^\infty \log\left|1 - \frac{z^2}{t^2}\right| d\sigma_n(t) \leqslant \beta_n|\Im z|.$$

Using this relation we can now show by a computation like the one involved in the proof of the *second* lemma from §C.5, Chapter VIII ($\sigma_n(t)$ being increasing) that

$$\frac{\sigma_n(t)}{t} \leqslant \frac{e\beta_n}{\pi} \quad \text{for } t > 0.$$

We proceed to the construction and examination of our function $\omega(x)$. By the first lemma of §B.1, $\log W(x)$ is *bounded above* for $|x| \leqslant 2A_1$; we *fix* any such bound and use it as our value for $\omega(0)$. Then we put

$$\omega(x) = \omega(0) + \sum_1^\infty G_n(x);$$

because the $G_n(x)$ are $\geqslant 0$ on $\mathbb{R}$ with $\log W(x) \leqslant G_n(x)$ for $2A_n \leqslant |x| \leqslant 2A_{n+1}$, and $\log W(x) \leqslant \omega(0)$ for $|x| \leqslant 2A_1$, we certainly have

$$\log W(x) \leqslant \omega(x), \quad x \in \mathbb{R}.$$

The function $\omega(x)$ is clearly even. Since $\beta_n \leqslant 4C(1+\eta_n)\alpha_n$ with $\eta_n \longrightarrow 0$ for $n \longrightarrow \infty$ and the sum of the α_n convergent, we have

$$\sum_1^\infty \beta_n < \infty,$$

whence, by the previous estimate of the integrals of the $G_n(t)/t^2$ and monotone convergence,

$$\int_{-\infty}^\infty \frac{\omega(x)}{1+x^2} dx < \infty.$$

Property (i) *thus holds for* ω.

To verify *infinite differentiability* for $\omega(x)$, we take

$$\sigma(t) = \sum_1^\infty \sigma_n(t)$$

and look at the increasing function $\sigma(t)$. The series on the right is surely convergent; we have indeed

$$\frac{\sigma(t)}{t} \;\leqslant\; \frac{e}{\pi}\sum_{1}^{\infty}\beta_n,$$

a finite quantity, for $t > 0$, thanks to the uniform bounds on the ratios $\sigma_n(t)/t$ found above. The function $\sigma(t)$ is actually *zero* for $0 \leqslant t \leqslant A_1/(\sqrt{2}\,(1+\eta_1))$, since all of the $\sigma_n(t)$ are (here is where we use the property that the η_n *decrease*). The summand $\sigma_n(t)$ is moreover different from zero only when $|t| \geqslant A_n/(\sqrt{2}\,(1+\eta_n))$, *a large number for large n.* Hence, since for any *given* real x, $\log|1 - (x^2/t^2)|$ is $\leqslant 0$ for $t \geqslant \sqrt{2}\,|x|$, we see – again by monotone convergence – that

$$\sum_{1}^{\infty} G_n(x) \;=\; -\int_{0}^{\infty} \log\left|1 \,-\, \frac{x^2}{t^2}\right| d\sigma(t), \qquad x \in \mathbb{R}.$$

Therefore

$$\omega(x) \;=\; \omega(0) \,-\, \int_{0}^{\infty} \log\left|1 \,-\, \frac{x^2}{t^2}\right| d\sigma(t).$$

Here, however, $\sigma(t)$ is infinitely differentiable on $\mathbb{R}$ and odd there; this is so because each individual $\sigma_n(t)$ is odd and $\mathscr{C}_\infty$ as noted above, and, on any given finite interval, *only finitely many* of the $\sigma_n(t)$ can be different from zero. We may therefore conclude that $\omega(x)$ is also $\mathscr{C}_\infty$ by invoking the result proved in the footnote to the theorem of §D.3.

Verification of property (ii) for the function $\omega(x)$ remains; that is more involved. What we have to do is look at the *size* of $\tilde{\omega}'(x)$ for $x > 0$.
   We have

$$\omega(0) \,-\, \omega(x) \;=\; \int_{0}^{\infty} \log\left|1 \,-\, \frac{x^2}{t^2}\right| d\sigma(t).$$

The ratio $\sigma(t)/t$ is bounded, and property (i) holds for ω. Therefore the *second* lemma from article 1 applies here, and tells us that

$$\tilde{\omega}(x) \;=\; \pi\sigma(x) \;=\; \pi\sum_{1}^{\infty}\sigma_n(x).$$

Thus,

$$\tilde{\omega}'(x) \;=\; \pi \sum_1^\infty \sigma_n'(x)$$

(with, on any *bounded* interval, only *finitely many terms* actually appearing in the sum on the right).

Let $\delta > 0$ be given. Corresponding to it, we have an N such that

$$e \sum_N^\infty \beta_n \;<\; \delta;$$

fixing such an N, we form the increasing function

$$s(t) \;=\; \sum_N^\infty \sigma_n(t)$$

and investigate its behaviour for $t \geqslant 0$.

That function is, in the first place, *zero* for $0 \leqslant t \leqslant A_N/(\sqrt{2}\,(1 + \eta_N))$. Also,

$$\frac{s(t)}{t} \;\leqslant\; \frac{e}{\pi} \sum_N^\infty \beta_n \;<\; \frac{\delta}{\pi} \qquad \text{for } t > 0$$

by the above uniform estimate on the ratios $\sigma_n(t)/t$. Therefore, if $x > 0$ is *sufficiently small*, the ratio

$$\frac{s(t) \;-\; s(x)}{t \;-\; x}$$

will be *zero* for $0 \leqslant t \leqslant A_N/(\sqrt{2}\,(1 + \eta_N)\,)$, and, for *larger* values of t,

$$\leqslant\; \frac{A_N}{A_N \;-\; \sqrt{2}\,(1 + \eta_N)x} \cdot \frac{s(t)}{t} \;\leqslant\; \frac{A_N}{A_N \;-\; \sqrt{2}\,(1 + \eta_N)x} \cdot \frac{e}{\pi} \sum_N^\infty \beta_n,$$

and hence $< \delta/\pi$. Doing, then, the F. Riesz construction on the graph of $s(x)$ vs. x for $x \geqslant 0$, and forming the open set

$$\mathcal{O} \;=\; \left\{ x > 0 : \frac{s(t) - s(x)}{t - x} \;>\; \frac{\delta}{\pi} \; \text{ for some } t > x \right\},$$

we see that $\mathcal{O}$ *can contain no points to the left of a certain* $a_0 > 0$.

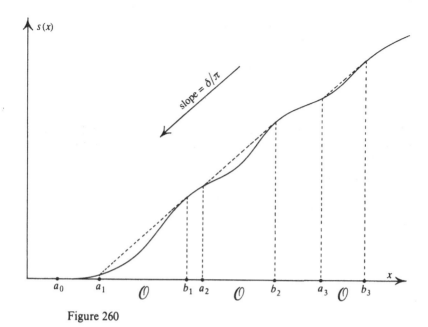

Figure 260

The set $\mathcal{O}$ is thus the union of a certain disjoint collection of intervals (a_k, b_k), $k = 1, 2, 3, \ldots$, with

$$a_k \geqslant a_0 > 0$$

for every k; these, of course, may be disposed in a much more complicated fashion than is shown in the diagram, there being no *a priori* lower limit to their lengths. *Every point $x > 0$ for which $s'(x) > \delta/\pi$ certainly belongs to $\mathcal{O}$*, so

$$s'(x) \;\leqslant\; \frac{\delta}{\pi} \quad \text{for } x \in [0, \infty) \sim \mathcal{O}.$$

For each $k \geqslant 1$,

$$\frac{s(b_k) - s(a_k)}{b_k - a_k} \;=\; \frac{\delta}{\pi}$$

as is clear from the figure.

It is now claimed that

$$\sum_1^\infty \left(\frac{b_k - a_k}{a_k} \right)^2 \;<\; \infty\,;$$

this we will show by an argument essentially the same as the one made in §D.1 of Chapter IX, using, however, the fifth lemma from the preceding article in place of the theorem of §C in Chapter IX. We work with the *subharmonic* function

$$U(z) \;=\; -\sum_N^\infty G_n(z) \;=\; \int_0^\infty \log\left|1 - \frac{z^2}{t^2}\right| \mathrm{d}s(t);$$

the two right-hand expressions are proved equal by what amounts to the reasoning used above in checking the analogous formula for $\omega(x)$ in terms of σ (monotone convergence).

Since the $G_n(x)$ are all $\geqslant 0$ on $\mathbb{R}$, we have

$$U(x) \;\geqslant\; \omega(0) - \omega(x), \qquad x \in \mathbb{R},$$

so, because ω has property (i),

$$\int_{-\infty}^\infty \frac{U(x)}{1+x^2}\,\mathrm{d}x \;>\; -\infty.$$

Writing

$$\beta \;=\; \sum_N^\infty \beta_n,$$

and recalling that $G_n(z) \geqslant -\beta_n|\Im z|$, we see, moreover, that

$$U(z) \;\leqslant\; \beta|\Im z|.$$

The convergence of the series $\sum_k((b_k - a_k)/a_k)^2$ will be deduced from the last two inequalities involving U and the fact that $\beta < \delta/e$ due to our choice of N. (It would in fact be enough if we merely had $\beta < \delta$; our having been somewhat crude in the estimation of the $\sigma_n(t)/t$ has required us to work with an extra margin of safety expressed by the factor $1/e$.)

Fixing our attention on any *particular* interval (a_k, b_k), let us denote its *midpoint* by c and its *length* by 2Δ, so as to have

$$(a_k, b_k) \;=\; (c - \Delta,\, c + \Delta).$$

The following discussion, corresponding to the one in §D.1 of Chapter IX, is actually quite simple; it may, however, at first appear complicated because of the changes of variable involved in it.

We take a certain quantity $R > 1$ (the same, in fact, for each of the intervals (a_k, b_k) – its exact *size* will be specified presently) and then, choosing a value for the parameter l,

$$\frac{2\Delta}{R + \dfrac{1}{R}} \;<\; l \;\leqslant\; \Delta,$$

apply the fifth lemma of article 1 to the subharmonic function

$$F(z) \;=\; U(lz + c).$$

Fix, for the moment, any number A large enough to ensure that

$$lA \gg \Delta,$$

and let us look at the Riesz representation for $F(z)$ (obtained by putting a minus sign in front of the one for the *superharmonic* function $-F(z)$!) in the disk $\{|z| < A\}$. In terms of the variable $w = lz + c$, we have this picture:

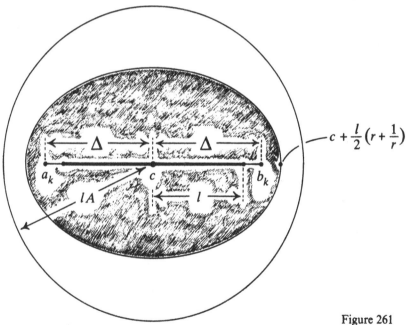

Figure 261

Because

$$U(w) \;=\; \int_0^\infty \log\left| 1 \;-\; \frac{w^2}{t^2} \right| \, ds(t),$$

we can, after making the change of variable $t = l\tau + c$, write

$$F(z) = U(lz + c) = \int_{-A}^{A} \log|z - \tau| \, ds(l\tau + c) + h(z),$$

with $h(z)$ a certain function *harmonic for* $|z| < A$. If, then, $r > 1$ and $\frac{1}{2}(r + 1/r) < A$, the *closed region* of the z-plane *bounded by the ellipse*

$$z = \frac{1}{2}\left(re^{i\vartheta} + \frac{e^{-i\vartheta}}{r}\right)$$

(*whose image in the w-plane is shown in the above figure*) *has mass* $M(r)$ *equal to*

$$\int_{-\frac{1}{2}(r + 1/r)}^{\frac{1}{2}(r + 1/r)} ds(l\tau + c) = s\left(c + \frac{l}{2}\left(r + \frac{1}{r}\right)\right) - s\left(c - \frac{l}{2}\left(r + \frac{1}{r}\right)\right)$$

assigned to it by the measure associated with the Riesz representation for $F(z)$ just given. By the fifth lemma of the last article we thus have

$$\int_{1}^{R} \frac{s(c + \frac{1}{2}l(r + r^{-1})) - s(c - \frac{1}{2}l(r + r^{-1}))}{r} \, dr = \int_{1}^{R} \frac{M(r)}{r} \, dr$$

$$= \frac{1}{2\pi}\int_{-\pi}^{\pi} U\left(c + \frac{l}{2}\left(Re^{i\vartheta} + \frac{e^{-i\vartheta}}{R}\right)\right) d\vartheta - \frac{1}{\pi}\int_{-1}^{1} \frac{U(lx + c)}{\sqrt{(1 - x^2)}} \, dx.$$

As in §D.1 of Chapter IX, it is convenient to now write

$$R = e^{\gamma}$$

(thus making γ a certain fixed quantity > 0), and to take a number $\varepsilon > 0$, considerably *smaller* than γ (corresponding to the quantity denoted by η in the passage referred to). If the parameter l is actually

$$\geqslant \frac{\Delta}{\cosh \varepsilon},$$

we will have

$$s\left(c + \frac{l}{2}\left(r + \frac{1}{r}\right)\right) - s\left(c - \frac{l}{2}\left(r + \frac{1}{r}\right)\right) \geqslant s(b_k) - s(a_k)$$

$$\text{for } r \geqslant e^{\varepsilon}$$

(see once more the preceding figure). By construction of our intervals (a_k, b_k), the quantity on the right is equal to

$$\frac{\delta}{\pi}(b_k - a_k) = \frac{2\delta}{\pi}\Delta.$$

The previous relation thus yields

$$\frac{2\delta}{\pi}\Delta\int_{e^{\varepsilon}}^{e^{\gamma}}\frac{dr}{r} \;\leqslant\; \int_{1}^{R}\frac{M(r)}{r}dr \;=\; \frac{1}{2\pi}\int_{-\pi}^{\pi} U(c \,+\, l\cosh(\gamma+i\vartheta))\,d\vartheta$$

$$-\;\frac{1}{\pi}\int_{-1}^{1}\frac{U(lx+c)}{\sqrt{(1-x^2)}}dx \qquad \text{for}\;\; \frac{\Delta}{\cosh\varepsilon} \leqslant l \leqslant \Delta.$$

Using the inequality

$$U(c \,+\, l\cosh(\gamma+i\vartheta)) \;\leqslant\; \beta l\,|\Im\cosh(\gamma+i\vartheta)| \;=\; \beta l\sinh\gamma\,|\sin\vartheta|$$

to estimate the next-to-the last integral on the right and making the change of variable $\xi = lx$ in the last one, one finds, after rearrangement, that

$$\int_{-l}^{l}\frac{U(\xi+c)}{\sqrt{(l^2-\xi^2)}}d\xi \;\leqslant\; 2\beta l\sinh\gamma \,-\, 2\delta(\gamma-\varepsilon)\Delta$$

$$\leqslant\; 2\sinh\gamma\left(\beta \,-\, \delta\frac{\gamma-\varepsilon}{\sinh\gamma}\right)\Delta, \qquad \text{for}\;\; \frac{\Delta}{\cosh\varepsilon} \leqslant l \leqslant \Delta.$$

Recall that in our present construction, we have

$$\beta \;<\; \frac{\delta}{e};$$

it is therefore certainly possible (and by far!) to choose $\gamma > 0$ *small enough* to make

$$\delta\frac{\gamma}{\sinh\gamma} \;>\; \beta$$

(*thus* is the value of $R = e^{\gamma}$ finally specified!), and then to fix an $\varepsilon > 0$ *yet smaller*, so as to *still have*

$$\delta\frac{\gamma-\varepsilon}{\sinh\gamma} \;>\; \beta.$$

These choices having been made, we write

$$a \;=\; \delta\frac{\gamma-\varepsilon}{\sinh\gamma} \,-\, \beta,$$

so that $a > 0$, and the above inequality becomes

$$\int_{-l}^{l}\frac{U(\xi+c)}{\sqrt{(l^2-\xi^2)}}d\xi \;\leqslant\; -2a\Delta\sinh\gamma, \qquad \frac{\Delta}{\cosh\varepsilon} \leqslant l \leqslant \Delta.$$

This relation is now multiplied by $l\,dl$, and both sides integrated over the range

$$\frac{\Delta}{\cosh\varepsilon} \;\leqslant\; l \;\leqslant\; \Delta.$$

In our circumstances, $U(t) \leqslant 0$ on the real axis, and the computation at this point is practically identical to the one in §D.1 of Chapter IX. It therefore suffices to merely give the result, which reads

$$\int_{a_k}^{b_k} U(t)\,dt \;=\; \int_{-\Delta}^{\Delta} U(\xi + c)\,d\xi \;\leqslant\; -a\Delta^2 \sinh\gamma\,\tanh\varepsilon$$

$$=\; -\frac{a}{4}(\sinh\gamma\,\tanh\varepsilon)(b_k - a_k)^2.$$

The last inequality, involving the quantities a, γ, and ε, now *fixed* (and > 0), holds for *any* of the intervals (a_k, b_k), $k \geqslant 1$. Since the a_k are $\geqslant a_0 > 0$ and $U(t) \leqslant 0$ on $\mathbb{R}$, we see from it that

$$\int_{a_k}^{b_k} \frac{U(t)}{t^2 + 1}\,dt \;\leqslant\; -\frac{aa_0^2}{4(a_0^2 + 1)}\sinh\gamma\,\tanh\varepsilon \left(\frac{b_k - a_k}{b_k}\right)^2.$$

Finally (using again the fact that $U(t) \leqslant 0$ on $\mathbb{R}$), we get

$$\sum_{k=1}^{\infty}\left(\frac{b_k - a_k}{b_k}\right)^2 \;\leqslant\; -\frac{4(a_0^2 + 1)}{aa_0^2 \sinh\gamma\,\tanh\varepsilon}\int_0^{\infty}\frac{U(t)}{t^2 + 1}\,dt \;<\; \infty,$$

and our claim that $\sum_k ((b_k - a_k)/a_k)^2 < \infty$ is thereby established*.

From this result, *property* (ii) *for the function* $\omega(x)$ readily follows. As we know,

$$\tilde{\omega}'(x) \;=\; \pi\sum_1^{\infty}\sigma_n'(x) \;=\; \pi\sum_1^{N-1}\sigma_n'(x) + \pi s'(x).$$

It has already been noted that *each one* of the derivatives $\sigma_n'(x)$ is *bounded* for $x > 0$; there is thus a number K such that

$$\pi\sum_1^{N-1}\sigma_n'(x) \;\leqslant\; K - \delta, \qquad x > 0.$$

The derivative $\pi s'(x)$ is, at the same time, $\leqslant \delta$ for all the $x > 0$ *outside*

$$\mathcal{O} \;=\; \bigcup_{k\geqslant 1}(a_k, b_k).$$

* for the preceding displayed relation implies in particular that the ratios b_k/a_k are *bounded* — cf. discussion, top of p. 81

Therefore

$$\tilde{\omega}'(x) \;-\; K \;\;=\;\; \pi \sum_{1}^{N-1} \sigma'_n(x) \;-\; (K-\delta) \;+\; \pi s'(x) \;-\; \delta$$

is $\leqslant 0$ for $x \in (0,\infty) \sim \mathcal{O}$.

On any of the components (a_k, b_k) of $\mathcal{O}$, we have

$$(\tilde{\omega}'(x) \;-\; K)^{+} \;\;\leqslant\;\; \left(\pi \sum_{1}^{N-1} \sigma'_n(x) \;-\; (K-\delta) \right)^{+} \;+\; (\pi s'(x) \;-\; \delta)^{+}$$

$$\leqslant \;\; 0 \;+\; \pi s'(x)$$

($s(x)$ being increasing!), so

$$\int_{a_k}^{b_k} (\tilde{\omega}'(x) \;-\; K)^{+} \, dx \;\leqslant\; \pi \int_{a_k}^{b_k} s'(x) \, dx \;=\; \pi(s(b_k) \;-\; s(a_k)) \;=\; \delta(b_k - a_k).$$

Property (ii) *therefore holds,* the sum $\sum_k ((b_k - a_k)/a_k)^2$ being convergent.

The *necessity* of our condition is thus proved.

2° The sufficiency. Suppose that there *is* an even $\mathscr{C}_\infty$ function

$$\omega(x) \;\geqslant\; \log W(x)$$

having the properties enumerated in the theorem's statement; we must show that W admits multipliers. Let, then, $\delta > 0$ be given; corresponding to it we have a K and an open subset $\mathcal{O}$ of $(0,\infty)$ with

$$\tilde{\omega}'(x) \;\leqslant\; K \quad \text{for } x \in (0,\infty) \;\sim\; \mathcal{O},$$

and, if $\mathcal{O}$ is the union of the disjoint intervals (a_k, b_k),

$$\sum_k \left(\frac{b_k - a_k}{a_k} \right)^2 \;<\; \infty,$$

while

$$\int_{a_k}^{b_k} (\tilde{\omega}'(x) \;-\; K)^{+} \, dx \;\leqslant\; \delta(b_k - a_k)$$

for each k.

We start by expressing $\tilde{\omega}(x)$ as the difference of two functions, each continuous and *increasing* on $[0,\infty)$. The given properties of $\tilde{\omega}'(x)$ make it possible for us to define a bounded measurable function $p(x)$ on $[0,\infty)$

with $0 \leqslant p(x) \leqslant \delta$ by taking

$$p(x) = \delta \quad \text{for } x \in [0, \infty) \sim \mathcal{O}$$

and then, on each of the interval components (a_k, b_k) of $\mathcal{O}$, having $p(x)$ assume the constant value needed to make

$$\int_{a_k}^{b_k} \{(\tilde{\omega}'(x) - K)^+ + p(x)\}\,dx = \delta(b_k - a_k).$$

Put now

$$\pi v_1(x) = \int_0^x \{(\tilde{\omega}'(t) - K)^+ + p(t)\}\,dt \qquad \text{for } x \geqslant 0$$

and

$$\pi v_2(x) = \int_0^x \{(K - \tilde{\omega}'(t))^+ + p(t)\}\,dt, \qquad x \geqslant 0.$$

We have $\tilde{\omega}(0) = 0$, for, since $\omega(x)$ is $\mathscr{C}_\infty$ and even, $\tilde{\omega}(x)$ is $\mathscr{C}_\infty$ and odd. Therefore, when $x \geqslant 0$,

$$\tilde{\omega}(x) - Kx = \int_0^x (\tilde{\omega}'(t) - K)\,dt$$

$$= \int_0^x \{(\tilde{\omega}'(t) - K)^+ - (\tilde{\omega}'(t) - K)^-\}\,dt = \pi v_1(x) - \pi v_2(x);$$

i.e.,

$$\pi v_1(x) + Kx - \pi v_2(x) = \tilde{\omega}(x), \qquad x \geqslant 0,$$

with $v_1(x)$ and $v_2(x)$ both *increasing* and *continuous* for $x \geqslant 0$.

The ratio $v_1(x)/x$ is *bounded* for $x > 0$. Indeed, if $x \in [0, \infty) \sim \mathcal{O}$, $\pi v_1(x) = \delta x$ by the definition of our function $p(x)$. And if $a_k < x < b_k$ for some k,

$$\frac{\pi v_1(x)}{x} < \frac{\pi v_1(b_k)}{a_k} = \delta \frac{b_k}{a_k}.$$

The ratios b_k/a_k are, however, *bounded above*, since $\sum_k ((b_k - a_k)/a_k)^2$ is convergent. Hence

$$\frac{v_1(x)}{x} \leqslant \text{const.} \qquad \text{for } x > 0.$$

The *odd* continuous increasing function $\mu(x)$ equal to $\pi v_1(x) + Kx$ for $x \geqslant 0$ is thus $O(x)$ on $[0, \infty)$, and $\tilde{\omega}(x) - \mu(x)$, *also odd* and equal, for $x \geqslant 0$, to $-\pi v_2(x)$ by the above formula, is *decreasing* on $\mathbb{R}$. The *fourth* lemma of article 1 can therefore be invoked (with $F(x) = -\omega(x)$, $\omega(x)$ being $\mathscr{C}_\infty$, $\geqslant 0$ and enjoying property (i) by hypothesis). This yields

$$|\tilde{\omega}(x)| \leqslant \text{const.}|x| \quad \text{for } x \in \mathbb{R},$$

which in turn makes

$$v_2(x) \leqslant \text{const.} x \quad \text{for } x \geqslant 0$$

in view of the preceding estimate on $v_1(x)$ and the formula just referred to.

All the conditions for application of the *third* lemma in article 1 (again with $F(x) = -\omega(x)$) are now verified. By that result we get, for $x \in \mathbb{R}$,

$$\omega(x) = \omega(0) - \frac{1}{\pi} \int_0^\infty \log\left|1 - \frac{x^2}{t^2}\right| d\tilde{\omega}(t)$$

$$= \omega(0) + \int_0^\infty \log\left|1 - \frac{x^2}{t^2}\right| dv_2(t)$$

$$- \int_0^\infty \log\left|1 - \frac{x^2}{t^2}\right| dv_1(t) - \frac{K}{\pi} \int_0^\infty \log\left|1 - \frac{x^2}{t^2}\right| dt.$$

The very last integral on the right is of course zero, so we have

$$\omega(x) + \int_0^\infty \log\left|1 - \frac{x^2}{t^2}\right| dv_1(t)$$

$$= \omega(0) + \int_0^\infty \log\left|1 - \frac{x^2}{t^2}\right| dv_2(t), \quad x \in \mathbb{R}.$$

The rest of our work here is based mainly on this formula.

Before looking more closely at the increasing function $v_1(t)$ and the expression

$$\int_0^\infty \log\left|1 - \frac{x^2}{t^2}\right| dv_1(t)$$

corresponding to it, we should attend to a detail regarding the *location* of the open set $\mathcal{O}$. We can, namely, arrange to *ensure that* $\mathcal{O} \subseteq (1, \infty)$

(which will turn out to be convenient later on) *by merely taking K large enough* to begin with. In the present circumstances, $\tilde{\omega}(x)$ (like $\omega(x)$) is $\mathscr{C}_\infty$ on $\mathbb{R}$,* so $\tilde{\omega}'(x)$ is *bounded on any finite interval*. Hence, if K is chosen large enough in the first place, we will have

$$\tilde{\omega}'(x) \leqslant K \qquad \text{for } 0 \leqslant x \leqslant 1,$$

so that *any component* (a_k, b_k) of the set $\mathcal{O}$ corresponding to this K which *lies entirely* in $(0, 1)$ may be simply *thrown away* (and $p(x) = \pi v_1'(x)$ just taken *equal* to δ thereon) without in any way affecting the properties of $v_1(x)$ and $v_2(x)$ used up to now. There may, however, *still* be a component (a_l, b_l) of $\mathcal{O}$ with $a_l < 1 < b_l$. In that event, b_l is certainly finite, and there is thus a $K' \geqslant K$ with

$$\tilde{\omega}'(x) \leqslant K' \qquad \text{for } 0 \leqslant x \leqslant b_l.$$

Then, if we *also throw away* (a_l, b_l), what *remains* of $\mathcal{O}$ will be a certain open set $\mathcal{O}' \subseteq (1, \infty)$ composed of the intervals (a_k, b_k) from $\mathcal{O}$ that *do not intersect* $[0, 1]$. We will have

$$\tilde{\omega}'(x) \leqslant K' \qquad \text{for } x \in (0, \infty) \sim \mathcal{O}',$$

and for each of the (a_k, b_k) making up $\mathcal{O}'$,

$$\int_{a_k}^{b_k} (\tilde{\omega}'(x) - K')^+ \, dx \leqslant \delta(b_k - a_k),$$

since the same relation holds with $K \leqslant K'$ standing in place of K'. *By increasing K to K', we thus ensure that none of the intervals* (a_k, b_k) *appearing in our construction intersect with* $[0, 1]$. This merely amounts to *choosing a larger value initially for the number K corresponding to our given* δ, which *we henceforth assume as having been done*. The intervals (a_k, b_k) involved in the formation of $v_1(x)$ and $v_2(x)$ are in such fashion guaranteed to *all lie in* $(1, \infty)$.

Having seen to this matter, we turn our attention to the behaviour of $\pi v_1(t)$ for $t \geqslant 0$. As we have already noted, $\pi v_1(t) = \delta t$ for $t \geqslant 0$ lying *outside* all the intervals (a_k, b_k). When $a_k < t < b_k$, we have, since $v_1(t)$ increases,

$$\delta a_k \leqslant \pi v_1(t) \leqslant \delta b_k.$$

At the same time,

$$\delta a_k < \delta t < \delta b_k,$$

* cf. initial footnote to third lemma of article 1

so

$$|\pi v_1(t) - \delta t| \leqslant \delta(b_k - a_k) \qquad \text{for } a_k < t < b_k.$$

Thence,

$$\int_{a_k}^{b_k} \frac{|\pi v_1(t) - \delta t|}{t^2} \, dt \;\leqslant\; \delta \left(\frac{b_k - a_k}{a_k} \right)^2$$

which, with the preceding observation, implies that

$$\int_0^\infty \frac{|\pi v_1(t) - \delta t|}{t^2} \, dt \;<\; \infty$$

on account of the convergence of $\sum_k ((b_k - a_k)/a_k)^2$; *it is here that we have made crucial use of that hypothesis.*

Let us, in the usual fashion, extend the increasing function $v_1(t)$ to all of $\mathbb{R}$ by making it *odd* there. Then the function

$$\Delta(t) \;=\; v_1(t) \;-\; \frac{\delta}{\pi} t$$

is *also* odd and, moreover, *zero* for $-1 < t < 1$ due to our having arranged that none of the intervals (a_k, b_k) intersect with $[0,1]$. According to what we have just seen,

$$\int_{-\infty}^\infty \frac{|\Delta(t)|}{t^2} \, dt \;<\; \infty;$$

$\Delta(t)$ *thus satisfies the hypothesis of the initial lemma in §B.2, Chapter X*, with δ/π playing the rôle of the number D figuring there. That result gives us a function $q(t)$, *zero* for $-1 < t < 1$, having the other properties of the one there denoted by $\delta(t)$, corresponding to a value δ/π of the parameter η. (Here we write $q(t)$ instead of $\delta(t)$ because the letter δ is already in service.) Since our present function $\Delta(t)$ is *odd*, the one furnished by the lemma referred to may be taken to be *odd also*, and

$$\lambda(t) \;=\; \frac{\delta}{\pi} t \;+\; q(t)$$

is then *odd*, besides being *increasing* on $\mathbb{R}$. We have

$$\lambda(t) \;=\; \frac{\delta}{\pi} t \qquad \text{for } -1 < t < 1$$

and moreover,

$$\frac{\lambda(t)}{t} \longrightarrow \frac{\delta}{\pi} \quad \text{as } t \longrightarrow \pm \infty.$$

One may now apply the *first* theorem of §B.2, Chapter X, to the present functions $\Delta(t)$ and $q(t)$, and then do a calculation like the one used to prove the lemma of §B.1 there. On account of the oddness of $v_1(t)$ and $\lambda(t)$, that computation simplifies quite a bit*, and the final result is that

$$\int_{-\infty}^{\infty} \frac{1}{1+x^2} \left| \int_0^{\infty} \log \left| 1 - \frac{x^2}{t^2} \right| d(v_1(t) + \lambda(t)) \right| dx \quad < \quad \infty.$$

Used with the previous boxed formula and the assumption (in the hypothesis) that $\omega(x) \geqslant 0$ enjoys property (i), this implies that

$$\int_{-\infty}^{\infty} \frac{1}{1+x^2} \left| \int_0^{\infty} \log \left| 1 - \frac{x^2}{t^2} \right| d(v_2(t) + \lambda(t)) \right| dx \quad < \quad \infty.$$

Put now

$$V(z) = \int_0^{\infty} \log \left| 1 - \frac{z^2}{t^2} \right| d(v_2(t) + \lambda(t));$$

the last relation can then be written

$$\int_{-\infty}^{\infty} \frac{|V(x)|}{1+x^2} dx \quad < \quad \infty.$$

We have

$$v_2(t) + \lambda(t) \quad \leqslant \quad \text{const.} \, t \quad \text{for } t \geqslant 0,$$

so V also satisfies an inequality of the form

$$V(z) \quad \leqslant \quad \text{const.} \, |z|.$$

These two properties of V imply that

$$V(z) = B|\Im z| + \frac{1}{\pi} \int_{-\infty}^{\infty} \frac{|\Im z| V(t)}{|z-t|^2} dt$$

with a suitable constant $B \geqslant 0$, according to a version of the result from §G.1, Chapter III – the use of such a version here can be justified by an

---

* The usual partial integration is carried out with $\Delta(t) + q(t)$ playing the rôle of $v(t)$; then the relation $\int_{-\infty}^{\infty} \log|1 - (x^2/t^2)| \, dt = 0$ is used.

argument like one made while proving the second theorem of §B.1.* From this formula and the first of the two relations for V preceding it, we get

$$\int_{-\infty}^{\infty} \frac{|V(x+i)|}{1+x^2}\, dx \;\; < \;\; \infty$$

in the usual way.

We desire to apply the *Theorem on the Multiplier* at this point, and for that an *entire function of exponential type* is needed. (It is not true here that $V(x) \geqslant 0$ on $\mathbb{R}$, so we are unable to directly adapt the *proof* of that theorem given in §C.2 to the function V.) Take, then, the entire function φ of exponential type given by the formula

$$\log|\varphi(z)| \;\; = \;\; \int_0^{\infty} \log\left|1 \,-\, \frac{z^2}{t^2}\right| d[v_2(t) + \lambda(t)].$$

By the lemma in §A.1, Chapter X,

$$\log|\varphi(x+i)| \;\; \leqslant \;\; V(x+i) \,+\, \log^+|x| \qquad \text{for } x \in \mathbb{R},$$

whence, by the preceding inequality,

$$\int_{-\infty}^{\infty} \frac{\log^+|\varphi(x+i)|}{1+x^2}\, dx \;\; < \;\; \infty.$$

The theorem on the multiplier thus gives us a non-zero entire function $\psi(z)$, of exponential type $\delta' \leqslant \delta$, bounded on $\mathbb{R}$ and with $|\varphi(x+i)\,\psi(x)|$ bounded on $\mathbb{R}$ as well. We may, of course, get such a ψ with $\delta' = \delta$ by simply multiplying the initial one by $\cos(\delta - \delta')z$.

We can also take $\psi(z)$ to be *even*, since φ is even, and, of course, can have $\psi(0) \neq 0$. The discussion following the first theorem of §B.1 shows furthermore that we can take $\psi(z)$ to have *real zeros only*, and thus be given in the form

$$\log|\psi(z)| \;\; = \;\; \int_0^{\infty} \log\left|1 \,-\, \frac{z^2}{t^2}\right| d\sigma(t),$$

with $\sigma(t)$ increasing, integer-valued, zero near the origin, and satisfying

$$\frac{\sigma(t)}{t} \;\; \longrightarrow \;\; \frac{\delta}{\pi} \qquad \text{for } t \longrightarrow \infty$$

---

* By its definition, $v_2(t)$ is absolutely continuous with $v_2'(t)$ bounded on finite intervals; $\lambda(t)$, on the other hand, has a graph similar to the one shown in fig. 226 (Chapter X, §B.2). These properties make $(V(z))^+$ continuous at the points of $\mathbb{R}$, and the arguments from §§E and G.1 of Chapter III may be used.

(by Levinson's theorem). By first dividing out four of the zeros of ψ if need be (it *has* infinitely many, being of exponential type $\delta > 0$ and bounded on $\mathbb{R}$!) we can finally ensure that in fact

$$|\varphi(x+i)\,\psi(x)| \;\leqslant\; \frac{\text{const.}}{(x^2+1)^2} \qquad \text{for } x \in \mathbb{R}$$

with (perhaps another) ψ of the kind described.

A relation between $V(x+i)$ and $\log|\varphi(x+i)|$ *opposite in sense to the above one* is now called for. To get it, observe that

$$V(z) \;=\; \int_0^\infty \log\left|1 \;-\; \frac{z^2}{t^2}\right| \, d(\,\min\,(v_2(t)+\lambda(t),\;1)\,)$$
$$+\; \int_0^\infty \log\left|1 \;-\; \frac{z^2}{t^2}\right| \, d(v_2(t)+\lambda(t)-1)^+.$$

Since $\lambda(t) = \delta t/\pi$ for $0 \leqslant t \leqslant 1$ and $v_2'(t)$ is certainly *bounded* there, the first integral on the right is

$$\leqslant\; 2\log^+|z| \;+\; \text{const.}.$$

Therefore, when $x \in \mathbb{R}$,

$$V(x+i) \;\leqslant\; 2\log^+|x| \;+\; \text{const.} \;+\; \int_0^\infty \log\left|1 \;-\; \frac{(x+i)^2}{t^2}\right| d(v_2(t)+\lambda(t)-1)^+.$$

However, $(v_2(t)+\lambda(t)-1)^+ \leqslant [v_2(t)+\lambda(t)]$ for $t \geqslant 0$, so, by reasoning identical to that used in proving the lemma of §A.1, Chapter X, we find that the last right-hand integral is

$$\leqslant\; \log|\varphi(x+i)| \;+\; \log^+|x|.$$

Thus,

$$V(x+i) \;\leqslant\; \log|\varphi(x+i)| \;+\; 3\log^+|x| \;+\; \text{const.}, \qquad x \in \mathbb{R}.$$

Referring to the previous relation involving $\varphi(x+i)$ and $\psi(x)$, we thence obtain

$$V(x+i) \;+\; \log|\psi(x)| \;\leqslant\; \text{const.}, \qquad x \in \mathbb{R}.$$

Clearly, $V(x) \leqslant V(x+i)$, so we have

$$V(x) \;+\; \int_0^\infty \log\left|1 \;-\; \frac{x^2}{t^2}\right| d\sigma(t) \;\leqslant\; \text{const.} \qquad \text{for } x \in \mathbb{R}$$

by our formula for $\log|\psi(z)|$.

Now by the previous boxed formula and our definition of the function V,

$$V(x) \;=\; \omega(x) \;-\; \omega(0) \;+\; \int_0^\infty \log\left|1 \;-\; \frac{x^2}{t^2}\right| \mathrm{d}(v_1(t) + \lambda(t)).$$

Combination of this with the preceding thus yields

$$\omega(x) \;+\; \int_0^\infty \log\left|1 \;-\; \frac{x^2}{t^2}\right| \mathrm{d}(v_1(t) + \lambda(t) + \sigma(t)) \;\leqslant\; \text{const.}, \qquad x \in \mathbb{R}$$

and hence, since $\log W(x) \leqslant \omega(x)$,

$$\log W(x) \;+\; \int_0^\infty \log\left|1 \;-\; \frac{x^2}{t^2}\right| \mathrm{d}(v_1(t) + \lambda(t) + \sigma(t)) \;\leqslant\; \text{const.}, \qquad x \in \mathbb{R}.$$

Here, $\lambda(t)/t$ and $\sigma(t)/t$ both tend to δ/π for $t \longrightarrow \infty$ as we have already noted. Again, since the ratios b_k/a_k corresponding to the intervals (a_k, b_k) used in the construction of $v_1(t)$ must tend to 1 for $k \longrightarrow \infty$, we also have*

$$\frac{v_1(t)}{t} \;\longrightarrow\; \frac{\delta}{\pi} \qquad \text{for } t \longrightarrow \infty$$

(look again at the above discussion of the behaviour of v_1). For the increasing function

$$\rho(t) \;=\; v_1(t) \;+\; \lambda(t) \;+\; \sigma(t)$$

it is thus true that

$$\frac{\rho(t)}{t} \;\longrightarrow\; \frac{3\delta}{\pi} \qquad \text{as } t \longrightarrow \infty,$$

and that

$$\log W(x) \;+\; \int_0^\infty \log\left|1 \;-\; \frac{x^2}{t^2}\right| \mathrm{d}\rho(t) \;\leqslant\; \text{const.}, \qquad x \in \mathbb{R}.$$

The quantity $\delta > 0$ was, however, arbitrary. Therefore, since $W(x)$, by hypothesis, meets the local regularity requirement of §B.1, *it admits multipliers* according to the *second theorem* of that §, and *sufficiency is now established*.

Our result is completely proved.

---

* although a_k need not $\longrightarrow \infty$ for $k \longrightarrow \infty$, all *sufficiently large* a_k certainly *do* have *arbitrarily large* indices k.

Remark 1 (added in proof). In the *sufficiency* proof, fulfilment of our local regularity requirement is only used at the end; in the absence of that requirement one still gets functions $\rho(t)$, increasing and $O(t)$ on $[0, \infty)$, with $\limsup_{t \to \infty} (\rho(t)/t)$ arbitrarily small and

$$\log W(x) \;+\; \int_0^\infty \log|1 - (x^2/t^2)|\,\mathrm{d}\rho(t)$$

bounded above on $\mathbb{R}$. The *necessity* proof, on the other hand, actually goes through – see the footnotes to its first part – whenever $W(x) \geqslant 1$ is *continuous* and such $\rho(t)$ exist. *The existence of a majorant $\omega(x)$ having the properties specified by the theorem is therefore equivalent to the existence of such increasing functions ρ for continuous weights W.* Our theorem thus holds, in particular, for continuous weights meeting the milder regularity requirement from the scholium at the end of §B.1. Continuity, indeed, need not even be assumed for such weights; that is evident after a little thought about the abovementioned footnotes and the passage they refer to.

Remark 2. The proof for the *necessity* shows that if $W(x)$ *does* admit multipliers, a majorant $\omega(x)$ for $\log W(x)$ having the properties asserted by the theorem exists, with

$$\omega(x) \;=\; \omega(0) \;-\; \int_0^\infty \log\left|1 - \frac{x^2}{t^2}\right|\mathrm{d}\sigma(t),$$

where $\sigma(t)$ is *increasing* on $[0, \infty)$, *zero* for t close enough to 0, and $O(t)$ for $t \longrightarrow \infty$. Now look again at the example in §D.4 and the discussion in §D.5!

Remark 3. It was by thinking about the above result that I came upon the method explained in §§B.2, B.3 and used in §C, being led to it by way of the construction in problem 55 (near end of §B.2).

Remark 4. It seems possible to tie the theorem's property (ii) more closely to the *local* behaviour of $\omega(x)$. Referring to the remark following the statement of the theorem, we see that

$$\tilde{\omega}'(x) \;=\; \frac{1}{\pi}\int_0^{Y(x)} \frac{2\omega(x) - \omega(x+t) - \omega(x-t)}{t^2}\,\mathrm{d}t$$
$$+\; \frac{1}{\pi}\int_{Y(x)}^\infty \frac{2\omega(x) - \omega(x+t) - \omega(x-t)}{t^2}\,\mathrm{d}t,$$

where for $Y(x)$ we can take *any* positive quantity, depending on x in any way we want.

Because $\omega \geqslant 0$, the *second* of the two integrals on the right is

$$\leqslant \quad \frac{2}{\pi Y(x)}\, \omega(x);$$

it is, on the other hand,

$$\geqslant \quad -\frac{2}{\pi} \int_{-\infty}^{\infty} \frac{\omega(t)}{(x-t)^2 + (Y(x))^2}\, dt.$$

For the present purpose this last expression's behaviour is adequately described by the 1967 lemma of Beurling and Malliavin given in §E.2 of Chapter IX. That result shows that for any given $\eta > 0$, the integral in question will lie between $-\eta$ and 0 for a function $Y(x) > 0$ with

$$\int_{-\infty}^{\infty} \int_{0}^{Y(x)} \frac{dy\,dx}{1 + x^2 + y^2} \quad < \quad \infty;$$

such a function is hence *not too large*.

Once a function $Y(x)$ is at hand, the set of $x > 0$ on which $\tilde{\omega}'(x)$ exceeds some large K seems to essentially be determined by the behaviour of $\omega(x)/Y(x)$ and of the integral

$$\frac{1}{\pi} \int_{0}^{Y(x)} \frac{2\omega(x) - \omega(x+t) - \omega(x-t)}{t^2}\, dt.$$

Both of these expressions involve *local behaviour* of ω.

I think an investigation along this line is worth trying, but have no time to undertake it now. *This book must go to press.*

Remark 5 (added in proof). We have been dealing with the notion of multiplier adopted in §B.1, using that term to desiquate a non-zero entire function of exponential type whose product with a given weight is *bounded* on $\mathbb{R}$. This specification of boundedness is largely responsible for our having had to introduce a local regularity requirement in §B.1.

Such requirements become to a certain extent irrelevant if we return to the broader interpretation of the term accepted in Chapter X and permit its use whenever the product in question belongs to some $L_p(\mathbb{R})$. This observation, already made by Beurling and Malliavin at the end of their 1962 article, is based on the following analogue of the second theorem in §B.2:

Lemma. Let $\Omega(x) \geqslant 1$ be Lebesgue measurable. Suppose, given $A > 0$, that there is a function $\rho(t)$, increasing and $O(t)$ on $[0, \infty)$, with

$$\limsup_{t \to \infty} (\rho(t)/t) \quad \leqslant \quad A/\pi$$

and

$$\log \Omega(x) \ + \ \int_0^\infty \log|1 \ - \ (x^2/t^2)| \, d\rho(t) \ \leqslant \ O(1) \qquad \text{a.e.}$$

on $\mathbb{R}$. *Then, if* $0 < p < \infty$, *there is a non-zero entire function* $\psi(z)$ *of exponential type* $\leqslant \ 4(p+2)A$ *such that*

$$\int_{-\infty}^\infty |\Omega(x)\psi(x)|^p \, dx \ < \ \infty.$$

Proof. We consider the case $p = 1$; treatment for the other values of p is similar.

Take, then, the increasing function $\rho(t)$ furnished by the hypothesis and put

$$\nu(t) \ = \ 4\rho(t),$$

making

$$4 \log \Omega(x) \ + \ \int_0^\infty \log|1 \ - \ (x^2/t^2)| \, d\nu(t) \ \leqslant \ C \qquad \text{a.e., } x \in \mathbb{R}.$$

Since $\limsup_{t \to \infty} (\nu(t)/t) \leqslant 4A/\pi$, the entire function $\varphi(z)$ given by the formula

$$\log|\varphi(z)| \ = \ \int_0^\infty \log|1 \ - \ (z^2/t^2)| \, d[\nu(t)]$$

is of exponential type $\leqslant 4A$; this may be checked by using partial integration to estimate $\log \varphi(|z|)$.

Putting

$$U(z) \ = \ \int_0^\infty \log|1 \ - \ (z^2/t^2)| \, d\nu(t),$$

we have

$$(\Omega(x))^4 \exp U(x) \ \leqslant \ C \qquad \text{a.e., } x \in \mathbb{R}.$$

The idea behind our proof is that $|\varphi(x)|$ cannot be too much larger than $\exp U(x)$.

The usual integration by parts yields

$$\log|\varphi(x)| \ - \ U(x) \ = \ \int_0^\infty \log|1 \ - \ (x^2/t^2)| \, d([\nu(t)] - \nu(t))$$

$$= \ \int_0^\infty \frac{2x^2}{x^2 - t^2} \cdot \frac{[\nu(t)] - \nu(t)}{t} \, dt$$

at every $x \in \mathbb{R}$ where $v'(x)$ exists and is finite, and hence almost everywhere (see the lemma in §B.1 of Chapter X). After extending v from $[0, \infty)$ to $\mathbb{R}$ by making it *odd* (which poses no problem, $v(t)$ being $O(t)$ for $t \geqslant 0$), we can rewrite the last integral as

$$\int_{-\infty}^{\infty} \frac{x}{x-t} \cdot \frac{[v(t)] - v(t)}{t}\, dt \;=\; \int_{-\infty}^{\infty} \left(\frac{1}{x-t} + \frac{1}{t} \right) ([v(t)] - v(t))\, dt$$

$$= \; b \; + \; \int_{-\infty}^{\infty} \left(\frac{1}{x-t} + \frac{t}{t^2+1} \right) ([v(t)] - v(t))\, dt,$$

where the quantity

$$b \;=\; \int_{-\infty}^{\infty} \frac{[v(t)] - v(t)}{t(t^2+1)}\, dt$$

is finite. Hence, aside from the additive constant b, $\log |\varphi(x)| - U(x)$ is just the Hilbert transform of $\pi([v(x)] - v(x))$ which is, however, *bounded by π in absolute value*. Referring now to problem 45(c)   (Chapter X, §F), we see that

$$\int_{-\infty}^{\infty} \frac{|\varphi(x)|^{1/4} e^{-U(x)/4}}{1+x^2}\, dx \;<\; \infty.$$

From this and the above relation involving $\Omega(x)$ and $U(x)$ we have, finally

$$\int_{-\infty}^{\infty} \frac{\Omega(x) |\varphi(x)|^{1/4}}{1+x^2}\, dx \;<\; \infty.$$

Write

$$\psi(z) \;=\; \left(\frac{\sin Az}{z} \right)^8 \varphi(z);$$

$\psi(z)$ is entire, of exponential type $\leqslant 12A$, with $|\psi(x)| \leqslant \text{const.} |\varphi(x)|/(x^2+1)^4$ on the real axis. It thence follows by the preceding inequality that

$$\int_{-\infty}^{\infty} \Omega(x) |\psi(x)|^{1/4}\, dx \;<\; \infty.$$

In order to conclude from this that

$$\int_{-\infty}^{\infty} \Omega(x) |\psi(x)|\, dx \;<\; \infty$$

(thus proving the lemma in the case $p = 1$), it is enough to show that $\psi(x)$ is *bounded* on $\mathbb{R}$.

For that purpose, we note that $\int_{-\infty}^{\infty} |\psi(x)|^{1/4} dx < \infty$ since $\Omega(x) \geqslant 1$, so surely

$$\int_{-\infty}^{\infty} \frac{\log^+ |\psi(x)|}{1 + x^2} dx < \infty.$$

This gives us the right to use the theorem from §G.1 of Chapter III (the easier one in that chapter's §E would do just as well) to get

$$\log |\psi(x + i)| \leqslant 12A + \frac{1}{\pi} \int_{-\infty}^{\infty} \frac{\log |\psi(t)|}{(x - t)^2 + 1} dt$$

for $x \in \mathbb{R}$. By the inequality between arithmetic and geometric means, the integral on the right is

$$\leqslant 4 \log \left(\frac{1}{\pi} \int_{-\infty}^{\infty} \frac{|\psi(t)|^{1/4}}{(x - t)^2 + 1} dt \right) \leqslant 4 \log \left(\frac{1}{\pi} \int_{-\infty}^{\infty} |\psi(t)|^{1/4} dt \right)$$

which, as we just observed, is finite. Therefore $\log |\psi(x + i)| \leqslant \text{const.}$, $x \in \mathbb{R}$. One can now conclude that $\psi(x)$ is bounded on $\mathbb{R}$, either by appealing to the third Phragmén–Lindelöf theorem from §C of Chapter III or by simply noting that $|\psi(x)| \leqslant |\psi(x + i)|$ on $\mathbb{R}$ for our function ψ (which has only real zeros). The proof is complete.

Let us now refer to Remark 1, and once more to the *sufficiency* proof for the above theorem. The argument made there furnished, for each $A > 0$, a function $\rho(t)$ satisfying the hypothesis of the lemma with the weight $\Omega(x) = \exp \omega(x)$; comparison of $\omega(x)$ with $\log W(x)$ did not take place until the very end. We can thereby conclude that *the existence, for* $\log W(x)$, *of an a.e. majorant* $\omega(x)$ *having the other properties enumerated in the theorem implies, for each* $p < \infty$, *the existence of entire functions* $\psi(z) \not\equiv 0$ *of arbitrarily small exponential type with*

$$\int_{-\infty}^{\infty} |W(x)\psi(x)|^p dx < \infty.$$

The function $\omega(x)$ with the stipulated properties does not even need to be an actual *majorant* of $\log W(x)$; as long as

$$\int_{-\infty}^{\infty} (e^{-\omega(x)} W(x))^{r_0} dx < \infty$$

for some $r_0 > 0$, *we will still, for each* $r < r_0$, *have entire functions* ψ *of the kind described with*

$$\int_{-\infty}^{\infty} |W(x)\psi(x)|^r \, dx \quad < \quad \infty.$$

This also follows from the lemma; it suffices to take $\Omega(x) = \exp \omega(x)$ and $p = r_0/(r_0 - r)$, and then use Hölder's inequality.

The first of these results should be confronted with one going in the opposite direction that was already pointed out in Remark 1. That says that, *for a continuous weight $W(x) \geqslant 1$, the existence of entire functions $\psi(z) \not\equiv 0$ of arbitrarily small exponential type making $W(x)\psi(x)$ bounded on $\mathbb{R}$ implies existence of a majorant $\omega(x)$ for $\log W(x)$ with the properties specified by the theorem.* Thus, insofar as *continuous* weights are concerned, our theorem's majorant criterion is at the same time a *necessary* condition for the admittance of multipliers (in the narrow L_∞ sense) and a *sufficient* one, albeit in the broader L_p sense only. No additional regularity of the weight (beyond continuity) is involved here.

A very similar observation can be made about the last theorem in §B.3. *Any continuous weight $W(x) \geqslant 1$ will admit multipliers in the L_p sense (with $p < \infty$) provided that, for each $A > 0$, the smallest superharmonic majorant of*

$$\frac{1}{\pi} \int_{-\infty}^{\infty} \frac{|\Im z| \log W(t)}{|z - t|^2} \, dt \quad - \quad A|\Im z|$$

is finite. This finiteness is, on the other hand, *necessary for the admittance of multipliers in the L_∞ sense by the weight W.* It is worthwhile in this connection to note, finally, the following fact: *for continuous weights W, finiteness of the smallest superharmonic majorants just mentioned is equivalent to the existence of an $\omega(x)$ enjoying all the properties described by the theorem.* That is an immediate consequence of the next-to-the-last theorem in §B.3 and Remark 1.

Scholium. One way of looking at the theorem on the multiplier is to view it as a *guarantee* of *admittance of multipliers* by smooth even weights $W(x) = e^{\omega(x)} \geqslant 1$ with

$$\int_{-\infty}^{\infty} \frac{\omega(x)}{1 + x^2} \, dx \quad < \quad \infty$$

under the subsidiary condition that $\tilde{\omega}(x) - Kx$ be decreasing on $\mathbb{R}$ for some K, i.e., that

$$\tilde{\omega}'(x) \leqslant K.$$

As long as the *growth* of $\tilde{\omega}(x)$ is *thus limited, convergence of the logarithmic integral* of W is *in itself sufficient.** Referring, however, to the very elementary Paley–Wiener multiplier theorem from §A.1, Chapter X, we see that the convergence is *also sufficient* subject to a *similar requirement* on $\omega(x)$ *itself*, namely that $\omega(x)$ be *increasing* for $x \geqslant 0$.

Part of what this article's theorem does is to *generalize* the first result. As long as $W(x)$ meets the local regularity requirement, *more growth of* $\tilde{\omega}(x)$ is in fact *permissible*; the theorem tells us exactly *how much*. Could not then the Paley–Wiener result be generalized in the same way, so as to allow for a *certain amount of decrease* in $\omega(x)$ for $x \geqslant 0$?

What comes to mind is that perhaps an *analogous generalization* of the second result would carry over. In that way one is led to consider the following conjecture:

Let $W(x) = e^{\omega(x)}$ with $\omega(x) \geqslant 0$, $\mathscr{C}_\infty$ and even. Suppose that

$$\int_{-\infty}^{\infty} \frac{\omega(x)}{1 + x^2} \, dx \;<\; \infty,$$

and that for a certain K,

$$\omega'(x) \;\geqslant\; -K$$

for all $x > 0$ outside a set of disjoint intervals $(a_k, b_k) \subseteq (0, \infty)$ with

$$\sum_k \left(\frac{b_k - a_k}{a_k} \right)^2 \;<\; \infty,$$

for each of which

$$\int_{a_k}^{b_k} (\omega'(x))^- \, dx \;\leqslant\; K(b_k - a_k).$$

Then $W(x)$ admits multipliers.

This conjecture is *true*. To prove it, one constructs a positive function $w(x)$, *uniformly* Lip 1 *on* $\mathbb{R}$, such that

$$w(x) \;\geqslant\; \omega(x)$$

---

* Without imposition of any local regularity requirement. Indeed, putting $Kt - \tilde{\omega}(t) = \pi\nu(t)$ and then $U(z) = \omega(0) + \int_0^\infty \log|1 - (z/t)^2|\,d\nu(t)$, we have $\omega(x) = U(x) \leqslant U(x + i)$ (see p. 503 and the lemmas, p. 516 and 521). If $\varphi(z)$ is the entire function given by $\log|\varphi(z)| = \int_0^\infty \log|1 - (z/t)^2|\,d[\nu(t)]$, $|\varphi(x + i)|$ *admits* multipliers in the present circumstances (see lemma, p. 521 and then pp. 546–7). But then $\exp U(x + i)$ does also (see p. 548), and so, finally, does $W(x) = \exp U(x)$.

there, and

$$\int_{-\infty}^{\infty} \frac{w(x)}{1+x^2}\,dx \ < \ \infty.$$

By the result in §C, Chapter X, it is *known* that $\exp w(x)$ *admits multipliers.* Hence $W(x) \ = \ \exp \omega(x)$ *must also.* The construction of $w(x)$ is outlined in the following two problems.

We may, first of all, ensure that all the intervals $(a_k, \ b_k)$ lie in $(1, \infty)$ by taking K large enough to begin with (see discussion in first half of the proof of sufficiency for the above theorem). This detail being settled, we take a function $\varphi(x) \geqslant 0$ defined on $[0, \infty)$ as follows:

$$\varphi(x) \ = \ K \ - \ (\omega'(x))^- \quad \text{for} \quad x \in [0, \infty) \ \sim \ \bigcup_k (a_k, b_k);$$

$$\varphi(x) \ = \ K \ - \ \frac{1}{b_k - a_k} \int_{a_k}^{b_k} (\omega'(t))^-\,dt \quad \text{for } a_k < x < b_k.$$

We then put

$$P(x) \ = \ \int_0^x \{(\omega'(t))^+ \ + \ \varphi(t)\}\,dt$$

and

$$N(x) \ = \ \int_0^x \{(\omega'(t))^- \ + \ \varphi(t)\}\,dt$$

getting, in this way, two continuous functions $P(x)$ and $N(x)$, both *increasing* on $[0, \infty]$, with

$$\omega(x) \ = \ P(x) \ - \ N(x), \quad x \geqslant 0.$$

Note that

$$N(x) \ = \ Kx \quad \text{for } x \in [0, \infty) \ \sim \ \bigcup_k (a_k, b_k);$$

in particular, $N(x) \ = \ Kx$ for $0 \leqslant x \leqslant 1$.

Fix now any number $M > K$ and consider the open set

$$\Omega \ = \ \left\{x > 0 \colon \ \frac{N(x) - N(\xi)}{x - \xi} \ > \ M(x - \xi)\right.$$

$$\left. \text{for some positive } \xi < x \ (sic!)\right\}.$$

Ω can be obtained by shining light *up from underneath* the graph of $N(x)$ vs x *from the left*, in a direction of slope M:

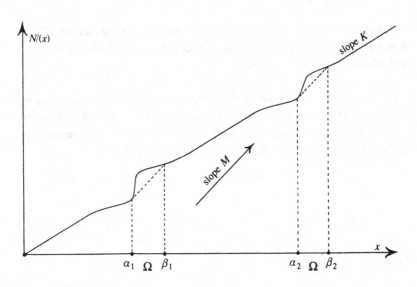

Figure 262

Ω is a *disjoint union* of certain open intervals $(\alpha_l, \beta_l) \subseteq (0, \infty)$ (not to be confounded with the given intervals (a_k, b_k)), and for $x \in (0, \infty) \sim \Omega$, $N'(x) \leqslant M$.

Problem 69

(a) Show that

$$\int_0^\infty \frac{|N(x) - Kx|}{x^2}\,dx \ < \ \infty.$$

(Hint: cf. the examination of $\pi v_1(t)$ in the proof of sufficiency for the above theorem.)

(b) Show that the intervals (α_l, β_l) actually lie in $(1, \infty)$.

For the rest of this problem, we make the following construction. Considering any one of the intervals (α_l, β_l), denote by $\mathscr{L}_l$ the line of slope M through the points $(\alpha_l, N(\alpha_l))$ and $(\beta_l, N(\beta_l))$. Then denote by γ_l the *abscissa* of the point where $\mathscr{L}_l$ and the *line of slope K through the origin* intersect (cf. proof of *third* lemma in §D.2, Chapter IX). Note that γ_l may well coincide with α_l or β_l, or even lie *outside* $[\alpha_l, \beta_l]$.

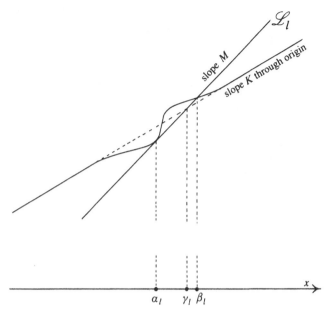

slope M

slope K through origin

$\mathscr{L}_l$

$\alpha_l \quad \gamma_l\, \beta_l$

x

Figure 263

Let R be the set of *indices l* for which γ_l *lies to the right of the midpoint* of (α_l, β_l), and S the set of those indices for which γ_l *lies to the left* of that midpoint.

(c) Show that $\sum_{l\in S}((\beta_l - \alpha_l)/\alpha_l)^2 < \infty$. (Hint: cf. proof of third lemma, Chapter IX, §D.2. Note that the *difference* between our present construction and the one used there is that *left and right have exchanged rôles*, as have *above and below*!)

(d) Show that if $\eta > 0$, there cannot be infinitely many indices l in R for which $\beta_l - \alpha_l > \eta\alpha_l$. (Hint: It is enough to consider η with

$$0 \;<\; \frac{M-K}{2K}\eta \;<\; 1.$$

If (α_l, β_l) is any interval corresponding to an $l \in R$ with $\beta_l - \alpha_l > \eta\alpha_l$, write

$$\alpha_l' \;=\; \left(1 - \frac{M-K}{2K}\eta\right)\alpha_l$$

and then estimate

$$\int_{\alpha_l'}^{\alpha_l} \frac{Kx - N(x)}{x^2}\,\mathrm{d}x$$

from below. Note that if this situation arises for infinitely many l in R, there must still be infinitely many of those indices for which the intervals (α'_l, α_l) are *disjoint.*)

(e*) Show that $\sum_{l\in R}((\beta_l - \alpha_l)/\alpha_l)^2 < \infty$. (Hint: by (d) we may wlog suppose that $((M-K)/2K)(\beta_l - \alpha_l) < \frac{1}{2}\alpha_l$ for all $l \in R$. For those l we then put

$$\alpha_l^* = \alpha_l - \frac{M-K}{2K}(\beta_l - \alpha_l)$$

and estimate each of the integrals

$$\int_{\alpha_l^*}^{\alpha_l} \frac{Kx - N(x)}{x^2}\,dx$$

from below. Starting, then, with an arbitrarily large *finite* subset R' of R, we go *first* to the *rightmost* of the (α_l, β_l) with $l \in R'$, and then make a covering argument like the one in the proof of the third lemma, §D.2, Chapter IX (used when considering the sums *over* S' figuring there), moving, however, *back towards the left instead of towards the right*, and working with the intervals (α_l^*, α_l). This gives a bound on

$$\sum_{l\in R'} \left(\frac{\beta_l - \alpha_l}{\alpha_l}\right)^2$$

independent of the size of R'.)

To finish this problem, we define a function $N_0(x)$ by putting

$$N_0(x) = N(x) \quad \text{for} \quad x \in [0, \infty) \sim \bigcup_l (\alpha_l, \beta_l)$$

and

$$N_0(x) = N(\alpha_l) + M(x - \alpha_l) \quad \text{for} \quad \alpha_l < x < \beta_l.$$

This makes

$$N_0(x) \leqslant N(x) \quad \text{for} \quad x \geqslant 0$$

and

$$N_0'(x) \leqslant M.$$

(f) Show that

$$\int_0^\infty \frac{N(x) - N_0(x)}{x^2}\,dx < \infty.$$

Carrying through the steps of the last problem has given us the increasing functions $P(x)$, $N(x)$ and $N_0(x)$, having the properties indicated above.

Let now

$$w_0(x) \;=\; P(x) - N_0(x) \qquad \text{for } x \geqslant 0.$$

Then

$$w_0(x) \;\geqslant\; P(x) - N(x) \;=\; \omega(x), \qquad x \geqslant 0$$

while

$$w_0'(x) \;\geqslant\; -N_0'(x) \;\geqslant\; -M.$$

At the same time, since

$$\int_0^\infty \frac{\omega(x)}{1+x^2}\,dx \;<\; \infty,$$

we have

$$\int_0^\infty \frac{w_0(x)}{1+x^2}\,dx \;<\; \infty$$

by part (f) of the problem, since

$$w_0(x) - \omega(x) \;=\; N(x) - N_0(x).$$

Problem 70

Denote by $w(x)$ the *smallest majorant* of $w_0(x)$ on $[0, \infty)$ having the property that

$$|w(x) - w(x')| \;\leqslant\; M|x - x'| \qquad \text{for } x \text{ and } x' \geqslant 0.$$

The object of this problem is to prove that

$$\int_0^\infty \frac{w(x)}{1+x^2}\,dx \;<\; \infty.$$

(a) Given $\eta > 0$, show that one cannot have $w_0(x) > \eta x$ for arbitrarily large x. (Hint: Given any such $x > 0$, estimate

$$\int_x^{(1+(\eta/2M))x} \frac{w_0(t)}{t^2}\,dt$$

from below. Cf. problem 69(d).)

(b) Hence show that $w(x) < \infty$ for $x \geqslant 0$ and that in $(0, \infty)$, $w(x) > w_0(x)$ on a certain set of disjoint *bounded* open intervals lying therein.

Continuing with this problem we take *just the intervals from* (b) *that lie in* $(1, \infty)$, and denote them by (A_n, B_n), with $n = 1, 2, 3, \ldots$. In order to verify the desired property of $w(x)$, it is enough to show that

$$\int_{A_0}^{\infty} \frac{w(x)}{x^2}\, dx \;\; < \;\; \infty,$$

where $A_0 = \inf_{n \geqslant 1} A_n$, a quantity $\geqslant 1$. In

$$(A_0, \infty) \sim \bigcup_{n=1}^{\infty} (A_n, B_n)$$

we have $w(x) = w_0(x)$, where, as we know

$$\int_1^{\infty} \frac{w_0(x)}{x^2}\, dx \;\; < \;\; \infty.$$

It is therefore only necessary for us to prove that

$$\sum_{n \geqslant 1} \int_{A_n}^{B_n} \frac{w(x)}{x^2}\, dx \;\; < \;\; \infty.$$

Note that for each $n \geqslant 1$, we have

$$w(A_n) = w_0(A_n),$$
$$w(B_n) = w_0(B_n)$$

and

$$w(x) \;\; = \;\; w_0(A_n) + M(x - A_n) \qquad \text{for } A_n \leqslant x \leqslant B_n.$$

(c) Show that $B_n/A_n \longrightarrow 1$ as $n \longrightarrow \infty$. (Hint: If $\eta > 0$ and there are infinitely many n with $B_n/A_n \geqslant 1 + \eta$, the corresponding A_n must tend to ∞ since the (A_n, B_n) are disjoint. Observe that for such n, since $w_0(x) \geqslant \omega(x) \geqslant 0$,

$$w_0(B_n) \;\; \geqslant \;\; M\eta B_n/(1 + \eta).$$

Refer to part (a).)

(d) For each $n \geqslant 1$, write

$$B_n^* \;\; = \;\; B_n + (B_n - A_n).$$

Show then that

$$\int_{A_n}^{B_n} \frac{w(x)}{x^2}\, dx \;\; \leqslant \;\; \left(\frac{B_n^*}{A_n}\right)^2 \int_{B_n}^{B_n^*} \frac{w_0(x)}{x^2}\, dx.$$

Hint:

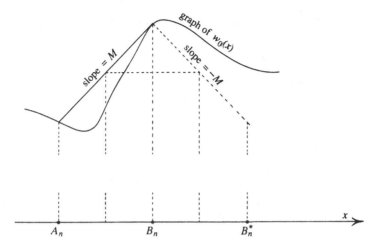

Figure 264

(e) Let us agree to call an interval (A_n, B_n) *special* if

$$w(A_n) \;\geq\; M(B_n - A_n).$$

Show then that if $(A_n, \; B_n)$ is special,

$$\int_{A_n}^{B_n} \frac{w(x)}{x^2}\,\mathrm{d}x \;\leq\; 3\left(\frac{B_n}{A_n}\right)^2 \int_{A_n}^{B_n} \frac{w_0(x)}{x^2}\,\mathrm{d}x.$$

Hint:

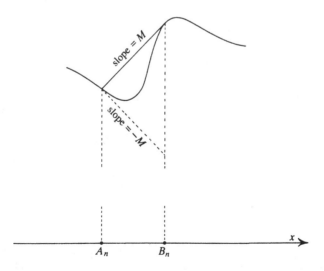

Figure 265

(f) Given any *finite* set T of integers $\geqslant 1$, obtain an upper bound *independent of T* on

$$\sum_{n\in T}\int_{A_n}^{B_n}\frac{w(x)}{x^2}\,dx,$$

hence showing that

$$\sum_{n\geqslant 1}\int_{A_n}^{B_n}\frac{w(x)}{x^2}\,dx \;<\; \infty.$$

(Procedure: Reindex the $(A_n,\,B_n)$ with $n\in T$ so as to have n *increase* from 1 up to some finite value as those intervals go *towards the right*. By (c), the ratios B_n^*/A_n must be bounded above by a quantity independent of T. Use then the result from (d) to estimate

$$\int_{A_1}^{B_1}\frac{w(x)}{x^2}\,dx.$$

Show next that *any interval $(A_n,\,B_n)$ entirely contained in $(B_1,\,B_1^*)$ must be special.* For such intervals, the result from (e) may be used to estimate

$$\int_{A_n}^{B_n}\frac{w(x)}{x^2}\,dx.$$

If there is an interval $(A_m,\,B_m)$ *intersecting* with $(B_1,\,B_1^*)$ but *not* lying *therein* $(m\in T)$, $(B_1,\,B_1^*)$ and $(B_m,\,B_m^*)$ are certainly *disjoint*, and we may again use the result of (d) to estimate

$$\int_{A_m}^{B_m}\frac{w(x)}{x^2}\,dx.$$

Then look to see if there are any $(A_n,\,B_n)$ entirely contained in $(B_m,\,B_m^*)$ and keep on going in this fashion, *moving steadily towards the right*, until all the $(A_n,\,B_n)$ with $n\in T$ are accounted for.)

The function $w(x)$ furnished by the constructions of these two problems is finally extended from $[0,\infty)$ to all of $\mathbb{R}$ by making it even. Then we will have

$$|w(x)-w(x')| \;\leqslant\; M|x-x'| \qquad \text{for x and x' in } \mathbb{R},$$

$$w(x) \;\geqslant\; \omega(x) \qquad \text{on } \mathbb{R},$$

and

$$\int_{-\infty}^{\infty}\frac{w(x)}{1+x^2}\,dx \;<\; \infty,$$

this last by problem 70. Our w thus has the properties we needed, and $W(x) = \exp\omega(x)$ admits multipliers, as explained at the beginning of this scholium.

One might hope to turn *around* the result just obtained and somehow show, in parallel to the necessity part of this article's theorem, that, for admittance of multipliers by a weight $W(x) \geqslant 1$ meeting the local regularity requirement, existence of a $\mathscr{C}_\infty$ even ω with

$$e^{\omega(x)} \geqslant W(x) \quad \text{on } \mathbb{R}$$

enjoying the other properties enumerated in the conjecture is *necessary*.

Problem 71

Show that such a proposition would be *false*. (Hint: Were such an ω to exist, the preceding constructions would give us an even uniformly Lip 1 $w(x) \geqslant \omega(x)$ for which

$$\int_{-\infty}^{\infty} \frac{w(x)}{1+x^2}\,dx < \infty.$$

Modify $w(x)$ in smooth fashion near 0 so as to obtain a new uniformly Lip 1 even function $w_1(x) \geqslant 0$, equal to zero at the origin and $O(x^2)$ *near* there, agreeing with $w(x)$ for $|x| \geqslant 1$, say. Then

$$\int_{0}^{\infty} \frac{w_1(x)}{x^2}\,dx < \infty.$$

Refer to problem 62 (end of §C.4) and then to the example of §D.4.)

August, 1983 – March 1986. Manuscript completed on March 2, 1986, in Outremont. Deeply affected by the assassination of Olof Palme, prime minister of Sweden, on the Friday, February 28th preceding.

Bibliography for volume II

Adamian, V.M., Arov, D.Z. and Krein, M.G. O beskonîechnykh Gankelîovykh matritsakh i obobschonnykh zadachakh Karateodori–Feĭera i F. Rissa. *Funkts. analiz i îevo prilozhenia*, 2:1 (1968), 1–19. Infinite Hankel matrices and generalized problems of Carathéodory–Fejér and F. Riesz. *Functl. Analysis and Appl.*, 2 (1968), 1–18.

Adamian, V.M., Arov, D.Z. and Krein, M.G. Beskonîechnye Gankelîovy matritsy i obobshchonnye zadachi Karateodori–Feĭera i I. Shura. *Funkts. analiz i îevo prilozhenia* 2:4 (1968), 1–17. Infinite Hankel matrices and generalized problems of Carathéodory–Fejér and I. Schur. *Functl. Analysis and Appl.*, 2 (1968), 269–81.

Adamian, V.M., Arov, D.Z. and Krein, M.G. Analiticheskie svoĭstva par Shmidta Gankelîova operatora i obobshchonnaîa zadacha Shura–Takagi. *Mat. Sbornik*, 86 (128), (1971), 34–75. Analytic properties of Schmidt pairs for a Hankel operator and the generalized Schur–Takagi problem. *Math. U.S.S.R. Sbornik*, 15 (1971), 31–73.

Adamian, V.M., Arov, D.Z., and Krein, M.G. Beskonîechnye blochno-Gankelîovy matritsy i svîazannye s nimi problemy prodolzheniîa. *Izvestia Akad. Nauk Armîan. S.S.R. Ser. Mat.*, 6 (1971), 87–112. Infinite Hankel block matrices and related extension problems. *A.M.S. Translations* (2), 111 (1978), 133–56.

Ahlfors, L.V. *Conformal Invariants: Topics in Geometric Function Theory*. McGraw-Hill, New York, 1973.

Ahlfors, L.V. and Beurling, A. Conformal invariants and function-theoretic nullsets. *Acta Math.* 83 (1950), 101–29.

Akhiezer, N.I. and Levin, B.Ia. Obobshchenie nîeravenstva S.N. Bernshteĭna dlîa proizvodnykh ot tselykh funktsiĭ. *Issledovaniîa po sovremennym problemam teorii funktsiĭ kompleksnovo peremennovo*, edited by A.I. Markushevich. Gosfizmatizdat., Moscow, 1960, pp 111–65. Généralisation de l'inégalité de S.N. Bernstein pour les dérivées des fonctions entières. *Fonctions d'une variable complexe. Problèmes contemporains*, edited by A.I. Marcouchevitch, translated by L. Nicolas. Gauthier-Villars, Paris, 1962, pp 109–61.

Arocena, R. and Cotlar, M. Generalized Herglotz–Bochner theorem and L^2-weighted inequalities with finite measures. *Conference on Harmonic Analysis* in *Honor of Antoni Zygmund*, edited by W. Beckner *et al.* Wadsworth, Belmont, 1983. Volume I, pp 258–69.

Arocena, R., Cotlar, M. and Sadosky, C. Weighted inequalities in L^2 and lifting' properties. *Mathematical Analysis and Applications*, Part A. Advances in Math. Supplemental Studies, 7a, edited by L. Nachbin. Academic Press, New York, 1981, pp 95–128.

Bateman Manuscript Project. *Tables of Integral Transforms*, Volume II, edited by A. Erdélyi. McGraw-Hill, New York, 1954.

Bernstein, V. *Leçons sur les progrès récents de la théorie des séries de Dirichlet.* Gauthier–Villars, Paris, 1933.

Beurling, A. *Etudes sur un problème de majoration.* Thesis, Uppsala, 1933.

Beurling, A. and Malliavin, P. On the zeros of entire functions of exponential type (I). Preprint, 1961.

Beurling, A. and Malliavin, P. On Fourier transforms of measures with compact support. *Acta Math.*, **107** (1962), 291–309.

Beurling, A. and Malliavin, P. On the closure of characters and the zeros of entire functions. *Acta Math.*, **118** (1967), 79–93.

Boas, R.P. *Entire Functions.* Academic press, New York, 1954.

Borichev, A.A. and Volberg, A.L. Teoremy ĩedinstvennosti dlĩa pochti analiticheskikh funktsiĭ. *Algebra i analiz*, **1** (1989), 146–77.

Carleson, L. *Selected Problems on Exceptional Sets.* Van Nostrand, Princeton, 1967.

Carleson, L. and Jones, P. Weighted norm inequalities and a theorem of Koosis. *Institut Mittag–Leffler*, Report no 2, 1981.

Coifman, R. A real variable characterization of H^p. *Studia Math.* **51** (1974), 269–74.

Cotlar, M. and Sadosky, C. On some L^p versions of the Helson–Szegő theorem. *Conference on Harmonic Analysis in Honor of Antoni Zygmund*, edited by W. Beckner, *et al.* Wadsworth, Belmont, 1983. Volume I, pp 306–17.

De Branges, L. The α-local operator problem. *Canadian J. Math.* **11** (1959), 583–92.

De Branges, L. *Hilbert Spaces of Entire Functions.* Prentice-Hall, Englewood Cliffs, 1968

Duren, P. *Theory of H^p Spaces.* Academic Press, New York, 1970.

Erdélyi, A. *et al.*—see under *Bateman Manuscript Project.*

Frostman, O. *Potential d'équilibre et capacité des ensembles avec quelques applications à la théorie des fonctions.* Thesis, Lund, 1935.

Fuchs, W. On the growth of functions of mean type. *Proc. Edinburgh Math. Soc.* Ser. 2, **9** (1954), 53–70.

Fuchs, W. *Topics in the Theory of Functions of one Complex Variable.* Van Nostrand, Princeton, 1967.

Gamelin, T. *Uniform Algebras and Jensen Measures.* L.M.S. lecture note series, 32. Cambridge Univ. Press, Cambridge, 1978.

García-Cuerva, J. and Rubio de Francia, J.L. *Weighted Norm Inequalities and Related Topics.* North-Holland, Amsterdam, 1985.

Garnett, J. *Analytic Capacity and Measure.* Lecture notes in math., 297. Springer, Berlin, 1972.

Garnett, J. *Bounded Analytic Functions.* Academic Press, New York, 1981.

Grötzsch, H. A series of papers, all in the *Berichte der Sächs. Akad. zu Leipzig.* Here are some of them:

Extremalprobleme der konformen Abblidung, in Volume 80 (1928), pp 367–76.

Über konforme Abbildung unendlich vielfach zusammenhängender schlichter Bereiche mit endlich vielen Häufungsrandkomponenten, in Volume 81 (1929), pp 51–86.

Zur konformen Abbildung mehrfach zusammenhängender schlichter Bereiche, in Volume 83 (1931), pp 67–76.

Zum Parallelschlitztheorem der konformen Abbildung schlichter unendlich-vielfach zusammenhängender Bereiche, in Volume 83 (1931), pp 185–200.

Über die Verzerrung bei schlichter konformer Abbildung mehrfach zusammenhängender schlichter Bereiche, in Volume 83 (1931), pp 283–97.

Haliste, K. Estimates of harmonic measures. *Arkiv för mat.* **6** (1967), 1–31.

Heins, M. *Selected Topics in the Classical Theory of Functions of a Complex Variable.* Holt, Rinehart and Winston, New York, 1962.

Helms, L. *Introduction to Potential Theory.* Wiley-Interscience, New York, 1969.

Helson, H. and Sarason, D. Past and future. *Math. Scand.* **21** (1967), 5–16.

Helson, H. and Szegő, G. A problem in prediction theory. *Annali mat. pura ed appl., ser.* 4, *Bologna.* **51** (1960), 107–38.

Hersch, J. Longueurs extrémales et théorie des fonctions. *Comm. Math. Helv.* **29** (1955), 301–37.

Herz, C. Bounded mean oscillation and regulated martingales. *Trans. A.M.S.,* **193** (1974), 199–215.

Hruščev, S.V. (Khrushchëv) and Nikolskiĭ, N.K. Funktsionalnaĩa model i nĩekotorye zadachi spektralnoĭ teorii funktsiĭ. *Trudy mat. inst. im. Steklova,* **176** (1987), 97–210. A function model and some problems in the spectral theory of functions. *Proc. Steklov Inst. of Math.,* A.M.S., Providence, **176** (1988), 101–214.

Hruščev, S.V. (Khrushchëv), Nikolskiĭ, N.K. and Pavlov, B.S. Unconditional bases of exponentials and of reproducing kernels. *Complex Analysis and Spectral Theory,* edited by V.P. Havin and N.K. Nikolskiĭ, Lecture notes in math., 864. Springer, Berlin, 1981, pp 214–335.

Kahane, J.P. Sur quelques problèmes d'unicité et de prolongement, relatifs aux fonctions approachables par des sommes d'exponentielles. *Annales Inst. Fourier Grenoble,* **5** (1955), 39–130.

Kahane, J.P. Travaux de Beurling et Malliavin. *Séminaire Bourbaki,* 1961/62, fasc 1, exposé no 225, 13 pages. Secrétariat mathématique, Inst. H. Poincaré, Paris, 1962. Reprinted by Benjamin, New York, 1966.

Kahane, J.P. and Salem, R. *Ensembles parfaits et séries trigonométriques.* Hermann, Paris, 1963.

Kellogg, O.D. *Foundations of Potential Theory.* Dover, New York, 1953.

Koosis, P. Sur la non-totalité de certaines suites d'exponentielles sur des intervalles assez longs. *Annales Ecole Norm. Sup.,* Sér. 3. **75** (1958), 125–52.

Koosis, P. Sur la totalite des systèmes d'exponentielles imaginaires. *C.R. Acad. Sci. Paris.* **250** (1960), 2102–3.

Koosis, P. Weighted quadratic means of Hilbert transforms. *Duke Math. J.*, **38** (1971), 609–34.

Koosis, P. Moyennes quadratiques de transformées de Hilbert et fonctions de type exponential. *C.R. Acad. Sci. Paris*, Sér. A, **276** (1973), 1201–4.

Koosis, P. Harmonic estimation in certain slit regions and a theorem of Beurling and Malliavin. *Acta Math.*, **142** (1979), 275–304.

Koosis, P. *Introduction to H_p Spaces*. L.M.S. lecture note series, 40. Cambridge Univ. Press, Cambridge, 1980.

Koosis, P. Moyennes quadratiques pondérées de fonctions périodiques et de leurs conjuguées harmoniques. *C.R. Acad. Sci. Paris*, Sér A, **291** (1980), 255–7.

Koosis, P. Entire functions of exponential type as multipliers for weight functions. *Pacific J. Math.*, **95** (1981), 105–23.

Koosis, P. Fonctions entières de type exponentiel comme multiplicateurs. Un exemple et une condition nécessaire et suffisante. *Annales Ecole Norm. Sup.*, Sér 4, **16** (1983), 375–407.

Koosis, P. La plus petite majorante surharmonique at son rapport avec l'existence des fonctions entières de type exponentiel jouant le rôle de multiplicateurs. *Annales Inst. Fourier Grenoble*, **33** (1983), 67–107.

Kriete, T. On the structure of certain $H^2(\mu)$ spaces. *Indiana Univ. Math. J.*, **28** (1979), 757–73.

Landkof, N.S. *Osnovy sovremennoĭ teorii potentsiala*. Nauka, Moscow, 1966. *Foundations of Modern Potential Theory*. Springer, New York, 1972.

Lebedev, N.A. *Printisip ploshchadeĭ v teorii odnolistnykh funktsiĭ*. Nauka, Moscow, 1975.

Leontiev, A.F. *Rīādy eksponent*. Nauka, Moscow, 1976.

Levinson, N. *Gap and Density Theorems*. Amer. Math. Soc. (Colloq. Publ., Volume 26), New York, 1940, reprinted 1968.

Lindelöf, E. *Le calcul des résidus et ses applications à la théorie des fonctions*. Gauthier-Villars, Paris, 1905.

Malliavin, P. Sur la croissance radiale d'une fonction méromorphe. *Illinois J. of Math.*, **1** (1957), 259–96.

Malliavin, P. The 1961 preprint. See Beurling, A. and Malliavin, P.

Malliavn, P. On the multiplier theorem for Fourier transforms of measures with compact support. *Arkiv för Mat.*, **17** (1979), 69–81.

Malliavin, P. and Rubel, L. On small entire functions of exponential type with given zeros. *Bull. Soc. Math. de France*, **89** (1961), 175–206.

Mandelbrojt, S. *Dirichlet Series*. Rice Institute Pamphlet, Volume 31, Houston, 1944.

Mandelbrojt, S. *Séries adhérentes. Régularisation des suites. Applications*. Gauthier-Villars, Paris, 1952.

Mandelbrojt, S. *Séries de Dirichlet – Principes et méthodes*. Gauthier-Villars, Paris, 1969. *Dirichlet Series. Principles and Methods*. Reidel, Dordrecht, 1972.

Markushevich, A.I. *Teoriīā analiticheskikh funktsiĭ*. Gostekhizdat, Moscow, 1950. Second augmented and corrected edition, Volumes I, II. Nauka, Moscow,

1967–8. *Theory of Functions of a Complex Variable*, Volumes I–III. Prentice-Hall, Englewood Cliffs, 1965–7. Second edition in one vol., Chelsea, New York, 1977.

Nagy, Béla Sz. and Foiaş, C. *Harmonic Analysis of Operators on Hilbert Space.* North-Holland, Amsterdam, 1970.

Nehari, Z. *Conformal Mapping.* McGraw-Hill, New York, 1952.

Neuwirth, J. and Newman, D.J. Positive $H^{1/2}$ functions are constant. *Proc. A.M.S.*, **18** (1967), 958.

Nevanlinna, R. *Eindeutige analytische Funktionen.* Second edition, Springer, Berlin, 1953. *Analytic Functions.* Springer, New York, 1970.

Nikolskiĭ, N.K. *Lektsii ob operatore sdviga.* Nauka, Moscow, 1980. *Treatise on the Shift Operator. Spectral Function Theory.* With appendix by S.V. Hruščev and V.V. Peller. Springer, Berlin, 1986.

Ohtsuka, M. *Dirichlet Problem, Extremal Length and Prime Ends.* Van Nostrand-Reinhold, New York, 1970.

Paley, R. and Wiener, N. *Fourier Transforms in the Complex Domain.* Amer. Math. Soc. (Colloq. Publ., Volume 19), Providence, 1934, reprinted 1960.

Pfluger, A. Extremallängen und Kapazität. *Comm. Math. Helv.* **29** (1955), 120–31.

Phelps, R. *Lectures on Choquet's Theorem.* Van Nostrand, Princeton, 1966.

Pólya, G. Untersuchungen über Lücken und Singularitäten von Potenzreihen. *Math. Zeitschr.*, **29** (1929), 549–640.

Proceedings of Symposia in Pure Mathematics, Volume VII. Convexity. Amer. Math. Soc., Providence, 1963.

Redheffer, R. On even entire functions with zeros having a density. *Trans. A.M.S.*, **77** (1954), 32–61.

Redheffer, R. Interpolation with entire functions having a regular distribution of zeros. *J. Analyse Math.*, **20** (1967), 353–70.

Redheffer, R. Eine Nevanlinna-Picardsche Theorie en miniature. *Arkiv för Mat.*, **7** (1967), 49–59.

Redheffer, R. Elementary remarks on completeness. *Duke Math. J.*, **35** (1968), 103–16.

Redheffer, R. Two consequences of the Beurling-Malliavin theory. *Proc. A.M.S.*, **36** (1972), 116–22.

Redheffer, R. Completeness of sets of complex exponentials. *Advances in Math.*, **24** (1977), 1–62.

Rubel, L. Necessary and sufficient conditions for Carlson's theorem on entire functions. *Trans. A.M.S.*, **83** (1956), 417–29.

Rubio de Francia, J.L. Boundedness of maximal functions and singular integrals in weighted L^p spaces. *Proc. A.M.S.*, **83** (1981), 673–9.

Sarason, D. *Function Theory on the Unit Circle.* Virginia Polytechnic Inst., Blacksburg, 1978.

Schwartz, L. *Etude des sommes d'exponentielles.* Hermann, Paris, 1959.

Titchmarsh, E.C. *The Theory of Functions.* Second edition, Oxford Univ. Press, Oxford, 1939. Reprinted 1952.

Treil, S.R. Geometrischeskiĭ podkhod k vesovym otsenkam preobrazovaniĭa Gilberta. *Funkts. analiz i ĭevo prilozh.*, 17:**4** (1983), 90–1. A geometric approach to the weighted estimates of Hilbert transforms. *Funct. Analysis and Appl.*, 17:**4** (1983), 319–21, Plenum Publ. Corp.

Treil, S.R. Operatornyĭ podkhod k vesovym otsenkam singuliarnykh integralov. *Akad. Nauk S.S.S.R., Mat. Inst. im. Steklova, Leningrad. Zapiski nauchnykh seminarov L.O.M.I.*, **135** (1984), 150–74.

Tsuji, M. *Potential Theory in Modern Function Theory.* Maruzen, Tokyo, 1959. Reprinted by Chelsea, New York, 1975.

Zalcman, L. *Analytic Capacity and Rational Approximation.* Lecture notes in math., 50. Springer, Berlin, 1968.

Index

Printed in the United States
By Bookmasters